# Residential Square Foot Costs

**Contractor's Pricing Guide 2002**

*Senior Editor*
Robert W. Mewis, CCC

*Contributing Editors*
Barbara Balboni
Robert A. Bastoni
Howard M. Chandler
John H. Chiang, PE
Jennifer L. Curran
Stephen E. Donnelly
J. Robert Lang
Robert C. McNichols
Melville J. Mossman, PE
John J. Moylan
Jeannene D. Murphy
Peter T. Nightingale
Stephen C. Plotner
Michael J. Regan
Marshall J. Stetson
Phillip R. Waier, PE

*Senior Engineering
Operations Manager*
John H. Ferguson, PE

*Vice President &
General Manager*
Roger J. Grant

*Senior Vice President*
Charles Spahr

*Vice President,
Sales & Marketing*
John M. Shea

*Production Manager*
Michael Kokernak

*Production Coordinator*
Marion E. Schofield

*Technical Support*
Thomas J. Dion
Jonathan Forgit
Mary Lou Geary
Gary L. Hoitt
Paula Reale-Camelio
Robin Richardson
Kathryn S. Rodriguez
Sheryl A. Rose
Elizabeth Testa

*Book & Cover Design*
Norman R. Forgit

**RSMeans
CMD** Understanding your craft.
Advancing your business.

# Residential Square Foot Costs

## Contractor's Pricing Guide 2002

- Residential Cost Models for All Standard Building Classes
- Costs for Modifications & Additions
- Costs for Hundreds of Residential Building Systems & Components
- Cost Adjustment Factors for Your Location
- Illustrations

Published by the
R.S. Means Company, Inc.

$39.95 per copy. (In United States).
Price subject to change without prior notice.

Copyright 2001

**R.S. Means Company, Inc.**

*Construction Publishers & Consultants*

*Construction Plaza*

*63 Smiths Lane*

*Kingston, MA 02364-0800*

*(781) 585-7880*

Special thanks to the following firms that provided the inspirations:
  Home Planners, Inc.
  Design Basics, Inc.
  Donald A. Gardner Architects, Inc.
  Larry E. Belk Designs
  Larry W. Garnett & Associates, Inc.

Printed in the United States of America
ISSN 1074-049X
ISBN 0-87629-647-9
**First Printing**

# Foreword

R.S. Means Co., Inc. is a subsidiary of CMD (Construction Market Data), a leading provider of construction information, products, and services in North America and globally. CMD's project information products include more than 100 regional editions, national construction data, sales leads, and over 70 local plan rooms in major business centers. CMD PlansDirect provides surveys, plans and specifications. The First Source suite of products consists of *First Source for Products,* SPEC-DATA™, MANU-SPEC™, CADBlocks, Manufacturer Catalogs and First Source Exchange (www.firstsourceexchange.com) for the selection of nationally available building products. CMD also publishes ProFile, a database of more then 20,000 U.S. architectural firms. R.S. Means provides construction cost data, training, and consulting services in print, CD-ROM and online. CMD is headquartered in Atlanta and has 1,400 employees worldwide. CMD is owned by Cahners Business Information (www.cahners.com), a leading provider of critical information and marketing solutions to business professionals in the media, manufacturing, electronics, construction and retail industries. Its market-leading properties include more than 135 business-to-business publications, over 125 Webzines and Web portals, as well as online services, custom publishing, directories, research and direct-marketing lists. Cahners is a member of the Reed Elsevier plc group (NYSE: RUK and ENL)—world-leading publisher and information provider operating in the science and medical, legal, education and business-to-business industry sectors.

## Our Mission

Since 1942, R.S. Means Company, Inc. has been actively engaged in construction cost publishing and consulting throughout North America.

Today, over 50 years after the company began, our primary objective remains the same: to provide you, the construction and facilities professional, with the most current and comprehensive construction cost data possible.

Whether you are a contractor, an owner, an architect, an engineer, a facilities manager, or anyone else who needs a fast and reliable construction cost estimate, you'll find this publication to be a highly useful and necessary tool.

Today, with the constant flow of new construction methods and materials, it's difficult to find the time to look at and evaluate all the different construction cost possibilities. In addition, because labor and material costs keep changing, last year's cost information is not a reliable basis for today's estimate or budget.

That's why so many construction professionals turn to R.S. Means. We keep track of the costs for you, along with a wide range of other key information, from city cost indexes . . . to productivity rates . . . to crew composition . . . to contractor's overhead and profit rates.

R.S. Means performs these functions by collecting data from all facets of the industry, and organizing it in a format that is instantly accessible to you. From the preliminary budget to the detailed unit price estimate, you'll find the data in this book useful for all phases of construction cost determination.

## The Staff, the Organization, and Our Services

When you purchase one of R.S. Means' publications, you are in effect hiring the services of a full-time staff of construction and engineering professionals.

Our thoroughly experienced and highly qualified staff works daily at collecting, analyzing, and disseminating comprehensive cost information for your needs. These staff members have years of practical construction experience and engineering training prior to joining the firm. As a result, you can count on them not only for the cost figures, but also for additional background reference information that will help you create a realistic estimate.

The Means organization is always prepared to help you solve construction problems through its five major divisions: Construction and Cost Data Publishing, Electronic Products and Services, Consulting Services, Insurance Services, and Educational Services.

Besides a full array of construction cost estimating books, Means also publishes a number of other reference works for the construction industry. Subjects include construction estimating and project and business management; special topics such as HVAC, roofing, plumbing, and hazardous waste remediation; and a library of facility management references.

In addition, you can access all of our construction cost data through your computer with Means CostWorks 2002 CD-ROM, an electronic tool that offers over 50,000 lines of Means detailed construction cost data, along with assembly and whole building cost data. You can also access Means cost information from our Web site at www.rsmeans.com

What's more, you can increase your knowledge and improve your construction estimating and management performance with a Means Construction Seminar or In-House Training Program. These two-day seminar programs offer unparalleled opportunities for everyone in your organization to get updated on a wide variety of construction-related issues.

Means also is a worldwide provider of construction cost management and analysis services for commercial and government owners and of claims and valuation services for insurers.

In short, R.S. Means can provide you with the tools and expertise for constructing accurate and dependable construction estimates and budgets in a variety of ways.

## Robert Snow Means Established a Tradition of Quality That Continues Today

Robert Snow Means spent years building his company, making certain he always delivered a quality product.

Today, at R.S. Means, we do more than talk about the quality of our data and the usefulness of our books. We stand behind all of our data, from historical cost indexes... to construction materials and techniques... to current costs.

If you have any questions about our products or services, please call us toll-free at 1-800-334-3509. Our customer service representatives will be happy to assist you or visit our Web site at www.rsmeans.com

# Table of Contents

# How the Book is Built: An Overview

## A Powerful Construction Tool

You have in your hands one of the most powerful construction tools available today. A successful project is built on the foundation of an accurate and dependable estimate. This book will enable you to construct just such an estimate.

For the casual user the book is designed to be:

- quickly and easily understood so you can get right to your estimate
- filled with valuable information so you can understand the necessary factors that go into the cost estimate

For the regular user, the book is designed to be:

- a handy desk reference that can be quickly referred to for key costs
- a comprehensive, fully reliable source of current construction costs so you'll be prepared to estimate any project
- a source book for preliminary project cost, product selections, and alternate materials and methods

To meet all of these requirements we have organized the book into the following clearly defined sections.

## Square Foot Cost Section

This section lists Square Foot costs for typical residential construction projects. The organizational format used divides the projects into basic building classes. These classes are defined at the beginning of the section. The individual projects are further divided into ten common components of construction. An outline of a typical page layout, an explanation of Square Foot prices, and a Table of Contents are located at the beginning of the section.

## Assemblies Cost Section

This section uses an "Assemblies" (sometimes referred to as "systems") format grouping all the functional elements of a building into nine construction divisions.

At the top of each "Assembly" cost table is an illustration, a brief description, and the design criteria used to develop the cost. Each of the components and its contributing cost to the system is shown.

**Material** These cost figures include a standard 10% markup for profit. They are national average material costs as of January of the current year and include delivery to the job site.

**Installation** The installation costs include labor and equipment, plus a markup for the installing contractor's overhead and profit.

For a complete breakdown and explanation of a typical "Assemblies" page, see "How To Use Assemblies Cost Tables" at the beginning of the Assembly Section.

**Location Factors:** Costs vary depending upon regional economy. You can adjust the "national average" costs in this book to over 930 major cities throughout the U.S. and Canada by using the data in this section.

**Abbreviations:** A listing of the abbreviations used throughout this book, along with the terms they represent, is included.

## Index

A comprehensive listing of all terms and subjects in this book to help you find what you need quickly.

## The Scope of This Book

This book is designed to be as comprehensive and as easy to use as possible. To that end we have made certain assumptions and limited its scope in three key ways:

1. We have established material prices based on a "national average."
2. We have computed labor costs based on a 7 major region average of open shop wage rates.
3. We have targeted the data for projects of a certain size range.

## Project Size

This book is intended for use by those involved primarily in Residential construction costing less than $750,000. This includes the construction of homes, row houses, townhouses, condominiums and apartments.

**With reasonable exercise of judgment the figures can be used for any building work.** For other types of projects, such as repair and remodeling or commercial buildings, consult the appropriate MEANS publication for more information.

# How to Use the Book: The Details

## What's Behind the Numbers? The Development of Cost Data

The staff at R.S. Means continuously monitors developments in the construction industry in order to ensure reliable, thorough and up-to-date cost information.

While **overall** construction costs may vary relative to general economic conditions, price fluctuations within the industry are dependent upon many factors. Individual price variations may, in fact, be opposite to overall economic trends. Therefore, costs are continually monitored and complete updates are published yearly. Also, new items are frequently added in response to changes in materials and methods.

## Costs – $ (U.S.)

All costs represent U.S. national averages and are given in U.S. dollars. The Means Location Factors can be used to adjust costs to a particular location. The Location Factors for Canada can be used to adjust U.S. national averages to local costs in Canadian dollars.

## Material Costs

The R.S. Means staff contacts manufacturers, dealers, distributors, and contractors all across the U.S. and Canada to determine national average material costs. If you have access to current material costs for your specific location, you may wish to make adjustments to reflect differences from the national average. Included within material costs are fasteners for a normal installation. R.S. Means engineers use manufacturers' recommendations, written specifications and/or standard construction practice for size and spacing of fasteners. Adjustments to material costs may be required for your specific application or location. Material costs do not include sales tax.

## Labor Costs

Labor costs are based on the average of open shop wages from across the U.S. for the current year. Rates along with overhead and profit markups are listed on the inside back cover of this book.

- If wage rates in your area vary from those used in this book, or if rate increases are expected within a given year, labor costs should be adjusted accordingly.

Labor costs reflect productivity based on actual working conditions. These figures include time spent during a normal workday on tasks other than actual installation, such as material receiving and handling, mobilization at site, site movement, breaks, and cleanup.

Productivity data is developed over an extended period so as not to be influenced by abnormal variations and reflects a typical average.

## Equipment Costs

Equipment costs include not only rental, but also operating costs for equipment under normal use. The operating costs include parts and labor for routine servicing such as repair and replacement of pumps, filters and worn lines. Normal operating expendables such as fuel, lubricants, tires and electricity (where applicable) are also included. Extraordinary operating expendables with highly variable wear patterns such as diamond bits and blades are excluded. These costs are included under materials. Equipment rental rates are obtained from industry sources throughout North America—contractors, suppliers, dealers, manufacturers, and distributors.

## Factors Affecting Costs

Costs can vary depending upon a number of variables. Here's how we have handled the main factors affecting costs.

**Quality**—The prices for materials and the workmanship upon which productivity is based represent sound construction work. They are also in line with U.S. government specifications.

**Overtime**—We have made no allowance for overtime. If you anticipate premium time or work beyond normal working hours, be sure to make an appropriate adjustment to your labor costs.

**Productivity**—The productivity, daily output, and labor-hour figures for each line item are based on working an eight-hour day in daylight hours in moderate temperatures. For work that extends beyond normal work hours or is performed under adverse conditions, productivity may decrease.

**Size of Project**—The size, scope of work, and type of construction project will have a significant impact on cost. Economies of scale can reduce costs for large projects. Unit costs can often run higher for small projects. Costs in this book are intended for the size and type of project as previously described in "How the Book Is Built: An Overview." Costs for projects of a significantly different size or type should be adjusted accordingly.

**Location**—Material prices in this book are for metropolitan areas. However, in dense urban areas, traffic and site storage limitations may increase costs. Beyond a 20-mile radius of large cities, extra trucking or transportation charges may also increase the material costs slightly. On the other hand, lower wage rates may be in effect. Be sure to consider both these factors when preparing an estimate, particularly if the job site is located in a central city or remote rural location.

In addition, highly specialized subcontract items may require travel and per diem expenses for mechanics.

**Other factors –**

- season of year
- contractor management
- weather conditions
- local union restrictions
- building code requirements

- availability of:
  - adequate energy
  - skilled labor
  - building materials
- owner's special requirements/ restrictions
- safety requirements
- environmental considerations

**General Conditions**—The "Square Foot" and "Assemblies" sections of this book use costs that include the installing contractor's overhead and profit (O&P). An allowance covering the general contractor's markup must be added to these figures. The general contractor can include this price in the bid with a normal markup ranging from 5% to 15%. The markup depends on economic conditions plus the supervision and troubleshooting expected by the general contractor. For purposes of this book, it is best for a general contractor to add an allowance of 10% to the figures in the Assemblies and Square Foot Cost sections.

**Overhead & Profit**—For systems costs and square foot costs, simply add 10% to the estimate for general contractor's profit.

**Unpredictable Factors**—General business conditions influence "in-place" costs of all items. Substitute materials and construction methods may have to be employed. These may affect the installed cost and/or life cycle costs. Such factors may be difficult to evaluate and cannot necessarily be predicted on the basis of the job's location in a particular section of the country. Thus, where these factors apply, you may find significant, but unavoidable cost variations for which you will have to apply a measure of judgment to your estimate.

## Rounding of Costs

In general, all unit prices in excess of $5.00 have been rounded to make them easier to use and still maintain adequate precision of the results. The rounding rules we have chosen are in the following table.

| Prices from . . . | Rounded to the nearest . . . |
|---|---|
| $.01 to $5.00 | $.01 |
| $5.01 to $20.00 | $.05 |
| $20.01 to $100.00 | $.50 |
| $100.01 to $300.00 | $1.00 |
| $300.01 to $1,000.00 | $5.00 |
| $1,000.01 to $10,000.00 | $25.00 |
| $10,000.01 to $50,000.00 | $100.00 |
| $50,000.01 and above | $500.00 |

## Final Checklist

Estimating can be a straightforward process provided you remember the basics. Here's a checklist of some of the items you should remember to do before completing your estimate.

Did you remember to . . .

- factor in the Location Factor for your locale
- take into consideration which items have been marked up and by how much
- mark up the entire estimate sufficiently for your purposes
- include all components of your project in the final estimate
- double check your figures to be sure of your accuracy
- call R.S. Means if you have any questions about your estimate or the data you've found in our publications

Remember, R.S. Means stands behind its publications. If you have any questions about your estimate . . . about the costs you've used from our books . . . or even about the technical aspects of the job that may affect your estimate, feel free to call the R.S. Means editors at 1-800-334-3509.

## Table of Contents

# How to Use the Square Foot Cost Pages

**Introduction:** This section contains costs per square foot for four classes of construction in seven building types. Costs are listed for various exterior wall materials which are typical of the class and building type. There are cost tables for Wings and Ells with modification tables to adjust the base cost of each class of building. Non-standard items can easily be added to the standard structures.

Accompanying each building type in each class is a list of components used in a typical residence. The components are divided into ten primary estimating divisions. The divisions correspond with the "Assemblies" section of this manual.

Cost estimating for a residence is a three step process:
(1) Identification
(2) Listing dimensions
(3) Calculations

Guidelines and a sample cost estimating form are shown on the following pages.

**Identification:** To properly identify a residential building, the class of construction, type, and exterior wall material must be determined. Located at the beginning of this section are drawings and guidelines for determining the class of construction. There are also detailed specifications accompanying each type of building along with additional drawings at the beginning of each set of tables to further aid in proper building class and type identification.

Sketches for seven types of residential buildings and their configurations are shown along with definitions of living area next to each sketch. Sketches and definitions of garage types follow the residential buildings.

**Living Area:** Base cost tables are prepared as costs per square foot of living area. The living area of a residence is that area which is suitable and normally designed for full time living. It does not include basement recreation rooms or finished attics, although these areas are often considered full time living areas by the owners.

Living area is calculated from the exterior dimensions without the need to adjust for exterior wall thickness. When calculating the living area of a 1-1/2 story, two story, three story or tri-level residence, overhangs and other differences in size and shape between floors must be considered.

Only the floor area with a ceiling height of six feet or more in a 1-1/2 story residence is considered living area. In bi-levels and tri-levels, the areas that are below grade are considered living area, even when these areas may not be completely finished.

**Base Tables and Modifications:** Base cost tables show the base cost per square foot without a basement, with one full bath and one full kitchen. Adjustments for finished and unfinished basements are part of the base cost tables. Adjustments for multi-family residences, additional bathrooms, townhouses, alternative roofs, air conditioning and heating systems are listed in Modifications, Adjustments and Alternatives tables below the base cost tables.

The component list for each residence type should also be consulted when preparing an estimate. If the components listed are not appropriate, modifications can be made by consulting the "Assemblies" section of this manual.

Costs for other modifications, adjustments, and alternatives, including garages, breezeways and site improvements, are in the modification tables at the end of this section. For additional information on contractor overhead and architectural fees, consult the "Reference" section of this manual.

**Listing of Dimensions:** To use this section of the manual, only the dimensions used to calculate the horizontal area of the building and additions and modifications are needed. The dimensions, normally the length and width, can come from drawings or field measurements. For ease in calculation, consider measuring in tenths of feet, i.e., 9 ft. 6 in. = 9.5 ft., 9 ft. 4 in. = 9.3 ft.

In all cases, make a sketch of the building. Any protrusions or other variations in shape should be noted on the sketch with dimensions.

**Calculations:** The calculations portion of the estimate is a two-step activity:
(1) The selection of appropriate costs from the tables
(2) Computations

**Selection of Appropriate Costs:** To select the appropriate cost from the base tables the following information is needed:
(1) Class of construction
(2) Type of residence
(3) Occupancy
(4) Building configuration
(5) Exterior wall construction
(6) Living area

Consult the tables and accompanying information to make the appropriate selections. Modifications, adjustments, and alternatives are classified by class, type and size. Further modifications can be made using the "Assemblies" Section.

**Computations:** The computation process should take the following sequence:
(1) Multiply the base cost by the area
(2) Add or subtract the modifications
(3) Apply the location modifier

When selecting costs, interpolate or use the cost that most nearly matches the structure under study. This applies to size, exterior wall construction and class.

# How to Use the Square Foot Cost Pages

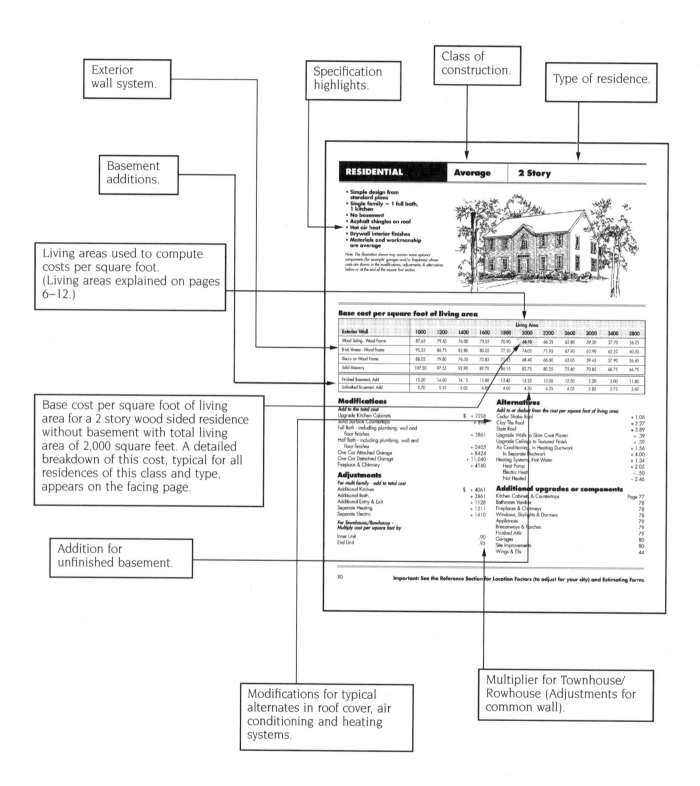

Exterior wall system.

Specification highlights.

Class of construction.

Type of residence.

Basement additions.

Living areas used to compute costs per square foot. (Living areas explained on pages 6–12.)

Base cost per square foot of living area for a 2 story wood sided residence without basement with total living area of 2,000 square feet. A detailed breakdown of this cost, typical for all residences of this class and type, appears on the facing page.

Addition for unfinished basement.

Modifications for typical alternates in roof cover, air conditioning and heating systems.

Multiplier for Townhouse/Rowhouse (Adjustments for common wall).


**RESIDENTIAL**  |  **Average**  |  **2 Story**

- Simple design from standard plans
- Single family — 1 full bath, 1 kitchen
- No basement
- Asphalt shingles on roof
- Hot air heat
- Drywall interior finishes
- Materials and workmanship are average

Note: The illustration shown may contain some optional components (for example garages and/or fireplaces) whose costs are shown in the modifications, adjustments, & alternatives below or at the end of the square foot section.

### Base cost per square foot of living area

| Exterior Wall | 1000 | 1200 | 1400 | 1600 | 1800 | 2000 | 2200 | 2600 | 3000 | 3400 | 3800 |
|---|---|---|---|---|---|---|---|---|---|---|---|
| Wood Siding - Wood Frame | 87.65 | 79.45 | 76.00 | 73.55 | 70.90 | 68.10 | 66.35 | 62.80 | 59.20 | 57.70 | 56.25 |
| Brick Veneer - Wood Frame | 95.55 | 86.75 | 82.80 | 80.05 | 77.10 | 74.05 | 71.95 | 67.90 | 63.90 | 62.20 | 60.50 |
| Stucco on Wood Frame | 88.05 | 79.80 | 76.30 | 73.85 | 71.25 | 68.40 | 66.60 | 63.05 | 59.45 | 57.90 | 56.45 |
| Solid Masonry | 107.20 | 97.55 | 92.90 | 89.70 | 86.15 | 82.75 | 80.25 | 75.40 | 70.85 | 68.75 | 66.75 |
| Finished Basement, Add | 15.20 | 14.60 | 14.15 | 13.80 | 13.45 | 13.25 | 13.00 | 12.50 | 12.20 | 12.00 | 11.80 |
| Unfinished Basement, Add | 5.70 | 5.35 | 5.05 | 4.85 | 4.60 | 4.50 | 4.35 | 4.05 | 3.85 | 3.75 | 3.60 |

### Modifications

Add to the total cost

| | |
|---|---|
| Upgrade Kitchen Cabinets | $ + 2206 |
| Solid Surface Countertops | + 898 |
| Full Bath - including plumbing, wall and floor finishes | + 3861 |
| Half Bath - including plumbing, wall and floor finishes | + 2405 |
| One Car Attached Garage | + 8424 |
| One Car Detached Garage | + 11,040 |
| Fireplace & Chimney | + 4160 |

### Adjustments

For multi family - add to total cost

| | |
|---|---|
| Additional Kitchen | $ + 4061 |
| Additional Bath | + 3861 |
| Additional Entry & Exit | + 1128 |
| Separate Heating | + 1211 |
| Separate Electric | + 1410 |

For Townhouse/Rowhouse -
Multiply cost per square foot by

| | |
|---|---|
| Inner Unit | .90 |
| End Unit | .95 |

### Alternatives

Add to or deduct from the cost per square foot of living area

| | |
|---|---|
| Cedar Shake Roof | + 1.06 |
| Clay Tile Roof | + 2.27 |
| Slate Roof | + 3.89 |
| Upgrade Walls to Skim Coat Plaster | + .39 |
| Upgrade Ceilings to Textured Finish | + .39 |
| Air Conditioning, in Heating Ductwork | + 1.56 |
| In Separate Ductwork | + 4.00 |
| Heating Systems, Hot Water | + 1.34 |
| Heat Pump | + 2.05 |
| Electric Heat | – .50 |
| Not Heated | – 2.46 |

### Additional upgrades or components

| | |
|---|---|
| Kitchen Cabinets & Countertops | Page 77 |
| Bathroom Vanities | 78 |
| Fireplaces & Chimneys | 78 |
| Windows, Skylights & Dormers | 78 |
| Appliances | 79 |
| Breezeways & Porches | 79 |
| Finished Attic | 79 |
| Garages | 80 |
| Site Improvements | 80 |
| Wings & Ells | 44 |

30    Important: See the Reference Section for Location Factors (to adjust for your city) and Estimating Forms

## Components
This page contains the ten components needed to develop the complete square foot cost of the typical dwelling specified. All components are defined with a description of the materials and/or task involved. Use cost figures from each component to estimate the cost per square foot of that section of the project.

## Specifications
The parameters for an example dwelling from the facing page are listed here. Included are the square foot dimensions of the proposed building. LIVING AREA takes into account the number of floors and other factors needed to define a building's TOTAL SQUARE FOOTAGE. Perimeter and partition dimensions are defined in terms of linear feet.

## Line Totals
The extreme right-hand column lists the sum of two figures. Use this total to determine the sum of MATERIAL COST plus INSTALLATION COST. The result is a convenient total cost for each of the ten components.

### Average 2 Story
Living Area - 2000 S.F.
Perimeter - 135 L.F.

| | | Labor-Hours | Cost Per Square Foot Of Living Area | | |
|---|---|---|---|---|---|
| | | | Mat. | Labor | Total |
| 1 Site Work | Site preparation for slab, 4' deep trench excavation for foundation wall. | .034 | | .51 | .51 |
| 2 Foundation | Continuous reinforced concrete footing 8" deep x 18" wide; dampproofed and insulated reinforced concrete foundation wall, 8" thick, 4' deep, 4" concrete slab on 4" crushed stone base and polyethylene vapor barrier, trowel finish. | .066 | 2.25 | 2.77 | 5.02 |
| 3 Framing | Exterior walls - 2" x 4" wood studs, 16" O.C.; 1/2" plywood sheathing; 2" x 6" rafters 16" O.C. with 1/2" plywood sheathing, 4 in 12 pitch; 2" x 6" ceiling joists 16" O.C.; 2" x 8" floor joists 16" O.C. with 5/8" plywood subfloor; 1/2" plywood subfloor on 1" x 2" wood sleepers 16" O.C. | .131 | 5.39 | 5.85 | 11.24 |
| 4 Exterior Walls | Beveled wood siding and #15 felt building paper on insulated wood frame walls; 6" attic insulation; double hung windows; 3 flush solid core wood exterior doors with storms. | .111 | 7.83 | 4.09 | 11.92 |
| 5 Roofing | 25 year asphalt shingles; #15 felt building paper; aluminum gutters, downspouts, drip edge and flashings. | .024 | .44 | .83 | 1.27 |
| 6 Interiors | Walls and ceilings, 1/2" taped and finished drywall, primed and painted with 2 coats; painted baseboard and trim, finished hardwood floor 40%, carpet with 1/2" underlayment 40%, vinyl tile with 1/2" underlayment 15%, ceramic tile with 1/2" underlayment 5%; hollow core and louvered interior doors. | .232 | 10.13 | 10.65 | 20.78 |
| 7 Specialties | Average grade kitchen cabinets - 14 L.F. wall and base with plastic laminate counter top and kitchen sink; 40 gallon electric water heater. | .021 | 1.15 | .48 | 1.63 |
| 8 Mechanical | 1 lavatory, white, wall hung; 1 water closet, white; 1 bathtub with shower; enameled steel, white; gas fired warm air heat. | .060 | 2.17 | 1.95 | 4.12 |
| 9 Electrical | 200 Amp. service; romex wiring; incandescent lighting fixtures, switches, receptacles. | .039 | .74 | 1.12 | 1.86 |
| 10 Overhead | Contractor's overhead and profit and plans. | | 5.03 | 4.72 | 9.75 |
| | Total | | 35.13 | 32.97 | 68.10 |

31

## Labor-hours
Use this column to determine the unit of measure in LABOR-HOURS needed to perform a task. This figure will give the builder LABOR-HOURS PER SQUARE FOOT of the building. The TOTAL LABOR-HOURS PER COMPONENT are determined by multiplying the LIVING AREA times the LABOR-HOURS listed on that line. (TOTAL LABOR-HOURS PER COMPONENT = LIVING AREA x LABOR-HOURS).

## Installation
The labor rates included here incorporate the overhead and profit costs for the installing contractor. The average mark-up used to create these figures is 70.3% over and above BARE LABOR COST including fringe benefits.

## Bottom Line Total
This figure is the complete square foot cost for the construction project. To determine TOTAL PROJECT COST, multiply the BOTTOM LINE TOTAL times the LIVING AREA. (TOTAL PROJECT COST = BOTTOM LINE TOTAL x LIVING AREA).

## Materials
This column gives the unit needed to develop the COST OF MATERIALS. Note: The figures given here are not BARE COSTS. Ten percent has been added to BARE MATERIAL COST to cover handling.

## *NOTE
The components listed on this page are typical of all the sizes of residences from the facing page. Specific quantities of components required would vary with the size of the dwelling and the exterior wall system.

# Building Classes

Given below are the four general definitions of building classes. Each building — Economy, Average, Custom and Luxury — is common in residential construction. All four are used in this book to determine costs per square foot.

## Economy Class

An economy class residence is usually mass-produced from stock plans. The materials and workmanship are sufficient only to satisfy minimum building codes. Low construction cost is more important than distinctive features. Design is seldom other than square or rectangular.

## Average Class

An average class residence is simple in design and is built from standard designer plans. Material and workmanship are average, but often exceed the minimum building codes. There are frequently special features that give the residence some distinctive characteristics.

## Custom Class

A custom class residence is usually built from a designer's plans which have been modified to give the building a distinction of design. Material and workmanship are generally above average with obvious attention given to construction details. Construction normally exceeds building code requirements.

## Luxury Class

A luxury class residence is built from an architect's plan for a specific owner. It is unique in design and workmanship. There are many special features, and construction usually exceeds all building codes. It is obvious that primary attention is placed on the owner's comfort and pleasure. Construction is supervised by an architect.

# Residential Building Types

## One Story

This is an example of a one-story dwelling. The living area of this type of residence is confined to the ground floor. The headroom in the attic is usually too low for use as a living area.

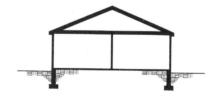

## One-and-a-half Story

The living area in the upper level of this type of residence is 50% to 90% of the ground floor. This is made possible by a combination of this design's high-peaked roof and/or dormers. Only the upper level area with a ceiling height of 6' or more is considered living area. The living area of this residence is the sum of the ground floor area plus the area on the second level with a ceiling height of 6' or more.

# Residential Building Types

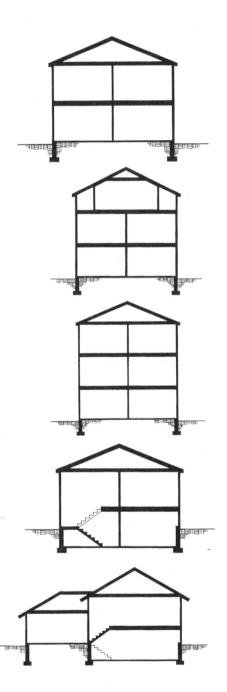

## Two Story

This type of residence has a second floor or upper level area which is equal or nearly equal to the ground floor area. The upper level of this type of residence can range from 90% to 110% of the ground floor area, depending on setbacks or overhangs. The living area is the sum of the ground floor area and the upper level floor area.

## Two-and-one-half Story

This type of residence has two levels of equal or nearly equal area and a third level which has a living area that is 50% to 90% of the ground floor. This is made possible by a high peaked roof, extended wall heights and/or dormers. Only the upper level area with a ceiling height of 6 feet or more is considered living area. The living area of this residence is the sum of the ground floor area, the second floor area and the area on the third level with a ceiling height of 6 feet or more.

## Three Story

This type of residence has three levels which are equal or nearly equal. As in the 2 story residence, the second and third floor areas may vary slightly depending on the setbacks or overhangs. The living area is the sum of the ground floor area and the two upper level floor areas.

## Bi-Level

This type of residence has two living areas, one above the other. One area is about 4 feet below grade and the second is about 4 feet above grade. Both are equal in size. The lower level in this type of residence is originally designed and built to serve as a living area and not as a basement. Both levels have full ceiling heights. The living area is the sum of the lower level area and the upper level area.

## Tri-Level

This type of residence has three levels of living area. One is at grade level, the second is about 4 feet below grade, and the third is about 4 feet above grade. All levels are originally designed to serve as living areas. All levels have full ceiling heights. The living area is the sum of the areas of each of the three levels.

# Exterior Wall Construction

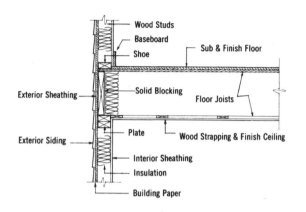

## Typical Frame Construction

Typical wood frame construction consists of wood studs with insulation between them. A typical exterior surface is made up of sheathing, building paper and exterior siding consisting of wood, vinyl, aluminum or stucco over the wood sheathing.

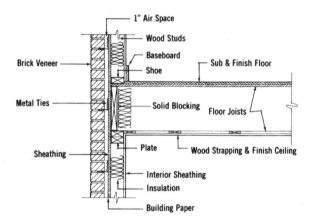

## Brick Veneer

Typical brick veneer construction consists of wood studs with insulation between them. A typical exterior surface is sheathing, building paper and an exterior of brick tied to the sheathing with metal strips.

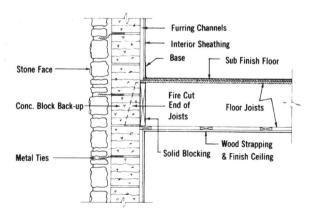

## Stone

Typical solid masonry construction consists of a stone or block wall covered on the exterior with brick, stone or other masonry.

# Residential Configurations

## Detached House

This category of residence is a freestanding separate building with or without an attached garage. It has four complete walls.

## Town/Row House

This category of residence has a number of attached units made up of inner units and end units. The units are joined by common walls. The inner units have only two exterior walls. The common walls are fireproof. The end units have three walls and a common wall. Town houses/row houses can be any of the five types.

## Semi-Detached House

This category of residence has two living units side-by-side. The common wall is a fireproof wall. Semi-detached residences can be treated as a row house with two end units. Semi-detached residences can be any of the five types.

# Residential Garage Types

## Attached Garages

Shares a common wall with the dwelling. Access is typically through a door between dwelling and garage.

## Built-In Garage

Constructed under the second floor living space and above basement level of dwelling. Reduces gross square feet of living area.

## Basement Garage

Constructed under the roof of the dwelling but below the living area.

## Detached Garage

Constructed apart from the main dwelling. Shares no common area or wall with the dwelling.

# RESIDENTIAL COST ESTIMATE

OWNER'S NAME: _____  APPRAISER: _____

RESIDENCE ADDRESS: _____  PROJECT: _____

CITY, STATE, ZIP CODE: _____  DATE: _____

| CLASS OF CONSTRUCTION | RESIDENCE TYPE | CONFIGURATION | EXTERIOR WALL SYSTEM |
|---|---|---|---|
| ☐ ECONOMY | ☐ 1 STORY | ☐ DETACHED | ☐ WOOD SIDING - WOOD FRAME |
| ☐ AVERAGE | ☐ 1-1/2 STORY | ☐ TOWN/ROW HOUSE | ☐ BRICK VENEER - WOOD FRAME |
| ☐ CUSTOM | ☐ 2 STORY | ☐ SEMI-DETACHED | ☐ STUCCO ON WOOD FRAME |
| ☐ LUXURY | ☐ 2-1/2 STORY | | ☐ PAINTED CONCRETE BLOCK |
| | ☐ 3 STORY | OCCUPANCY | ☐ SOLID MASONRY (AVERAGE & CUSTOM) |
| | ☐ BI-LEVEL | ☐ ONE FAMILY | ☐ STONE VENEER - WOOD FRAME |
| | ☐ TRI-LEVEL | ☐ TWO FAMILY | ☐ SOLID BRICK (LUXURY) |
| | | ☐ THREE FAMILY | ☐ SOLID STONE (LUXURY) |
| | | ☐ OTHER _____ | |

| * LIVING AREA (Main Building) | | * LIVING AREA (Wing or Ell) ( ) | | * LIVING AREA ( WING or ELL ) ( ) | |
|---|---|---|---|---|---|
| First Level | _____ S.F. | First Level | _____ S.F. | First Level | _____ S.F. |
| Second level | _____ S.F. | Second level | _____ S.F. | Second level | _____ S.F. |
| Third Level | _____ S.F. | Third Level | _____ S.F. | Third Level | _____ S.F. |
| Total | _____ S.F. | Total | _____ S.F. | Total | _____ S.F. |

\* Basement Area is not part of living area.

| MAIN BUILDING | COSTS PER S.F. LIVING AREA |
|---|---|
| Cost per Square Foot of Living Area, from Page _____ | $ |
| Basement Addition: _____ % Finished, _____ % Unfinished | + |
| Roof Cover Adjustment: _____ Type, Page _____ (Add or Deduct) | ( ) |
| Central Air Conditioning: ☐ Separate Ducts ☐ Heating Ducts, Page _____ | + |
| Heating System Adjustment: _____ Type, Page _____ (Add or Deduct) | ( ) |
| Main Building: Adjusted Cost per S.F. of Living Area | $ |

| MAIN BUILDING TOTAL COST | $ _____ /S.F. Cost per S.F. Living Area | X | _____ S.F. Living Area | X | _____ Town/Row House Multiplier (Use 1 for Detached) | = | $ _____ TOTAL COST |
|---|---|---|---|---|---|---|---|

| WING OR ELL ( ) _____ STORY | COSTS PER S.F. LIVING AREA |
|---|---|
| Cost per Square Foot of Living Area, from Page _____ | $ |
| Basement Addition: _____ % Finished, _____ % Unfinished | + |
| Roof Cover Adjustment: _____ Type, Page _____ (Add or Deduct) | ( ) |
| Central Air Conditioning: ☐ Separate Ducts ☐ Heating Ducts, Page _____ | + |
| Heating System Adjustment: _____ Type, Page _____ (Add or Deduct) | ( ) |
| Wing or Ell ( ): Adjusted Cost per S.F. of Living Area | $ |

| WING OR ELL ( ) TOTAL COST | $ _____ /S.F. Cost per S.F. Living Area | X | _____ S.F. Living Area | X | | = | $ _____ TOTAL COST |
|---|---|---|---|---|---|---|---|

| WING OR ELL ( ) _____ STORY | COSTS PER S.F. LIVING AREA |
|---|---|
| Cost per Square Foot of Living Area, from Page _____ | $ |
| Basement Addition: _____ % Finished, _____ % Unfinished | + |
| Roof Cover Adjustment: _____ Type, Page _____ (Add or Deduct) | ( ) |
| Central Air Conditioning: ☐ Separate Ducts ☐ Heating Ducts, Page _____ | + |
| Heating System Adjustment: _____ Type, Page _____ (Add or Deduct) | ( ) |
| Wing or Ell ( ): Adjusted Cost per S.F. of Living Area | $ |

| WING OR ELL ( ) TOTAL COST | $ _____ /S.F. Cost per S.F. Living Area | X | _____ S.F. Living Area | X | | = | $ _____ TOTAL COST |
|---|---|---|---|---|---|---|---|

TOTAL THIS PAGE  [ _____ ]

Page 1 of 2

11

## RESIDENTIAL
## COST ESTIMATE

| Total Page 1 | | | $ |
|---|---|---|---|
| | QUANTITY | UNIT COST | |
| Additional Bathrooms: _____ Full _____ Half | | | |
| Finished Attic: _____ Ft. x _____ Ft. | S.F. | | + |
| Breezeway: ☐ Open ☐ Enclosed _____ Ft. x _____ Ft. | S.F. | | + |
| Covered Porch: ☐ Open ☐ Enclosed _____ Ft. x _____ Ft. | S.F. | | + |
| Fireplace: ☐ Interior Chimney ☐ Exterior Chimney | | | |
| ☐ No. of Flues ☐ Additional Fireplaces | | | + |
| Appliances: | | | + |
| Kitchen Cabinets Adjustments: (±) | | | |
| ☐ Garage ☐ Carport: _____ Car(s) Description _____ (±) | | | |
| Miscellaneous: | | | + |

ADJUSTED TOTAL BUILDING COST  | $ |

| REPLACEMENT COST | |
|---|---|
| ADJUSTED TOTAL BUILDING COST | $ _____ |
| Site Improvements | |
|    (A) Paving & Sidewalks | $ _____ |
|    (B) Landscaping | $ _____ |
|    (C) Fences | $ _____ |
|    (D) Swimming Pools | $ _____ |
|    (E) Miscellaneous | $ _____ |
| TOTAL | $ _____ |
| Location Factor | x _____ |
| Location Replacement Cost | $ _____ |
| Depreciation | - $ _____ |
| LOCAL DEPRECIATED COST | $ _____ |

| INSURANCE COST | |
|---|---|
| ADJUSTED TOTAL BUILDING COST | $ _____ |
| Insurance Exclusions | |
|    (A) Footings, Site work, Underground Piping | - $ _____ |
|    (B) Architects Fees | - $ _____ |
| Total Building Cost Less Exclusion | $ _____ |
| Location Factor | x _____ |
| LOCAL INSURABLE REPLACEMENT COST | $ _____ |

---

### SKETCH AND ADDITIONAL CALCULATIONS

**SQUARE FOOT COSTS**

## 1 Story

© Home Planners, Inc.

## 1-1/2 Story

## 2 Story

## Bi-Level

## Tri-Level

©Design Basics, Inc.

# RESIDENTIAL | Economy | 1 Story

- **Mass produced from stock plans**
- **Single family — 1 full bath, 1 kitchen**
- **No basement**
- **Asphalt shingles on roof**
- **Hot air heat**
- **Drywall interior finishes**
- **Materials and workmanship are sufficient to meet codes**

*Note: The illustration shown may contain some optional components (for example: garages and/or fireplaces) whose costs are shown in the modifications, adjustments, & alternatives below or at the end of the square foot section.*

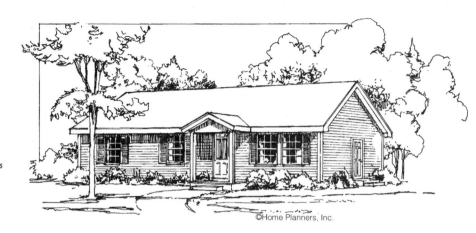

©Home Planners, Inc.

## Base cost per square foot of living area

| Exterior Wall | Living Area | | | | | | | | | | |
|---|---|---|---|---|---|---|---|---|---|---|---|
| | 600 | 800 | 1000 | 1200 | 1400 | 1600 | 1800 | 2000 | 2400 | 2800 | 3200 |
| Wood Siding - Wood Frame | 77.65 | 70.55 | 65.15 | 60.70 | 56.75 | 54.30 | 53.15 | 51.50 | 48.05 | 45.55 | 43.95 |
| Brick Veneer - Wood Frame | 84.30 | 76.45 | 70.50 | 65.55 | 61.20 | 58.40 | 57.15 | 55.25 | 51.50 | 48.75 | 46.90 |
| Stucco on Wood Frame | 75.80 | 68.90 | 63.65 | 59.35 | 55.55 | 53.15 | 52.05 | 50.45 | 47.10 | 44.70 | 43.15 |
| Painted Concrete Block | 78.95 | 71.70 | 66.15 | 61.65 | 57.65 | 55.05 | 53.95 | 52.20 | 48.75 | 46.15 | 44.50 |
| Finished Basement, Add | 20.85 | 19.65 | 18.75 | 18.00 | 17.35 | 16.95 | 16.70 | 16.40 | 15.90 | 15.55 | 15.25 |
| Unfinished Basement, Add | 9.45 | 8.45 | 7.75 | 7.15 | 6.65 | 6.35 | 6.15 | 5.90 | 5.50 | 5.25 | 5.00 |

## Modifications

*Add to the total cost*

| | |
|---|---|
| Upgrade Kitchen Cabinets | $ + 287 |
| Solid Surface Countertops | + 665 |
| Full Bath - including plumbing, wall and floor finishes | + 3076 |
| Half Bath - including plumbing, wall and floor finishes | + 1916 |
| One Car Attached Garage | + 8010 |
| One Car Detached Garage | + 10,343 |
| Fireplace & Chimney | + 3655 |

## Adjustments

*For multi family - add to total cost*

| | |
|---|---|
| Additional Kitchen | $ + 2167 |
| Additional Bath | + 3076 |
| Additional Entry & Exit | + 1128 |
| Separate Heating | + 1211 |
| Separate Electric | + 748 |

*For Townhouse/Rowhouse - Multiply cost per square foot by*

| | |
|---|---|
| Inner Unit | .95 |
| End Unit | .97 |

## Alternatives

*Add to or deduct from the cost per square foot of living area*

| | |
|---|---|
| Composition Roll Roofing | – .5 |
| Cedar Shake Roof | + 2.4 |
| Upgrade Walls and Ceilings to Skim Coat Plaster | + .5 |
| Upgrade Ceilings to Textured Finish | + .3 |
| Air Conditioning, in Heating Ductwork | + 2.5 |
| In Separate Ductwork | + 4.9 |
| Heating Systems, Hot Water | + 1.4 |
| Heat Pump | + 1.7 |
| Electric Heat | – 1.2 |
| Not Heated | – 3.1 |

## Additional upgrades or components

| | |
|---|---|
| Kitchen Cabinets & Countertops | Page 7 |
| Bathroom Vanities | 7 |
| Fireplaces & Chimneys | 7 |
| Windows, Skylights & Dormers | 7 |
| Appliances | 7 |
| Breezeways & Porches | 7 |
| Finished Attic | 7 |
| Garages | 8 |
| Site Improvements | 8 |
| Wings & Ells | 2 |

**Important: See the Reference Section for Location Factors (to adjust for your city) and Estimating Form**

| | | Labor-Hours | Cost Per Square Foot Of Living Area | | |
|---|---|---|---|---|---|
| | | | Mat. | Labor | Total |
| **1** Site Work | Site preparation for slab; 4' deep trench excavation for foundation wall. | .060 | | .84 | .84 |
| **2** Foundation | Continuous reinforced concrete footing, 8" deep x 18" wide; dampproofed and insulated 8" thick reinforced concrete block foundation wall, 4' deep; 4" concrete slab on 4" crushed stone base and polyethylene vapor barrier, trowel finish. | .131 | 4.12 | 4.95 | 9.07 |
| **3** Framing | Exterior walls - 2" x 4" wood studs, 16" O.C.; 1/2" insulation board sheathing; wood truss roof frame, 24" O.C. with 1/2" plywood sheathing, 4 in 12 pitch. | .098 | 3.89 | 4.23 | 8.12 |
| **4** Exterior Walls | Metal lath reinforced stucco exterior on insulated wood frame walls; 6" attic insulation; sliding sash wood windows; 2 flush solid core wood exterior doors with storms. | .110 | 4.38 | 4.44 | 8.82 |
| **5** Roofing | 20 year asphalt shingles; #15 felt building paper; aluminum gutters, downspouts, drip edge and flashings. | .047 | .85 | 1.62 | 2.47 |
| **6** Interiors | Walls and ceilings, 1/2" taped and finished drywall, primed and painted with 2 coats; painted baseboard and trim; rubber backed carpeting 80%, asphalt tile 20%; hollow core wood interior doors. | .243 | 5.85 | 8.08 | 13.93 |
| **7** Specialties | Economy grade kitchen cabinets - 6 L.F. wall and base with plastic laminate counter top and kitchen sink; 30 gallon electric water heater. | .004 | 1.15 | .62 | 1.77 |
| **8** Mechanical | 1 lavatory, white, wall hung; 1 water closet, white; 1 bathtub, enameled steel, white; gas fired warm air heat. | .086 | 2.75 | 2.27 | 5.02 |
| **9** Electrical | 100 Amp. service; romex wiring; incandescent lighting fixtures, switches, receptacles. | .036 | .56 | 1.00 | 1.56 |
| **10** Overhead | Contractor's overhead and profit | | 3.54 | 4.21 | 7.75 |
| | **Total** | | 27.09 | 32.26 | **59.35** |

**SQUARE FOOT COSTS**

- **Mass produced from stock plans**
- **Single family — 1 full bath, 1 kitchen**
- **No basement**
- **Asphalt shingles on roof**
- **Hot air heat**
- **Drywall interior finishes**
- **Materials and workmanship are sufficient to meet codes**

*Note: The illustration shown may contain some optional components (for example: garages and/or fireplaces) whose costs are shown in the modifications, adjustments, & alternatives below or at the end of the square foot section.*

## Base cost per square foot of living area

| Exterior Wall | Living Area | | | | | | | | | | |
|---|---|---|---|---|---|---|---|---|---|---|---|
| | 600 | 800 | 1000 | 1200 | 1400 | 1600 | 1800 | 2000 | 2400 | 2800 | 3200 |
| Wood Siding - Wood Frame | 87.25 | 73.00 | 65.20 | 61.70 | 59.25 | 55.30 | 53.45 | 51.40 | 47.25 | 45.70 | 44.10 |
| Brick Veneer - Wood Frame | 96.55 | 79.75 | 71.45 | 67.55 | 64.85 | 60.35 | 58.25 | 56.00 | 51.25 | 49.50 | 47.60 |
| Stucco on Wood Frame | 84.70 | 71.10 | 63.50 | 60.15 | 57.75 | 53.90 | 52.15 | 50.15 | 46.10 | 44.65 | 43.10 |
| Painted Concrete Block | 89.05 | 74.30 | 66.45 | 62.85 | 60.35 | 56.30 | 54.40 | 52.30 | 48.05 | 46.40 | 44.75 |
| Finished Basement, Add | 16.00 | 13.60 | 13.00 | 12.55 | 12.20 | 11.75 | 11.45 | 11.20 | 10.70 | 10.50 | 10.25 |
| Unfinished Basement, Add | 8.20 | 6.35 | 5.85 | 5.50 | 5.20 | 4.85 | 4.65 | 4.45 | 4.05 | 3.90 | 3.70 |

## Modifications

*Add to the total cost*

| | |
|---|---|
| Upgrade Kitchen Cabinets | $ + 287 |
| Solid Surface Countertops | + 665 |
| Full Bath - including plumbing, wall and floor finishes | + 3076 |
| Half Bath - including plumbing, wall and floor finishes | + 1916 |
| One Car Attached Garage | + 8010 |
| One Car Detached Garage | + 10,343 |
| Fireplace & Chimney | + 3655 |

## Adjustments

*For multi family - add to total cost*

| | |
|---|---|
| Additional Kitchen | $ + 2167 |
| Additional Bath | + 3076 |
| Additional Entry & Exit | + 1128 |
| Separate Heating | + 1211 |
| Separate Electric | + 748 |

*For Townhouse/Rowhouse -*
*Multiply cost per square foot by*

| | |
|---|---|
| Inner Unit | .95 |
| End Unit | .97 |

## Alternatives

*Add to or deduct from the cost per square foot of living area*

| | |
|---|---|
| Composition Roll Roofing | – .3 |
| Cedar Shake Roof | + 1.7 |
| Upgrade Walls and Ceilings to Skim Coat Plaster | + .5 |
| Upgrade Ceilings to Textured Finish | + .3 |
| Air Conditioning, in Heating Ductwork | + 1.9 |
| In Separate Ductwork | + 4.2 |
| Heating Systems, Hot Water | + 1.3 |
| Heat Pump | + 1.8 |
| Electric Heat | – .9 |
| Not Heated | – 2.9 |

## Additional upgrades or components

| | |
|---|---|
| Kitchen Cabinets & Countertops | Page 7 |
| Bathroom Vanities | 7 |
| Fireplaces & Chimneys | 7 |
| Windows, Skylights & Dormers | 7 |
| Appliances | 7 |
| Breezeways & Porches | 7 |
| Finished Attic | 7 |
| Garages | 8 |
| Site Improvements | 8 |
| Wings & Ells | 2 |

| | | Labor-Hours | Cost Per Square Foot Of Living Area | | |
|---|---|---|---|---|---|
| | | | Mat. | Labor | Total |
| **1 Site Work** | Site preparation for slab; 4' deep trench excavation for foundation wall. | .041 | | .63 | .63 |
| **2 Foundation** | Continuous reinforced concrete footing, 8" deep x 18" wide; dampproofed and insulated 8" thick reinforced concrete block foundation wall, 4' deep; 4" concrete slab on 4" crushed stone base and polyethylene vapor barrier, trowel finish. | .073 | 2.74 | 3.36 | 6.10 |
| **3 Framing** | Exterior walls - 2" x 4" wood studs, 16" O.C.; 1/2" insulation board sheathing; 2" x 6" rafters, 16" O.C. with 1/2" plywood sheathing, 8 in 12 pitch; 2" x 8" floor joists 16" O.C. with bridging and 5/8" plywood subfloor. | .090 | 4.06 | 4.72 | 8.78 |
| **4 Exterior Walls** | Beveled wood siding and #15 felt building paper on insulated wood frame walls; 6" attic insulation; double hung windows; 2 flush solid core wood exterior doors with storms. | .077 | 5.89 | 3.70 | 9.59 |
| **5 Roofing** | 20 year asphalt shingles; #15 felt building paper; aluminum gutters, downspouts, drip edge and flashings. | .029 | .53 | 1.01 | 1.54 |
| **6 Interiors** | Walls and ceilings, 1/2" taped and finished drywall, primed and painted with 2 coats; painted baseboard and trim; rubber backed carpeting 80%, asphalt tile 20%; hollow core wood interior doors. | .204 | 6.01 | 8.39 | 14.40 |
| **7 Specialties** | Economy grade kitchen cabinets - 6 L.F. wall and base with plastic laminate counter top and kitchen sink; 30 gallon electric water heater. | .020 | .86 | .47 | 1.33 |
| **8 Mechanical** | 1 lavatory, white, wall hung; 1 water closet, white; 1 bathtub, enameled steel, white; gas fired warm air heat. | .079 | 2.29 | 2.03 | 4.32 |
| **9 Electrical** | 100 Amp. service; romex wiring; incandescent lighting fixtures, switches, receptacles. | .033 | .50 | .90 | 1.40 |
| **10 Overhead** | Contractor's overhead and profit. | | 3.43 | 3.78 | 7.21 |
| | **Total** | | 26.31 | 28.99 | **55.30** |

# RESIDENTIAL | Economy | 2 Story

- **Mass produced from stock plans**
- **Single family — 1 full bath, 1 kitchen**
- **No basement**
- **Asphalt shingles on roof**
- **Hot air heat**
- **Drywall interior finishes**
- **Materials and workmanship are sufficient to meet codes**

*Note: The illustration shown may contain some optional components (for example: garages and/or fireplaces) whose costs are shown in the modifications, adjustments, & alternatives below or at the end of the square foot section.*

## Base cost per square foot of living area

| Exterior Wall | Living Area | | | | | | | | | | |
|---|---|---|---|---|---|---|---|---|---|---|---|
| | 1000 | 1200 | 1400 | 1600 | 1800 | 2000 | 2200 | 2600 | 3000 | 3400 | 3800 |
| Wood Siding - Wood Frame | 69.95 | 63.20 | 60.25 | 58.20 | 56.20 | 53.70 | 52.20 | 49.30 | 46.25 | 44.95 | 43.85 |
| Brick Veneer - Wood Frame | 77.15 | 69.85 | 66.50 | 64.15 | 61.80 | 59.10 | 57.35 | 53.90 | 50.55 | 49.05 | 47.70 |
| Stucco on Wood Frame | 67.95 | 61.35 | 58.55 | 56.60 | 54.70 | 52.20 | 50.80 | 48.00 | 45.05 | 43.85 | 42.75 |
| Painted Concrete Block | 71.35 | 64.50 | 61.50 | 59.35 | 57.30 | 54.75 | 53.20 | 50.20 | 47.10 | 45.75 | 44.60 |
| Finished Basement, Add | 10.90 | 10.45 | 10.10 | 9.85 | 9.55 | 9.40 | 9.20 | 8.85 | 8.60 | 8.45 | 8.30 |
| Unfinished Basement, Add | 5.10 | 4.70 | 4.40 | 4.20 | 4.00 | 3.90 | 3.70 | 3.45 | 3.25 | 3.10 | 3.00 |

## Modifications

*Add to the total cost*

| | |
|---|---|
| Upgrade Kitchen Cabinets | $ + 287 |
| Solid Surface Countertops | + 665 |
| Full Bath - including plumbing, wall and floor finishes | + 3076 |
| Half Bath - including plumbing, wall and floor finishes | + 1916 |
| One Car Attached Garage | + 8010 |
| One Car Detached Garage | + 10,343 |
| Fireplace & Chimney | + 4035 |

## Adjustments

*For multi family - add to total cost*

| | |
|---|---|
| Additional Kitchen | $ + 2167 |
| Additional Bath | + 3076 |
| Additional Entry & Exit | + 1128 |
| Separate Heating | + 1211 |
| Separate Electric | + 748 |

*For Townhouse/Rowhouse -*
*Multiply cost per square foot by*

| | |
|---|---|
| Inner Unit | .93 |
| End Unit | .96 |

## Alternatives

*Add to or deduct from the cost per square foot of living area*

| | |
|---|---|
| Composition Roll Roofing | – .2 |
| Cedar Shake Roof | + 1.2 |
| Upgrade Walls and Ceilings to Skim Coat Plaster | + .5 |
| Upgrade Ceilings to Textured Finish | + .3 |
| Air Conditioning, in Heating Ductwork | + 1.5 |
| In Separate Ductwork | + 3.8 |
| Heating Systems, Hot Water | + 1.3 |
| Heat Pump | + 1.9 |
| Electric Heat | – .7 |
| Not Heated | – 2.7 |

## Additional upgrades or components

| | |
|---|---|
| Kitchen Cabinets & Countertops | Page 7 |
| Bathroom Vanities | 7 |
| Fireplaces & Chimneys | 7 |
| Windows, Skylights & Dormers | 7 |
| Appliances | 7 |
| Breezeways & Porches | 7 |
| Finished Attic | 7 |
| Garages | 8 |
| Site Improvements | 8 |
| Wings & Ells | 2 |

**Important: See the Reference Section for Location Factors (to adjust for your city) and Estimating Form**

# Economy 2 Story

**Living Area - 2000 S.F.**
**Perimeter - 135 L.F.**

| | | Labor-Hours | Cost Per Square Foot Of Living Area | | |
|---|---|---|---|---|---|
| | | | Mat. | Labor | Total |
| **1 Site Work** | Site preparation for slab; 4' deep trench excavation for foundation wall. | .034 | | .50 | .50 |
| **2 Foundation** | Continuous reinforced concrete footing, 8" deep x 18" wide; dampproofed and insulated 8" thick reinforced concrete block foundation wall, 4' deep; 4" concrete slab on 4" crushed stone base and polyethylene vapor barrier, trowel finish. | .069 | 2.19 | 2.69 | 4.88 |
| **3 Framing** | Exterior walls - 2" x 4" wood studs, 16" O.C.; 1/2" insulation board sheathing; wood truss roof frame, 24" O.C. with 1/2" plywood sheathing, 4 in 12 pitch; 2" x 8" floor joists 16" O.C. with bridging and 5/8" plywood subfloor. | .112 | 4.17 | 4.97 | 9.14 |
| **4 Exterior Walls** | Beveled wood siding and #15 felt building paper on insulated wood frame walls; 6" attic insulation; double hung windows; 2 flush solid core wood exterior doors with storms. | .107 | 5.96 | 3.78 | 9.74 |
| **5 Roofing** | 20 year asphalt shingles; #15 felt building paper; aluminum gutters, downspouts, drip edge and flashings. | .024 | .43 | .81 | 1.24 |
| **6 Interiors** | Walls and ceilings, 1/2" taped and finished drywall, primed and painted with 2 coats; painted baseboard and trim; rubber backed carpeting 80%, asphalt tile 20%; hollow core wood interior doors. | .219 | 6.23 | 8.69 | 14.92 |
| **7 Specialties** | Economy grade kitchen cabinets - 6 L.F. wall and base with plastic laminate counter top and kitchen sink; 30 gallon electric water heater. | .017 | .69 | .37 | 1.06 |
| **8 Mechanical** | 1 lavatory, white, wall hung; 1 water closet, white; 1 bathtub, enameled steel, white; gas fired warm air heat. | .061 | 2.02 | 1.90 | 3.92 |
| **9 Electrical** | 100 Amp. service; romex wiring; incandescent lighting fixtures; switches, receptacles. | .030 | .47 | .84 | 1.31 |
| **10 Overhead** | Contractor's overhead and profit | | 3.32 | 3.67 | 6.99 |
| | **Total** | | 25.48 | 28.23 | **53.70** |

19

# RESIDENTIAL | Economy | Bi-Level

- **Mass produced from stock plans**
- **Single family — 1 full bath, 1 kitchen**
- **No basement**
- **Asphalt shingles on roof**
- **Hot air heat**
- **Drywall interior finishes**
- **Materials and workmanship are sufficient to meet codes**

*Note: The illustration shown may contain some optional components (for example: garages and/or fireplaces) whose costs are shown in the modifications, adjustments, & alternatives below or at the end of the square foot section.*

## Base cost per square foot of living area

| Exterior Wall | Living Area | | | | | | | | | | |
|---|---|---|---|---|---|---|---|---|---|---|---|
| | 1000 | 1200 | 1400 | 1600 | 1800 | 2000 | 2200 | 2600 | 3000 | 3400 | 3800 |
| Wood Siding - Wood Frame | 65.25 | 58.85 | 56.20 | 54.35 | 52.55 | 50.20 | 48.90 | 46.25 | 43.40 | 42.30 | 41.30 |
| Brick Veneer - Wood Frame | 70.65 | 63.85 | 60.90 | 58.80 | 56.75 | 54.25 | 52.70 | 49.75 | 46.65 | 45.35 | 44.20 |
| Stucco on Wood Frame | 63.75 | 57.50 | 54.95 | 53.10 | 51.40 | 49.10 | 47.85 | 45.30 | 42.55 | 41.45 | 40.55 |
| Painted Concrete Block | 66.30 | 59.85 | 57.10 | 55.20 | 53.40 | 50.95 | 49.60 | 46.95 | 44.05 | 42.90 | 41.90 |
| Finished Basement, Add | 10.90 | 10.45 | 10.10 | 9.85 | 9.55 | 9.40 | 9.20 | 8.85 | 8.60 | 8.45 | 8.30 |
| Unfinished Basement, Add | 5.10 | 4.70 | 4.40 | 4.20 | 4.00 | 3.90 | 3.70 | 3.45 | 3.25 | 3.10 | 3.00 |

## Modifications

*Add to the total cost*

| | |
|---|---|
| Upgrade Kitchen Cabinets | $  + 287 |
| Solid Surface Countertops | + 665 |
| Full Bath - including plumbing, wall and floor finishes | + 3076 |
| Half Bath - including plumbing, wall and floor finishes | + 1916 |
| One Car Attached Garage | + 8010 |
| One Car Detached Garage | + 10,343 |
| Fireplace & Chimney | + 3655 |

## Adjustments

*For multi family - add to total cost*

| | |
|---|---|
| Additional Kitchen | $  + 2167 |
| Additional Bath | + 3076 |
| Additional Entry & Exit | + 1128 |
| Separate Heating | + 1211 |
| Separate Electric | + 748 |

*For Townhouse/Rowhouse -*
*Multiply cost per square foot by*

| | |
|---|---|
| Inner Unit | .94 |
| End Unit | .97 |

## Alternatives

*Add to or deduct from the cost per square foot of living area*

| | |
|---|---|
| Composition Roll Roofing | – .2 |
| Cedar Shake Roof | + 1.2 |
| Upgrade Walls and Ceilings to Skim Coat Plaster | + .5 |
| Upgrade Ceilings to Textured Finish | + .3 |
| Air Conditioning, in Heating Ductwork | + 1.5 |
| In Separate Ductwork | + 3.8 |
| Heating Systems, Hot Water | + 1.3 |
| Heat Pump | + 1.9 |
| Electric Heat | – .7 |
| Not Heated | – 2.7 |

## Additional upgrades or components

| | |
|---|---|
| Kitchen Cabinets & Countertops | Page 7 |
| Bathroom Vanities | 7 |
| Fireplaces & Chimneys | 7 |
| Windows, Skylights & Dormers | 7 |
| Appliances | 7 |
| Breezeways & Porches | 7 |
| Finished Attic | 7 |
| Garages | 8 |
| Site Improvements | 8 |
| Wings & Ells | 2 |

**Important: See the Reference Section for Location Factors (to adjust for your city) and Estimating Form**

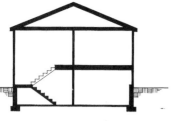

| | | Labor-Hours | Cost Per Square Foot Of Living Area | | |
|---|---|---|---|---|---|
| | | | Mat. | Labor | Total |
| **1 Site Work** | Excavation for lower level, 4' deep. Site preparation for slab. | .029 | | .50 | .50 |
| **2 Foundation** | Continuous reinforced concrete footing, 8" deep x 18" wide; dampproofed and insulated 8" thick reinforced concrete block foundation wall, 4' deep; 4" concrete slab on 4" crushed stone base and polyethylene vapor barrier, trowel finish. | .069 | 2.19 | 2.69 | 4.88 |
| **3 Framing** | Exterior walls - 2" x 4" wood studs, 16" O.C.; 1/2" insulation board sheathing; wood truss roof frame, 24" O.C. with 1/2" plywood sheathing, 4 in 12 pitch; 2" x 8" floor joists 16" O.C. with bridging and 5/8" plywood subfloor. | .107 | 3.91 | 4.67 | 8.58 |
| **4 Exterior Walls** | Beveled wood siding and #15 felt building paper on insulated wood frame walls; 6" attic insulation; double hung windows; 2 flush solid core wood exterior doors with storms. | .089 | 4.71 | 2.96 | 7.67 |
| **5 Roofing** | 20 year asphalt shingles; #15 felt building paper; aluminum gutters, downspouts, drip edge and flashings. | .024 | .43 | .81 | 1.24 |
| **6 Interiors** | Walls and ceilings, 1/2" taped and finished drywall, primed and painted with 2 coats; painted baseboard and trim, rubber backed carpeting 80%, asphalt tile 20%; hollow core wood interior doors. | .213 | 6.08 | 8.41 | 14.49 |
| **7 Specialties** | Economy grade kitchen cabinets - 6 L.F. wall and base with plastic laminate counter top and kitchen sink; 30 gallon electric water heater. | .018 | .69 | .37 | 1.06 |
| **8 Mechanical** | 1 lavatory, white, wall hung; 1 water closet, white; 1 bathtub, enameled steel, white; gas fired warm air heat. | .061 | 2.02 | 1.90 | 3.92 |
| **9 Electrical** | 100 Amp. service; romex wiring; incandescent lighting fixtures; switches, receptacles. | .030 | .47 | .84 | 1.31 |
| **10 Overhead** | Contractor's overhead and profit. | | 3.08 | 3.47 | 6.55 |
| | **Total** | | 23.58 | 26.62 | **50.20** |

**SQUARE FOOT COSTS**

- **Mass produced from stock plans**
- **Single family — 1 full bath, 1 kitchen**
- **No basement**
- **Asphalt shingles on roof**
- **Hot air heat**
- **Drywall interior finishes**
- **Materials and workmanship are sufficient to meet codes**

*Note: The illustration shown may contain some optional components (for example: garages and/or fireplaces) whose costs are shown in the modifications, adjustments, & alternatives below or at the end of the square foot section.*

©Design Basics, Inc.

## Base cost per square foot of living area

| Exterior Wall | Living Area | | | | | | | | | | |
|---|---|---|---|---|---|---|---|---|---|---|---|
| | 1200 | 1500 | 1800 | 2000 | 2200 | 2400 | 2800 | 3200 | 3600 | 4000 | 4400 |
| Wood Siding - Wood Frame | 60.25 | 55.50 | 51.95 | 50.50 | 48.40 | 46.65 | 45.35 | 43.60 | 41.40 | 40.70 | 39.00 |
| Brick Veneer - Wood Frame | 65.20 | 59.95 | 56.00 | 54.35 | 52.10 | 50.10 | 48.65 | 46.65 | 44.25 | 43.45 | 41.60 |
| Stucco on Wood Frame | 58.90 | 54.25 | 50.80 | 49.40 | 47.35 | 45.70 | 44.40 | 42.70 | 40.60 | 39.90 | 38.30 |
| Solid Masonry | 61.25 | 56.35 | 52.70 | 51.25 | 49.10 | 47.30 | 45.95 | 44.15 | 41.95 | 41.25 | 39.50 |
| Finished Basement, Add* | 13.05 | 12.50 | 11.95 | 11.75 | 11.50 | 11.30 | 11.10 | 10.85 | 10.60 | 10.50 | 10.30 |
| Unfinished Basement, Add* | 5.60 | 5.15 | 4.75 | 4.60 | 4.40 | 4.20 | 4.05 | 3.85 | 3.65 | 3.60 | 3.45 |

*Basement under middle level only.

## Modifications

*Add to the total cost*

| | |
|---|---|
| Upgrade Kitchen Cabinets | $ + 287 |
| Solid Surface Countertops | + 665 |
| Full Bath - including plumbing, wall and floor finishes | + 3076 |
| Half Bath - including plumbing, wall and floor finishes | + 1916 |
| One Car Attached Garage | + 8010 |
| One Car Detached Garage | + 10,343 |
| Fireplace & Chimney | + 3655 |

## Adjustments

*For multi family - add to total cost*

| | |
|---|---|
| Additional Kitchen | $ + 2167 |
| Additional Bath | + 3076 |
| Additional Entry & Exit | + 1128 |
| Separate Heating | + 1211 |
| Separate Electric | + 748 |

*For Townhouse/Rowhouse - Multiply cost per square foot by*

| | |
|---|---|
| Inner Unit | .93 |
| End Unit | .96 |

## Alternatives

*Add to or deduct from the cost per square foot of living area*

| | |
|---|---|
| Composition Roll Roofing | – . |
| Cedar Shake Roof | + 1. |
| Upgrade Walls and Ceilings to Skim Coat Plaster | + . |
| Upgrade Ceilings to Textured Finish | + . |
| Air Conditioning, in Heating Ductwork | + 1. |
| In Separate Ductwork | + 3. |
| Heating Systems, Hot Water | + 1. |
| Heat Pump | + 2. |
| Electric Heat | – . |
| Not Heated | – 2. |

## Additional upgrades or components

| | |
|---|---|
| Kitchen Cabinets & Countertops | Page |
| Bathroom Vanities | |
| Fireplaces & Chimneys | |
| Windows, Skylights & Dormers | |
| Appliances | |
| Breezeways & Porches | |
| Finished Attic | |
| Garages | |
| Site Improvements | |
| Wings & Ells | |

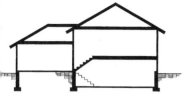

**Living Area - 2400 S.F.**
**Perimeter - 163 L.F.**

**SQUARE FOOT COSTS**

| | | Labor-Hours | Cost Per Square Foot Of Living Area | | |
|---|---|---|---|---|---|
| | | | Mat. | Labor | Total |
| **1 Site Work** | Site preparation for slab; 4' deep trench excavation for foundation wall, excavation for lower level, 4' deep. | .027 | | .42 | .42 |
| **2 Foundation** | Continuous reinforced concrete footing, 8" deep x 18" wide; dampproofed and insulated 8" thick reinforced concrete block foundation wall, 4' deep; 4" concrete slab on 4" crushed stone base and polyethylene vapor barrier, trowel finish. | .071 | 2.51 | 2.93 | 5.44 |
| **3 Framing** | Exterior walls - 2" x 4" wood studs, 16" O.C.; 1/2" insulation board sheathing; wood truss roof frame, 24" O.C. with 1/2" plywood sheathing, 4 in 12 pitch; 2" x 8" floor joists 16" O.C. with bridging and 5/8" plywood subfloor. | .094 | 3.64 | 4.21 | 7.85 |
| **4 Exterior Walls** | Beveled wood siding and #15 felt building paper on insulated wood frame walls; 6" attic insulation; double hung windows; 2 flush solid core wood exterior doors with storms. | .081 | 4.09 | 2.56 | 6.65 |
| **5 Roofing** | 20 year asphalt shingles; #15 felt building paper; aluminum gutters, downspouts, drip edge and flashings. | .032 | .57 | 1.08 | 1.65 |
| **6 Interiors** | Walls and ceilings, 1/2" taped and finished drywall, primed and painted with 2 coats; painted baseboard and trim, rubber backed carpeting 80%, asphalt tile 20%; hollow core wood interior doors. | .177 | 5.33 | 7.44 | 12.77 |
| **7 Specialties** | Economy grade kitchen cabinets - 6 L.F. wall and base with plastic laminate counter top and kitchen sink; 30 gallon electric water heater. | .014 | .58 | .31 | .89 |
| **8 Mechanical** | 1 lavatory, white, wall hung; 1 water closet, white; 1 bathtub, enameled steel, white; gas fired warm air heat. | .057 | 1.84 | 1.80 | 3.64 |
| **9 Electrical** | 100 Amp. service; romex wiring; incandescent lighting fixtures, switches, receptacles. | .029 | .44 | .80 | 1.24 |
| **10 Overhead** | Contractor's overhead and profit | | 2.85 | 3.25 | 6.10 |
| | **Total** | | 21.85 | 24.78 | **46.65** |

# RESIDENTIAL | Economy | Wings & Ells

## 1 Story — Base cost per square foot of living area

| Exterior Wall | Living Area | | | | | | | |
|---|---|---|---|---|---|---|---|---|
| | 50 | 100 | 200 | 300 | 400 | 500 | 600 | 700 |
| Wood Siding - Wood Frame | 107.55 | 82.15 | 71.05 | 59.70 | 56.10 | 53.90 | 52.50 | 53.00 |
| Brick Veneer - Wood Frame | 124.35 | 94.15 | 81.05 | 66.40 | 62.10 | 59.50 | 57.80 | 58.15 |
| Stucco on Wood Frame | 102.90 | 78.85 | 68.30 | 57.85 | 54.45 | 52.40 | 51.00 | 51.60 |
| Painted Concrete Block | 110.80 | 84.50 | 73.00 | 61.05 | 57.25 | 55.00 | 53.55 | 54.00 |
| Finished Basement, Add | 31.40 | 25.70 | 23.30 | 19.40 | 18.60 | 18.10 | 17.80 | 17.55 |
| Unfinished Basement, Add | 17.35 | 13.00 | 11.15 | 8.15 | 7.55 | 7.15 | 6.95 | 6.75 |

## 1-1/2 Story — Base cost per square foot of living area

| Exterior Wall | Living Area | | | | | | | |
|---|---|---|---|---|---|---|---|---|
| | 100 | 200 | 300 | 400 | 500 | 600 | 700 | 800 |
| Wood Siding - Wood Frame | 84.05 | 67.45 | 57.30 | 51.65 | 48.60 | 47.15 | 45.25 | 44.75 |
| Brick Veneer - Wood Frame | 99.05 | 79.40 | 67.30 | 59.45 | 55.80 | 53.95 | 51.70 | 51.15 |
| Stucco on Wood Frame | 79.90 | 64.10 | 54.55 | 49.50 | 46.60 | 45.30 | 43.50 | 43.00 |
| Painted Concrete Block | 86.95 | 69.75 | 59.25 | 53.15 | 50.00 | 48.45 | 46.55 | 46.00 |
| Finished Basement, Add | 21.00 | 18.65 | 17.10 | 15.35 | 14.85 | 14.55 | 14.25 | 14.20 |
| Unfinished Basement, Add | 10.75 | 8.95 | 7.75 | 6.40 | 6.05 | 5.80 | 5.55 | 5.55 |

## 2 Story — Base cost per square foot of living area

| Exterior Wall | Living Area | | | | | | | |
|---|---|---|---|---|---|---|---|---|
| | 100 | 200 | 400 | 600 | 800 | 1000 | 1200 | 1400 |
| Wood Siding - Wood Frame | 85.25 | 63.40 | 53.75 | 44.85 | 41.70 | 39.85 | 38.60 | 39.25 |
| Brick Veneer - Wood Frame | 102.05 | 75.40 | 63.75 | 51.55 | 47.70 | 45.45 | 43.90 | 44.40 |
| Stucco on Wood Frame | 80.65 | 60.10 | 51.00 | 43.05 | 40.05 | 38.30 | 37.10 | 37.85 |
| Painted Concrete Block | 88.50 | 65.75 | 55.70 | 46.20 | 42.85 | 40.95 | 39.65 | 40.25 |
| Finished Basement, Add | 15.70 | 12.90 | 11.70 | 9.70 | 9.30 | 9.05 | 8.90 | 8.80 |
| Unfinished Basement, Add | 8.70 | 6.50 | 5.60 | 4.05 | 3.75 | 3.60 | 3.45 | 3.40 |

Base costs do not include bathroom or kitchen facilities. Use Modifications/Adjustments/Alternatives on pages 77-80 where appropriate.

## 1 Story

## 1-1/2 Story

## 2 Story

## 2-1/2 Story

## Bi-Level

## Tri-Level

- **Simple design from standard plans**
- **Single family — 1 full bath, 1 kitchen**
- **No basement**
- **Asphalt shingles on roof**
- **Hot air heat**
- **Drywall interior finishes**
- **Materials and workmanship are average**

*Note: The illustration shown may contain some optional components (for example: garages and/or fireplaces) whose costs are shown in the modifications, adjustments, & alternatives below or at the end of the square foot section.*

©Home Planners, Inc.

## Base cost per square foot of living area

| Exterior Wall | Living Area | | | | | | | | | | |
|---|---|---|---|---|---|---|---|---|---|---|---|
| | 600 | 800 | 1000 | 1200 | 1400 | 1600 | 1800 | 2000 | 2400 | 2800 | 3200 |
| Wood Siding - Wood Frame | 99.05 | 90.40 | 83.80 | 78.35 | 73.85 | 70.85 | 69.30 | 67.30 | 63.30 | 60.35 | 58.35 |
| Brick Veneer - Wood Frame | 113.30 | 103.85 | 96.60 | 90.60 | 85.60 | 82.30 | 80.60 | 78.35 | 73.95 | 70.75 | 68.50 |
| Stucco on Wood Frame | 106.30 | 97.60 | 90.95 | 85.50 | 81.00 | 77.95 | 76.40 | 74.40 | 70.35 | 67.45 | 65.40 |
| Solid Masonry | 123.15 | 112.65 | 104.60 | 97.80 | 92.15 | 88.40 | 86.50 | 83.90 | 79.05 | 75.45 | 72.85 |
| Finished Basement, Add | 26.25 | 24.70 | 23.55 | 22.55 | 21.70 | 21.15 | 20.85 | 20.40 | 19.75 | 19.30 | 18.85 |
| Unfinished Basement, Add | 10.45 | 9.45 | 8.75 | 8.15 | 7.65 | 7.35 | 7.15 | 6.90 | 6.55 | 6.25 | 6.00 |

## Modifications

*Add to the total cost*

| | |
|---|---|
| Upgrade Kitchen Cabinets | $ + 2206 |
| Solid Surface Countertops | + 896 |
| Full Bath - including plumbing, wall and floor finishes | + 3861 |
| Half Bath - including plumbing, wall and floor finishes | + 2405 |
| One Car Attached Garage | + 8424 |
| One Car Detached Garage | + 11,040 |
| Fireplace & Chimney | + 3730 |

## Adjustments

*For multi family - add to total cost*

| | |
|---|---|
| Additional Kitchen | $ + 4061 |
| Additional Bath | + 3861 |
| Additional Entry & Exit | + 1128 |
| Separate Heating | + 1211 |
| Separate Electric | + 1410 |

*For Townhouse/Rowhouse -
Multiply cost per square foot by*

| | |
|---|---|
| Inner Unit | .92 |
| End Unit | .96 |

## Alternatives

*Add to or deduct from the cost per square foot of living area*

| | |
|---|---|
| Cedar Shake Roof | + 2.1 |
| Clay Tile Roof | + 4.5 |
| Slate Roof | + 7.7 |
| Upgrade Walls to Skim Coat Plaster | + .3 |
| Upgrade Ceilings to Textured Finish | + .3 |
| Air Conditioning, in Heating Ductwork | + 2.6 |
|    In Separate Ductwork | + 5.0 |
| Heating Systems, Hot Water | + 1.4 |
|    Heat Pump | + 1.7 |
|    Electric Heat | – .6 |
|    Not Heated | – 2.6 |

## Additional upgrades or components

| | |
|---|---|
| Kitchen Cabinets & Countertops | Page 7 |
| Bathroom Vanities | 7 |
| Fireplaces & Chimneys | 7 |
| Windows, Skylights & Dormers | 7 |
| Appliances | |
| Breezeways & Porches | |
| Finished Attic | |
| Garages | 8 |
| Site Improvements | 8 |
| Wings & Ells | |

**Important: See the Reference Section for Location Factors (to adjust for your city) and Estimating Form**

**SQUARE FOOT COSTS**

| | | Labor-Hours | Cost Per Square Foot Of Living Area | | |
|---|---|---|---|---|---|
| | | | Mat. | Labor | Total |
| **1** Site Work | Site preparation for slab; 4' deep trench excavation for foundation wall. | .048 | | .65 | .65 |
| **2** Foundation | Continuous reinforced concrete footing 8" deep x 18" wide; dampproofed and insulated reinforced concrete foundation wall, 8" thick, 4' deep; 4" concrete slab on 4" crushed stone base and polyethylene vapor barrier, trowel finish. | .113 | 3.87 | 4.50 | 8.37 |
| **3** Framing | Exterior walls - 2" x 4" wood studs, 16" O.C.; 1/2" plywood sheathing; 2" x 6" rafters 16" O.C. with 1/2" plywood sheathing, 4 in 12 pitch; 2" x 6" ceiling joists 16" O.C.; 1/2" plywood subfloor on 1" x 2" wood sleepers 16" O.C. | .136 | 5.52 | 6.24 | 11.76 |
| **4** Exterior Walls | Beveled wood siding and #15 felt building paper on insulated wood frame walls; 6" attic insulation; double hung windows; 3 flush solid core wood exterior doors with storms. | .098 | 6.64 | 3.46 | 10.10 |
| **5** Roofing | 25 year asphalt shingles; #15 felt building paper; aluminum gutters, downspouts, drip edge and flashings. | .047 | .87 | 1.67 | 2.54 |
| **6** Interiors | Walls and ceilings, 1/2" taped and finished drywall, primed and painted with 2 coats; painted baseboard and trim, finished hardwood floor 40%, carpet with 1/2" underlayment 40%, vinyl tile with 1/2" underlayment 15%, ceramic tile with 1/2" underlayment 5%; hollow core and louvered interior doors. | .251 | 9.23 | 9.39 | 18.62 |
| **7** Specialties | Average grade kitchen cabinets - 14 L.F. wall and base with plastic laminate counter top and kitchen sink; 40 gallon electric water heater. | .009 | 1.44 | .61 | 2.05 |
| **8** Mechanical | 1 lavatory, white, wall hung; 1 water closet, white; 1 bathtub with shower, enameled steel, white; gas fired warm air heat. | .098 | 2.47 | 2.09 | 4.56 |
| **9** Electrical | 200 Amp. service; romex wiring; incandescent lighting fixtures, switches, receptacles. | .041 | .82 | 1.22 | 2.04 |
| **10** Overhead | Contractor's overhead and profit and plans. | | 5.17 | 4.99 | 10.16 |
| **Total** | | | 36.03 | 34.82 | 70.85 |

- **Simple design from standard plans**
- **Single family — 1 full bath, 1 kitchen**
- **No basement**
- **Asphalt shingles on roof**
- **Hot air heat**
- **Drywall interior finishes**
- **Materials and workmanship are average**

*Note: The illustration shown may contain some optional components (for example: garages and/or fireplaces) whose costs are shown in the modifications, adjustments, & alternatives below or at the end of the square foot section.*

©By Designer

## Base cost per square foot of living area

| Exterior Wall | Living Area | | | | | | | | | | |
|---|---|---|---|---|---|---|---|---|---|---|---|
| | 600 | 800 | 1000 | 1200 | 1400 | 1600 | 1800 | 2000 | 2400 | 2800 | 3200 |
| Wood Siding - Wood Frame | 109.40 | 92.55 | 83.10 | 78.95 | 75.95 | 71.25 | 69.05 | 66.55 | 61.70 | 59.85 | 57.80 |
| Brick Veneer - Wood Frame | 119.60 | 99.95 | 89.95 | 85.35 | 82.10 | 76.80 | 74.30 | 71.55 | 66.10 | 64.05 | 61.65 |
| Stucco on Wood Frame | 109.90 | 92.90 | 83.45 | 79.25 | 76.25 | 71.50 | 69.30 | 66.80 | 61.90 | 60.05 | 57.95 |
| Solid Masonry | 133.40 | 109.95 | 99.20 | 94.00 | 90.35 | 84.30 | 81.40 | 78.35 | 72.05 | 69.65 | 66.90 |
| Finished Basement, Add | 21.40 | 18.25 | 17.45 | 16.85 | 16.45 | 15.80 | 15.45 | 15.15 | 14.50 | 14.20 | 13.85 |
| Unfinished Basement, Add | 8.90 | 7.05 | 6.55 | 6.20 | 5.90 | 5.55 | 5.35 | 5.15 | 4.75 | 4.60 | 4.40 |

## Modifications

*Add to the total cost*

| | |
|---|---|
| Upgrade Kitchen Cabinets | $ + 2206 |
| Solid Surface Countertops | + 896 |
| Full Bath - including plumbing, wall and floor finishes | + 3861 |
| Half Bath - including plumbing, wall and floor finishes | + 2405 |
| One Car Attached Garage | + 8424 |
| One Car Detached Garage | + 11,040 |
| Fireplace & Chimney | + 3730 |

## Adjustments

*For multi family - add to total cost*

| | |
|---|---|
| Additional Kitchen | $ + 4061 |
| Additional Bath | + 3861 |
| Additional Entry & Exit | + 1128 |
| Separate Heating | + 1211 |
| Separate Electric | + 1410 |

*For Townhouse/Rowhouse -*
*Multiply cost per square foot by*

| | |
|---|---|
| Inner Unit | .92 |
| End Unit | .96 |

## Alternatives

*Add to or deduct from the cost per square foot of living area*

| | |
|---|---|
| Cedar Shake Roof | + 1.5 |
| Clay Tile Roof | + 3.2 |
| Slate Roof | + 5.6 |
| Upgrade Walls to Skim Coat Plaster | + .3 |
| Upgrade Ceilings to Textured Finish | + .3 |
| Air Conditioning, in Heating Ductwork | + 1.9 |
| In Separate Ductwork | + 4.3 |
| Heating Systems, Hot Water | + 1.3 |
| Heat Pump | + 1.9 |
| Electric Heat | − .5 |
| Not Heated | − 2.5 |

## Additional upgrades or components

| | |
|---|---|
| Kitchen Cabinets & Countertops | Page 7 |
| Bathroom Vanities | 7 |
| Fireplaces & Chimneys | 7 |
| Windows, Skylights & Dormers | 7 |
| Appliances | 7 |
| Breezeways & Porches | 7 |
| Finished Attic | 7 |
| Garages | 8 |
| Site Improvements | 8 |
| Wings & Ells | 4 |

| | | Labor-Hours | Cost Per Square Foot Of Living Area | | |
|---|---|---|---|---|---|
| | | | Mat. | Labor | Total |
| **1 Site Work** | Site preparation for slab; 4' deep trench excavation for foundation wall. | .037 | | .58 | .58 |
| **2 Foundation** | Continuous reinforced concrete footing 8" deep x 18" wide; dampproofed and insulated reinforced concrete foundation wall, 8" thick, 4' deep; 4" concrete slab on 4" crushed stone base and polyethylene vapor barrier, trowel finish. | .073 | 2.72 | 3.31 | 6.03 |
| **3 Framing** | Exterior walls - 2" x 4" wood studs, 16" O.C.; 1/2" plywood sheathing; 2" x 6" rafters 16" O.C. with 1/2" plywood sheathing, 8 in 12 pitch; 2" x 8" floor joists 16" O.C. with 5/8" plywood subfloor; 1/2" plywood subfloor on 1" x 2" wood sleepers 16" O.C. | .098 | 5.33 | 5.71 | 11.04 |
| **4 Exterior Walls** | Beveled wood siding and #15 felt building paper on insulated wood frame walls; 6" attic insulation; double hung windows; 3 flush solid core wood exterior doors with storms. | .078 | 7.21 | 3.74 | 10.95 |
| **5 Roofing** | 25 year asphalt shingles; #15 felt building paper; aluminum gutters, downspouts, drip edge and flashings. | .029 | .54 | 1.04 | 1.58 |
| **6 Interiors** | Walls and ceilings, 1/2" taped and finished drywall, primed and painted with 2 coats; painted baseboard and trim, finished hardwood floor 40%, carpet with 1/2" underlayment 40%, vinyl tile with 1/2" underlayment 15%, ceramic tile with 1/2" underlayment 5%; hollow core and louvered interior doors. | .225 | 10.21 | 10.68 | 20.89 |
| **7 Specialties** | Average grade kitchen cabinets - 14 L.F. wall and base with plastic laminate counter top and kitchen sink; 40 gallon electric water heater. | .022 | 1.27 | .53 | 1.80 |
| **8 Mechanical** | 1 lavatory, white, wall hung; 1 water closet, white; 1 bathtub with shower, enameled steel, white; gas fired warm air heat. | .049 | 2.31 | 2.01 | 4.32 |
| **9 Electrical** | 200 Amp. service; romex wiring; incandescent lighting fixtures, switches, receptacles. | .039 | .78 | 1.17 | 1.95 |
| **10 Overhead** | Contractor's overhead and profit and plans. | | 5.09 | 4.82 | 9.91 |
| **Total** | | | 35.46 | 33.59 | **69.05** |

**SQUARE FOOT COSTS**

- **Simple design from standard plans**
- **Single family — 1 full bath, 1 kitchen**
- **No basement**
- **Asphalt shingles on roof**
- **Hot air heat**
- **Drywall interior finishes**
- **Materials and workmanship are average**

*Note: The illustration shown may contain some optional components (for example: garages and/or fireplaces) whose costs are shown in the modifications, adjustments, & alternatives below or at the end of the square foot section.*

## Base cost per square foot of living area

| Exterior Wall | Living Area | | | | | | | | | | |
|---|---|---|---|---|---|---|---|---|---|---|---|
| | 1000 | 1200 | 1400 | 1600 | 1800 | 2000 | 2200 | 2600 | 3000 | 3400 | 3800 |
| Wood Siding - Wood Frame | 87.65 | 79.45 | 76.00 | 73.55 | 70.90 | 68.10 | 66.35 | 62.80 | 59.20 | 57.70 | 56.25 |
| Brick Veneer - Wood Frame | 95.55 | 86.75 | 82.80 | 80.05 | 77.10 | 74.05 | 71.95 | 67.90 | 63.90 | 62.20 | 60.50 |
| Stucco on Wood Frame | 88.05 | 79.80 | 76.30 | 73.85 | 71.25 | 68.40 | 66.60 | 63.05 | 59.45 | 57.90 | 56.45 |
| Solid Masonry | 107.20 | 97.55 | 92.90 | 89.70 | 86.15 | 82.75 | 80.25 | 75.40 | 70.85 | 68.75 | 66.75 |
| Finished Basement, Add | 15.20 | 14.60 | 14.15 | 13.80 | 13.45 | 13.25 | 13.00 | 12.50 | 12.20 | 12.00 | 11.80 |
| Unfinished Basement, Add | 5.70 | 5.35 | 5.05 | 4.85 | 4.60 | 4.50 | 4.35 | 4.05 | 3.85 | 3.75 | 3.60 |

## Modifications

*Add to the total cost*

| | |
|---|---|
| Upgrade Kitchen Cabinets | $ + 2206 |
| Solid Surface Countertops | + 896 |
| Full Bath - including plumbing, wall and floor finishes | + 3861 |
| Half Bath - including plumbing, wall and floor finishes | + 2405 |
| One Car Attached Garage | + 8424 |
| One Car Detached Garage | + 11,040 |
| Fireplace & Chimney | + 4160 |

## Adjustments

*For multi family - add to total cost*

| | |
|---|---|
| Additional Kitchen | $ + 4061 |
| Additional Bath | + 3861 |
| Additional Entry & Exit | + 1128 |
| Separate Heating | + 1211 |
| Separate Electric | + 1410 |

*For Townhouse/Rowhouse - Multiply cost per square foot by*

| | |
|---|---|
| Inner Unit | .90 |
| End Unit | .95 |

## Alternatives

*Add to or deduct from the cost per square foot of living area*

| | |
|---|---|
| Cedar Shake Roof | + 1. |
| Clay Tile Roof | + 2. |
| Slate Roof | + 3. |
| Upgrade Walls to Skim Coat Plaster | + . |
| Upgrade Ceilings to Textured Finish | + . |
| Air Conditioning, in Heating Ductwork | + 1. |
| In Separate Ductwork | + 4. |
| Heating Systems, Hot Water | + 1. |
| Heat Pump | + 2. |
| Electric Heat | – . |
| Not Heated | – 2. |

## Additional upgrades or components

| | |
|---|---|
| Kitchen Cabinets & Countertops | Page |
| Bathroom Vanities | |
| Fireplaces & Chimneys | |
| Windows, Skylights & Dormers | |
| Appliances | |
| Breezeways & Porches | |
| Finished Attic | |
| Garages | |
| Site Improvements | |
| Wings & Ells | |

**Important: See the Reference Section for Location Factors (to adjust for your city) and Estimating Forr**

**Living Area - 2000 S.F.**
**Perimeter - 135 L.F.**

| | | Labor-Hours | Cost Per Square Foot Of Living Area | | |
|---|---|---|---|---|---|
| | | | Mat. | Labor | Total |
| **1 Site Work** | Site preparation for slab; 4' deep trench excavation for foundation wall. | .034 | | .51 | .51 |
| **2 Foundation** | Continuous reinforced concrete footing 8" deep x 18" wide; dampproofed and insulated reinforced concrete foundation wall, 8" thick, 4' deep, 4" concrete slab on 4" crushed stone base and polyethylene vapor barrier, trowel finish. | .066 | 2.25 | 2.77 | 5.02 |
| **3 Framing** | Exterior walls - 2" x 4" wood studs, 16" O.C.; 1/2" plywood sheathing; 2" x 6" rafters 16" O.C. with 1/2" plywood sheathing, 4 in 12 pitch; 2" x 6" ceiling joists 16" O.C.; 2" x 8" floor joists 16" O.C. with 5/8" plywood subfloor; 1/2" plywood subfloor on 1" x 2" wood sleepers 16" O.C. | .131 | 5.39 | 5.85 | 11.24 |
| **4 Exterior Walls** | Beveled wood siding and #15 felt building paper on insulated wood frame walls; 6" attic insulation; double hung windows; 3 flush solid core wood exterior doors with storms. | .111 | 7.83 | 4.09 | 11.92 |
| **5 Roofing** | 25 year asphalt shingles; #15 felt building paper; aluminum gutters, downspouts, drip edge and flashings. | .024 | .44 | .83 | 1.27 |
| **6 Interiors** | Walls and ceilings, 1/2" taped and finished drywall, primed and painted with 2 coats; painted baseboard and trim, finished hardwood floor 40%, carpet with 1/2" underlayment 40%, vinyl tile with 1/2" underlayment 15%, ceramic tile with 1/2" underlayment 5%; hollow core and louvered interior doors. | .232 | 10.13 | 10.65 | 20.78 |
| **7 Specialties** | Average grade kitchen cabinets - 14 L.F. wall and base with plastic laminate counter top and kitchen sink; 40 gallon electric water heater. | .021 | 1.15 | .48 | 1.63 |
| **8 Mechanical** | 1 lavatory, white, wall hung; 1 water closet, white; 1 bathtub with shower; enameled steel, white; gas fired warm air heat. | .060 | 2.17 | 1.95 | 4.12 |
| **9 Electrical** | 200 Amp. service; romex wiring; incandescent lighting fixtures, switches, receptacles. | .039 | .74 | 1.12 | 1.86 |
| **10 Overhead** | Contractor's overhead and profit and plans. | | 5.03 | 4.72 | 9.75 |
| | **Total** | | 35.13 | 32.97 | **68.10** |

**SQUARE FOOT COSTS**

- **Simple design from standard plans**
- **Single family — 1 full bath, 1 kitchen**
- **No basement**
- **Asphalt shingles on roof**
- **Hot air heat**
- **Drywall interior finishes**
- **Materials and workmanship are average**

*Note: The illustration shown may contain some optional components (for example: garages and/or fireplaces) whose costs are shown in the modifications, adjustments, & alternatives below or at the end of the square foot section.*

## Base cost per square foot of living area

| Exterior Wall | Living Area | | | | | | | | | | |
|---|---|---|---|---|---|---|---|---|---|---|---|
| | 1200 | 1400 | 1600 | 1800 | 2000 | 2400 | 2800 | 3200 | 3600 | 4000 | 4400 |
| Wood Siding - Wood Frame | 86.60 | 81.80 | 74.60 | 73.50 | 71.00 | 67.05 | 63.95 | 60.70 | 59.10 | 56.10 | 55.20 |
| Brick Veneer - Wood Frame | 95.05 | 89.35 | 81.60 | 80.50 | 77.60 | 73.00 | 69.60 | 65.85 | 63.95 | 60.60 | 59.60 |
| Stucco on Wood Frame | 87.05 | 82.15 | 74.90 | 73.80 | 71.30 | 67.30 | 64.20 | 60.90 | 59.35 | 56.30 | 55.40 |
| Solid Masonry | 106.40 | 99.65 | 91.00 | 89.95 | 86.50 | 81.00 | 77.30 | 72.75 | 70.55 | 66.75 | 65.50 |
| Finished Basement, Add | 12.90 | 12.30 | 11.85 | 11.80 | 11.50 | 11.00 | 10.80 | 10.45 | 10.25 | 10.05 | 9.90 |
| Unfinished Basement, Add | 4.75 | 4.40 | 4.10 | 4.05 | 3.90 | 3.60 | 3.50 | 3.25 | 3.15 | 3.00 | 2.95 |

## Modifications

*Add to the total cost*

| | |
|---|---|
| Upgrade Kitchen Cabinets | $ + 2206 |
| Solid Surface Countertops | + 896 |
| Full Bath - including plumbing, wall and floor finishes | + 3861 |
| Half Bath - including plumbing, wall and floor finishes | + 2405 |
| One Car Attached Garage | + 8424 |
| One Car Detached Garage | + 11,040 |
| Fireplace & Chimney | + 4730 |

## Adjustments

*For multi family - add to total cost*

| | |
|---|---|
| Additional Kitchen | $ + 4061 |
| Additional Bath | + 3861 |
| Additional Entry & Exit | + 1128 |
| Separate Heating | + 1211 |
| Separate Electric | + 1410 |

*For Townhouse/Rowhouse -*
*Multiply cost per square foot by*

| | |
|---|---|
| Inner Unit | .90 |
| End Unit | .95 |

## Alternatives

*Add to or deduct from the cost per square foot of living area*

| | |
|---|---|
| Cedar Shake Roof | + . |
| Clay Tile Roof | + 1. |
| Slate Roof | + 3. |
| Upgrade Walls to Skim Coat Plaster | + . |
| Upgrade Ceilings to Textured Finish | + . |
| Air Conditioning, in Heating Ductwork | + 1. |
| In Separate Ductwork | + 3. |
| Heating Systems, Hot Water | + 1. |
| Heat Pump | + 2. |
| Electric Heat | – . |
| Not Heated | – 2. |

## Additional upgrades or components

| | |
|---|---|
| Kitchen Cabinets & Countertops | Page |
| Bathroom Vanities | |
| Fireplaces & Chimneys | |
| Windows, Skylights & Dormers | |
| Appliances | |
| Breezeways & Porches | |
| Finished Attic | |
| Garages | |
| Site Improvements | |
| Wings & Ells | |

**Important: See the Reference Section for Location Factors (to adjust for your city) and Estimating Form**

SQUARE FOOT COSTS

| | | Labor-Hours | Cost Per Square Foot Of Living Area | | |
|---|---|---|---|---|---|
| | | | Mat. | Labor | Total |
| **1 Site Work** | Site preparation for slab; 4' deep trench excavation for foundation wall. | .046 | | .33 | .33 |
| **2 Foundation** | Continuous reinforced concrete footing 8" deep x 18" wide, dampproofed and insulated reinforced concrete foundation wall, 8" thick, 4' deep; 4" concrete slab on 4" crushed stone base and polyethylene vapor barrier, trowel finish. | .061 | 1.62 | 1.96 | 3.58 |
| **3 Framing** | Exterior walls - 2" x 4" wood studs, 16" O.C.; 1/2" plywood sheathing; 2" x 6" rafters 16" O.C. with 1/2" plywood sheathing, 4 in 12 pitch; 2" x 6" ceiling joists 16" O.C.; 2" x 8" floor joists 16" O.C. with 5/8" plywood subfloor; 1/2" plywood subfloor on 1" x 2" wood sleepers 16" O.C. | .127 | 5.17 | 5.68 | 10.85 |
| **4 Exterior Walls** | Beveled wood siding and #15 felt building paper on insulated wood frame walls; 6" attic insulation; double hung windows; 3 flush solid core wood exterior doors with storms. | .136 | 6.57 | 3.43 | 10.00 |
| **5 Roofing** | 25 year asphalt shingles; #15 felt building paper; aluminum gutters, downspouts, drip edge and flashings. | .018 | .34 | .64 | .98 |
| **6 Interiors** | Walls and ceilings, 1/2" taped and finished drywall, primed and painted with 2 coats; painted baseboard and trim, finished hardwood floor 40%, carpet with 1/2" underlayment 40%, vinyl tile with 1/2" underlayment 15%, ceramic tile with 1/2" underlayment 5%; hollow core and louvered interior doors. | .286 | 9.86 | 10.30 | 20.16 |
| **7 Specialties** | Average grade kitchen cabinets - 14 L.F. wall and base with plastic laminate counter top and kitchen sink; 40 gallon electric water heater. | .030 | .73 | .30 | 1.03 |
| **8 Mechanical** | 1 lavatory, white, wall hung; 1 water closet, white; 1 bathtub with shower, enameled steel, white; gas fired warm air heat. | .072 | 1.74 | 1.72 | 3.46 |
| **9 Electrical** | 200 Amp. service; romex wiring; incandescent lighting fixtures, switches, receptacles. | .046 | .62 | .98 | 1.60 |
| **10 Overhead** | Contractor's overhead and profit and plans. | | 4.46 | 4.25 | 8.71 |
| | **Total** | | 31.11 | 29.59 | **60.70** |

# RESIDENTIAL | Average | 3 Story

- **Simple design from standard plans**
- **Single family — 1 full bath, 1 kitchen**
- **No basement**
- **Asphalt shingles on roof**
- **Hot air heat**
- **Drywall interior finishes**
- **Materials and workmanship are average**

*Note: The illustration shown may contain some optional components (for example: garages and/or fireplaces) whose costs are shown in the modifications, adjustments, & alternatives below or at the end of the square foot section.*

## Base cost per square foot of living area

| Exterior Wall | Living Area | | | | | | | | | | |
|---|---|---|---|---|---|---|---|---|---|---|---|
| | 1500 | 1800 | 2100 | 2500 | 3000 | 3500 | 4000 | 4500 | 5000 | 5500 | 6000 |
| Wood Siding - Wood Frame | 80.75 | 73.35 | 70.35 | 68.00 | 63.25 | 61.30 | 58.45 | 55.25 | 54.30 | 53.15 | 52.00 |
| Brick Veneer - Wood Frame | 88.65 | 80.65 | 77.20 | 74.50 | 69.20 | 66.90 | 63.55 | 59.95 | 58.85 | 57.55 | 56.10 |
| Stucco on Wood Frame | 81.10 | 73.65 | 70.65 | 68.30 | 63.50 | 61.55 | 58.70 | 55.45 | 54.50 | 53.35 | 52.20 |
| Solid Masonry | 99.30 | 90.50 | 86.40 | 83.30 | 77.15 | 74.45 | 70.45 | 66.30 | 65.00 | 63.45 | 61.65 |
| Finished Basement, Add | 11.10 | 10.70 | 10.40 | 10.15 | 9.80 | 9.60 | 9.30 | 9.10 | 9.00 | 8.85 | 8.75 |
| Unfinished Basement, Add | 3.85 | 3.65 | 3.45 | 3.30 | 3.05 | 2.95 | 2.80 | 2.65 | 2.60 | 2.50 | 2.45 |

## Modifications

*Add to the total cost*

| | |
|---|---|
| Upgrade Kitchen Cabinets | $ + 2206 |
| Solid Surface Countertops | + 896 |
| Full Bath - including plumbing, wall and floor finishes | + 3861 |
| Half Bath - including plumbing, wall and floor finishes | + 2405 |
| One Car Attached Garage | + 8424 |
| One Car Detached Garage | + 11,040 |
| Fireplace & Chimney | + 4730 |

## Adjustments

*For multi family - add to total cost*

| | |
|---|---|
| Additional Kitchen | $ + 4061 |
| Additional Bath | + 3861 |
| Additional Entry & Exit | + 1128 |
| Separate Heating | + 1211 |
| Separate Electric | + 1410 |

*For Townhouse/Rowhouse - Multiply cost per square foot by*

| | |
|---|---|
| Inner Unit | .88 |
| End Unit | .94 |

## Alternatives

*Add to or deduct from the cost per square foot of living area*

| | |
|---|---|
| Cedar Shake Roof | + . |
| Clay Tile Roof | + 1. |
| Slate Roof | + 2. |
| Upgrade Walls to Skim Coat Plaster | + . |
| Upgrade Ceilings to Textured Finish | + . |
| Air Conditioning, in Heating Ductwork | + 1.4 |
| In Separate Ductwork | + 3.8 |
| Heating Systems, Hot Water | + 1.2 |
| Heat Pump | + 2.1 |
| Electric Heat | – . |
| Not Heated | – 2. |

## Additional upgrades or components

| | |
|---|---|
| Kitchen Cabinets & Countertops | Page 7 |
| Bathroom Vanities | 7 |
| Fireplaces & Chimneys | 7 |
| Windows, Skylights & Dormers | 7 |
| Appliances | 7 |
| Breezeways & Porches | 7 |
| Finished Attic | 7 |
| Garages | 8 |
| Site Improvements | 8 |
| Wings & Ells | 4 |

**Important: See the Reference Section for Location Factors (to adjust for your city) and Estimating Form**

| | Labor-Hours | Cost Per Square Foot Of Living Area | | |
|---|---|---|---|---|
| | | Mat. | Labor | Total |
| **1 Site Work** Site preparation for slab; 4' deep trench excavation for foundation wall. | .038 | | .35 | .35 |
| **2 Foundation** Continuous reinforced concrete footing 8" deep x 18" wide, dampproofed and insulated reinforced concrete foundation wall, 8" thick, 4' deep; 4" concrete slab on 4" crushed stone base and polyethylene vapor barrier, trowel finish. | .053 | 1.50 | 1.84 | 3.34 |
| **3 Framing** Exterior walls - 2" x 4" wood studs, 16" O.C.; 1/2" plywood sheathing; 2" x 6" rafters 16" O.C. with 1/2" plywood sheathing, 4 in 12 pitch; 2" x 6" ceiling joists 16" O.C.; 2" x 8" floor joists 16" O.C. with 5/8" plywood subfloor; 1/2" plywood subfloor on 1" x 2" wood sleepers 16" O.C. | .128 | 5.28 | 5.85 | 11.13 |
| **4 Exterior Walls** Horizontal beveled wood siding; #15 felt building paper; 3-1/2" batt insulation; wood double hung windows; 3 flush solid core wood exterior doors; storms and screens. | .139 | 7.50 | 3.92 | 11.42 |
| **5 Roofing** 25 year asphalt shingles; #15 felt building paper; aluminum gutters, downspouts, drip edge and flashings. | .014 | .29 | .56 | .85 |
| **6 Interiors** Walls and ceilings, 1/2" taped and finished drywall, primed and painted with 2 coats; painted baseboard and trim, finished hardwood floor 40%, carpet with 1/2" underlayment 40%, vinyl tile with 1/2" underlayment 15%, ceramic tile with 1/2" underlayment 5%; hollow core and louvered interior doors. | .280 | 10.17 | 10.71 | 20.88 |
| **7 Specialties** Average grade kitchen cabinets - 14 L.F. wall and base with plastic laminate counter top and kitchen sink; 40 gallon electric water heater. | .025 | .77 | .32 | 1.09 |
| **8 Mechanical** 1 lavatory, white, wall hung; 1 water closet, white; 1 bathtub with shower, enameled steel, white; gas fired warm air heat. | .065 | 1.77 | 1.75 | 3.52 |
| **9 Electrical** 200 Amp. service; romex wiring; incandescent lighting fixtures, switches, receptacles. | .042 | .63 | 1.00 | 1.63 |
| **10 Overhead** Contractor's overhead and profit and plans. | | 4.67 | 4.37 | 9.04 |
| **Total** | | 32.58 | 30.67 | **63.25** |

**SQUARE FOOT COSTS**

- **Simple design from standard plans**
- **Single family — 1 full bath, 1 kitchen**
- **No basement**
- **Asphalt shingles on roof**
- **Hot air heat**
- **Drywall interior finishes**
- **Materials and workmanship are average**

*Note: The illustration shown may contain some optional components (for example: garages and/or fireplaces) whose costs are shown in the modifications, adjustments, & alternatives below or at the end of the square foot section.*

## Base cost per square foot of living area

| Exterior Wall | Living Area | | | | | | | | | | |
|---|---|---|---|---|---|---|---|---|---|---|---|
| | 1000 | 1200 | 1400 | 1600 | 1800 | 2000 | 2200 | 2600 | 3000 | 3400 | 3800 |
| Wood Siding - Wood Frame | 82.05 | 74.25 | 71.10 | 68.85 | 66.55 | 63.90 | 62.35 | 59.15 | 55.85 | 54.50 | 53.25 |
| Brick Veneer - Wood Frame | 87.95 | 79.75 | 76.25 | 73.75 | 71.15 | 68.35 | 66.55 | 63.00 | 59.35 | 57.85 | 56.40 |
| Stucco on Wood Frame | 82.30 | 74.50 | 71.35 | 69.10 | 66.75 | 64.10 | 62.55 | 59.35 | 56.00 | 54.65 | 53.40 |
| Solid Masonry | 95.95 | 87.10 | 83.15 | 80.35 | 77.40 | 74.35 | 72.20 | 68.15 | 64.15 | 62.40 | 60.70 |
| Finished Basement, Add | 15.20 | 14.60 | 14.15 | 13.80 | 13.45 | 13.25 | 13.00 | 12.50 | 12.20 | 12.00 | 11.80 |
| Unfinished Basement, Add | 5.70 | 5.35 | 5.05 | 4.85 | 4.60 | 4.50 | 4.35 | 4.05 | 3.85 | 3.75 | 3.60 |

## Modifications

*Add to the total cost*

| | |
|---|---|
| Upgrade Kitchen Cabinets | $ + 2206 |
| Solid Surface Countertops | + 896 |
| Full Bath - including plumbing, wall and floor finishes | + 3861 |
| Half Bath - including plumbing, wall and floor finishes | + 2405 |
| One Car Attached Garage | + 8424 |
| One Car Detached Garage | + 11,040 |
| Fireplace & Chimney | + 3730 |

## Adjustments

*For multi family - add to total cost*

| | |
|---|---|
| Additional Kitchen | $ + 4061 |
| Additional Bath | + 3861 |
| Additional Entry & Exit | + 1128 |
| Separate Heating | + 1211 |
| Separate Electric | + 1410 |

*For Townhouse/Rowhouse - Multiply cost per square foot by*

| | |
|---|---|
| Inner Unit | .91 |
| End Unit | .96 |

## Alternatives

*Add to or deduct from the cost per square foot of living area*

| | |
|---|---|
| Cedar Shake Roof | + 2.1 |
| Clay Tile Roof | + 2.2 |
| Slate Roof | + 3.8 |
| Upgrade Walls to Skim Coat Plaster | + .3 |
| Upgrade Ceilings to Textured Finish | + .3 |
| Air Conditioning, in Heating Ductwork | + 1.5 |
| In Separate Ductwork | + 4.0 |
| Heating Systems, Hot Water | + 1.3 |
| Heat Pump | + 2.0 |
| Electric Heat | − .5 |
| Not Heated | − 2.4 |

## Additional upgrades or components

| | |
|---|---|
| Kitchen Cabinets & Countertops | Page 7 |
| Bathroom Vanities | 7 |
| Fireplaces & Chimneys | 7 |
| Windows, Skylights & Dormers | 7 |
| Appliances | 7 |
| Breezeways & Porches | 7 |
| Finished Attic | 7 |
| Garages | 8 |
| Site Improvements | 8 |
| Wings & Ells | 4 |

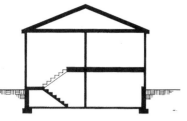

| | | Labor-Hours | Cost Per Square Foot Of Living Area | | |
|---|---|---|---|---|---|
| | | | Mat. | Labor | Total |
| **1 Site Work** | Excavation for lower level, 4' deep. Site preparation for slab. | .029 | | .51 | .51 |
| **2 Foundation** | Continuous reinforced concrete footing 8" deep x 18" wide, dampproofed and insulated reinforced concrete foundation wall, 8" thick, 4' deep; 4" concrete slab on 4" crushed stone base and polyethylene vapor barrier, trowel finish. | .066 | 2.25 | 2.77 | 5.02 |
| **3 Framing** | Exterior walls - 2" x 4" wood studs, 16" O.C.; 1/2" plywood sheathing; 2" x 6" rafters 16" O.C. with 1/2" plywood sheathing, 4 in 12 pitch; 2" x 6" ceiling joists 16" O.C.; 2" x 8" floor joists 16" O.C. with 5/8" plywood subfloor; 1/2" plywood subfloor on 1" x 2" wood sleepers 16" O.C. | .118 | 5.12 | 5.54 | 10.66 |
| **4 Exterior Walls** | Horizontal beveled wood siding; #15 felt building paper; 3-1/2" batt insulation; wood double hung windows; 3 flush solid core wood exterior doors; storms and screens. | .091 | 6.13 | 3.20 | 9.33 |
| **5 Roofing** | 25 year asphalt shingles; #15 felt building paper; aluminum gutters, downspouts, drip edge and flashings. | .024 | .44 | .83 | 1.27 |
| **6 Interiors** | Walls and ceilings, 1/2" taped and finished drywall, primed and painted with 2 coats; painted baseboard and trim, finished hardwood floor 40%, carpet with 1/2" underlayment 40%, vinyl tile with 1/2" underlayment 15%, ceramic tile with 1/2" underlayment 5%; hollow core and louvered interior doors. | .217 | 9.97 | 10.36 | 20.33 |
| **7 Specialties** | Average grade kitchen cabinets - 14 L.F. wall and base with plastic laminate counter top and kitchen sink; 40 gallon electric water heater. | .021 | 1.15 | .48 | 1.63 |
| **8 Mechanical** | 1 lavatory, white, wall hung; 1 water closet, white; 1 bathtub with shower, enameled steel, white; gas fired warm air heat. | .061 | 2.17 | 1.95 | 4.12 |
| **9 Electrical** | 200 Amp. service; romex wiring; incandescent lighting fixtures, switches, receptacles. | .039 | .74 | 1.12 | 1.86 |
| **10 Overhead** | Contractor's overhead and profit and plans. | | 4.68 | 4.49 | 9.17 |
| | **Total** | | 32.65 | 31.25 | **63.90** |

# RESIDENTIAL | Average | Tri-Level

- Simple design from standard plans
- Single family — 1 full bath, 1 kitchen
- No basement
- Asphalt shingles on roof
- Hot air heat
- Drywall interior finishes
- Materials and workmanship are average

*Note: The illustration shown may contain some optional components (for example: garages and/or fireplaces) whose costs are shown in the modifications, adjustments, & alternatives below or at the end of the square foot section.*

*©Design Basics, Inc.*

## Base cost per square foot of living area

| Exterior Wall | Living Area | | | | | | | | | | |
|---|---|---|---|---|---|---|---|---|---|---|---|
| | 1200 | 1500 | 1800 | 2100 | 2400 | 2700 | 3000 | 3400 | 3800 | 4200 | 4600 |
| Wood Siding - Wood Frame | 78.05 | 72.30 | 68.05 | 64.35 | 61.90 | 60.50 | 59.00 | 57.55 | 55.10 | 53.15 | 52.10 |
| Brick Veneer - Wood Frame | 83.45 | 77.25 | 72.50 | 68.40 | 65.65 | 64.15 | 62.45 | 60.90 | 58.20 | 56.05 | 54.90 |
| Stucco on Wood Frame | 78.30 | 72.55 | 68.25 | 64.50 | 62.05 | 60.70 | 59.15 | 57.70 | 55.25 | 53.30 | 52.25 |
| Solid Masonry | 90.80 | 83.90 | 78.50 | 73.85 | 70.80 | 69.10 | 67.10 | 65.35 | 62.35 | 59.95 | 58.70 |
| Finished Basement, Add* | 17.40 | 16.65 | 15.95 | 15.40 | 15.05 | 14.85 | 14.55 | 14.35 | 14.05 | 13.80 | 13.70 |
| Unfinished Basement, Add* | 6.35 | 5.90 | 5.50 | 5.15 | 4.95 | 4.80 | 4.65 | 4.55 | 4.35 | 4.20 | 4.15 |

*Basement under middle level only.

## Modifications

**Add to the total cost**

| | |
|---|---|
| Upgrade Kitchen Cabinets | $ + 2206 |
| Solid Surface Countertops | + 896 |
| Full Bath - including plumbing, wall and floor finishes | + 3861 |
| Half Bath - including plumbing, wall and floor finishes | + 2405 |
| One Car Attached Garage | + 8424 |
| One Car Detached Garage | + 11,040 |
| Fireplace & Chimney | + 4160 |

## Adjustments

**For multi family - add to total cost**

| | |
|---|---|
| Additional Kitchen | $ + 4061 |
| Additional Bath | + 3861 |
| Additional Entry & Exit | + 1128 |
| Separate Heating | + 1211 |
| Separate Electric | + 1410 |

**For Townhouse/Rowhouse - Multiply cost per square foot by**

| | |
|---|---|
| Inner Unit | .90 |
| End Unit | .95 |

## Alternatives

**Add to or deduct from the cost per square foot of living area**

| | |
|---|---|
| Cedar Shake Roof | + 1.50 |
| Clay Tile Roof | + 3.22 |
| Slate Roof | + 5.61 |
| Upgrade Walls to Skim Coat Plaster | + .31 |
| Upgrade Ceilings to Textured Finish | + .39 |
| Air Conditioning, in Heating Ductwork | + 1.31 |
| In Separate Ductwork | + 3.74 |
| Heating Systems, Hot Water | + 1.29 |
| Heat Pump | + 2.13 |
| Electric Heat | – .41 |
| Not Heated | – 2.37 |

## Additional upgrades or components

| | |
|---|---|
| Kitchen Cabinets & Countertops | Page 77 |
| Bathroom Vanities | 78 |
| Fireplaces & Chimneys | 78 |
| Windows, Skylights & Dormers | 78 |
| Appliances | 79 |
| Breezeways & Porches | 79 |
| Finished Attic | 79 |
| Garages | 80 |
| Site Improvements | 80 |
| Wings & Ells | 44 |

**Important: See the Reference Section for Location Factors (to adjust for your city) and Estimating Forms**

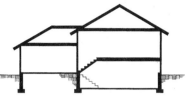

| | | Labor-Hours | Cost Per Square Foot Of Living Area | | |
|---|---|---|---|---|---|
| | | | Mat. | Labor | Total |
| **1 Site Work** | Site preparation for slab; 4' deep trench excavation for foundation wall, excavation for lower level, 4' deep. | .029 | | .43 | .43 |
| **2 Foundation** | Continuous reinforced concrete footing 8" deep x 18" wide; dampproofed and insulated reinforced concrete foundation wall, 8" thick, 4' deep; 4" concrete slab on 4" crushed stone base and polyethylene vapor barrier, trowel finish. | .080 | 2.58 | 3.01 | 5.59 |
| **3 Framing** | Exterior walls - 2" x 4" wood studs, 16" O.C.; 1/2" plywood sheathing; 2" x 6" rafters 16" O.C. with 1/2" plywood sheathing, 4 in 12 pitch; 2" x 6" ceiling joists 16" O.C.; 2" x 8" floor joists 16" O.C. with 5/8" plywood subfloor; 1/2" plywood subfloor on 1" x 2" wood sleepers 16" O.C. | .124 | 4.98 | 5.27 | 10.25 |
| **4 Exterior Walls** | Horizontal beveled wood siding: #15 felt building paper; 3-1/2" batt insulation; wood double hung windows; 3 flush solid core wood exterior doors; storms and screens. | .083 | 5.29 | 2.76 | 8.05 |
| **5 Roofing** | 25 year asphalt shingles; #15 felt building paper; aluminum gutters, downspouts, drip edge and flashings. | .032 | .59 | 1.11 | 1.70 |
| **6 Interiors** | Walls and ceilings, 1/2" taped and finished drywall, primed and painted with 2 coats; painted baseboard and trim, finished hardwood floor 40%, carpet with 1/2" underlayment 40%, vinyl tile with 1/2" underlayment 15%, ceramic tile with 1/2" underlayment 5%; hollow core and louvered interior doors. | .186 | 10.07 | 10.04 | 20.11 |
| **7 Specialties** | Average grade kitchen cabinets - 14 L.F. wall and base with plastic laminate counter top and kitchen sink; 40 gallon electric water heater. | .012 | .97 | .40 | 1.37 |
| **8 Mechanical** | 1 lavatory, white, wall hung; 1 water closet, white; 1 bathtub with shower, enameled steel, white; gas fired warm air heat. | .059 | 1.91 | 1.85 | 3.76 |
| **9 Electrical** | 200 Amp. service; romex wiring; incandescent lighting fixtures, switches, receptacles. | .036 | .69 | 1.06 | 1.75 |
| **10 Overhead** | Contractor's overhead and profit and plans. | | 4.52 | 4.37 | 8.89 |
| | **Total** | | 31.60 | 30.30 | **61.90** |

- **Post and beam frame**
- **Log exterior walls**
- **Simple design from standard plans**
- **Single family — 1 full bath, 1 kitchen**
- **No basement**
- **Asphalt shingles on roof**
- **Hot air heat**
- **Drywall interior finishes**
- **Materials and workmanship are average**

*Note: The illustration shown may contain some optional components (for example: garages and/or fireplaces) whose costs are shown in the modifications, adjustments, & alternatives below or at the end of the square foot section.*

## Base cost per square foot of living area

| Exterior Wall | Living Area | | | | | | | | | | |
|---|---|---|---|---|---|---|---|---|---|---|---|
| | 600 | 800 | 1000 | 1200 | 1400 | 1600 | 1800 | 2000 | 2400 | 2800 | 3200 |
| 6" Log - Solid Wall | 97.00 | 87.70 | 80.40 | 74.70 | 70.10 | 66.10 | 64.20 | 62.80 | 58.50 | 55.20 | 53.40 |
| 8" Log - Solid Wall | 96.30 | 87.00 | 79.80 | 74.10 | 69.50 | 65.60 | 63.80 | 62.40 | 58.20 | 54.60 | 52.90 |
| Finished Basement, Add | 22.30 | 20.90 | 20.00 | 19.10 | 18.50 | 18.00 | 17.70 | 17.30 | 16.80 | 16.40 | 16.10 |
| Unfinished Basement, Add | 9.50 | 8.70 | 8.10 | 7.50 | 7.10 | 6.70 | 6.60 | 6.40 | 6.00 | 5.70 | 5.50 |

## Modifications

*Add to the total cost*

| | |
|---|---|
| Upgrade Kitchen Cabinets | $ + 2206 |
| Solid Surface Countertops | + 896 |
| Full Bath - including plumbing, wall and floor finishes | + 3861 |
| Half Bath - including plumbing, wall and floor finishes | + 2405 |
| One Car Attached Garage | + 8424 |
| One Car Detached Garage | + 11,040 |
| Fireplace & Chimney | + 3730 |

## Adjustments

*For multi family - add to total cost*

| | |
|---|---|
| Additional Kitchen | $ + 4061 |
| Additional Bath | + 3861 |
| Additional Entry & Exit | + 1128 |
| Separate Heating | + 1211 |
| Separate Electric | + 1410 |

*For Townhouse/Rowhouse - Multiply cost per square foot by*

| | |
|---|---|
| Inner Unit | .92 |
| End Unit | .96 |

## Alternatives

*Add to or deduct from the cost per square foot of living area*

| | |
|---|---|
| Cedar Shake Roof | + 2.1 |
| Air Conditioning, in Heating Ductwork | + 2.6 |
| In Separate Ductwork | + 5.0 |
| Heating Systems, Hot Water | + 1.4 |
| Heat Pump | + 1.7 |
| Electric Heat | – .6 |
| Not Heated | – 2.6 |

## Additional upgrades or components

| | |
|---|---|
| Kitchen Cabinets & Countertops | Page 7 |
| Bathroom Vanities | 7 |
| Fireplaces & Chimneys | 7 |
| Windows, Skylights & Dormers | 7 |
| Appliances | 7 |
| Breezeways & Porches | 7 |
| Finished Attic | 7 |
| Garages | 8 |
| Site Improvements | 8 |
| Wings & Ells | 4 |

**Important: See the Reference Section for Location Factors (to adjust for your city) and Estimating Form**

| | | Labor-Hours | Cost Per Square Foot Of Living Area | | |
|---|---|---|---|---|---|
| | | | Mat. | Labor | Total |
| **1 Site Work** | Site preparation for slab; 4' deep trench excavation for foundation wall. | .048 | | .78 | .78 |
| **2 Foundation** | Continuous reinforced concrete footing 8" deep x 18" wide; dampproofed and insulated reinforced concrete foundation wall, 8" thick, 4' deep; 4" concrete slab on 4" crushed stone base and polyethylene vapor barrier, trowel finish. | .113 | 3.64 | 4.55 | 8.19 |
| **3 Framing** | Exterior walls - Precut traditional log home. Handicrafted white cedar or pine logs. Delivery included. | .201 | 10.89 | 7.73 | 18.62 |
| **4 Exterior Walls** | | | | | |
| **5 Roofing** | 25 year asphalt shingles; #15 felt building paper; aluminum gutters, downspouts, drip edge and flashings. | .047 | .90 | 1.59 | 2.49 |
| **6 Interiors** | Walls and ceilings, 1/2" taped and finished drywall, primed and painted with 2 coats; painted baseboard and trim, finished hardwood floor 40%, carpet with 1/2" underlayment 40%, vinyl tile with 1/2" underlayment 15%, ceramic tile with 1/2" underlayment 5%; hollow core and louvered interior doors. | .232 | 7.92 | 7.78 | 15.70 |
| **7 Specialties** | Average grade kitchen cabinets - 14 L.F. wall and base with plastic laminate counter top and kitchen sink; 40 gallon electric water heater. | .009 | 1.65 | .56 | 2.21 |
| **8 Mechanical** | 1 lavatory, white, wall hung; 1 water closet, white; 1 bathtub with shower, enameled steel, white; gas fired warm air heat. | .098 | 2.60 | 2.02 | 4.62 |
| **9 Electrical** | 200 Amp. service; romex wiring; incandescent lighting fixtures, switches, receptacles. | .041 | .83 | 1.21 | 2.04 |
| **10 Overhead** | Contractor's overhead and profit and plans. | | 6.75 | 4.70 | 11.45 |
| **Total** | | | 35.18 | 30.92 | **66.10** |

# RESIDENTIAL | Solid Wall | 2 Story

- Post and beam frame
- Log exterior walls
- Simple design from standard plans
- Single family — 1 full bath, 1 kitchen
- No basement
- Asphalt shingles on roof
- Hot air heat
- Drywall interior finishes
- Materials and workmanship are average

*Note: The illustration shown may contain some optional components (for example: garages and/or fireplaces) whose costs are shown in the modifications, adjustments, & alternatives below or at the end of the square foot section.*

## Base cost per square foot of living area

| Exterior Wall | Living Area | | | | | | | | | | |
|---|---|---|---|---|---|---|---|---|---|---|---|
| | 1000 | 1200 | 1400 | 1600 | 1800 | 2000 | 2200 | 2600 | 3000 | 3400 | 3800 |
| 6" Log-Solid | 97.20 | 80.10 | 76.40 | 73.70 | 70.70 | **67.90** | 65.90 | 62.00 | 57.50 | 56.00 | 54.50 |
| 8" Log-Solid | 93.30 | 83.70 | 79.60 | 77.00 | 73.90 | 70.70 | 68.80 | 64.60 | 60.20 | 58.60 | 56.80 |
| Finished Basement, Add | 12.90 | 12.50 | 12.00 | 11.80 | 11.50 | 11.30 | 11.10 | 10.70 | 10.40 | 10.30 | 10.10 |
| Unfinished Basement, Add | 5.20 | 4.90 | 4.60 | 4.40 | 4.30 | 4.00 | 4.00 | 3.70 | 3.50 | 3.40 | 3.30 |

## Modifications

*Add to the total cost*

| | |
|---|---|
| Upgrade Kitchen Cabinets | $ + 2206 |
| Solid Surface Countertops | + 896 |
| Full Bath - including plumbing, wall and floor finishes | + 3861 |
| Half Bath - including plumbing, wall and floor finishes | + 2405 |
| One Car Attached Garage | + 8424 |
| One Car Detached Garage | + 11,040 |
| Fireplace & Chimney | + 4160 |

## Adjustments

*For multi family - add to total cost*

| | |
|---|---|
| Additional Kitchen | $ + 4061 |
| Additional Bath | + 3861 |
| Additional Entry & Exit | + 1128 |
| Separate Heating | + 1211 |
| Separate Electric | + 1410 |

*For Townhouse/Rowhouse - Multiply cost per square foot by*

| | |
|---|---|
| Inner Unit | .92 |
| End Unit | .96 |

## Alternatives

*Add to or deduct from the cost per square foot of living area*

| | |
|---|---|
| Cedar Shake Roof | + 1.50 |
| Air Conditioning, in Heating Ductwork | + 1.56 |
| In Separate Ductwork | + 4.00 |
| Heating Systems, Hot Water | + 1.34 |
| Heat Pump | + 2.05 |
| Electric Heat | – .50 |
| Not Heated | – 2.46 |

## Additional upgrades or components

| | |
|---|---|
| Kitchen Cabinets & Countertops | Page 77 |
| Bathroom Vanities | 78 |
| Fireplaces & Chimneys | 78 |
| Windows, Skylights & Dormers | 78 |
| Appliances | 79 |
| Breezeways & Porches | 79 |
| Finished Attic | 79 |
| Garages | 80 |
| Site Improvements | 80 |
| Wings & Ells | 44 |

**Important: See the Reference Section for Location Factors (to adjust for your city) and Estimating Forms**

# Solid Wall 2 Story

**Living Area - 2000 S.F.**
**Perimeter - 135 L.F.**

| | | Labor-Hours | Cost Per Square Foot Of Living Area | | |
| --- | --- | --- | --- | --- | --- |
| | | | Mat. | Labor | Total |
| **1** Site Work | Site preparation for slab; 4' deep trench excavation for foundation wall. | .034 | | .58 | .58 |
| **2** Foundation | Continuous reinforced concrete footing 8" deep x 18" wide; dampproofed and insulated reinforced concrete foundation wall, 8" thick, 4' deep; 4" concrete slab on 4" crushed stone base and polyethylene vapor barrier, trowel finish. | .066 | 2.18 | 2.77 | 4.95 |
| **3** Framing | Exterior walls - Precut traditional log home. Handicrafted white cedar or pine logs. Delivery included. | .232 | 13.25 | 8.80 | 22.05 |
| **4** Exterior Walls | | | | | |
| **5** Roofing | 25 year asphalt shingles; #15 felt building paper; aluminum gutters, downspouts, drip edge and flashings. | .024 | .45 | .82 | 1.27 |
| **6** Interiors | Walls and ceilings, 1/2" taped and finished drywall, primed and painted with 2 coats; painted baseboard and trim, finished hardwood floor 40%, carpet with 1/2" underlayment 40%, vinyl tile with 1/2" underlayment 15%, ceramic tile with 1/2" underlayment 5%; hollow core and louvered interior doors. | .225 | 9.30 | 9.12 | 18.42 |
| **7** Specialties | Average grade kitchen cabinets - 14 L.F. wall and base with plastic laminate counter top and kitchen sink; 40 gallon electric water heater. | .021 | 1.30 | .46 | 1.76 |
| **8** Mechanical | 1 lavatory, white, wall hung; 1 water closet, white; 1 bathtub with shower, enameled steel, white; gas fired warm air heat. | .060 | 2.29 | 1.87 | 4.16 |
| **9** Electrical | 200 Amp. service; romex wiring; incandescent lighting fixtures, switches, receptacles. | .039 | .75 | 1.10 | 1.85 |
| **10** Overhead | Contractor's overhead and profit and plans. | | 6.56 | 6.30 | 12.86 |
| | **Total** | | 36.08 | 31.82 | **67.90** |

# RESIDENTIAL    Average    Wings & Ells

## 1 Story    Base cost per square foot of living area

| Exterior Wall | Living Area | | | | | | | |
|---|---|---|---|---|---|---|---|---|
| | 50 | 100 | 200 | 300 | 400 | 500 | 600 | 700 |
| Wood Siding - Wood Frame | 132.10 | 101.55 | 88.10 | 75.25 | 70.85 | 68.20 | 66.45 | 67.40 |
| Brick Veneer - Wood Frame | 143.60 | 107.80 | 92.15 | 75.65 | 70.55 | 67.45 | 65.40 | 66.15 |
| Stucco on Wood Frame | 132.80 | 102.00 | 88.45 | 75.40 | 71.00 | 68.35 | 66.55 | 67.50 |
| Solid Masonry | 175.40 | 132.45 | 113.85 | 92.40 | 86.30 | 82.65 | 81.30 | 81.65 |
| Finished Basement, Add | 41.35 | 33.45 | 30.20 | 24.70 | 23.60 | 22.95 | 22.50 | 22.20 |
| Unfinished Basement, Add | 18.35 | 14.00 | 12.20 | 9.15 | 8.55 | 8.20 | 7.95 | 7.75 |

## 1-1/2 Story    Base cost per square foot of living area

| Exterior Wall | Living Area | | | | | | | |
|---|---|---|---|---|---|---|---|---|
| | 100 | 200 | 300 | 400 | 500 | 600 | 700 | 800 |
| Wood Siding - Wood Frame | 107.85 | 86.30 | 73.60 | 66.95 | 63.15 | 61.40 | 59.05 | 58.30 |
| Brick Veneer - Wood Frame | 146.55 | 110.60 | 91.95 | 81.10 | 75.50 | 72.60 | 69.25 | 68.05 |
| Stucco on Wood Frame | 130.85 | 98.05 | 81.50 | 72.95 | 67.95 | 65.45 | 62.55 | 61.35 |
| Solid Masonry | 168.75 | 128.35 | 106.75 | 92.65 | 86.15 | 82.65 | 78.80 | 77.45 |
| Finished Basement, Add | 28.00 | 24.70 | 22.55 | 20.10 | 19.45 | 19.00 | 18.65 | 18.55 |
| Unfinished Basement, Add | 11.60 | 9.80 | 8.60 | 7.25 | 6.90 | 6.65 | 6.40 | 6.40 |

## 2 Story    Base cost per square foot of living area

| Exterior Wall | Living Area | | | | | | | |
|---|---|---|---|---|---|---|---|---|
| | 100 | 200 | 400 | 600 | 800 | 1000 | 1200 | 1400 |
| Wood Siding - Wood Frame | 106.50 | 79.65 | 67.70 | 57.40 | 53.50 | 51.15 | 49.60 | 50.70 |
| Brick Veneer - Wood Frame | 148.90 | 104.80 | 84.65 | 68.70 | 63.05 | 59.70 | 57.45 | 58.05 |
| Stucco on Wood Frame | 131.35 | 92.25 | 74.20 | 61.70 | 56.80 | 53.85 | 51.85 | 52.70 |
| Solid Masonry | 173.75 | 122.55 | 99.45 | 78.55 | 71.95 | 68.00 | 65.35 | 65.70 |
| Finished Basement, Add | 22.15 | 18.20 | 16.55 | 13.80 | 13.30 | 12.95 | 12.75 | 12.60 |
| Unfinished Basement, Add | 9.30 | 7.10 | 6.20 | 4.70 | 4.40 | 4.20 | 4.05 | 4.00 |

Base costs do not include bathroom or kitchen facilities. Use Modifications/Adjustments/Alternatives on pages 77-80 where appropriate.

**Important: See the Reference Section for Location Factors (to adjust for your city) and Estimating Forms**

## 1 Story

## 1-1/2 Story

## 2 Story

## 2-1/2 Story

## Bi-Level

## Tri-Level

- **A distinct residence from designer's plans**
- **Single family — 1 full bath, 1 half bath, 1 kitchen**
- **No basement**
- **Asphalt shingles on roof**
- **Forced hot air heat/air conditioning**
- **Drywall interior finishes**
- **Materials and workmanship are above average**

*Note: The illustration shown may contain some optional components (for example: garages and/or fireplaces) whose costs are shown in the modifications, adjustments, & alternatives below or at the end of the square foot section.*

eDesign Basics, Inc.

## Base cost per square foot of living area

| Exterior Wall | Living Area | | | | | | | | | | |
|---|---|---|---|---|---|---|---|---|---|---|---|
| | 800 | 1000 | 1200 | 1400 | 1600 | 1800 | 2000 | 2400 | 2800 | 3200 | 3600 |
| Wood Siding - Wood Frame | 119.90 | 109.40 | 100.90 | 94.05 | 89.35 | 86.90 | 83.85 | 78.00 | 73.80 | 70.80 | 67.65 |
| Brick Veneer - Wood Frame | 132.65 | 121.65 | 112.65 | 105.40 | 100.40 | 97.85 | 94.55 | 88.40 | 83.95 | 80.75 | 77.40 |
| Stone Veneer - Wood Frame | 137.95 | 126.45 | 117.00 | 109.35 | 104.15 | 101.40 | 97.90 | 91.45 | 86.80 | 83.40 | 79.85 |
| Solid Masonry | 138.45 | 126.90 | 117.40 | 109.70 | 104.45 | 101.75 | 98.15 | 91.75 | 87.05 | 83.60 | 80.05 |
| Finished Basement, Add | 37.45 | 35.65 | 34.00 | 32.75 | 31.90 | 31.45 | 30.70 | 29.75 | 29.05 | 28.40 | 27.85 |
| Unfinished Basement, Add | 14.30 | 13.45 | 12.70 | 12.05 | 11.65 | 11.40 | 11.10 | 10.60 | 10.25 | 9.95 | 9.65 |

## Modifications

*Add to the total cost*

| | |
|---|---|
| Upgrade Kitchen Cabinets | $ + 526 |
| Solid Surface Countertops | + 1280 |
| Full Bath - including plumbing, wall and floor finishes | + 4575 |
| Half Bath - including plumbing, wall and floor finishes | + 2850 |
| Two Car Attached Garage | + 17,118 |
| Two Car Detached Garage | + 19,464 |
| Fireplace & Chimney | + 3925 |

## Adjustments

*For multi family - add to total cost*

| | |
|---|---|
| Additional Kitchen | $ + 8491 |
| Additional Full Bath & Half Bath | + 7425 |
| Additional Entry & Exit | + 1128 |
| Separate Heating & Air Conditioning | + 4133 |
| Separate Electric | + 1410 |

*For Townhouse/Rowhouse - Multiply cost per square foot by*

| | |
|---|---|
| Inner Unit | .90 |
| End Unit | .95 |

## Alternatives

*Add to or deduct from the cost per square foot of living area*

| | |
|---|---|
| Cedar Shake Roof | + 1.78 |
| Clay Tile Roof | + 4.19 |
| Slate Roof | + 7.43 |
| Upgrade Ceilings to Textured Finish | + .39 |
| Air Conditioning, in Heating Ductwork | Base System |
| Heating Systems, Hot Water | + 1.48 |
| Heat Pump | + 1.77 |
| Electric Heat | – 1.98 |
| Not Heated | – 3.30 |

## Additional upgrades or components

| | |
|---|---|
| Kitchen Cabinets & Countertops | Page 77 |
| Bathroom Vanities | 78 |
| Fireplaces & Chimneys | 78 |
| Windows, Skylights & Dormers | 78 |
| Appliances | 79 |
| Breezeways & Porches | 79 |
| Finished Attic | 79 |
| Garages | 80 |
| Site Improvements | 80 |
| Wings & Ells | 60 |

**Important: See the Reference Section for Location Factors (to adjust for your city) and Estimating Forms**

| | | | Labor-Hours | Cost Per Square Foot Of Living Area | | |
|---|---|---|---|---|---|---|
| | | | | Mat. | Labor | Total |
| **1** | **Site Work** | Site preparation for slab; 4' deep trench excavation for foundation wall. | .028 | | .51 | .51 |
| **2** | **Foundation** | Continuous reinforced concrete footing 8" deep x 18" wide; dampproofed and insulated reinforced concrete foundation wall, 8" thick, 4' deep; 4" concrete slab on 4" crushed stone base and polyethylene vapor barrier, trowel finish. | .113 | 4.30 | 4.93 | 9.23 |
| **3** | **Framing** | Exterior walls - 2" x 6" wood studs, 16" O.C.; 1/2" plywood sheathing; 2" x 8" rafters 16" O.C. with 1/2" plywood sheathing, 4 in 12 pitch; 2" x 6" ceiling joists 16" O.C.; 5/8" plywood subfloor on 1" x 3" wood sleepers 16" O.C. | .190 | 3.61 | 4.52 | 8.13 |
| **4** | **Exterior Walls** | Horizontal beveled wood siding; #15 felt building paper; 6" batt insulation; wood double hung windows; 3 solid core wood exterior doors; storms and screens. | .085 | 6.98 | 2.33 | 9.31 |
| **5** | **Roofing** | 30 year asphalt shingles; #15 felt building paper; aluminum gutters, downspouts and drip edge; copper flashings. | .082 | 2.79 | 2.39 | 5.18 |
| **6** | **Interiors** | Walls and ceilings - 5/8" drywall, skim coat plaster, painted with primer and 2 coats; hardwood baseboard and trim, sanded and finished; hardwood floor 70%, ceramic tile with underlayment 20%, vinyl tile with underlayment 10%; wood panel interior doors, primed and painted with 2 coats. | .292 | 11.55 | 9.47 | 21.02 |
| **7** | **Specialties** | Custom grade kitchen cabinets - 20 L.F. wall and base with plastic laminate counter top and kitchen sink; 4 L.F. bathroom vanity; 75 gallon electric water heater, medicine cabinet. | .019 | 3.03 | .61 | 3.64 |
| **8** | **Mechanical** | Gas fired warm air heat/air conditioning; one full bath including: bathtub, corner shower, built in lavatory and water closet; one 1/2 bath including: built in lavatory and water closet. | .092 | 4.09 | 2.12 | 6.21 |
| **9** | **Electrical** | 200 Amp. service; romex wiring; fluorescent and incandescent lighting fixtures, switches, receptacles. | .039 | .77 | 1.20 | 1.97 |
| **10** | **Overhead** | Contractor's overhead and profit and design. | | 7.27 | 5.53 | 12.80 |
| | **Total** | | | 44.39 | 33.61 | **78.00** |

# RESIDENTIAL | Custom | 1-1/2 Story

- **A distinct residence from designer's plans**
- **Single family — 1 full bath, 1 half bath, 1 kitchen**
- **No basement**
- **Asphalt shingles on roof**
- **Forced hot air heat/air conditioning**
- **Drywall interior finishes**
- **Materials and workmanship are above average**

*Note: The illustration shown may contain some optional components (for example: garages and/or fireplaces) whose costs are shown in the modifications, adjustments, & alternatives below or at the end of the square foot section.*

•Donald A. Gardner Architects, Inc.

## Base cost per square foot of living area

| Exterior Wall | Living Area | | | | | | | | | | |
|---|---|---|---|---|---|---|---|---|---|---|---|
| | 1000 | 1200 | 1400 | 1600 | 1800 | 2000 | 2400 | 2800 | 3200 | 3600 | 4000 |
| Wood Siding - Wood Frame | 108.25 | 101.45 | 96.60 | 90.30 | 86.90 | 83.35 | 76.65 | 73.75 | 70.95 | 68.75 | 65.65 |
| Brick Veneer - Wood Frame | 114.10 | 107.00 | 101.85 | 95.10 | 91.45 | 87.65 | 80.45 | 77.35 | 74.25 | 71.95 | 68.65 |
| Stone Veneer - Wood Frame | 119.70 | 112.20 | 106.80 | 99.60 | 95.70 | 91.70 | 84.05 | 80.75 | 77.40 | 75.00 | 71.45 |
| Solid Masonry | 120.15 | 112.65 | 107.25 | 100.00 | 96.10 | 92.10 | 84.35 | 81.05 | 77.70 | 75.25 | 71.70 |
| Finished Basement, Add | 24.90 | 23.95 | 23.25 | 22.30 | 21.75 | 21.25 | 20.25 | 19.75 | 19.20 | 18.95 | 18.50 |
| Unfinished Basement, Add | 9.65 | 9.20 | 8.90 | 8.40 | 8.15 | 7.90 | 7.40 | 7.20 | 6.90 | 6.80 | 6.60 |

## Modifications

*Add to the total cost*

| | |
|---|---|
| Upgrade Kitchen Cabinets | $ + 526 |
| Solid Surface Countertops | + 1280 |
| Full Bath - including plumbing, wall and floor finishes | + 4575 |
| Half Bath - including plumbing, wall and floor finishes | + 2850 |
| Two Car Attached Garage | + 17,118 |
| Two Car Detached Garage | + 19,464 |
| Fireplace & Chimney | + 3925 |

## Adjustments

*For multi family - add to total cost*

| | |
|---|---|
| Additional Kitchen | $ + 8491 |
| Additional Full Bath & Half Bath | + 7425 |
| Additional Entry & Exit | + 1128 |
| Separate Heating & Air Conditioning | + 4133 |
| Separate Electric | + 1410 |

*For Townhouse/Rowhouse - Multiply cost per square foot by*

| | |
|---|---|
| Inner Unit | .90 |
| End Unit | .95 |

## Alternatives

*Add to or deduct from the cost per square foot of living area*

| | |
|---|---|
| Cedar Shake Roof | + 1.25 |
| Clay Tile Roof | + 3.02 |
| Slate Roof | + 5.37 |
| Upgrade Ceilings to Textured Finish | + .39 |
| Air Conditioning, in Heating Ductwork | Base System |
| Heating Systems, Hot Water | + 1.40 |
|    Heat Pump | + 1.87 |
|    Electric Heat | – 1.72 |
|    Not Heated | – 3.04 |

## Additional upgrades or components

| | |
|---|---|
| Kitchen Cabinets & Countertops | Page 77 |
| Bathroom Vanities | 78 |
| Fireplaces & Chimneys | 78 |
| Windows, Skylights & Dormers | 78 |
| Appliances | 79 |
| Breezeways & Porches | 79 |
| Finished Attic | 79 |
| Garages | 80 |
| Site Improvements | 80 |
| Wings & Ells | 60 |

**Important: See the Reference Section for Location Factors (to adjust for your city) and Estimating Forms**

**SQUARE FOOT COSTS**

| | | Labor-Hours | Cost Per Square Foot Of Living Area | | |
|---|---|---|---|---|---|
| | | | Mat. | Labor | Total |
| **1 Site Work** | Site preparation for slab; 4' deep trench excavation for foundation wall. | .028 | | .44 | .44 |
| **2 Foundation** | Continuous reinforced concrete footing 8" deep x 18" wide; dampproofed and insulated reinforced concrete foundation wall, 8" thick, 4' deep; 4" concrete slab on 4" crushed stone base and polyethylene vapor barrier, trowel finish. | .065 | 2.97 | 3.52 | 6.49 |
| **3 Framing** | Exterior walls - 2" x 6" wood studs, 16" O.C.; 1/2" plywood sheathing; 2" x 8" rafters 16" O.C. with 1/2" plywood sheathing, 8 in 12 pitch; 2" x 10" floor joists 16" O.C. with 5/8" plywood subfloor; 5/8" plywood subfloor on 1" x 3" wood sleepers 16" O.C. | .192 | 4.10 | 4.64 | 8.74 |
| **4 Exterior Walls** | Horizontal beveled wood siding; #15 felt building paper; 6" batt insulation; wood double hung windows; 3 solid core wood exterior doors; storms and screens. | .064 | 7.03 | 2.34 | 9.37 |
| **5 Roofing** | 30 year asphalt shingles; #15 felt building paper; aluminum gutters, downspouts and drip edge; copper flashings. | .048 | 1.75 | 1.49 | 3.24 |
| **6 Interiors** | Walls and ceilings - 5/8" drywall, skim coat plaster, painted with primer and 2 coats; hardwood baseboard and trim, sanded and finished; hardwood floor 70%, ceramic tile with underlayment 20%, vinyl tile with underlayment 10%; wood panel interior doors, primed and painted with 2 coats. | .259 | 12.33 | 10.52 | 22.85 |
| **7 Specialties** | Custom grade kitchen cabinets - 20 L.F. wall and base with plastic laminate counter top and kitchen sink; 4 L.F. bathroom vanity; 75 gallon electric water heater, medicine cabinet. | .030 | 2.60 | .51 | 3.11 |
| **8 Mechanical** | Gas fired warm air heat/air conditioning; one full bath including: bathtub, corner shower, built in lavatory and water closet; one 1/2 bath including: built in lavatory and water closet. | .084 | 3.56 | 1.98 | 5.54 |
| **9 Electrical** | 200 Amp. service; romex wiring; fluorescent and incandescent lighting fixtures, switches, receptacles. | .038 | .73 | 1.16 | 1.89 |
| **10 Overhead** | Contractor's overhead and profit and design. | | 6.86 | 5.22 | 12.08 |
| | **Total** | | 41.93 | 31.82 | **73.75** |

# RESIDENTIAL | Custom | 2 Story

- **A distinct residence from designer's plans**
- **Single family — 1 full bath, 1 half bath, 1 kitchen**
- **No basement**
- **Asphalt shingles on roof**
- **Forced hot air heat/air conditioning**
- **Drywall interior finishes**
- **Materials and workmanship are above average**

*Note: The illustration shown may contain some optional components (for example: garages and/or fireplaces) whose costs are shown in the modifications, adjustments, & alternatives below or at the end of the square foot section.*

## Base cost per square foot of living area

| Exterior Wall | Living Area | | | | | | | | | | |
|---|---|---|---|---|---|---|---|---|---|---|---|
| | 1200 | 1400 | 1600 | 1800 | 2000 | 2400 | 2800 | 3200 | 3600 | 4000 | 4400 |
| Wood Siding - Wood Frame | 102.10 | 96.60 | 92.65 | 89.05 | 85.10 | 79.40 | 74.60 | 71.35 | 69.45 | 67.55 | 65.75 |
| Brick Veneer - Wood Frame | 108.40 | 102.45 | 98.25 | 94.30 | 90.20 | 84.00 | 78.75 | 75.25 | 73.25 | 71.10 | 69.20 |
| Stone Veneer - Wood Frame | 114.35 | 108.00 | 103.55 | 99.30 | 95.00 | 88.35 | 82.75 | 79.00 | 76.80 | 74.45 | 72.45 |
| Solid Masonry | 114.85 | 108.50 | 104.05 | 99.75 | 95.45 | 88.70 | 83.05 | 79.30 | 77.15 | 74.75 | 72.75 |
| Finished Basement, Add | 20.00 | 19.25 | 18.75 | 18.20 | 17.85 | 17.05 | 16.40 | 16.00 | 15.75 | 15.40 | 15.20 |
| Unfinished Basement, Add | 7.75 | 7.40 | 7.15 | 6.90 | 6.75 | 6.35 | 6.05 | 5.85 | 5.70 | 5.55 | 5.45 |

## Modifications

*Add to the total cost*

| | |
|---|---|
| Upgrade Kitchen Cabinets | $ + 526 |
| Solid Surface Countertops | + 1280 |
| Full Bath - including plumbing, wall and floor finishes | + 4575 |
| Half Bath - including plumbing, wall and floor finishes | + 2850 |
| Two Car Attached Garage | + 17,118 |
| Two Car Detached Garage | + 19,464 |
| Fireplace & Chimney | + 4430 |

## Adjustments

*For multi family - add to total cost*

| | |
|---|---|
| Additional Kitchen | $ + 8491 |
| Additional Full Bath & Half Bath | + 7425 |
| Additional Entry & Exit | + 1128 |
| Separate Heating & Air Conditioning | + 4133 |
| Separate Electric | + 1410 |

*For Townhouse/Rowhouse -*
*Multiply cost per square foot by*

| | |
|---|---|
| Inner Unit | .87 |
| End Unit | .93 |

## Alternatives

*Add to or deduct from the cost per square foot of living area*

| | |
|---|---|
| Cedar Shake Roof | + .8? |
| Clay Tile Roof | + 2.1? |
| Slate Roof | + 3.7? |
| Upgrade Ceilings to Textured Finish | + .3? |
| Air Conditioning, in Heating Ductwork | Base System |
| Heating Systems, Hot Water | + 1.3? |
|    Heat Pump | + 2.0? |
|    Electric Heat | − 1.7? |
|    Not Heated | − 2.8? |

## Additional upgrades or components

| | |
|---|---|
| Kitchen Cabinets & Countertops | Page 7? |
| Bathroom Vanities | 7? |
| Fireplaces & Chimneys | 7? |
| Windows, Skylights & Dormers | 7? |
| Appliances | 7? |
| Breezeways & Porches | 7? |
| Finished Attic | 7? |
| Garages | 8? |
| Site Improvements | 8? |
| Wings & Ells | 6? |

**Important: See the Reference Section for Location Factors (to adjust for your city) and Estimating Form**

**SQUARE FOOT COSTS**

| | | Labor-Hours | Cost Per Square Foot Of Living Area | | |
|---|---|---|---|---|---|
| | | | Mat. | Labor | Total |
| **1 Site Work** | Site preparation for slab; 4' deep trench excavation for foundation wall. | .024 | | .44 | .44 |
| **2 Foundation** | Continuous reinforced concrete footing 8" deep x 18" wide; dampproofed and insulated reinforced concrete foundation wall, 8" thick, 4' deep; 4" concrete slab on 4" crushed stone base and polyethylene vapor barrier, trowel finish. | .058 | 2.51 | 3.02 | 5.53 |
| **3 Framing** | Exterior walls - 2" x 6" wood studs, 16" O.C.; 1/2" plywood sheathing; 2" x 8" rafters 16" O.C. with 1/2" plywood sheathing, 6 in 12 pitch; 2" x 8" ceiling joists 16" O.C.; 2" x 10" floor joists 16" O.C. with 5/8" plywood subfloor; 5/8" plywood subfloor on 1" x 3" wood sleepers 16" O.C. | .159 | 4.37 | 4.76 | 9.13 |
| **4 Exterior Walls** | Horizontal beveled wood siding; #15 felt building paper; 6" batt insulation; wood double hung windows; 3 solid core wood exterior doors; storms and screens. | .091 | 8.01 | 2.66 | 10.67 |
| **5 Roofing** | 30 year asphalt shingles; #15 felt building paper; aluminum gutters, downspouts and drip edge; copper flashings. | .042 | 1.40 | 1.20 | 2.60 |
| **6 Interiors** | Walls and ceilings - 5/8" drywall, skim coat plaster, painted with primer and 2 coats; hardwood baseboard and trim, sanded and finished; hardwood floor 70%, ceramic tile with underlayment 20%, vinyl tile with underlayment 10%; wood panel interior doors, primed and painted with 2 coats. | .271 | 12.51 | 10.82 | 23.33 |
| **7 Specialties** | Custom grade kitchen cabinets - 20 L.F. wall and base with plastic laminate counter top and kitchen sink; 4 L.F. bathroom vanity; 75 gallon electric water heater, medicine cabinet. | .028 | 2.60 | .51 | 3.11 |
| **8 Mechanical** | Gas fired warm air heat/air conditioning; one full bath including: bathtub, corner shower; built in lavatory and water closet; one 1/2 bath including: built in lavatory and water closet. | .078 | 3.67 | 2.02 | 5.69 |
| **9 Electrical** | 200 Amp. service; romex wiring; fluorescent and incandescent lighting fixtures, switches, receptacles. | .038 | .73 | 1.16 | 1.89 |
| **10 Overhead** | Contractor's overhead and profit and design. | | 7.01 | 5.20 | 12.21 |
| **Total** | | | 42.81 | 31.79 | **74.60** |

# RESIDENTIAL  **Custom**  **2-1/2 Story**

- **A distinct residence from designer's plans**
- **Single family — 1 full bath, 1 half bath, 1 kitchen**
- **No basement**
- **Asphalt shingles on roof**
- **Forced hot air heat/air conditioning**
- **Drywall interior finishes**
- **Materials and workmanship are above average**

*Note: The illustration shown may contain some optional components (for example: garages and/or fireplaces) whose costs are shown in the modifications, adjustments, & alternatives below or at the end of the square foot section.*

## Base cost per square foot of living area

| Exterior Wall | Living Area | | | | | | | | | | |
|---|---|---|---|---|---|---|---|---|---|---|---|
| | 1500 | 1800 | 2100 | 2400 | 2800 | 3200 | 3600 | 4000 | 4500 | 5000 | 5500 |
| Wood Siding - Wood Frame | 101.35 | 91.35 | 85.85 | 82.45 | 78.20 | 73.85 | 71.60 | 67.75 | 65.90 | 64.15 | 62.25 |
| Brick Veneer - Wood Frame | 107.85 | 97.40 | 91.25 | 87.55 | 83.05 | 78.25 | 75.80 | 71.65 | 69.60 | 67.60 | 65.60 |
| Stone Veneer - Wood Frame | 114.05 | 103.10 | 96.35 | 92.40 | 87.70 | 82.45 | 79.75 | 75.35 | 73.05 | 70.95 | 68.75 |
| Solid Masonry | 114.60 | 103.60 | 96.75 | 92.80 | 88.10 | 82.80 | 80.10 | 75.65 | 73.40 | 71.25 | 69.05 |
| Finished Basement, Add | 15.90 | 15.20 | 14.40 | 14.00 | 13.65 | 13.10 | 12.80 | 12.45 | 12.20 | 11.95 | 11.75 |
| Unfinished Basement, Add | 6.25 | 5.90 | 5.50 | 5.30 | 5.15 | 4.85 | 4.75 | 4.55 | 4.40 | 4.35 | 4.20 |

## Modifications

*Add to the total cost*

| | |
|---|---|
| Upgrade Kitchen Cabinets | $ + 526 |
| Solid Surface Countertops | + 1280 |
| Full Bath - including plumbing, wall and floor finishes | + 4575 |
| Half Bath - including plumbing, wall and floor finishes | + 2850 |
| Two Car Attached Garage | + 17,118 |
| Two Car Detached Garage | + 19,464 |
| Fireplace & Chimney | + 5005 |

## Adjustments

*For multi family - add to total cost*

| | |
|---|---|
| Additional Kitchen | $ + 8491 |
| Additional Full Bath & Half Bath | + 7425 |
| Additional Entry & Exit | + 1128 |
| Separate Heating & Air Conditioning | + 4133 |
| Separate Electric | + 1410 |

*For Townhouse/Rowhouse -
Multiply cost per square foot by*

| | |
|---|---|
| Inner Unit | .87 |
| End Unit | .94 |

## Alternatives

*Add to or deduct from the cost per square foot of living area*

| | |
|---|---|
| Cedar Shake Roof | + .7 |
| Clay Tile Roof | + 1.8 |
| Slate Roof | + 3.2 |
| Upgrade Ceilings to Textured Finish | + .3 |
| Air Conditioning, in Heating Ductwork | Base System |
| Heating Systems, Hot Water | + 1.2 |
|     Heat Pump | + 2.1 |
|     Electric Heat | – 3.0 |
|     Not Heated | – 2.8 |

## Additional upgrades or components

| | |
|---|---|
| Kitchen Cabinets & Countertops | Page 7 |
| Bathroom Vanities | 7 |
| Fireplaces & Chimneys | 7 |
| Windows, Skylights & Dormers | 7 |
| Appliances | 7 |
| Breezeways & Porches | 7 |
| Finished Attic | 7 |
| Garages | 8 |
| Site Improvements | 8 |
| Wings & Ells | 6 |

**Important: See the Reference Section for Location Factors (to adjust for your city) and Estimating Form**

**SQUARE FOOT COSTS**

| | | Labor-Hours | Cost Per Square Foot Of Living Area | | |
|---|---|---|---|---|---|
| | | | Mat. | Labor | Total |
| **1 Site Work** | Site preparation for slab; 4' deep trench excavation for foundation wall. | .048 | | .39 | .39 |
| **2 Foundation** | Continuous reinforced concrete footing 8" deep x 18" wide; dampproofed and insulated reinforced concrete foundation wall, 8" thick, 4' deep; 4" concrete slab on 4" crushed stone base and polyethylene vapor barrier, trowel finish. | .063 | 2.05 | 2.51 | 4.56 |
| **3 Framing** | Exterior walls - 2" x 6" wood studs, 16" O.C.; 1/2" plywood sheathing; 2" x 8" rafters 16" O.C. with 1/2" plywood sheathing, 6 in 12 pitch; 2" x 8" ceiling joists 16" O.C.; 2" x 10" floor joists 16" O.C. with 5/8" plywood subfloor; 5/8" plywood subfloor on 1" x 3" wood sleepers 16" O.C. | .177 | 4.62 | 4.90 | 9.52 |
| **4 Exterior Walls** | Horizontal beveled wood siding; #15 felt building paper; 6" batt insulation; wood double hung windows; 3 solid core wood exterior doors; storms and screens. | .134 | 8.28 | 2.76 | 11.04 |
| **5 Roofing** | 30 year asphalt shingles; #15 felt building paper; aluminum gutters, downspouts and drip edge; copper flashings. | .032 | 1.08 | .92 | 2.00 |
| **6 Interiors** | Walls and ceilings - 5/8" drywall, skim coat plaster, painted with primer and 2 coats; hardwood baseboard and trim, sanded and finished; hardwood floor 70%, ceramic tile with underlayment 20%, vinyl tile with underlayment 10%; wood panel interior doors, primed and painted with 2 coats. | .354 | 12.93 | 11.46 | 24.39 |
| **7 Specialties** | Custom grade kitchen cabinets - 20 L.F. wall and base with plastic laminate counter top and kitchen sink; 4 L.F. bathroom vanity; 75 gallon electric water heater, medicine cabinet. | .053 | 2.28 | .46 | 2.74 |
| **8 Mechanical** | Gas fired warm air heat/air conditioning; one full bath including: bathtub, corner shower; built in lavatory and water closet; one 1/2 bath including: built in lavatory and water closet. | .104 | 3.36 | 1.93 | 5.29 |
| **9 Electrical** | 200 Amp. service; romex wiring; fluorescent and incandescent lighting fixtures, switches, receptacles. | .048 | .70 | 1.12 | 1.82 |
| **10 Overhead** | Contractor's overhead and profit and design. | | 6.91 | 5.19 | 12.10 |
| **Total** | | | 42.21 | 31.64 | **73.85** |

# RESIDENTIAL | Custom | 3 Story

- **A distinct residence from designer's plans**
- **Single family — 1 full bath, 1 half bath, 1 kitchen**
- **No basement**
- **Asphalt shingles on roof**
- **Forced hot air heat/air conditioning**
- **Drywall interior finishes**
- **Materials and workmanship are above average**

*Note: The illustration shown may contain some optional components (for example: garages and/or fireplaces) whose costs are shown in the modifications, adjustments, & alternatives below or at the end of the square foot section.*

## Base cost per square foot of living area

| Exterior Wall | Living Area | | | | | | | | | | |
|---|---|---|---|---|---|---|---|---|---|---|---|
| | 1500 | 1800 | 2100 | 2500 | 3000 | 3500 | 4000 | 4500 | 5000 | 5500 | 6000 |
| Wood Siding - Wood Frame | 100.90 | 91.05 | 86.65 | 83.05 | 77.00 | 74.15 | 70.35 | 66.35 | 65.05 | 63.50 | 62.05 |
| Brick Veneer - Wood Frame | 107.65 | 97.30 | 92.50 | 88.65 | 82.10 | 79.00 | 74.75 | 70.40 | 68.95 | 67.30 | 65.60 |
| Stone Veneer - Wood Frame | 114.10 | 103.25 | 98.05 | 93.95 | 86.90 | 83.55 | 78.90 | 74.25 | 72.65 | 70.85 | 68.95 |
| Solid Masonry | 114.65 | 103.80 | 98.60 | 94.40 | 87.35 | 83.95 | 79.25 | 74.60 | 72.95 | 71.20 | 69.25 |
| Finished Basement, Add | 14.00 | 13.35 | 12.85 | 12.45 | 11.90 | 11.55 | 11.15 | 10.80 | 10.65 | 10.45 | 10.25 |
| Unfinished Basement, Add | 5.45 | 5.15 | 4.95 | 4.75 | 4.50 | 4.35 | 4.10 | 3.95 | 3.90 | 3.80 | 3.70 |

## Modifications

*Add to the total cost*

| | |
|---|---|
| Upgrade Kitchen Cabinets | $ + 526 |
| Solid Surface Countertops | + 1280 |
| Full Bath - including plumbing, wall and floor finishes | + 4575 |
| Half Bath - including plumbing, wall and floor finishes | + 2850 |
| Two Car Attached Garage | + 17,118 |
| Two Car Detached Garage | + 19,464 |
| Fireplace & Chimney | + 5005 |

## Adjustments

*For multi family - add to total cost*

| | |
|---|---|
| Additional Kitchen | $ + 8491 |
| Additional Full Bath & Half Bath | + 7425 |
| Additional Entry & Exit | + 1128 |
| Separate Heating & Air Conditioning | + 4133 |
| Separate Electric | + 1410 |

*For Townhouse/Rowhouse -*
*Multiply cost per square foot by*

| | |
|---|---|
| Inner Unit | .85 |
| End Unit | .93 |

## Alternatives

*Add to or deduct from the cost per square foot of living area*

| | |
|---|---|
| Cedar Shake Roof | + .5 |
| Clay Tile Roof | + 1.4 |
| Slate Roof | + 2.4 |
| Upgrade Ceilings to Textured Finish | + .3 |
| Air Conditioning, in Heating Ductwork | Base Syste |
| Heating Systems, Hot Water | + 1.2 |
| Heat Pump | + 2.1 |
| Electric Heat | – 3.0 |
| Not Heated | – 2.7 |

## Additional upgrades or components

| | |
|---|---|
| Kitchen Cabinets & Countertops | Page 7 |
| Bathroom Vanities | 7 |
| Fireplaces & Chimneys | 7 |
| Windows, Skylights & Dormers | 7 |
| Appliances | 7 |
| Breezeways & Porches | 7 |
| Finished Attic | 7 |
| Garages | 8 |
| Site Improvements | 8 |
| Wings & Ells | 6 |

**Important: See the Reference Section for Location Factors (to adjust for your city) and Estimating Form**

# Custom 3 Story

**Living Area - 3000 S.F.**
**Perimeter - 135 L.F.**

| | | Labor-Hours | Cost Per Square Foot Of Living Area | | |
|---|---|---|---|---|---|
| | | | Mat. | Labor | Total |
| **1 Site Work** | Site preparation for slab; 4' deep trench excavation for foundation wall. | .048 | | .41 | .41 |
| **2 Foundation** | Continuous reinforced concrete footing 8" deep x 18" wide; dampproofed and insulated reinforced concrete foundation wall, 8" thick, 4' deep; 4" concrete slab on 4" crushed stone base and polyethylene vapor barrier, trowel finish. | .060 | 1.91 | 2.36 | 4.27 |
| **3 Framing** | Exterior walls - 2" x 6" wood studs, 16" O.C.; 1/2" plywood sheathing; 2" x 8" rafters 16" O.C. with 1/2" plywood sheathing, 6 in 12 pitch; 2" x 8" ceiling joists 16" O.C.; 2" x 10" floor joists 16" O.C. with 5/8" plywood subfloor; 5/8" plywood subfloor on 1" x 3" wood sleepers 16" O.C. | .191 | 4.83 | 5.05 | 9.88 |
| **4 Exterior Walls** | Horizontal beveled wood siding; #15 felt building paper; 6" batt insulation; wood double hung windows; 3 solid core wood exterior doors; storms and screens. | .150 | 9.47 | 3.19 | 12.66 |
| **5 Roofing** | 30 year asphalt shingles; #15 felt building paper; aluminum gutters, downspouts and drip edge; copper flashings. | .028 | .93 | .79 | 1.72 |
| **6 Interiors** | Walls and ceilings - 5/8" drywall, skim coat plaster, painted with primer and 2 coats; hardwood baseboard and trim, sanded and finished; hardwood floor 70%, ceramic tile with underlayment 20%, vinyl tile with underlayment 10%; wood panel interior doors, primed and painted with 2 coats. | .409 | 13.30 | 11.93 | 25.23 |
| **7 Specialties** | Custon grade kitchen cabinets - 20 L.F. wall and base with plastic laminate counter top and kitchen sink; 4 L.F. bathroom vanity; 75 gallon electric water heater, medicine cabinet. | .053 | 2.41 | .49 | 2.90 |
| **8 Mechanical** | Gas fired warm air heat/air conditioning; one full bath including: bathtub, corner shower; built in lavatory and water closet; one 1/2 bath including: built in lavatory and water closet. | .105 | 3.50 | 1.97 | 5.47 |
| **9 Electrical** | 200 Amp. service; romex wiring; fluorescent and incandescent lighting fixtures, switches, receptacles. | .048 | .71 | 1.14 | 1.85 |
| **10 Overhead** | Contractor's overhead and profit and design. | | 7.27 | 5.34 | 12.61 |
| **Total** | | | 44.33 | 32.67 | **77.00** |

# RESIDENTIAL | Custom | Bi-Level

- **A distinct residence from designer's plans**
- **Single family — 1 full bath, 1 half bath, 1 kitchen**
- **No basement**
- **Asphalt shingles on roof**
- **Forced hot air heat/air conditioning**
- **Drywall interior finishes**
- **Materials and workmanship are above average**

*Note: The illustration shown may contain some optional components (for example: garages and/or fireplaces) whose costs are shown in the modifications, adjustments, & alternatives below or at the end of the square foot section.*

## Base cost per square foot of living area

| Exterior Wall | Living Area | | | | | | | | | | |
|---|---|---|---|---|---|---|---|---|---|---|---|
| | 1200 | 1400 | 1600 | 1800 | 2000 | 2400 | 2800 | 3200 | 3600 | 4000 | 4400 |
| Wood Siding - Wood Frame | 96.35 | 91.15 | 87.45 | 84.20 | 80.40 | 75.15 | 70.75 | 67.80 | 66.05 | 64.30 | 62.60 |
| Brick Veneer - Wood Frame | 101.05 | 95.50 | 91.65 | 88.15 | 84.20 | 78.60 | 73.90 | 70.75 | 68.90 | 66.95 | 65.15 |
| Stone Veneer - Wood Frame | 105.50 | 99.70 | 95.65 | 91.90 | 87.85 | 81.85 | 76.85 | 73.50 | 71.55 | 69.50 | 67.60 |
| Solid Masonry | 105.90 | 100.05 | 96.00 | 92.20 | 88.15 | 82.10 | 77.10 | 73.75 | 71.80 | 69.70 | 67.80 |
| Finished Basement, Add | 20.00 | 19.25 | 18.75 | 18.20 | 17.85 | 17.05 | 16.40 | 16.00 | 15.75 | 15.40 | 15.20 |
| Unfinished Basement, Add | 7.75 | 7.40 | 7.15 | 6.90 | 6.75 | 6.35 | 6.05 | 5.85 | 5.70 | 5.55 | 5.45 |

## Modifications

*Add to the total cost*

| | |
|---|---|
| Upgrade Kitchen Cabinets | $ + 526 |
| Solid Surface Countertops | + 1280 |
| Full Bath - including plumbing, wall and floor finishes | + 4575 |
| Half Bath - including plumbing, wall and floor finishes | + 2850 |
| Two Car Attached Garage | + 17,118 |
| Two Car Detached Garage | + 19,464 |
| Fireplace & Chimney | + 3925 |

## Adjustments

*For multi family - add to total cost*

| | |
|---|---|
| Additional Kitchen | $ + 8491 |
| Additional Full Bath & Half Bath | + 7425 |
| Additional Entry & Exit | + 1128 |
| Separate Heating & Air Conditioning | + 4133 |
| Separate Electric | + 1410 |

*For Townhouse/Rowhouse -*
*Multiply cost per square foot by*

| | |
|---|---|
| Inner Unit | .89 |
| End Unit | .95 |

## Alternatives

*Add to or deduct from the cost per square foot of living area*

| | |
|---|---|
| Cedar Shake Roof | + .8 |
| Clay Tile Roof | + 2.1 |
| Slate Roof | + 3.7 |
| Upgrade Ceilings to Textured Finish | + .3 |
| Air Conditioning, in Heating Ductwork | Base Syste |
| Heating Systems, Hot Water | + 1.3 |
| Heat Pump | + 2.0 |
| Electric Heat | – 1.7 |
| Not Heated | – 2.7 |

## Additional upgrades or components

| | |
|---|---|
| Kitchen Cabinets & Countertops | Page 7 |
| Bathroom Vanities | 7 |
| Fireplaces & Chimneys | 7 |
| Windows, Skylights & Dormers | 7 |
| Appliances | 7 |
| Breezeways & Porches | 7 |
| Finished Attic | 7 |
| Garages | 8 |
| Site Improvements | 8 |
| Wings & Ells | 6 |

**Important: See the Reference Section for Location Factors (to adjust for your city) and Estimating Form**

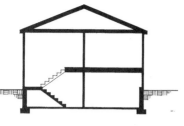

**SQUARE FOOT COSTS**

| | | Labor-Hours | Cost Per Square Foot Of Living Area | | |
|---|---|---|---|---|---|
| | | | Mat. | Labor | Total |
| **1 Site Work** | Excavation for lower level, 4' deep. Site preparation for slab. | .024 | | .44 | .44 |
| **2 Foundation** | Continuous reinforced concrete footing 8" deep x 18" wide; dampproofed and insulated reinforced concrete foundation wall, 8" thick, 4' deep; 4" concrete slab on 4" crushed stone base and polyethylene vapor barrier, trowel finish. | .058 | 2.51 | 3.02 | 5.53 |
| **3 Framing** | Exterior walls - 2" x 6" wood studs, 16" O.C.; 1/2" plywood sheathing; 2" x 8" rafters 16" O.C. with 1/2" plywood sheathing, 6 in 12 pitch; 2" x 8" ceiling joists 16" O.C.; 2" x 10" floor joists 16" O.C. with 5/8" plywood subfloor; 5/8" plywood subfloor on 1" x 3" wood sleepers 16" O.C. | .147 | 4.11 | 4.53 | 8.64 |
| **4 Exterior Walls** | Horizontal beveled wood siding; #15 felt building paper; 6" batt insulation; wood double hung windows; 3 solid core wood exterior doors; storms and screens. | .079 | 6.25 | 2.07 | 8.32 |
| **5 Roofing** | 30 year asphalt shingles; #15 felt building paper; aluminum gutters, downspouts and drip edge; copper flashings. | .033 | 1.40 | 1.20 | 2.60 |
| **6 Interiors** | Walls and ceilings - 5/8" drywall, skim coat plaster, painted with primer and 2 coats; hardwood baseboard and trim, sanded and finished; hardwood floor 70%, ceramic tile with underlayment 20%, vinyl tile with underlayment 10%; wood panel interior doors, primed and painted with 2 coats. | .257 | 12.39 | 10.57 | 22.96 |
| **7 Specialties** | Custom grade kitchen cabinets - 20 L.F. wall and base with plastic laminate counter top and kitchen sink; 4 L.F. bathroom vanity; 75 gallon electric water heater, medicine cabinet. | .028 | 2.60 | .51 | 3.11 |
| **8 Mechanical** | Gas fired warm air heat/air conditioning; one full bath including: bathtub, corner shower, built in lavatory and water closet; one 1/2 bath including: built in lavatory and water closet. | .078 | 3.67 | 2.02 | 5.69 |
| **9 Electrical** | 200 Amp. service; romex wiring; fluorescent and incandescent lighting fixtures, switches, receptacles. | .038 | .73 | 1.16 | 1.89 |
| **10 Overhead** | Contractor's overhead and profit and design. | | 6.60 | 4.97 | 11.57 |
| **Total** | | 40.26 | 30.49 | | **70.75** |

- **A distinct residence from designer's plans**
- **Single family — 1 full bath, 1 half bath, 1 kitchen**
- **No basement**
- **Asphalt shingles on roof**
- **Forced hot air heat/air conditioning**
- **Drywall interior finishes**
- **Materials and workmanship are above average**

*Note: The illustration shown may contain some optional components (for example: garages and/or fireplaces) whose costs are shown in the modifications, adjustments, & alternatives below or at the end of the square foot section.*

©Design Basics, Inc.

## Base cost per square foot of living area

| Exterior Wall | Living Area | | | | | | | | | | |
|---|---|---|---|---|---|---|---|---|---|---|---|
| | 1200 | 1500 | 1800 | 2100 | 2400 | 2800 | 3200 | 3600 | 4000 | 4500 | 5000 |
| Wood Siding - Wood Frame | 99.30 | 90.85 | 84.45 | 79.10 | 75.45 | 72.95 | 69.80 | 66.40 | 65.10 | 61.80 | 60.05 |
| Brick Veneer - Wood Frame | 104.00 | 95.10 | 88.25 | 82.60 | 78.70 | 76.05 | 72.70 | 69.10 | 67.75 | 64.20 | 62.35 |
| Stone Veneer - Wood Frame | 108.40 | 99.10 | 91.90 | 85.90 | 81.80 | 79.05 | 75.45 | 71.65 | 70.20 | 66.50 | 64.50 |
| Solid Masonry | 108.80 | 99.45 | 92.15 | 86.20 | 82.10 | 79.30 | 75.70 | 71.85 | 70.40 | 66.70 | 64.70 |
| Finished Basement, Add* | 24.90 | 23.70 | 22.65 | 21.80 | 21.25 | 20.85 | 20.30 | 19.85 | 19.60 | 19.15 | 18.85 |
| Unfinished Basement, Add* | 9.55 | 8.95 | 8.45 | 8.05 | 7.75 | 7.60 | 7.30 | 7.10 | 7.00 | 6.75 | 6.60 |

*Basement under middle level only.

## Modifications

*Add to the total cost*

| | |
|---|---|
| Upgrade Kitchen Cabinets | $  + 526 |
| Solid Surface Countertops | + 1280 |
| Full Bath - including plumbing, wall and floor finishes | + 4575 |
| Half Bath - including plumbing, wall and floor finishes | + 2850 |
| Two Car Attached Garage | + 17,118 |
| Two Car Detached Garage | + 19,464 |
| Fireplace & Chimney | + 4430 |

## Adjustments

*For multi family - add to total cost*

| | |
|---|---|
| Additional Kitchen | $  + 8491 |
| Additional Full Bath & Half Bath | + 7425 |
| Additional Entry & Exit | + 1128 |
| Separate Heating & Air Conditioning | + 4133 |
| Separate Electric | + 1410 |

*For Townhouse/Rowhouse - Multiply cost per square foot by*

| | |
|---|---|
| Inner Unit | .87 |
| End Unit | .94 |

## Alternatives

*Add to or deduct from the cost per square foot of living area*

| | |
|---|---|
| Cedar Shake Roof | + 1.2 |
| Clay Tile Roof | + 3.0 |
| Slate Roof | + 5.3 |
| Upgrade Ceilings to Textured Finish | + .3 |
| Air Conditioning, in Heating Ductwork | Base System |
| Heating Systems, Hot Water | + 1.3 |
| Heat Pump | + 2.1 |
| Electric Heat | − 1.5 |
| Not Heated | − 2.7 |

## Additional upgrades or components

| | |
|---|---|
| Kitchen Cabinets & Countertops | Page 7 |
| Bathroom Vanities | 7 |
| Fireplaces & Chimneys | 7 |
| Windows, Skylights & Dormers | 7 |
| Appliances | 7 |
| Breezeways & Porches | 7 |
| Finished Attic | 7 |
| Garages | 8 |
| Site Improvements | 8 |
| Wings & Ells | 6 |

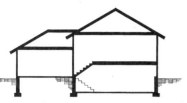

**Living Area - 3200 S.F.**
**Perimeter - 198 L.F.**

**SQUARE FOOT COSTS**

| | | Labor-Hours | Cost Per Square Foot Of Living Area | | |
|---|---|---|---|---|---|
| | | | Mat. | Labor | Total |
| **1 Site Work** | Site preparation for slab; 4' deep trench excavation for foundation wall, excavation for lower level, 4' deep. | .023 | | .39 | .39 |
| **2 Foundation** | Continuous reinforced concrete footing 8" deep x 18" wide; dampproofed and insulated reinforced concrete foundation wall, 8" thick, 4' deep; 4" concrete slab on 4" crushed stone base and polyethylene vapor barrier, trowel finish. | .073 | 2.99 | 3.49 | 6.48 |
| **3 Framing** | Exterior walls - 2" x 6" wood studs, 16" O.C.; 1/2" plywood sheathing; 2" x 8" rafters 16" O.C. with 1/2" plywood sheathing, 6 in 12 pitch; 2" x 8" ceiling joists 16" O.C.; 2" x 10" floor joists 16" O.C. with 5/8" plywood subfloor; 5/8" plywood subfloor on 1" x 3" wood sleepers 16" O.C. | .162 | 3.88 | 4.50 | 8.38 |
| **4 Exterior Walls** | Horizontal beveled wood siding; #15 felt building paper; 6" batt insulation; wood double hung windows; 3 solid core wood exterior doors; storms and screens. | .076 | 6.04 | 2.01 | 8.05 |
| **5 Roofing** | 30 year asphalt shingles; #15 felt building paper; aluminum gutters, downspouts and drip edge; copper flashings. | .045 | 1.86 | 1.60 | 3.46 |
| **6 Interiors** | Walls and ceilings - 5/8" drywall, skim coat plaster, painted with primer and 2 coats; hardwood baseboard and trim, sanded and finished; hardwood floor 70%, ceramic tile with underlayment 20%, vinyl tile with underlayment 10%; wood panel interior doors, primed and painted with 2 coats. | .242 | 11.85 | 9.93 | 21.78 |
| **7 Specialties** | Custom grade kitchen cabinets - 20 L.F. wall and base with plastic laminate counter top and kitchen sink; 4 L.F. bathroom vanity; 75 gallon electric water heater, medicine cabinet. | .026 | 2.28 | .46 | 2.74 |
| **8 Mechanical** | Gas fired warm air heat/air conditioning; one full bath including: bathtub, corner shower, built in lavatory and water closet; one 1/2 bath including: built in lavatory and water closet. | .073 | 3.36 | 1.93 | 5.29 |
| **9 Electrical** | 200 Amp. service; romex wiring; fluorescent and incandescent lighting fixtures, switches, receptacles. | .036 | .70 | 1.12 | 1.82 |
| **10 Overhead** | Contractor's overhead and profit and design. | | 6.45 | 4.96 | 11.41 |
| | **Total** | | 39.41 | 30.39 | **69.80** |

# RESIDENTIAL | Custom | Wings & Ells

## 1 Story | Base cost per square foot of living area

| Exterior Wall | Living Area | | | | | | | |
|---|---|---|---|---|---|---|---|---|
| | 50 | 100 | 200 | 300 | 400 | 500 | 600 | 700 |
| Wood Siding - Wood Frame | 158.55 | 121.75 | 105.70 | 89.85 | 84.60 | 81.40 | 79.30 | 80.40 |
| Brick Veneer - Wood Frame | 174.35 | 133.10 | 115.15 | 96.15 | 90.20 | 86.70 | 84.30 | 85.20 |
| Stone Veneer - Wood Frame | 189.35 | 143.80 | 124.05 | 102.10 | 95.60 | 91.70 | 89.05 | 89.80 |
| Solid Masonry | 190.65 | 144.75 | 124.85 | 102.60 | 96.00 | 92.15 | 89.50 | 90.20 |
| Finished Basement, Add | 60.35 | 49.05 | 44.35 | 36.55 | 34.95 | 34.00 | 33.40 | 32.90 |
| Unfinished Basement, Add | 42.00 | 28.40 | 22.15 | 16.80 | 15.35 | 14.45 | 13.90 | 13.45 |

## 1-1/2 Story | Base cost per square foot of living area

| Exterior Wall | Living Area | | | | | | | |
|---|---|---|---|---|---|---|---|---|
| | 100 | 200 | 300 | 400 | 500 | 600 | 700 | 800 |
| Wood Siding - Wood Frame | 128.15 | 103.75 | 89.15 | 81.30 | 76.85 | 74.85 | 72.10 | 71.20 |
| Brick Veneer - Wood Frame | 142.30 | 115.05 | 98.55 | 88.65 | 83.60 | 81.25 | 78.15 | 77.20 |
| Stone Veneer - Wood Frame | 155.65 | 125.75 | 107.50 | 95.60 | 90.05 | 87.30 | 83.90 | 82.90 |
| Solid Masonry | 156.85 | 126.70 | 108.30 | 96.20 | 90.60 | 87.85 | 84.40 | 83.40 |
| Finished Basement, Add | 40.15 | 35.50 | 32.35 | 28.90 | 27.95 | 27.35 | 26.70 | 26.65 |
| Unfinished Basement, Add | 24.75 | 18.45 | 15.55 | 13.15 | 12.25 | 11.70 | 11.20 | 11.00 |

## 2 Story | Base cost per square foot of living area

| Exterior Wall | Living Area | | | | | | | |
|---|---|---|---|---|---|---|---|---|
| | 100 | 200 | 400 | 600 | 800 | 1000 | 1200 | 1400 |
| Wood Siding - Wood Frame | 127.70 | 96.20 | 82.30 | 70.05 | 65.55 | 62.75 | 60.95 | 62.25 |
| Brick Veneer - Wood Frame | 143.55 | 107.50 | 91.70 | 76.35 | 71.15 | 68.05 | 66.00 | 67.10 |
| Stone Veneer - Wood Frame | 158.55 | 118.20 | 100.65 | 82.30 | 76.55 | 73.05 | 70.75 | 71.70 |
| Solid Masonry | 159.85 | 119.15 | 101.45 | 82.80 | 76.95 | 73.50 | 71.15 | 72.05 |
| Finished Basement, Add | 30.20 | 24.55 | 22.20 | 18.25 | 17.50 | 17.05 | 16.75 | 16.50 |
| Unfinished Basement, Add | 21.00 | 14.20 | 11.10 | 8.40 | 7.70 | 7.25 | 6.95 | 6.75 |

Base costs do not include bathroom or kitchen facilities. Use Modifications/Adjustments/Alternatives on pages 77-80 where appropriate.

## 1 Story

## 1-1/2 Story

## 2 Story

## 2-1/2 Story

## Bi-Level

## Tri-Level

**SQUARE FOOT COSTS**

- **Unique residence built from an architect's plan**
- **Single family — 1 full bath, 1 half bath, 1 kitchen**
- **No basement**
- **Cedar shakes on roof**
- **Forced hot air heat/air conditioning**
- **Double drywall interior**
- **Many special features**
- **Extraordinary materials and workmanship**

*Note: The illustration shown may contain some optional components (for example: garages and/or fireplaces) whose costs are shown in the modifications, adjustments, & alternatives below or at the end of the square foot section.*

©Home Planners, Inc.

## Base cost per square foot of living area

| Exterior Wall | Living Area | | | | | | | | | | |
|---|---|---|---|---|---|---|---|---|---|---|---|
| | 1000 | 1200 | 1400 | 1600 | 1800 | 2000 | 2400 | 2800 | 3200 | 3600 | 4000 |
| Wood Siding - Wood Frame | 138.45 | 128.45 | 120.35 | 114.90 | 111.85 | 108.30 | 101.50 | **96.65** | 93.20 | 89.60 | 86.55 |
| Brick Veneer - Wood Frame | 144.30 | 133.70 | 125.20 | 119.40 | 116.20 | 112.40 | 105.25 | 100.10 | 96.40 | 92.55 | 89.35 |
| Solid Brick | 154.20 | 142.65 | 133.30 | 127.00 | 123.55 | 119.30 | 111.55 | 105.90 | 101.75 | 97.60 | 94.05 |
| Solid Stone | 155.15 | 143.50 | 134.10 | 127.70 | 124.25 | 119.95 | 112.15 | 106.45 | 102.30 | 98.05 | 94.50 |
| Finished Basement, Add | 42.40 | 40.40 | 38.80 | 37.75 | 37.20 | 36.35 | 35.15 | 34.20 | 33.45 | 32.75 | 32.15 |
| Unfinished Basement, Add | 15.20 | 14.35 | 13.65 | 13.20 | 12.95 | 12.55 | 12.05 | 11.60 | 11.25 | 10.95 | 10.70 |

## Modifications

*Add to the total cost*

| | |
|---|---|
| Upgrade Kitchen Cabinets | $ + 695 |
| Solid Surface Countertops | + 1284 |
| Full Bath - including plumbing, wall and floor finishes | + 5307 |
| Half Bath - including plumbing, wall and floor finishes | + 3306 |
| Two Car Attached Garage | + 19,788 |
| Two Car Detached Garage | + 22,357 |
| Fireplace & Chimney | + 5545 |

## Adjustments

*For multi family - add to total cost*

| | |
|---|---|
| Additional Kitchen | $ + 10,887 |
| Additional Full Bath & Half Bath | + 8613 |
| Additional Entry & Exit | + 1718 |
| Separate Heating & Air Conditioning | + 4133 |
| Separate Electric | + 1410 |

*For Townhouse/Rowhouse -
Multiply cost per square foot by*

| | |
|---|---|
| Inner Unit | .90 |
| End Unit | .95 |

## Alternatives

*Add to or deduct from the cost per square foot of living area*

| | |
|---|---|
| Heavyweight Asphalt Shingles | – 1.7 |
| Clay Tile Roof | + 2.4 |
| Slate Roof | + 5.6 |
| Upgrade Ceilings to Textured Finish | + .3 |
| Air Conditioning, in Heating Ductwork | Base Syste |
| Heating Systems, Hot Water | + 1.5 |
| Heat Pump | + 1.9 |
| Electric Heat | – 1.7 |
| Not Heated | – 3.5 |

## Additional upgrades or components

| | |
|---|---|
| Kitchen Cabinets & Countertops | Page 7 |
| Bathroom Vanities | 7 |
| Fireplaces & Chimneys | 7 |
| Windows, Skylights & Dormers | 7 |
| Appliances | 7 |
| Breezeways & Porches | 7 |
| Finished Attic | 7 |
| Garages | 8 |
| Site Improvements | 8 |
| Wings & Ells | 7 |

| | | Labor-Hours | Cost Per Square Foot Of Living Area | | |
|---|---|---|---|---|---|
| | | | Mat. | Labor | Total |
| **1** Site Work | Site preparation for slab; 4' deep trench excavation for foundation wall. | .028 | | .47 | .47 |
| **2** Foundation | Continuous reinforced concrete footing 8" deep x 18" wide; dampproofed and insulated reinforced concrete foundation wall, 12" thick, 4' deep; 4" concrete slab on 4" crushed stone base and polyethylene vapor barrier, trowel finish. | .098 | 5.04 | 5.13 | 10.17 |
| **3** Framing | Exterior walls - 2" x 6" wood studs, 16" O.C.; 5/8" plywood sheathing; 2" x 10" rafters 16" O.C. with 5/8" plywood sheathing, 6 in 12 pitch; 2" x 8" ceiling joists 16" O.C.; 5/8" plywood subfloor on 1" x 3" wood sleepers 16" O.C. | .260 | 10.55 | 9.22 | 19.77 |
| **4** Exterior Walls | Horizontal beveled wood siding; #15 felt building paper; 6" batt insulation; wood double hung windows; 3 solid core wood exterior doors; storms and screens. | .204 | 6.49 | 2.26 | 8.75 |
| **5** Roofing | Red cedar shingles; #15 felt building paper; aluminum gutters, downspouts and drip edge; copper flashings. | .082 | 3.26 | 2.79 | 6.05 |
| **6** Interiors | Walls and ceilings - 5/8" drywall, skim coat plaster, painted with primer and 2 coats; hardwood baseboard and trim, sanded and finished; hardwood floor 70%, ceramic tile with underlayment 20%, vinyl tile with underlayment 10%; wood panel interior doors, primed and painted with 2 coats. | .287 | 9.83 | 10.32 | 20.15 |
| **7** Specialties | Luxury grade kitchen cabinets - 25 L.F. wall and base with plastic laminate counter top and kitchen sink; 6 L.F. bathroom vanity; 75 gallon electric water heater; medicine cabinet. | .052 | 3.39 | .63 | 4.02 |
| **8** Mechanical | Gas fired warm air heat/air conditioning; one full bath including: bathtub, corner shower; built in lavatory and water closet; one 1/2 bath including: built in lavatory and water closet. | .078 | 4.18 | 2.22 | 6.40 |
| **9** Electrical | 200 Amp. service; romex wiring; fluorescent and incandescent lighting fixtures; intercom, switches, receptacles. | .044 | .86 | 1.39 | 2.25 |
| **10** Overhead | Contractor's overhead and profit and architect's fees. | | 10.40 | 8.22 | 18.62 |
| | **Total** | | 54.00 | 42.65 | **96.65** |

**SQUARE FOOT COSTS**

- **Unique residence built from an architect's plan**
- **Single family — 1 full bath, 1 half bath, 1 kitchen**
- **No basement**
- **Cedar shakes on roof**
- **Forced hot air heat/air conditioning**
- **Double drywall interior**
- **Many special features**
- **Extraordinary materials and workmanship**

*Note: The illustration shown may contain some optional components (for example: garages and/or fireplaces) whose costs are shown in the modifications, adjustments, & alternatives below or at the end of the square foot section.*

©Larry E. Belk Designs

## Base cost per square foot of living area

| Exterior Wall | Living Area | | | | | | | | | | |
|---|---|---|---|---|---|---|---|---|---|---|---|
| | 1000 | 1200 | 1400 | 1600 | 1800 | 2000 | 2400 | 2800 | 3200 | 3600 | 4000 |
| Wood Siding - Wood Frame | 126.60 | 118.45 | 112.50 | 105.25 | 101.20 | 97.00 | 89.25 | 85.80 | 82.50 | 79.95 | 76.40 |
| Brick Veneer - Wood Frame | 133.40 | 124.80 | 118.60 | 110.75 | 106.40 | 101.95 | 93.60 | 89.95 | 86.35 | 83.65 | 79.80 |
| Solid Brick | 144.80 | 135.50 | 128.80 | 120.00 | 115.20 | 110.30 | 100.95 | 96.90 | 92.80 | 89.85 | 85.60 |
| Solid Stone | 145.95 | 136.50 | 129.75 | 120.90 | 116.00 | 111.10 | 101.70 | 97.60 | 93.40 | 90.45 | 86.15 |
| Finished Basement, Add | 29.75 | 28.60 | 27.75 | 26.55 | 25.85 | 25.25 | 23.95 | 23.40 | 22.75 | 22.35 | 21.90 |
| Unfinished Basement, Add | 10.90 | 10.40 | 10.05 | 9.50 | 9.20 | 8.95 | 8.40 | 8.15 | 7.85 | 7.70 | 7.45 |

## Modifications

*Add to the total cost*

| | |
|---|---|
| Upgrade Kitchen Cabinets | $  + 695 |
| Solid Surface Countertops | + 1284 |
| Full Bath - including plumbing, wall and floor finishes | + 5307 |
| Half Bath - including plumbing, wall and floor finishes | + 3306 |
| Two Car Attached Garage | + 19,788 |
| Two Car Detached Garage | + 22,357 |
| Fireplace & Chimney | + 5545 |

## Adjustments

*For multi family - add to total cost*

| | |
|---|---|
| Additional Kitchen | $  + 10,887 |
| Additional Full Bath & Half Bath | + 8613 |
| Additional Entry & Exit | + 1718 |
| Separate Heating & Air Conditioning | + 4133 |
| Separate Electric | + 1410 |

*For Townhouse/Rowhouse - Multiply cost per square foot by*

| | |
|---|---|
| Inner Unit | .90 |
| End Unit | .95 |

## Alternatives

*Add to or deduct from the cost per square foot of living area*

| | |
|---|---|
| Heavyweight Asphalt Shingles | – 1.2 |
| Clay Tile Roof | + 1.7 |
| Slate Roof | + 4.1 |
| Upgrade Ceilings to Textured Finish | + .3 |
| Air Conditioning, in Heating Ductwork | Base Syste |
| Heating Systems, Hot Water | + 1.5 |
|    Heat Pump | + 2.1 |
|    Electric Heat | – 1.7 |
|    Not Heated | – 3.2 |

## Additional upgrades or components

| | |
|---|---|
| Kitchen Cabinets & Countertops | Page 7 |
| Bathroom Vanities | 7 |
| Fireplaces & Chimneys | 7 |
| Windows, Skylights & Dormers | 7 |
| Appliances | 7 |
| Breezeways & Porches | 7 |
| Finished Attic | 7 |
| Garages | 8 |
| Site Improvements | 8 |
| Wings & Ells | 7 |

**Important: See the Reference Section for Location Factors (to adjust for your city) and Estimating Form**

**SQUARE FOOT COSTS**

| | | | Labor-Hours | Cost Per Square Foot Of Living Area | | |
|---|---|---|---|---|---|---|
| | | | | Mat. | Labor | Total |
| **1** | **Site Work** | Site preparation for slab; 4' deep trench excavation for foundation wall. | .025 | | .47 | .47 |
| **2** | **Foundation** | Continuous reinforced concrete footing 8" deep x 18" wide; dampproofed and insulated reinforced concrete foundation wall, 12" thick, 4' deep; 4" concrete slab on 4" crushed stone base and polyethylene vapor barrier, trowel finish. | .066 | 3.58 | 3.88 | 7.46 |
| **3** | **Framing** | Exterior walls - 2" x 6" wood studs, 16" O.C.; 5/8" plywood sheathing; 2" x 10" rafters 16" O.C. with 5/8" plywood sheathing, 8 in 12 pitch; 2" x 8" ceiling joists 16" O.C.; 2" x 12" floor joists 16" O.C. with 5/8" plywood subfloor; 5/8" plywood subfloor on 1" x 3" wood sleepers 16" O.C. | .189 | 6.36 | 5.95 | 12.31 |
| **4** | **Exterior Walls** | Horizontal beveled wood siding; #15 felt building paper; 6" batt insulation; wood double hung windows; 3 solid core wood exterior doors; storms and screens. | .174 | 7.14 | 2.52 | 9.66 |
| **5** | **Roofing** | Red cedar shingles; #15 felt building paper; aluminum gutters, downspouts and drip edge; copper flashings. | .065 | 2.03 | 1.75 | 3.78 |
| **6** | **Interiors** | Walls and ceilings - 5/8" drywall, skim coat plaster, painted with primer and 2 coats; hardwood baseboard and trim, sanded and finished; hardwood floor 70%, ceramic tile with underlayment 20%, vinyl tile with underlayment 10%; wood panel interior doors, primed and painted with 2 coats. | .260 | 10.96 | 11.93 | 22.89 |
| **7** | **Specialties** | Luxury grade kitchen cabinets - 25 L.F. wall and base with plastic laminate counter top and kitchen sink; 6 L.F. bathroom vanity; 75 gallon electric water heater; medicine cabinet. | .062 | 3.39 | .63 | 4.02 |
| **8** | **Mechanical** | Gas fired warm air heat/air conditioning; one full bath including: bathtub, corner shower; built in lavatory and water closet; one 1/2 bath including: built in lavatory and water closet. | .080 | 4.18 | 2.22 | 6.40 |
| **9** | **Electrical** | 200 Amp. service; romex wiring; fluorescent and incandescent lighting fixtures; intercom, switches, receptacles. | .044 | .86 | 1.39 | 2.25 |
| **10** | **Overhead** | Contractor's overhead and profit and architect's fees. | | 9.22 | 7.34 | 16.56 |
| | | **Total** | | 47.72 | 38.08 | **85.80** |

**SQUARE FOOT COSTS**

- **Unique residence built from an architect's plan**
- **Single family — 1 full bath, 1 half bath, 1 kitchen**
- **No basement**
- **Cedar shakes on roof**
- **Forced hot air heat/air conditioning**
- **Double drywall interior**
- **Many special features**
- **Extraordinary materials and workmanship**

*Note: The illustration shown may contain some optional components (for example: garages and/or fireplaces) whose costs are shown in the modifications, adjustments, & alternatives below or at the end of the square foot section.*

## Base cost per square foot of living area

| Exterior Wall | Living Area | | | | | | | | | | |
|---|---|---|---|---|---|---|---|---|---|---|---|
| | 1200 | 1400 | 1600 | 1800 | 2000 | 2400 | 2800 | 3200 | 3600 | 4000 | 4400 |
| Wood Siding - Wood Frame | 118.15 | 111.65 | 107.00 | 102.65 | 98.20 | 91.50 | 85.95 | 82.25 | 80.05 | 77.80 | 75.75 |
| Brick Veneer - Wood Frame | 125.40 | 118.40 | 113.45 | 108.75 | 104.05 | 96.80 | 90.75 | 86.75 | 84.40 | 81.90 | 79.70 |
| Solid Brick | 137.60 | 129.80 | 124.35 | 119.00 | 113.95 | 105.75 | 98.90 | 94.35 | 91.70 | 88.80 | 86.35 |
| Solid Stone | 138.80 | 130.90 | 125.40 | 120.00 | 114.85 | 106.55 | 99.65 | 95.05 | 92.45 | 89.40 | 87.00 |
| Finished Basement, Add | 23.90 | 22.95 | 22.35 | 21.70 | 21.25 | 20.25 | 19.45 | 18.95 | 18.60 | 18.25 | 17.95 |
| Unfinished Basement, Add | 8.75 | 8.35 | 8.10 | 7.80 | 7.60 | 7.20 | 6.85 | 6.60 | 6.50 | 6.30 | 6.20 |

## Modifications

*Add to the total cost*

| | |
|---|---|
| Upgrade Kitchen Cabinets | $ + 695 |
| Solid Surface Countertops | + 1284 |
| Full Bath - including plumbing, wall and floor finishes | + 5307 |
| Half Bath - including plumbing, wall and floor finishes | + 3306 |
| Two Car Attached Garage | + 19,788 |
| Two Car Detached Garage | + 22,357 |
| Fireplace & Chimney | + 6075 |

## Adjustments

*For multi family - add to total cost*

| | |
|---|---|
| Additional Kitchen | $ + 10,887 |
| Additional Full Bath & Half Bath | + 8613 |
| Additional Entry & Exit | + 1718 |
| Separate Heating & Air Conditioning | + 4133 |
| Separate Electric | + 1410 |

*For Townhouse/Rowhouse - Multiply cost per square foot by*

| | |
|---|---|
| Inner Unit | .86 |
| End Unit | .93 |

## Alternatives

*Add to or deduct from the cost per square foot of living area*

| | |
|---|---|
| Heavyweight Asphalt Shingles | – .8 |
| Clay Tile Roof | + 1.2 |
| Slate Roof | + 2.8 |
| Upgrade Ceilings to Textured Finish | + .3 |
| Air Conditioning, in Heating Ductwork | Base Syste |
| Heating Systems, Hot Water | + 1.4 |
| Heat Pump | + 2.2 |
| Electric Heat | – 1.5 |
| Not Heated | – 3.1 |

## Additional upgrades or components

| | |
|---|---|
| Kitchen Cabinets & Countertops | Page 7 |
| Bathroom Vanities | 7 |
| Fireplaces & Chimneys | 7 |
| Windows, Skylights & Dormers | 7 |
| Appliances | 7 |
| Breezeways & Porches | 7 |
| Finished Attic | 7 |
| Garages | 8 |
| Site Improvements | 8 |
| Wings & Ells | 7 |

**Important: See the Reference Section for Location Factors (to adjust for your city) and Estimating Form**

# Luxury 2 Story

**Living Area - 3200 S.F.**
**Perimeter - 163 L.F.**

| | | Labor-Hours | Cost Per Square Foot Of Living Area | | |
|---|---|---|---|---|---|
| | | | Mat. | Labor | Total |
| **1 Site Work** | Site preparation for slab; 4' deep trench excavation for foundation wall. | .024 | | .42 | .42 |
| **2 Foundation** | Continuous reinforced concrete footing 8" deep x 18" wide; dampproofed and insulated reinforced concrete foundation wall, 12" thick, 4' deep; 4" concrete slab on 4" crushed stone base and polyethylene vapor barrier, trowel finish. | .058 | 2.92 | 3.15 | 6.07 |
| **3 Framing** | Exterior walls - 2" x 6" wood studs, 16" O.C.; 5/8" plywood sheathing; 2" x 10" rafters 16" O.C. with 5/8" plywood sheathing, 6 in 12 pitch; 2" x 8" ceiling joists 16" O.C.; 2" x 12" floor joists 16" O.C. with 5/8" plywood subfloor; 5/8" plywood subfloor on 1" x 3" wood sleepers 16" O.C. | .193 | 6.42 | 5.91 | 12.33 |
| **4 Exterior Walls** | Horizontal beveled wood siding, #15 felt building paper; 6" batt insulation; wood double hung windows; 3 solid core wood exterior doors; storms and screens. | .247 | 7.53 | 2.67 | 10.20 |
| **5 Roofing** | Red cedar shingles; #15 felt building paper; aluminum gutters, downspouts and drip edge; copper flashings. | .049 | 1.63 | 1.40 | 3.03 |
| **6 Interiors** | Walls and ceilings - 5/8" drywall, skim coat plaster, painted with primer and 2 coats; hardwood baseboard and trim, sanded and finished; hardwood floor 70%, ceramic tile with underlayment 20%, vinyl tile with underlayment 10%; wood panel interior doors, primed and painted with 2 coats. | .252 | 10.84 | 11.86 | 22.70 |
| **7 Specialties** | Luxury grade kitchen cabinets - 25 L.F. wall and base with plastic laminate counter top and kitchen sink; 6 L.F. bathroom vanity; 75 gallon electric water heater; medicine cabinet. | .057 | 2.96 | .57 | 3.53 |
| **8 Mechanical** | Gas fired warm air heat/air conditioning; one full bath including: bathtub, corner shower; built in lavatory and water closet; one 1/2 bath including: built in lavatory and water closet. | .071 | 3.82 | 2.14 | 5.96 |
| **9 Electrical** | 200 Amp. service; romex wiring; fluorescent and incandescent lighting fixtures; intercom, switches, receptacles. | .042 | .83 | 1.35 | 2.18 |
| **10 Overhead** | Contractor's overhead and profit and architect's fee. | | 8.80 | 7.03 | 15.83 |
| | **Total** | | 45.75 | 36.50 | **82.25** |

**SQUARE FOOT COSTS**

- **Unique residence built from an architect's plan**
- **Single family — 1 full bath, 1 half bath, 1 kitchen**
- **No basement**
- **Cedar shakes on roof**
- **Forced hot air heat/air conditioning**
- **Double drywall interior**
- **Many special features**
- **Extraordinary materials and workmanship**

*Note: The illustration shown may contain some optional components (for example: garages and/or fireplaces) whose costs are shown in the modifications, adjustments, & alternatives below or at the end of the square foot section.*

©Larry W. Garnett & Associates, Inc

## Base cost per square foot of living area

| Exterior Wall | Living Area | | | | | | | | | | |
|---|---|---|---|---|---|---|---|---|---|---|---|
| | 1500 | 1800 | 2100 | 2500 | 3000 | 3500 | 4000 | 4500 | 5000 | 5500 | 6000 |
| Wood Siding - Wood Frame | 116.10 | 104.65 | 98.30 | 93.60 | 87.00 | 82.20 | 77.40 | 75.25 | 73.20 | 71.05 | 69.00 |
| Brick Veneer - Wood Frame | 123.65 | 111.60 | 104.50 | 99.45 | 92.35 | 87.00 | 81.90 | 79.45 | 77.25 | 74.90 | 72.70 |
| Solid Brick | 136.35 | 123.25 | 114.95 | 109.30 | 101.35 | 95.20 | 89.45 | 86.65 | 84.05 | 81.40 | 78.90 |
| Solid Stone | 137.55 | 124.40 | 115.95 | 110.25 | 102.15 | 95.95 | 90.15 | 87.30 | 84.75 | 82.05 | 79.50 |
| Finished Basement, Add | 19.05 | 18.20 | 17.15 | 16.60 | 15.85 | 15.20 | 14.75 | 14.40 | 14.15 | 13.85 | 13.65 |
| Unfinished Basement, Add | 7.05 | 6.65 | 6.20 | 6.00 | 5.65 | 5.35 | 5.15 | 5.00 | 4.90 | 4.80 | 4.70 |

## Modifications

*Add to the total cost*

| | |
|---|---|
| Upgrade Kitchen Cabinets | $ + 695 |
| Solid Surface Countertops | + 1284 |
| Full Bath - including plumbing, wall and floor finishes | + 5307 |
| Half Bath - including plumbing, wall and floor finishes | + 3306 |
| Two Car Attached Garage | + 19,788 |
| Two Car Detached Garage | + 22,357 |
| Fireplace & Chimney | + 6645 |

## Adjustments

*For multi family - add to total cost*

| | |
|---|---|
| Additional Kitchen | $ + 10,887 |
| Additional Full Bath & Half Bath | + 8613 |
| Additional Entry & Exit | + 1718 |
| Separate Heating & Air Conditioning | + 4133 |
| Separate Electric | + 1410 |

*For Townhouse/Rowhouse -
Multiply cost per square foot by*

| | |
|---|---|
| Inner Unit | .86 |
| End Unit | .93 |

## Alternatives

*Add to or deduct from the cost per square foot of living area*

| | |
|---|---|
| Heavyweight Asphalt Shingles | – .7 |
| Clay Tile Roof | + 1.0 |
| Slate Roof | + 2.4 |
| Upgrade Ceilings to Textured Finish | + .3 |
| Air Conditioning, in Heating Ductwork | Base System |
| Heating Systems, Hot Water | + 1.3 |
| Heat Pump | + 2.3 |
| Electric Heat | – 3.0 |
| Not Heated | – 3.1 |

## Additional upgrades or components

| | |
|---|---|
| Kitchen Cabinets & Countertops | Page 7 |
| Bathroom Vanities | 7 |
| Fireplaces & Chimneys | 7 |
| Windows, Skylights & Dormers | 7 |
| Appliances | 7 |
| Breezeways & Porches | 7 |
| Finished Attic | 7 |
| Garages | 8 |
| Site Improvements | 8 |
| Wings & Ells | 7 |

| | | Labor-Hours | Cost Per Square Foot Of Living Area | | |
|---|---|---|---|---|---|
| | | | Mat. | Labor | Total |
| **1 Site Work** | Site preparation for slab; 4' deep trench excavation for foundation wall. | .055 | | .44 | .44 |
| **2 Foundation** | Continuous reinforced concrete footing 8" deep x 18" wide; dampproofed and insulated reinforced concrete foundation wall, 12" thick, 4' deep; 4" concrete slab on 4" crushed stone base and polyethylene vapor barrier, trowel finish. | .067 | 2.52 | 2.85 | 5.37 |
| **3 Framing** | Exterior walls - 2" x 6" wood studs, 16" O.C.; 5/8" plywood sheathing; 2" x 10" rafters 16" O.C. with 5/8" plywood sheathing, 6 in 12 pitch; 2" x 8" ceiling joists 16" O.C.; 2" x 12" floor joists 16" O.C. with 5/8" plywood subfloor; 5/8" plywood subfloor on 1" x 3" wood sleepers 16" O.C. | .209 | 6.65 | 6.06 | 12.71 |
| **4 Exterior Walls** | Horizontal beveled wood siding; #15 felt building paper; 6" batt insulation; wood double hung windows; 3 solid core wood exterior doors; storms and screens. | .405 | 8.82 | 3.12 | 11.94 |
| **5 Roofing** | Red cedar shingles; #15 felt building paper; aluminum gutters, downspouts and drip edge; copper flashings. | .039 | 1.25 | 1.07 | 2.32 |
| **6 Interiors** | Walls and ceilings - 5/8" drywall, skim coat plaster, painted with primer and 2 coats; hardwood baseboard and trim, sanded and finished; hardwood floor 70%, ceramic tile with underlayment 20%, vinyl tile with underlayment 10%; wood panel interior doors, primed and painted with 2 coats. | .341 | 11.99 | 13.37 | 25.36 |
| **7 Specialties** | Luxury grade kitchen cabinets - 25 L.F. wall and base with plastic laminate counter top and kitchen sink; 6 L.F. bathroom vanity; 75 gallon electric water heater; medicine cabinet. | .119 | 3.15 | .60 | 3.75 |
| **8 Mechanical** | Gas fired warm air heat/air conditioning; one full bath including: bathtub, corner shower; built in lavatory and water closet; one 1/2 bath including: built in lavatory and water closet. | .103 | 3.99 | 2.17 | 6.16 |
| **9 Electrical** | 200 Amp. service; romex wiring; fluorescent and incandescent lighting fixtures; intercom, switches, receptacles. | .054 | .84 | 1.37 | 2.21 |
| **10 Overhead** | Contractor's overhead and profit and architect's fee. | | 9.35 | 7.39 | 16.74 |
| | **Total** | | 48.56 | 38.44 | **87.00** |

# RESIDENTIAL    Luxury    3 Story

- **Unique residence built from an architect's plan**
- **Single family — 1 full bath, 1 half bath, 1 kitchen**
- **No basement**
- **Cedar shakes on roof**
- **Forced hot air heat/air conditioning**
- **Double drywall interior**
- **Many special features**
- **Extraordinary materials and workmanship**

*Note: The illustration shown may contain some optional components (for example: garages and/or fireplaces) whose costs are shown in the modifications, adjustments, & alternatives below or at the end of the square foot section.*

**SQUARE FOOT COSTS**

## Base cost per square foot of living area

| Exterior Wall | Living Area | | | | | | | | | | |
|---|---|---|---|---|---|---|---|---|---|---|---|
| | 1500 | 1800 | 2100 | 2500 | 3000 | 3500 | 4000 | 4500 | 5000 | 5500 | 6000 |
| Wood Siding - Wood Frame | 115.35 | 103.95 | 98.90 | 94.60 | 87.65 | 84.30 | 79.95 | 75.45 | 73.90 | 72.15 | 70.50 |
| Brick Veneer - Wood Frame | 123.15 | 111.20 | 105.65 | 101.05 | 93.55 | 89.90 | 85.05 | 80.10 | 78.40 | 76.50 | 74.60 |
| Solid Brick | 136.30 | 123.40 | 117.05 | 111.95 | 103.40 | 99.20 | 93.55 | 88.00 | 86.00 | 83.85 | 81.50 |
| Solid Stone | 137.60 | 124.60 | 118.15 | 113.00 | 104.35 | 100.10 | 94.35 | 88.75 | 86.75 | 84.55 | 82.10 |
| Finished Basement, Add | 16.75 | 15.95 | 15.30 | 14.85 | 14.15 | 13.75 | 13.20 | 12.80 | 12.60 | 12.40 | 12.15 |
| Unfinished Basement, Add | 6.15 | 5.85 | 5.55 | 5.40 | 5.05 | 4.90 | 4.65 | 4.45 | 4.40 | 4.30 | 4.20 |

## Modifications

*Add to the total cost*

| | |
|---|---|
| Upgrade Kitchen Cabinets | $ + 695 |
| Solid Surface Countertops | + 1284 |
| Full Bath - including plumbing, wall and floor finishes | + 5307 |
| Half Bath - including plumbing, wall and floor finishes | + 3306 |
| Two Car Attached Garage | + 19,788 |
| Two Car Detached Garage | + 22,357 |
| Fireplace & Chimney | + 6645 |

## Adjustments

*For multi family - add to total cost*

| | |
|---|---|
| Additional Kitchen | $ + 10,887 |
| Additional Full Bath & Half Bath | + 8613 |
| Additional Entry & Exit | + 1718 |
| Separate Heating & Air Conditioning | + 4133 |
| Separate Electric | + 1410 |

*For Townhouse/Rowhouse - Multiply cost per square foot by*

| | |
|---|---|
| Inner Unit | .84 |
| End Unit | .92 |

## Alternatives

*Add to or deduct from the cost per square foot of living area*

| | |
|---|---|
| Heavyweight Asphalt Shingles | – .59 |
| Clay Tile Roof | + .80 |
| Slate Roof | + 1.88 |
| Upgrade Ceilings to Textured Finish | + .39 |
| Air Conditioning, in Heating Ductwork | Base System |
| Heating Systems, Hot Water | + 1.33 |
|     Heat Pump | + 2.32 |
|     Electric Heat | – 3.05 |
|     Not Heated | – 3.01 |

## Additional upgrades or components

| | |
|---|---|
| Kitchen Cabinets & Countertops | Page 77 |
| Bathroom Vanities | 78 |
| Fireplaces & Chimneys | 78 |
| Windows, Skylights & Dormers | 78 |
| Appliances | 79 |
| Breezeways & Porches | 79 |
| Finished Attic | 79 |
| Garages | 80 |
| Site Improvements | 80 |
| Wings & Ells | 76 |

**Important: See the Reference Section for Location Factors (to adjust for your city) and Estimating Forms**

| | | | Labor-Hours | Cost Per Square Foot Of Living Area | | |
|---|---|---|---|---|---|---|
| | | | | Mat. | Labor | Total |
| **1** | **Site Work** | Site preparation for slab; 4' deep trench excavation for foundation wall. | .055 | | .44 | .44 |
| **2** | **Foundation** | Continuous reinforced concrete footing 8" deep x 18" wide; dampproofed and insulated reinforced concrete foundation wall, 12" thick, 4' deep; 4" concrete slab on 4" crushed stone base and polyethylene vapor barrier, trowel finish. | .063 | 2.27 | 2.59 | 4.86 |
| **3** | **Framing** | Exterior walls - 2" x 6" wood studs, 16" O.C.; 5/8" plywood sheathing; 2" x 10" rafters 16" O.C. with 5/8" plywood sheathing, 6 in 12 pitch; 2" x 8" ceiling joists 16" O.C.; 2" x 12" floor joists 16" O.C. with 5/8" plywood subfloor; 5/8" plywood subfloor on 1" x 3" wood sleepers 16" O.C. | .225 | 6.75 | 6.10 | 12.85 |
| **4** | **Exterior Walls** | Horizontal beveled wood siding; #15 felt building paper; 6" batt insulation; wood double hung windows; 3 solid core wood exterior doors; storms and screens. | .454 | 9.61 | 3.43 | 13.04 |
| **5** | **Roofing** | Red cedar shingles; #15 felt building paper; aluminum gutters, downspouts and drip edge; copper flashings. | .034 | 1.09 | .93 | 2.02 |
| **6** | **Interiors** | Walls and ceilings - 5/8" drywall, skim coat plaster, painted with primer and 2 coats; hardwood baseboard and trim, sanded and finished; hardwood floor 70%, ceramic tile with underlayment 20%, vinyl tile with underlayment 10%; wood panel interior doors, primed and painted with 2 coats. | .390 | 12.02 | 13.45 | 25.47 |
| **7** | **Specialties** | Luxury grade kitchen cabinets - 25 L.F. wall and base with plastic laminate counter top and kitchen sink; 6 L.F. bathroom vanity; 75 gallon electric water heater; medicine cabinet. | .119 | 3.15 | .60 | 3.75 |
| **8** | **Mechanical** | Gas fired warm air heat/air conditioning; one full bath including: bathtub, corner shower; built in lavatory and water closet; one 1/2 bath including: built in lavatory and water closet. | .103 | 3.99 | 2.17 | 6.16 |
| **9** | **Electrical** | 200 Amp. service; romex wiring; fluorescent and incandescent lighting fixtures; intercom, switches, receptacles. | .053 | .84 | 1.37 | 2.21 |
| **10** | **Overhead** | Contractor's overhead and profit and architect's fees. | | 9.46 | 7.39 | 16.85 |
| | | **Total** | | 49.18 | 38.47 | **87.65** |

# RESIDENTIAL | Luxury | Bi-Level

- **Unique residence built from an architect's plan**
- **Single family — 1 full bath, 1 half bath, 1 kitchen**
- **No basement**
- **Cedar shakes on roof**
- **Forced hot air heat/air conditioning**
- **Double drywall interior**
- **Many special features**
- **Extraordinary materials and workmanship**

*Note: The illustration shown may contain some optional components (for example: garages and/or fireplaces) whose costs are shown in the modifications, adjustments, & alternatives below or at the end of the square foot section.*

## Base cost per square foot of living area

| Exterior Wall | Living Area | | | | | | | | | | |
|---|---|---|---|---|---|---|---|---|---|---|---|
| | 1200 | 1400 | 1600 | 1800 | 2000 | 2400 | 2800 | 3200 | 3600 | 4000 | 4400 |
| Wood Siding - Wood Frame | 111.80 | 105.65 | 101.25 | 97.35 | 92.95 | 86.85 | 81.80 | **78.30** | 76.25 | 74.20 | 72.30 |
| Brick Veneer - Wood Frame | 117.20 | 110.70 | 106.10 | 101.90 | 97.40 | 90.80 | 85.40 | 81.70 | 79.50 | 77.25 | 75.25 |
| Solid Brick | 126.40 | 119.25 | 114.25 | 109.55 | 104.80 | 97.50 | 91.45 | 87.40 | 85.00 | 82.45 | 80.25 |
| Solid Stone | 127.25 | 120.05 | 115.05 | 110.30 | 105.50 | 98.15 | 92.05 | 87.90 | 85.55 | 82.95 | 80.70 |
| Finished Basement, Add | 23.90 | 22.95 | 22.35 | 21.70 | 21.25 | 20.25 | 19.45 | 18.95 | 18.60 | 18.25 | 17.95 |
| Unfinished Basement, Add | 8.75 | 8.35 | 8.10 | 7.80 | 7.60 | 7.20 | 6.85 | 6.60 | 6.50 | 6.30 | 6.20 |

## Modifications

**Add to the total cost**

| | |
|---|---|
| Upgrade Kitchen Cabinets | $ + 695 |
| Solid Surface Countertops | + 1284 |
| Full Bath - including plumbing, wall and floor finishes | + 5307 |
| Half Bath - including plumbing, wall and floor finishes | + 3306 |
| Two Car Attached Garage | + 19,788 |
| Two Car Detached Garage | + 22,357 |
| Fireplace & Chimney | + 5545 |

## Adjustments

**For multi family - add to total cost**

| | |
|---|---|
| Additional Kitchen | $ + 10,887 |
| Additional Full Bath & Half Bath | + 8613 |
| Additional Entry & Exit | + 1718 |
| Separate Heating & Air Conditioning | + 4133 |
| Separate Electric | + 1410 |

**For Townhouse/Rowhouse - Multiply cost per square foot by**

| | |
|---|---|
| Inner Unit | .89 |
| End Unit | .94 |

## Alternatives

**Add to or deduct from the cost per square foot of living area**

| | |
|---|---|
| Heavyweight Asphalt Shingles | – .8 |
| Clay Tile Roof | + 1.2 |
| Slate Roof | + 5.6 |
| Upgrade Ceilings to Textured Finish | + .3 |
| Air Conditioning, in Heating Ductwork | Base System |
| Heating Systems, Hot Water | + 1.4 |
|     Heat Pump | + 2.2 |
|     Electric Heat | – 1.5 |
|     Not Heated | – 3.1 |

## Additional upgrades or components

| | |
|---|---|
| Kitchen Cabinets & Countertops | Page 7 |
| Bathroom Vanities | 7 |
| Fireplaces & Chimneys | 7 |
| Windows, Skylights & Dormers | 7 |
| Appliances | 79 |
| Breezeways & Porches | 79 |
| Finished Attic | 79 |
| Garages | 80 |
| Site Improvements | 80 |
| Wings & Ells | 76 |

**Important: See the Reference Section for Location Factors (to adjust for your city) and Estimating Forms**

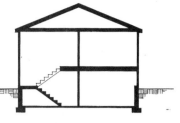

| | | Labor-Hours | Cost Per Square Foot Of Living Area | | |
|---|---|---|---|---|---|
| | | | Mat. | Labor | Total |
| **1 Site Work** | Excavation for lower level, 4' deep. Site preparation for slab. | .024 | | .42 | .42 |
| **2 Foundation** | Continuous reinforced concrete footing 8" deep x 18" wide; dampproofed and insulated reinforced concrete foundation wall, 12" thick, 4' deep; 4" concrete slab on 4" crushed stone base and polyethylene vapor barrier, trowel finish. | .058 | 2.92 | 3.15 | 6.07 |
| **3 Framing** | Exterior walls - 2" x 6" wood studs, 16" O.C.; 5/8" plywood sheathing; 2" x 10" rafters 16" O.C. with 5/8" plywood sheathing, 6 in 12 pitch; 2" x 8" ceiling joists 16" O.C.; 2" x 12" floor joists 16" O.C. with 5/8" plywood subfloor; 5/8" plywood subfloor on 1" x 3" wood sleepers 16" O.C. | .232 | 6.13 | 5.64 | 11.77 |
| **4 Exterior Walls** | Horizontal beveled wood siding; #15 felt building paper; 6" batt insulation; wood double hung windows; 3 solid core wood exterior doors; storms and screens. | .185 | 5.90 | 2.08 | 7.98 |
| **5 Roofing** | Red cedar shingles: #15 felt building paper; aluminum gutters, downspouts and drip edge; copper flashings. | .042 | 1.63 | 1.40 | 3.03 |
| **6 Interiors** | Walls and ceilings - 5/8" drywall, skim coat plaster, painted with primer and 2 coats; hardwood baseboard and trim, sanded and finished; hardwood floor 70%, ceramic tile with underlayment 20%; vinyl tile with underlayment 10%; wood panel interior doors, primed and painted with 2 coats. | .238 | 10.70 | 11.60 | 22.30 |
| **7 Specialties** | Luxury grade kitchen cabinets - 25 L.F. wall and base with plastic laminate counter top and kitchen sink; 6 L.F. bathroom vanity; 75 gallon electric water heater; medicine cabinet. | .056 | 2.96 | .57 | 3.53 |
| **8 Mechanical** | Gas fired warm air heat/air conditioning; one full bath including: bathtub, corner shower; built in lavatory and water closet; one 1/2 bath including: built in lavatory and water closet. | .071 | 3.82 | 2.14 | 5.96 |
| **9 Electrical** | 200 Amp. service; romex wiring; fluorescent and incandescent lighting fixtures; intercom, switches, receptacles. | .042 | .83 | 1.35 | 2.18 |
| **10 Overhead** | Contractor's overhead and profit and architect's fees. | | 8.32 | 6.74 | 15.06 |
| | **Total** | | 43.21 | 35.09 | **78.30** |

# RESIDENTIAL | Luxury | Tri-Level

- **Unique residence built from an architect's plan**
- **Single family — 1 full bath, 1 half bath, 1 kitchen**
- **No basement**
- **Cedar shakes on roof**
- **Forced hot air heat/air conditioning**
- **Double drywall interior**
- **Many special features**
- **Extraordinary materials and workmanship**

*Note: The illustration shown may contain some optional components (for example: garages and/or fireplaces) whose costs are shown in the modifications, adjustments, & alternatives below or at the end of the square foot section.*

©Home Planners, Inc.

## Base cost per square foot of living area

| Exterior Wall | Living Area | | | | | | | | | | |
|---|---|---|---|---|---|---|---|---|---|---|---|
| | 1500 | 1800 | 2100 | 2400 | 2800 | 3200 | 3600 | 4000 | 4500 | 5000 | 5500 |
| Wood Siding - Wood Frame | 105.90 | 98.40 | 92.25 | 87.95 | 84.95 | 81.35 | **77.40** | 75.85 | 72.05 | 70.10 | 67.80 |
| Brick Veneer - Wood Frame | 110.80 | 102.80 | 96.25 | 91.70 | 88.55 | 84.65 | 80.50 | 78.85 | 74.85 | 72.70 | 70.30 |
| Solid Brick | 119.05 | 110.20 | 103.00 | 98.00 | 94.60 | 90.30 | 85.75 | 83.95 | 79.55 | 77.15 | 74.50 |
| Solid Stone | 119.85 | 110.90 | 103.65 | 98.65 | 95.20 | 90.85 | 86.25 | 84.45 | 80.00 | 77.55 | 74.85 |
| Finished Basement, Add* | 28.25 | 26.95 | 25.85 | 25.20 | 24.70 | 24.05 | 23.40 | 23.15 | 22.55 | 22.15 | 21.80 |
| Unfinished Basement, Add* | 10.15 | 9.55 | 9.10 | 8.80 | 8.60 | 8.30 | 8.05 | 7.90 | 7.65 | 7.45 | 7.30 |

*Basement under middle level only.

## Modifications

**Add to the total cost**

| | |
|---|---|
| Upgrade Kitchen Cabinets | $ + 695 |
| Solid Surface Countertops | + 1284 |
| Full Bath - including plumbing, wall and floor finishes | + 5307 |
| Half Bath - including plumbing, wall and floor finishes | + 3306 |
| Two Car Attached Garage | + 19,788 |
| Two Car Detached Garage | + 22,357 |
| Fireplace & Chimney | + 6075 |

## Adjustments

**For multi family - add to total cost**

| | |
|---|---|
| Additional Kitchen | $ + 10,887 |
| Additional Full Bath & Half Bath | + 8613 |
| Additional Entry & Exit | + 1718 |
| Separate Heating & Air Conditioning | + 4133 |
| Separate Electric | + 1410 |

**For Townhouse/Rowhouse - Multiply cost per square foot by**

| | |
|---|---|
| Inner Unit | .86 |
| End Unit | .93 |

## Alternatives

**Add to or deduct from the cost per square foot of living area**

| | |
|---|---|
| Heavyweight Asphalt Shingles | – 1.2. |
| Clay Tile Roof | + 1.7. |
| Slate Roof | + 4.1. |
| Upgrade Ceilings to Textured Finish | + .3. |
| Air Conditioning, in Heating Ductwork | Base System |
| Heating Systems, Hot Water | + 1.4. |
|    Heat Pump | + 2.3. |
|    Electric Heat | – 1.3. |
|    Not Heated | – 3.0. |

## Additional upgrades or components

| | |
|---|---|
| Kitchen Cabinets & Countertops | Page 7. |
| Bathroom Vanities | 78 |
| Fireplaces & Chimneys | 78 |
| Windows, Skylights & Dormers | 78 |
| Appliances | 79 |
| Breezeways & Porches | 79 |
| Finished Attic | 79 |
| Garages | 80 |
| Site Improvements | 80 |
| Wings & Ells | 76 |

**Important: See the Reference Section for Location Factors (to adjust for your city) and Estimating Forms**

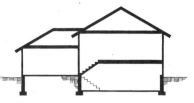

| | | Labor-Hours | Cost Per Square Foot Of Living Area | | |
|---|---|---|---|---|---|
| | | | Mat. | Labor | Total |
| **1 Site Work** | Site preparation for slab; 4' deep trench excavation for foundation wall, excavation for lower level, 4' deep. | .021 | | .37 | .37 |
| **2 Foundation** | Continuous reinforced concrete footing 8" deep x 18" wide; dampproofed and insulated reinforced concrete foundation wall, 12" thick, 4' deep; 4" concrete slab on 4" crushed stone base and polyethylene vapor barrier, trowel finish. | .109 | 3.50 | 3.63 | 7.13 |
| **3 Framing** | Exterior walls - 2" x 6" wood studs, 16" O.C.; 5/8" plywood sheathing; 2" x 10" rafters 16" O.C. with 5/8" plywood sheathing, 6 in 12 pitch; 2" x 8" ceiling joists 16" O.C.; 2" x 12" floor joists 16" O.C. with 5/8" plywood subfloor; 5/8" plywood subfloor on 1" x 3" wood sleepers 16" O.C. | .204 | 6.06 | 5.65 | 11.71 |
| **4 Exterior Walls** | Horizontal beveled wood siding; #15 felt building paper; 6" batt insulation; wood double hung windows; 3 solid core wood exterior doors; storms and screens. | .181 | 5.66 | 1.98 | 7.64 |
| **5 Roofing** | Red cedar shingles; #15 felt building paper; aluminum gutters, downspouts and drip edge; copper flashings. | .056 | 2.17 | 1.87 | 4.04 |
| **6 Interiors** | Walls and ceilings - 5/8" drywall, skim coat plaster, painted with primer and 2 coats; hardwood baseboard and trim, sanded and finished; hardwood floor 70%, ceramic tile with underlayment 20%, vinyl tile with underlayment 10%; wood panel interior doors, primed and painted with 2 coats. | .217 | 10.05 | 10.69 | 20.74 |
| **7 Specialties** | Luxury grade kitchen cabinets - 25 L.F. wall and base with plastic laminate counter top and kitchen sink; 6 L.F. bathroom vanity; 75 gallon electric water heater; medicine cabinet. | .048 | 2.62 | .51 | 3.13 |
| **8 Mechanical** | Gas fired warm air heat/air conditioning; one full bath including: bathtub, corner shower; built in lavatory and water closet; one 1/2 bath including: built in lavatory and water closet. | .057 | 3.55 | 2.07 | 5.62 |
| **9 Electrical** | 200 Amp. service; romex wiring; fluorescent and incandescent lighting fixtures; intercom, switches, receptacles. | .039 | .79 | 1.32 | 2.11 |
| **10 Overhead** | Contractor's overhead and profit and architect's fees. | | 8.22 | 6.69 | 14.91 |
| | **Total** | | 42.62 | 34.78 | **77.40** |

**SQUARE FOOT COSTS**

## 1 Story — Base cost per square foot of living area

| Exterior Wall | Living Area | | | | | | | |
| --- | --- | --- | --- | --- | --- | --- | --- | --- |
| | 50 | 100 | 200 | 300 | 400 | 500 | 600 | 700 |
| Wood Siding - Wood Frame | 178.10 | 137.30 | 119.50 | 101.95 | 96.10 | 92.55 | 90.20 | 91.45 |
| Brick Veneer - Wood Frame | 196.40 | 150.35 | 130.35 | 109.20 | 102.60 | 98.60 | 96.00 | 97.00 |
| Solid Brick | 227.15 | 172.35 | 148.65 | 121.40 | 113.65 | 108.90 | 105.75 | 106.45 |
| Solid Stone | 230.10 | 174.45 | 150.40 | 122.55 | 114.60 | 109.85 | 106.70 | 107.35 |
| Finished Basement, Add | 74.25 | 59.90 | 53.90 | 43.90 | 41.90 | 40.70 | 39.90 | 39.35 |
| Unfinished Basement, Add | 30.25 | 23.60 | 20.85 | 16.20 | 15.25 | 14.70 | 14.35 | 14.10 |

## 1-1/2 Story — Base cost per square foot of living area

| Exterior Wall | Living Area | | | | | | | |
| --- | --- | --- | --- | --- | --- | --- | --- | --- |
| | 100 | 200 | 300 | 400 | 500 | 600 | 700 | 800 |
| Wood Siding - Wood Frame | 142.20 | 115.05 | 98.90 | 90.20 | 85.25 | 83.05 | 80.05 | 79.00 |
| Brick Veneer - Wood Frame | 158.50 | 128.10 | 109.75 | 98.70 | 93.10 | 90.45 | 87.00 | 85.95 |
| Solid Brick | 185.95 | 150.10 | 128.00 | 112.95 | 106.25 | 102.90 | 98.80 | 97.60 |
| Solid Stone | 188.55 | 152.20 | 129.80 | 114.30 | 107.55 | 104.10 | 99.90 | 98.75 |
| Finished Basement, Add | 49.10 | 43.10 | 39.10 | 34.70 | 33.50 | 32.70 | 31.95 | 31.85 |
| Unfinished Basement, Add | 19.45 | 16.65 | 14.80 | 12.75 | 12.20 | 11.85 | 11.50 | 11.45 |

## 2 Story — Base cost per square foot of living area

| Exterior Wall | Living Area | | | | | | | |
| --- | --- | --- | --- | --- | --- | --- | --- | --- |
| | 100 | 200 | 400 | 600 | 800 | 1000 | 1200 | 1400 |
| Wood Siding - Wood Frame | 140.15 | 105.20 | 89.80 | 76.30 | 71.30 | 68.20 | 66.20 | 67.65 |
| Brick Veneer - Wood Frame | 158.40 | 118.25 | 100.65 | 83.55 | 77.80 | 74.30 | 72.00 | 73.25 |
| Solid Brick | 189.20 | 140.20 | 118.95 | 95.75 | 88.85 | 84.60 | 81.75 | 82.65 |
| Solid Stone | 192.10 | 142.30 | 120.75 | 96.95 | 89.80 | 85.55 | 82.70 | 83.55 |
| Finished Basement, Add | 37.20 | 30.00 | 27.00 | 22.00 | 21.00 | 20.40 | 20.00 | 19.70 |
| Unfinished Basement, Add | 15.15 | 11.80 | 10.45 | 8.10 | 7.65 | 7.35 | 7.20 | 7.05 |

Base costs do not include bathroom or kitchen facilities. Use Modifications/Adjustments/Alternatives on pages 77-80 where appropriate.

**Important: See the Reference Section for Location Factors (to adjust for your city) and Estimating Forms**

## Kitchen cabinets - Base units, hardwood *(Cost per Unit)*

| | Economy | Average | Custom | Luxury |
|---|---|---|---|---|
| **24″ deep, 35″ high,** | | | | |
| One top drawer, | | | | |
| One door below | | | | |
| 12″ wide | $122 | $162 | $215 | $285 |
| 15″ wide | 161 | 215 | 285 | 375 |
| 18″ wide | 186 | 248 | 330 | 435 |
| 21″ wide | 178 | 237 | 315 | 415 |
| 24″ wide | 204 | 272 | 360 | 475 |
| Four drawers | | | | |
| 12″ wide | 274 | 365 | 485 | 640 |
| 15″ wide | 236 | 315 | 420 | 550 |
| 18″ wide | 240 | 320 | 425 | 560 |
| 24″ wide | 259 | 345 | 460 | 605 |
| Two top drawers, | | | | |
| Two doors below | | | | |
| 27″ wide | 229 | 305 | 405 | 535 |
| 30″ wide | 244 | 325 | 430 | 570 |
| 33″ wide | 244 | 325 | 430 | 570 |
| 36″ wide | 255 | 340 | 450 | 595 |
| 42″ wide | 274 | 365 | 485 | 640 |
| 48″ wide | 289 | 385 | 510 | 675 |
| **Range or sink base** | | | | |
| *(Cost per unit)* | | | | |
| Two doors below | | | | |
| 30″ wide | 200 | 266 | 355 | 465 |
| 33″ wide | 212 | 283 | 375 | 495 |
| 36″ wide | 222 | 296 | 395 | 520 |
| 42″ wide | 236 | 315 | 420 | 550 |
| 48″ wide | 248 | 330 | 440 | 580 |
| **Corner Base Cabinet** | | | | |
| *(Cost per unit)* | | | | |
| 36″ wide | 202 | 269 | 360 | 470 |
| **Lazy Susan** *(Cost per unit)* | | | | |
| With revolving door | 311 | 415 | 550 | 725 |

## Kitchen cabinets - Wall cabinets, hardwood *(Cost per Unit)*

| | Economy | Average | Custom | Luxury |
|---|---|---|---|---|
| **12″ deep, 2 doors** | | | | |
| 12″ high | | | | |
| 30″ wide | $130 | $173 | $230 | $ 305 |
| 36″ wide | 149 | 198 | 265 | 345 |
| 15″ high | | | | |
| 30″ wide | 138 | 184 | 245 | 320 |
| 33″ wide | 152 | 202 | 270 | 355 |
| 36″ wide | 157 | 209 | 280 | 365 |
| 24″ high | | | | |
| 30″ wide | 164 | 219 | 290 | 385 |
| 36″ wide | 180 | 240 | 320 | 420 |
| 42″ wide | 197 | 263 | 350 | 460 |
| 30″ high, 1 door | | | | |
| 12″ wide | 119 | 158 | 210 | 275 |
| 15″ wide | 132 | 176 | 235 | 310 |
| 18″ wide | 144 | 192 | 255 | 335 |
| 24″ wide | 158 | 210 | 280 | 370 |
| 30″ high, 2 doors | | | | |
| 27″ wide | 194 | 258 | 345 | 450 |
| 30″ wide | 188 | 251 | 335 | 440 |
| 36″ wide | 214 | 285 | 380 | 500 |
| 42″ wide | 229 | 305 | 405 | 535 |
| 48″ wide | 255 | 340 | 450 | 595 |
| Corner wall, 30″ high | | | | |
| 24″ wide | 138 | 184 | 245 | 320 |
| 30″ wide | 162 | 216 | 285 | 380 |
| 36″ wide | 175 | 233 | 310 | 410 |
| **Broom closet** | | | | |
| 84″ high, 24″ deep | | | | |
| 18″ wide | 345 | 460 | 610 | 805 |
| **Oven Cabinet** | | | | |
| 84″ high, 24″ deep | | | | |
| 27″ wide | 499 | 665 | 885 | 1165 |

## Kitchen countertops *(Cost per L.F.)*

| | Economy | Average | Custom | Luxury |
|---|---|---|---|---|
| Solid Surface | | | | |
| 24″ wide, no backsplash | $83 | $111 | $150 | $195 |
| with backsplash | 91 | 121 | 160 | 210 |
| Stock plastic laminate, 24″ wide | | | | |
| with backsplash | 14 | 18 | 25 | 30 |
| Custom plastic laminate, no splash | | | | |
| 7/8″ thick, alum. molding | 20 | 27 | 35 | 45 |
| 1-1/4″ thick, no splash | 24 | 32 | 45 | 55 |
| Marble | | | | |
| 1/2″ - 3/4″ thick w/splash | 37 | 50 | 65 | 85 |
| Maple, laminated | | | | |
| 1-1/2″ thick w/splash | 36 | 49 | 65 | 85 |
| Stainless steel | | | | |
| (per S.F.) | 70 | 93 | 125 | 165 |
| Cutting blocks, recessed | | | | |
| 16″ x 20″ x 1″ (each) | 61 | 81 | 110 | 140 |

**ADJUSTMENTS**

## Vanity bases  (Cost per Unit)

|  | Economy | Average | Custom | Luxury |
|---|---|---|---|---|
| **2 door, 30" high, 21" deep** | | | | |
| 24" wide | 155 | 206 | 275 | 360 |
| 30" wide | 180 | 240 | 320 | 420 |
| 36" wide | 236 | 315 | 420 | 550 |
| 48" wide | 281 | 375 | 500 | 655 |

## Solid surface vanity tops  (Cost Each)

|  | Economy | Average | Custom | Luxury |
|---|---|---|---|---|
| **Center bowl** | | | | |
| 22" x 25" | $281 | $303 | $328 | $354 |
| 22" x 31" | 325 | 351 | 379 | 409 |
| 22" x 37" | 370 | 400 | 432 | 466 |
| 22" x 49" | 460 | 497 | 537 | 579 |

## Fireplaces & chimneys  (Cost per Unit)

|  | 1-1/2 Story | 2 Story | 3 Story |
|---|---|---|---|
| **Economy (prefab metal)** | | | |
| Exterior chimney & 1 fireplace | $3655 | $4035 | $4420 |
| Interior chimney & 1 fireplace | 3495 | 3890 | 4075 |
| **Average (masonry)** | | | |
| Exterior chimney & 1 fireplace | 3730 | 4160 | 4730 |
| Interior chimney & 1 fireplace | 3495 | 3920 | 4270 |
| For more than 1 flue, add | 275 | 455 | 765 |
| For more than 1 fireplace, add | 2645 | 2645 | 2645 |
| **Custom (masonry)** | | | |
| Exterior chimney & 1 fireplace | 3925 | 4430 | 5005 |
| Interior chimney & 1 fireplace | 3685 | 4165 | 4490 |
| For more than 1 flue, add | 310 | 530 | 725 |
| For more than 1 fireplace, add | 2820 | 2820 | 2820 |
| **Luxury (masonry)** | | | |
| Exterior chimney & 1 fireplace | 5545 | 6075 | 6645 |
| Interior chimney & 1 fireplace | 5295 | 5790 | 6120 |
| For more than 1 flue, add | 455 | 765 | 1070 |
| For more than 1 fireplace, add | 4365 | 4365 | 4365 |

## Windows and skylights  (Cost Each)

|  | Economy | Average | Custom | Luxury |
|---|---|---|---|---|
| **Fixed Picture Windows** | | | | |
| 3'-6" x 4'-0" | $ 390 | $ 421 | $ 455 | $ 491 |
| 4'-0" x 6'-0" | 617 | 667 | 720 | 778 |
| 5'-0" x 6'-0" | 793 | 856 | 925 | 999 |
| 6'-0" x 6'-0" | 806 | 870 | 940 | 1015 |
| **Bay/Bow Windows** | | | | |
| 8'-0" x 5'-0" | 1157 | 1250 | 1350 | 1458 |
| 10'-0" x 5'-0" | 1629 | 1759 | 1900 | 2052 |
| 10'-0" x 6'-0" | 1715 | 1852 | 2000 | 2160 |
| 12'-0" x 6'-0" | 2036 | 2199 | 2375 | 2565 |
| **Palladian Windows** | | | | |
| 3'-2" x 6'-4" | | 1435 | 1550 | 1674 |
| 4'-0" x 6'-0" | | 1481 | 1600 | 1728 |
| 5'-5" x 6'-10" | | 1852 | 2000 | 2160 |
| 8'-0" x 6'-0" | | 2269 | 2450 | 2646 |
| **Skylights** | | | | |
| 46" x 21-1/2" | 342 | 369 | 513 | 554 |
| 46" x 28" | 451 | 487 | 637 | 688 |
| 57" x 44" | 559 | 604 | 789 | 852 |

## Dormers  (Cost/S.F. of plan area)

|  | Economy | Average | Custom | Luxury |
|---|---|---|---|---|
| **Framing and Roofing Only** | | | | |
| Gable dormer, 2" x 6" roof frame | $19 | $21 | $24 | $39 |
| 2" x 8" roof frame | 20 | 23 | 25 | 41 |
| Shed dormer, 2" x 6" roof frame | 12 | 14 | 15 | 25 |
| 2" x 8" roof frame | 13 | 15 | 17 | 26 |
| 2" x 10" roof frame | 14 | 16 | 18 | 27 |

## Appliances *(Cost per Unit)*

| | Economy | Average | Custom | Luxury |
|---|---|---|---|---|
| **Range** | | | | |
| 30″ free standing, 1 oven | $ 320 | $ 873 | $1149 | $ 1425 |
| 2 oven | 1475 | 1563 | 1606 | 1650 |
| 30″ built-in, 1 oven | 400 | 925 | 1188 | 1450 |
| 2 oven | 1200 | 1600 | 1800 | 2000 |
| 21″ free standing | | | | |
| 1 oven | 355 | 493 | 561 | 630 |
| **Counter Top Ranges** | | | | |
| 4 burner standard | 262 | 426 | 508 | 590 |
| As above with griddle | 485 | 588 | 639 | 690 |
| **Microwave Oven** | 174 | 385 | 490 | 595 |
| **Combination Range,** | | | | |
| **Refrigerator, Sink** | | | | |
| 30″ wide | 995 | 1498 | 1749 | 2000 |
| 60″ wide | 2925 | 3364 | 3584 | 3803 |
| 72″ wide | 3325 | 3824 | 4074 | 4323 |
| **Comb. Range, Refrig., Sink,** | | | | |
| **Microwave Oven & Ice Maker** | 5147 | 5919 | 6305 | 66910 |
| **Compactor** | | | | |
| 4 to 1 compaction | 445 | 483 | 501 | 520 |
| **Deep Freeze** | | | | |
| 15 to 23 C.F. | 500 | 590 | 635 | 680 |
| 30 C.F. | 930 | 1053 | 1114 | 1175 |
| **Dehumidifier,** portable, auto. | | | | |
| 15 pint | 223 | 257 | 273 | 290 |
| 30 pint | 247 | 284 | 303 | 321 |
| **Washing Machine,** automatic | 470 | 748 | 886 | 1025 |
| **Water Heater** | | | | |
| **Electric,** glass lined | | | | |
| 30 gal. | 325 | 400 | 438 | 475 |
| 80 gal. | 680 | 865 | 958 | 1050 |
| **Water Heater, Gas,** glass lined | | | | |
| 30 gal. | 455 | 560 | 613 | 665 |
| 50 gal. | 685 | 855 | 940 | 1025 |
| **Water Softener,** automatic | | | | |
| 30 grains/gal. | 585 | 673 | 717 | 761 |
| 100 grains/gal. | 760 | 874 | 931 | 988 |
| **Dishwasher,** built-in | | | | |
| 2 cycles | 375 | 440 | 473 | 505 |
| 4 or more cycles | 415 | 468 | 494 | 520 |
| **Dryer,** automatic | 440 | 655 | 763 | 870 |
| **Garage Door Opener** | 273 | 324 | 350 | 375 |
| **Garbage Disposal** | 82 | 161 | 201 | 240 |
| **Heater, Electric,** built-in | | | | |
| 1250 watt ceiling type | 155 | 192 | 211 | 229 |
| 1250 watt wall type | 187 | 218 | 233 | 248 |
| Wall type w/blower | | | | |
| 1500 watt | 213 | 232 | 241 | 250 |
| 3000 watt | 380 | 420 | 440 | 460 |
| **Hood For Range,** 2 speed | | | | |
| 30″ wide | 112 | 364 | 489 | 615 |
| 42″ wide | 262 | 606 | 778 | 950 |
| **Humidifier,** portable | | | | |
| 7 gal. per day | 111 | 128 | 136 | 144 |
| 15 gal. per day | 180 | 207 | 221 | 234 |
| **Ice Maker,** automatic | | | | |
| 13 lb. per day | 650 | 748 | 796 | 845 |
| 51 lb. per day | 1050 | 1208 | 1286 | 1365 |
| **Refrigerator,** no frost | | | | |
| 10-12 C.F. | 535 | 683 | 756 | 830 |
| 14-16 C.F. | 550 | 615 | 648 | 680 |
| 18-20 C.F. | 605 | 758 | 834 | 910 |
| 21-29 C.F. | 805 | 2003 | 2601 | 3200 |
| **Sump Pump,** 1/3 H.P. | 202 | 281 | 321 | 360 |

## Breezeway *(Cost per S.F.)*

| Class | Type | Area (S.F.) | | | |
|---|---|---|---|---|---|
| | | 50 | 100 | 150 | 200 |
| Economy | Open | $ 16.60 | $14.15 | $11.85 | $11.60 |
| | Enclosed | 79.65 | 61.55 | 51.10 | 44.75 |
| Average | Open | 21.10 | 18.55 | 16.20 | 14.75 |
| | Enclosed | 89.55 | 66.55 | 54.55 | 48.00 |
| Custom | Open | 30.40 | 26.70 | 23.25 | 21.30 |
| | Enclosed | 124.80 | 92.80 | 75.90 | 66.70 |
| Luxury | Open | 31.55 | 27.70 | 25.00 | 24.20 |
| | Enclosed | 127.25 | 94.35 | 76.30 | 68.35 |

## Porches *(Cost per S.F.)*

| Class | Type | Area (S.F.) | | | | |
|---|---|---|---|---|---|---|
| | | 25 | 50 | 100 | 200 | 300 |
| Economy | Open | $ 47.60 | $ 31.90 | $24.85 | $21.10 | $18.00 |
| | Enclosed | 95.15 | 66.30 | 50.15 | 39.10 | 33.50 |
| Average | Open | 59.25 | 37.75 | 28.95 | 24.10 | 24.10 |
| | Enclosed | 117.05 | 79.30 | 60.00 | 46.50 | 39.35 |
| Custom | Open | 77.55 | 51.45 | 38.80 | 33.85 | 30.35 |
| | Enclosed | 154.20 | 105.35 | 80.20 | 62.35 | 53.80 |
| Luxury | Open | 83.65 | 54.70 | 40.40 | 36.30 | 32.30 |
| | Enclosed | 163.75 | 115.00 | 85.40 | 66.25 | 57.10 |

## Finished attic *(Cost per S.F.)*

| Class | Area (S.F.) | | | | |
|---|---|---|---|---|---|
| | 400 | 500 | 600 | 800 | 1000 |
| Economy | $12.75 | $12.30 | $11.75 | $11.55 | $11.20 |
| Average | 20.10 | 19.55 | 19.15 | 18.85 | 18.35 |
| Custom | 24.75 | 24.15 | 23.70 | 23.25 | 22.80 |
| Luxury | 31.35 | 30.70 | 29.95 | 29.25 | 28.75 |

## Alarm system *(Cost per System)*

| Class | Burglar Alarm | Smoke Detector |
|---|---|---|
| Economy | $ 350 | $ 52 |
| Average | 405 | 64 |
| Custom | 684 | 133 |
| Luxury | 1025 | 164 |

## Sauna, prefabricated
*(Cost per unit, including heater and controls—7′ high)*

| Size | Cost |
|---|---|
| 6′ x 4′ | $4600 |
| 6′ x 5′ | 4075 |
| 6′ x 6′ | 5375 |
| 6′ x 9′ | 5275 |
| 8′ x 10′ | 6825 |
| 8′ x 12′ | 8275 |
| 10′ x 12′ | 9025 |

**ADJUSTMENTS**

## Garages *

(Costs include exterior wall systems comparable with the quality of the residence. Included in the cost is an allowance for one personnel door, manual overhead door(s) and electrical fixture.)

| Class | Type | | | | | | | | | |
|---|---|---|---|---|---|---|---|---|---|---|
| | Detached | | | Attached | | | Built-in | | Basement | |
| | One Car | Two Car | Three Car | One Car | Two Car | Three Car | One Car | Two Car | One Car | Two Car |
| **Economy** | | | | | | | | | | |
| Wood | $10,343 | $15,719 | $21,095 | $ 8010 | $13,807 | $19,183 | $-1374 | $-2747 | $1018 | $1280 |
| Masonry | 14,108 | 20,388 | 26,667 | 10,343 | 17,031 | 23,311 | -1913 | -3827 | | |
| **Average** | | | | | | | | | | |
| Wood | 11,040 | 16,548 | 22,056 | 8424 | 14,340 | 19,848 | -1537 | -3073 | 1242 | 1674 |
| Masonry | 14,200 | 20,503 | 26,806 | 10,401 | 17,112 | 23,415 | -1925 | -3357 | | |
| **Custom** | | | | | | | | | | |
| Wood | 12,841 | 19,464 | 26,088 | 9957 | 17,118 | 23,741 | -3731 | -4172 | 1771 | 2732 |
| Masonry | 15,531 | 22,783 | 30,036 | 11,616 | 19,393 | 26,645 | -4078 | -4875 | | |
| **Luxury** | | | | | | | | | | |
| Wood | 14,496 | 22,357 | 30,218 | 11,402 | 19,788 | 27,649 | -3808 | -4336 | 2399 | 3675 |
| Masonry | 18,754 | 27,671 | 36,587 | 14,059 | 23,500 | 32,417 | -4333 | -5386 | | |

*See the Introduction to this section for definitions of garage types.

## Swimming pools  (Cost per S.F.)

| | |
|---|---|
| **Residential**  (includes equipment) | |
| In-ground | $  18 - 45 |
| **Deck equipment** | 1.30 |
| **Paint pool,** preparation & 3 coats (epoxy) | 2.78 |
| Rubber base paint | 2.56 |
| **Pool Cover** | .66 |
| **Swimming Pool Heaters**   (Cost per unit) | |
| (not including wiring, external piping, base or pad) | |
| **Gas** | |
| 155 MBH | $  2600 |
| 190 MBH | 3500 |
| 500 MBH | 8500 |
| **Electric** | |
| 15 KW    7200 gallon pool | 2050 |
| 24 KW    9600 gallon pool | 2700 |
| 54 KW    24,000 gallon pool | 4025 |

## Wood and coal stoves

| | |
|---|---|
| **Wood Only** | |
| Free Standing (minimum) | $ 1250 |
| Fireplace Insert (minimum) | 1260 |
| **Coal Only** | |
| Free Standing | $ 1424 |
| Fireplace Insert | 1559 |
| **Wood and Coal** | |
| Free Standing | $ 2928 |
| Fireplace Insert | 3000 |

## Sidewalks  (Cost per S.F.)

| | | |
|---|---|---|
| Concrete, 3000 psi with wire mesh | 4" thick | $ 2.52 |
| | 5" thick | 3.07 |
| | 6" thick | 3.45 |
| Precast concrete patio blocks (natural) | 2" thick | 7.50 |
| Precast concrete patio blocks (colors) | 2" thick | 8.00 |
| Flagstone, bluestone | 1" thick | 10.87 |
| Flagstone, bluestone | 1-1/2" thick | 14.35 |
| Slate (natural, irregular) | 3/4" thick | 10.46 |
| Slate (random rectangular) | 1/2" thick | 15.57 |
| **Seeding** | | |
| Fine grading & seeding includes lime, fertilizer & seed | per S.Y. | 1.71 |
| **Lawn Sprinkler System** | per S.F. | .68 |

## Fencing  (Cost per L.F.)

| | |
|---|---|
| Chain Link, 4' high, galvanized | $ 10.80 |
| Gate, 4' high (each) | 116.00 |
| Cedar Picket, 3' high, 2 rail | 9.25 |
| Gate (each) | 117.00 |
| 3 Rail, 4' high | 10.25 |
| Gate (each) | 126.00 |
| Cedar Stockade, 3 Rail, 6' high | 10.50 |
| Gate (each) | 125.00 |
| Board & Battens, 2 sides 6' high, pine | 15.35 |
| 6' high, cedar | 22.00 |
| No. 1 Cedar, basketweave, 6' high | 12.15 |
| Gate, 6' high (each) | 145.00 |

## Carport  (Cost per S.F.)

| | |
|---|---|
| Economy | $ 5.99 |
| Average | 9.31 |
| Custom | 14.03 |
| Luxury | 16.06 |

# Assemblies Section

## Table of Contents

# How to Use the Assemblies Section

**Illustration**
Each building assembly system is accompanied by a detailed, illustrated description. Each individual component is labeled. Every element involved in the total system function is shown.

**Description**
Each page includes a brief outline of any special conditions to be used when pricing a system. All units of measure are defined here.

**System Definition**
Not only are all components broken down for each system, but alternative components can be found on the opposite page. Simply insert any chosen new element into the chart to develop a custom system.

**Labor-hours**
Total labor-hours for a system can be found by simply multiplying the quantity of the system required times LABOR-HOURS. The resulting figure is the total labor-hours needed to complete the system.
(QUANTITY OF SYSTEM x LABOR-HOURS = TOTAL SYSTEM LABOR-HOURS)

**Materials**
This column contains the MATERIAL COST of each element. These cost figures include 10% for profit.

**Installation**
Labor rates include both the INSTALLATION COST of the contractor and the standard contractor's O&P. On the average, the LABOR COST will be 70.3% over the above BARE LABOR COST.

**Totals**
This row provides the necessary system cost totals. TOTAL SYSTEM COST can be derived by multiplying the TOTAL times each system's SQUARE FOOT ESTIMATE (TOTAL x SQUARE FEET = TOTAL SYSTEM COST).

**Work Sheet**
Using the SELECTIVE PRICE SHEET on the page opposite each system, it is possible to create estimates with alternative items for any number of systems.

**Note:**
Throughout this section, the words assembly and system are used interchangeably.

**Quantities**
Each material in a system is shown with the quantity required for the system unit. For example, the rafters in this system have 1.170 L.F. per S.F. of ceiling area.

**Unit of Measure**
In the three right-hand columns, each cost figure is adjusted to agree with the unit of measure for the entire system. In this case, COST PER SQUARE FOOT (S.F.) is the common unit of measure. NOTE: In addition, under the UNIT heading, all the elements of each system are defined in relation to the product as a selling commodity. For example, "fascia board" is defined in linear feet, instead of in board feet.

## 3 | FRAMING    12 | Gable End Roof Framing Systems

Labeled: Ridge Board, Sheathing, Rafters, Rafter Tie, Ceiling Joists, Furring Strips, Soffit Nailer, Fascia Board

| System Description | QUAN. | UNIT | LABOR HOURS | MAT. | INST. | TOTAL |
|---|---|---|---|---|---|---|
| **2" X 6" RAFTERS, 16" O.C., 4/12 PITCH** | | | | | | |
| Rafters, 2" x 6", 16" O.C., 4/12 pitch | 1.170 | L.F. | .019 | .73 | .67 | 1.40 |
| Ceiling joists, 2" x 4", 16" O.C. | 1.000 | L.F. | .013 | .39 | .46 | .85 |
| Ridge board, 2" x 6" | .050 | L.F. | .002 | .03 | .06 | .09 |
| Fascia board, 2" x 6" | .100 | L.F. | .005 | .07 | .19 | .26 |
| Rafter tie, 1" x 4", 4' O.C. | .060 | L.F. | .001 | .02 | .04 | .06 |
| Soffit nailer (outrigger), 2" x 4", 24" O.C. | .170 | L.F. | .004 | .07 | .16 | .23 |
| Sheathing, exterior, plywood, CDX, 1/2" thick | 1.170 | S.F. | .013 | .63 | .48 | 1.11 |
| Furring strips, 1" x 3", 16" O.C. | 1.000 | L.F. | .023 | .21 | .81 | 1.02 |
| TOTAL | | | .080 | 2.15 | 2.87 | 5.02 |
| **2" X 8" RAFTERS, 16" O.C., 4/12 PITCH** | | | | | | |
| Rafters, 2" x 8", 16" O.C., 4/12 pitch | 1.170 | L.F. | .020 | 1.11 | .70 | 1.81 |
| Ceiling joists, 2" x 6", 16" O.C. | 1.000 | L.F. | .013 | .62 | .46 | 1.08 |
| Ridge board, 2" x 8" | .050 | L.F. | .002 | .05 | .06 | .11 |
| Fascia board, 2" x 8" | .100 | L.F. | .007 | .10 | .25 | .35 |
| Rafter tie, 1" x 4", 4' O.C. | .060 | L.F. | .001 | .02 | .04 | .06 |
| Soffit nailer (outrigger), 2" x 4", 24" O.C. | .170 | L.F. | .004 | .07 | .16 | .23 |
| Sheathing, exterior, plywood, CDX, 1/2" thick | 1.170 | S.F. | .013 | .63 | .48 | 1.11 |
| Furring strips, 1" x 3", 16" O.C. | 1.000 | L.F. | .023 | .21 | .81 | 1.02 |
| TOTAL | | | .083 | 2.81 | 2.96 | 5.77 |

The cost of this system is based on the square foot of plan area.
All quantities have been adjusted accordingly.

| Description | QUAN. | UNIT | LABOR HOURS | MAT. | INST. | TOTAL |
|---|---|---|---|---|---|---|
| | | | | | | |
| | | | | | | |
| | | | | | | |
| | | | | | | |
| | | | | | | |
| | | | | | | |

120    **Important: See the Reference Section for critical supporting data - Reference Nos., Crews & Location Factors**

**Total**
MATERIAL COST + INSTALLATION COST = TOTAL. Work on the table from left to right across cost columns to derive totals.

# Division 1
# Site Work

Backfill

Excavate

| System Description | QUAN. | UNIT | LABOR HOURS | COST EACH | | |
|---|---|---|---|---|---|---|
| | | | | MAT. | INST. | TOTAL |
| **BUILDING, 24' X 38', 4' DEEP** | | | | | | |
| Clear and strip, dozer, light trees, 30' from building | .190 | Acre | 9.120 | | 456.00 | 456.00 |
| Excavate, backhoe | 174.000 | C.Y. | 2.319 | | 264.48 | 264.48 |
| Backfill, dozer, 4" lifts, no compaction | 87.000 | C.Y. | .580 | | 87.00 | 87.00 |
| Rough grade, dozer, 30' from building | 87.000 | C.Y. | .580 | | 87.00 | 87.00 |
| TOTAL | | | 12.599 | | 894.48 | 894.48 |
| **BUILDING, 26' X 46', 4' DEEP** | | | | | | |
| Clear and strip, dozer, light trees, 30' from building | .210 | Acre | 10.080 | | 504.00 | 504.00 |
| Excavate, backhoe | 201.000 | C.Y. | 2.679 | | 305.52 | 305.52 |
| Backfill, dozer, 4" lifts, no compaction | 100.000 | C.Y. | .667 | | 100.00 | 100.00 |
| Rough grade, dozer, 30' from building | 100.000 | C.Y. | .667 | | 100.00 | 100.00 |
| TOTAL | | | 14.093 | | 1009.52 | 1009.52 |
| **BUILDING, 26' X 60', 4' DEEP** | | | | | | |
| Clear and strip, dozer, light trees, 30' from building | .240 | Acre | 11.520 | | 576.00 | 576.00 |
| Excavate, backhoe | 240.000 | C.Y. | 3.199 | | 364.80 | 364.80 |
| Backfill, dozer, 4" lifts, no compaction | 120.000 | C.Y. | .800 | | 120.00 | 120.00 |
| Rough grade, dozer, 30' from building | 120.000 | C.Y. | .800 | | 120.00 | 120.00 |
| TOTAL | | | 16.319 | | 1180.80 | 1180.80 |
| **BUILDING, 30' X 66', 4' DEEP** | | | | | | |
| Clear and strip, dozer, light trees, 30' from building | .260 | Acre | 12.480 | | 624.00 | 624.00 |
| Excavate, backhoe | 268.000 | C.Y. | 3.572 | | 407.36 | 407.36 |
| Backfill, dozer, 4" lifts, no compaction | 134.000 | C.Y. | .894 | | 134.00 | 134.00 |
| Rough grade, dozer, 30' from building | 134.000 | C.Y. | .894 | | 134.00 | 134.00 |
| TOTAL | | | 17.840 | | 1299.36 | 1299.36 |

The costs in this system are on a cost each basis.
Quantities are based on 1'-0" clearance on each side of footing.

| Description | QUAN. | UNIT | LABOR HOURS | COST EACH | | |
|---|---|---|---|---|---|---|
| | | | | MAT. | INST. | TOTAL |
| | | | | | | |
| | | | | | | |
| | | | | | | |

**Important: See the Reference Section for critical supporting data - Reference Nos., Crews & Location Factors**

SITE WORK 1

## Footing Excavation Price Sheet

| | QUAN. | UNIT | LABOR HOURS | COST EACH MAT. | COST EACH INST. | COST EACH TOTAL |
|---|---|---|---|---|---|---|
| Clear and grub, medium brush, 30' from building, 24' x 38' | .190 | Acre | 9.120 | | 455.00 | 455.00 |
| 26' x 46' | .210 | Acre | 10.080 | | 505.00 | 505.00 |
| 26' x 60' | .240 | Acre | 11.520 | | 575.00 | 575.00 |
| 30' x 66' | .260 | Acre | 12.480 | | 625.00 | 625.00 |
| Light trees, to 6" dia. cut & chip, 24' x 38' | .190 | Acre | 9.120 | | 455.00 | 455.00 |
| 26' x 46' | .210 | Acre | 10.080 | | 505.00 | 505.00 |
| 26' x 60' | .240 | Acre | 11.520 | | 575.00 | 575.00 |
| 30' x 66' | .260 | Acre | 12.480 | | 625.00 | 625.00 |
| Medium trees, to 10" dia. cut & chip, 24' x 38' | .190 | Acre | 13.029 | | 650.00 | 650.00 |
| 26' x 46' | .210 | Acre | 14.400 | | 720.00 | 720.00 |
| 26' x 60' | .240 | Acre | 16.457 | | 820.00 | 820.00 |
| 30' x 66' | .260 | Acre | 17.829 | | 890.00 | 890.00 |
| Excavation, footing, 24' x 38', 2' deep | 68.000 | C.Y. | .906 | | 104.00 | 104.00 |
| 4' deep | 174.000 | C.Y. | 2.319 | | 265.00 | 265.00 |
| 8' deep | 384.000 | C.Y. | 5.119 | | 585.00 | 585.00 |
| 26' x 46', 2' deep | 79.000 | C.Y. | 1.053 | | 120.00 | 120.00 |
| 4' deep | 201.000 | C.Y. | 2.679 | | 305.00 | 305.00 |
| 8' deep | 404.000 | C.Y. | 5.385 | | 615.00 | 615.00 |
| 26' x 60', 2' deep | 94.000 | C.Y. | 1.253 | | 143.00 | 143.00 |
| 4' deep | 240.000 | C.Y. | 3.199 | | 365.00 | 365.00 |
| 8' deep | 483.000 | C.Y. | 6.438 | | 730.00 | 730.00 |
| 30' x 66', 2' deep | 105.000 | C.Y. | 1.400 | | 160.00 | 160.00 |
| 4' deep | 268.000 | C.Y. | 3.572 | | 405.00 | 405.00 |
| 8' deep | 539.000 | C.Y. | 7.185 | | 820.00 | 820.00 |
| Backfill, 24' x 38', 2" lifts, no compaction | 34.000 | C.Y. | .227 | | 34.00 | 34.00 |
| Compaction, air tamped | 34.000 | C.Y. | 2.267 | | 228.00 | 228.00 |
| 4" lifts, no compaction | 87.000 | C.Y. | .580 | | 87.00 | 87.00 |
| Compaction, air tamped | 87.000 | C.Y. | 5.800 | | 585.00 | 585.00 |
| 8" lifts, no compaction | 192.000 | C.Y. | 1.281 | | 192.00 | 192.00 |
| Compaction, air tamped | 192.000 | C.Y. | 12.801 | | 1300.00 | 1300.00 |
| 26' x 46', 2" lifts, no compaction | 40.000 | C.Y. | .267 | | 40.00 | 40.00 |
| Compaction, air tamped | 40.000 | C.Y. | 2.667 | | 268.00 | 268.00 |
| 4" lifts, no compaction | 100.000 | C.Y. | .667 | | 100.00 | 100.00 |
| Compaction, air tamped | 100.000 | C.Y. | 6.667 | | 670.00 | 670.00 |
| 8" lifts, no compaction | 202.000 | C.Y. | 1.347 | | 203.00 | 203.00 |
| Compaction, air tamped | 202.000 | C.Y. | 13.467 | | 1350.00 | 1350.00 |
| 26' x 60', 2" lifts, no compaction | 47.000 | C.Y. | .313 | | 47.00 | 47.00 |
| Compaction, air tamped | 47.000 | C.Y. | 3.133 | | 315.00 | 315.00 |
| 4" lifts, no compaction | 120.000 | C.Y. | .800 | | 120.00 | 120.00 |
| Compaction, air tamped | 120.000 | C.Y. | 8.000 | | 805.00 | 805.00 |
| 8" lifts, no compaction | 242.000 | C.Y. | 1.614 | | 242.00 | 242.00 |
| Compaction, air tamped | 242.000 | C.Y. | 16.134 | | 1625.00 | 1625.00 |
| 30' x 66', 2" lifts, no compaction | 53.000 | C.Y. | .354 | | 53.00 | 53.00 |
| Compaction, air tamped | 53.000 | C.Y. | 3.534 | | 355.00 | 355.00 |
| 4" lifts, no compaction | 134.000 | C.Y. | .894 | | 134.00 | 134.00 |
| Compaction, air tamped | 134.000 | C.Y. | 8.934 | | 900.00 | 900.00 |
| 8" lifts, no compaction | 269.000 | C.Y. | 1.794 | | 269.00 | 269.00 |
| Compaction, air tamped | 269.000 | C.Y. | 17.934 | | 1800.00 | 1800.00 |
| Rough grade, 30' from building, 24' x 38' | 87.000 | C.Y. | .580 | | 87.00 | 87.00 |
| 26' x 46' | 100.000 | C.Y. | .667 | | 100.00 | 100.00 |
| 26' x 60' | 120.000 | C.Y. | .800 | | 120.00 | 120.00 |
| 30' x 66' | 134.000 | C.Y. | .894 | | 134.00 | 134.00 |

**SITE WORK**

**1**

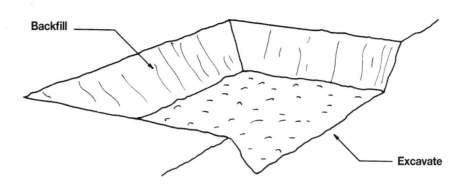

Backfill

Excavate

| System Description | QUAN. | UNIT | LABOR HOURS | COST EACH | | |
|---|---|---|---|---|---|---|
| | | | | MAT. | INST. | TOTAL |
| **BUILDING, 24' X 38', 8' DEEP** | | | | | | |
| Clear & grub, dozer, medium brush, 30' from building | .190 | Acre | 2.027 | | 177.65 | 177.65 |
| Excavate, track loader, 1-1/2 C.Y. bucket | 550.000 | C.Y. | 7.860 | | 566.50 | 566.50 |
| Backfill, dozer, 8" lifts, no compaction | 180.000 | C.Y. | 1.201 | | 180.00 | 180.00 |
| Rough grade, dozer, 30' from building | 280.000 | C.Y. | 1.868 | | 280.00 | 280.00 |
| TOTAL | | | 12.956 | | 1204.15 | 1204.15 |
| **BUILDING, 26' X 46', 8' DEEP** | | | | | | |
| Clear & grub, dozer, medium brush, 30' from building | .210 | Acre | 2.240 | | 196.35 | 196.35 |
| Excavate, track loader, 1-1/2 C.Y. bucket | 672.000 | C.Y. | 9.603 | | 692.16 | 692.16 |
| Backfill, dozer, 8" lifts, no compaction | 220.000 | C.Y. | 1.467 | | 220.00 | 220.00 |
| Rough grade, dozer, 30' from building | 340.000 | C.Y. | 2.268 | | 340.00 | 340.00 |
| TOTAL | | | 15.578 | | 1448.51 | 1448.51 |
| **BUILDING, 26' X 60', 8' DEEP** | | | | | | |
| Clear & grub, dozer, medium brush, 30' from building | .240 | Acre | 2.560 | | 224.40 | 224.40 |
| Excavate, track loader, 1-1/2 C.Y. bucket | 829.000 | C.Y. | 11.846 | | 853.87 | 853.87 |
| Backfill, dozer, 8" lifts, no compaction | 270.000 | C.Y. | 1.801 | | 270.00 | 270.00 |
| Rough grade, dozer, 30' from building | 420.000 | C.Y. | 2.801 | | 420.00 | 420.00 |
| TOTAL | | | 19.008 | | 1768.27 | 1768.27 |
| **BUILDING, 30' X 66', 8' DEEP** | | | | | | |
| Clear & grub, dozer, medium brush, 30' from building | .260 | Acre | 2.773 | | 243.10 | 243.10 |
| Excavate, track loader, 1-1/2 C.Y. bucket | 990.000 | C.Y. | 14.147 | | 1019.70 | 1019.70 |
| Backfill dozer, 8" lifts, no compaction | 320.000 | C.Y. | 2.134 | | 320.00 | 320.00 |
| Rough grade, dozer, 30' from building | 500.000 | C.Y. | 3.335 | | 500.00 | 500.00 |
| TOTAL | | | 22.389 | | 2082.80 | 2082.80 |

The costs in this system are on a cost each basis.
Quantities are based on 1'-0" clearance beyond footing projection.

| Description | QUAN. | UNIT | LABOR HOURS | COST EACH | | |
|---|---|---|---|---|---|---|
| | | | | MAT. | INST. | TOTAL |
| | | | | | | |
| | | | | | | |
| | | | | | | |

**Important: See the Reference Section for critical supporting data - Reference Nos., Crews & Location Factors**

SITE WORK 1

# Foundation Excavation Price Sheet

| | QUAN. | UNIT | LABOR HOURS | COST EACH MAT. | COST EACH INST. | COST EACH TOTAL |
|---|---|---|---|---|---|---|
| Clear & grub, medium brush, 30' from building, 24' x 38' | .190 | Acre | 2.027 | | 178.00 | 178.00 |
| 26' x 46' | .210 | Acre | 2.240 | | 197.00 | 197.00 |
| 26' x 60' | .240 | Acre | 2.560 | | 224.00 | 224.00 |
| 30' x 66' | .260 | Acre | 2.773 | | 243.00 | 243.00 |
| Light trees, to 6" dia. cut & chip, 24' x 38' | .190 | Acre | 9.120 | | 455.00 | 455.00 |
| 26' x 46' | .210 | Acre | 10.080 | | 505.00 | 505.00 |
| 26' x 60' | .240 | Acre | 11.520 | | 575.00 | 575.00 |
| 30' x 66' | .260 | Acre | 12.480 | | 625.00 | 625.00 |
| Medium trees, to 10" dia. cut & chip, 24' x 38' | .190 | Acre | 13.029 | | 650.00 | 650.00 |
| 26' x 46' | .210 | Acre | 14.400 | | 720.00 | 720.00 |
| 26' x 60' | .240 | Acre | 16.457 | | 820.00 | 820.00 |
| 30' x 66' | .260 | Acre | 17.829 | | 890.00 | 890.00 |
| Excavation, basement, 24' x 38', 2' deep | 98.000 | C.Y. | 1.400 | | 101.00 | 101.00 |
| 4' deep | 220.000 | C.Y. | 3.144 | | 226.00 | 226.00 |
| 8' deep | 550.000 | C.Y. | 7.860 | | 565.00 | 565.00 |
| 26' x 46', 2' deep | 123.000 | C.Y. | 1.758 | | 127.00 | 127.00 |
| 4' deep | 274.000 | C.Y. | 3.915 | | 282.00 | 282.00 |
| 8' deep | 672.000 | C.Y. | 9.603 | | 695.00 | 695.00 |
| 26' x 60', 2' deep | 157.000 | C.Y. | 2.244 | | 162.00 | 162.00 |
| 4' deep | 345.000 | C.Y. | 4.930 | | 355.00 | 355.00 |
| 8' deep | 829.000 | C.Y. | 11.846 | | 855.00 | 855.00 |
| 30' x 66', 2' deep | 192.000 | C.Y. | 2.744 | | 198.00 | 198.00 |
| 4' deep | 419.000 | C.Y. | 5.988 | | 430.00 | 430.00 |
| 8' deep | 990.000 | C.Y. | 14.147 | | 1025.00 | 1025.00 |
| Backfill, 24' x 38', 2" lifts, no compaction | 32.000 | C.Y. | .213 | | 32.00 | 32.00 |
| Compaction, air tamped | 32.000 | C.Y. | 2.133 | | 215.00 | 215.00 |
| 4" lifts, no compaction | 72.000 | C.Y. | .480 | | 72.00 | 72.00 |
| Compaction, air tamped | 72.000 | C.Y. | 4.800 | | 485.00 | 485.00 |
| 8" lifts, no compaction | 180.000 | C.Y. | 1.201 | | 180.00 | 180.00 |
| Compaction, air tamped | 180.000 | C.Y. | 12.001 | | 1200.00 | 1200.00 |
| 26' x 46', 2" lifts, no compaction | 40.000 | C.Y. | .267 | | 40.00 | 40.00 |
| Compaction, air tamped | 40.000 | C.Y. | 2.667 | | 268.00 | 268.00 |
| 4" lifts, no compaction | 90.000 | C.Y. | .600 | | 90.00 | 90.00 |
| Compaction, air tamped | 90.000 | C.Y. | 6.000 | | 605.00 | 605.00 |
| 8" lifts, no compaction | 220.000 | C.Y. | 1.467 | | 220.00 | 220.00 |
| Compacton, air tamped | 220.000 | C.Y. | 14.667 | | 1475.00 | 1475.00 |
| 26' x 60', 2" lifts, no compaction | 50.000 | C.Y. | .334 | | 50.00 | 50.00 |
| Compaction, air tamped | 50.000 | C.Y. | 3.334 | | 335.00 | 335.00 |
| 4" lifts, no compaction | 110.000 | C.Y. | .734 | | 110.00 | 110.00 |
| Compaction, air tamped | 110.000 | C.Y. | 7.334 | | 740.00 | 740.00 |
| 8" lifts, no compaction | 270.000 | C.Y. | 1.801 | | 270.00 | 270.00 |
| Compaction, air tamped | 270.000 | C.Y. | 18.001 | | 1800.00 | 1800.00 |
| 30' x 66', 2" lifts, no compaction | 60.000 | C.Y. | .400 | | 60.00 | 60.00 |
| Compaction, air tamped | 60.000 | C.Y. | 4.000 | | 405.00 | 405.00 |
| 4" lifts, no compaction | 130.000 | C.Y. | .867 | | 130.00 | 130.00 |
| Compaction, air tamped | 130.000 | C.Y. | 8.667 | | 870.00 | 870.00 |
| 8" lifts, no compaction | 320.000 | C.Y. | 2.134 | | 320.00 | 320.00 |
| Compaction, air tamped | 320.000 | C.Y. | 21.334 | | 2150.00 | 2150.00 |
| Rough grade, 30' from building, 24' x 38' | 280.000 | C.Y. | 1.868 | | 280.00 | 280.00 |
| 26' x 46' | 340.000 | C.Y. | 2.268 | | 340.00 | 340.00 |
| 26' x 60' | 420.000 | C.Y. | 2.801 | | 420.00 | 420.00 |
| 30' x 66' | 500.000 | C.Y. | 3.335 | | 500.00 | 500.00 |

**SITE WORK**

**1**

SITE WORK

| System Description | QUAN. | UNIT | LABOR HOURS | COST PER L.F. | | |
|---|---|---|---|---|---|---|
| | | | | MAT. | INST. | TOTAL |
| **2′ DEEP** | | | | | | |
| Excavation, backhoe | .296 | C.Y. | .032 | | 1.36 | 1.36 |
| Bedding, sand | .111 | C.Y. | .044 | 1.45 | 1.35 | 2.80 |
| Utility, sewer, 6″ cast iron | 1.000 | L.F. | .283 | 13.12 | 9.59 | 22.71 |
| Backfill, incl. compaction | .185 | C.Y. | .044 | | 1.13 | 1.13 |
| TOTAL | | | .403 | 14.57 | 13.43 | 28.00 |
| **4′ DEEP** | | | | | | |
| Excavation, backhoe | .889 | C.Y. | .095 | | 4.08 | 4.08 |
| Bedding, sand | .111 | C.Y. | .044 | 1.45 | 1.35 | 2.80 |
| Utility, sewer, 6″ cast iron | 1.000 | L.F. | .283 | 13.12 | 9.59 | 22.71 |
| Backfill, incl. compaction | .778 | C.Y. | .183 | | 4.75 | 4.75 |
| TOTAL | | | .605 | 14.57 | 19.77 | 34.34 |
| **6′ DEEP** | | | | | | |
| Excavation, backhoe | 1.770 | C.Y. | .189 | | 8.13 | 8.13 |
| Bedding, sand | .111 | C.Y. | .044 | 1.45 | 1.35 | 2.80 |
| Utility, sewer, 6″ cast iron | 1.000 | L.F. | .283 | 13.12 | 9.59 | 22.71 |
| Backfill, incl. compaction | 1.660 | C.Y. | .391 | | 10.13 | 10.13 |
| TOTAL | | | .907 | 14.57 | 29.20 | 43.77 |
| **8′ DEEP** | | | | | | |
| Excavation, backhoe | 2.960 | C.Y. | .316 | | 13.59 | 13.59 |
| Bedding, sand | .111 | C.Y. | .044 | 1.45 | 1.35 | 2.80 |
| Utility, sewer, 6″ cast iron | 1.000 | L.F. | .283 | 13.12 | 9.59 | 22.71 |
| Backfill, incl. compaction | 2.850 | C.Y. | .671 | | 17.39 | 17.39 |
| TOTAL | | | 1.314 | 14.57 | 41.92 | 56.49 |

The costs in this system are based on a cost per linear foot of trench,
and based on 2′ wide at bottom of trench up to 6′ deep.

| Description | QUAN. | UNIT | LABOR HOURS | COST PER L.F. | | |
|---|---|---|---|---|---|---|
| | | | | MAT. | INST. | TOTAL |
| | | | | | | |
| | | | | | | |
| | | | | | | |

| Utility Trenching Price Sheet | QUAN. | UNIT | LABOR HOURS | COST PER L.F. | | |
|---|---|---|---|---|---|---|
| | | | | MAT. | INST. | TOTAL |
| Excavation, bottom of trench 2' wide, 2' deep | .296 | C.Y. | .032 | | 1.36 | 1.36 |
| 4' deep | .889 | C.Y. | .095 | | 4.08 | 4.08 |
| 6' deep | 1.770 | C.Y. | .142 | | 6.45 | 6.45 |
| 8' deep | 2.960 | C.Y. | .105 | | 10.50 | 10.50 |
| Bedding, sand, bottom of trench 2' wide, no compaction, pipe, 2" diameter | .070 | C.Y. | .028 | .91 | .86 | 1.77 |
| 4" diameter | .084 | C.Y. | .034 | 1.10 | 1.02 | 2.12 |
| 6" diameter | .105 | C.Y. | .042 | 1.37 | 1.28 | 2.65 |
| 8" diameter | .122 | C.Y. | .049 | 1.59 | 1.48 | 3.07 |
| Compacted, pipe, 2" diameter | .074 | C.Y. | .030 | .97 | .90 | 1.87 |
| 4" diameter | .092 | C.Y. | .037 | 1.20 | 1.12 | 2.32 |
| 6" diameter | .111 | C.Y. | .044 | 1.45 | 1.35 | 2.80 |
| 8" diameter | .129 | C.Y. | .052 | 1.68 | 1.57 | 3.25 |
| 3/4" stone, bottom of trench 2' wide, pipe, 4" diameter | .082 | C.Y. | .033 | 1.07 | 1.00 | 2.07 |
| 6" diameter | .099 | C.Y. | .040 | 1.29 | 1.21 | 2.50 |
| 3/8" stone, bottom of trench 2' wide, pipe, 4" diameter | .084 | C.Y. | .034 | 1.10 | 1.02 | 2.12 |
| 6" diameter | .102 | C.Y. | .041 | 1.33 | 1.24 | 2.57 |
| Utilities, drainage & sewerage, corrugated plastic, 6" diameter | 1.000 | L.F. | .069 | 2.73 | 1.85 | 4.58 |
| 8" diameter | 1.000 | L.F. | .072 | 4.84 | 1.93 | 6.77 |
| Bituminous fiber, 4" diameter | 1.000 | L.F. | .064 | 1.57 | 1.73 | 3.30 |
| 6" diameter | 1.000 | L.F. | .069 | 2.73 | 1.85 | 4.58 |
| 8" diameter | 1.000 | L.F. | .072 | 4.84 | 1.93 | 6.77 |
| Concrete, non-reinforced, 6" diameter | 1.000 | L.F. | .181 | 3.85 | 5.80 | 9.65 |
| 8" diameter | 1.000 | L.F. | .214 | 4.24 | 6.85 | 11.09 |
| PVC, SDR 35, 4" diameter | 1.000 | L.F. | .064 | 1.57 | 1.73 | 3.30 |
| 6" diameter | 1.000 | L.F. | .069 | 2.73 | 1.85 | 4.58 |
| 8" diameter | 1.000 | L.F. | .072 | 4.84 | 1.93 | 6.77 |
| Vitrified clay, 4" diameter | 1.000 | L.F. | .091 | 1.83 | 2.44 | 4.27 |
| 6" diameter | 1.000 | L.F. | .120 | 3.06 | 3.24 | 6.30 |
| 8" diameter | 1.000 | L.F. | .140 | 4.36 | 4.84 | 9.20 |
| Gas & service, polyethylene, 1-1/4" diameter | 1.000 | L.F. | .059 | .76 | 1.85 | 2.61 |
| Steel sched.40, 1" diameter | 1.000 | L.F. | .107 | 2.33 | 4.26 | 6.59 |
| 2" diameter | 1.000 | L.F. | .114 | 3.66 | 4.56 | 8.22 |
| Sub-drainage, PVC, perforated, 3" diameter | 1.000 | L.F. | .064 | 1.57 | 1.73 | 3.30 |
| 4" diameter | 1.000 | L.F. | .064 | 1.57 | 1.73 | 3.30 |
| 5" diameter | 1.000 | L.F. | .069 | 2.73 | 1.85 | 4.58 |
| 6" diameter | 1.000 | L.F. | .069 | 2.73 | 1.85 | 4.58 |
| Porous wall concrete, 4" diameter | 1.000 | L.F. | .072 | 2.02 | 1.93 | 3.95 |
| Vitrified clay, perforated, 4" diameter | 1.000 | L.F. | .120 | 2.20 | 3.82 | 6.02 |
| 6" diameter | 1.000 | L.F. | .152 | 3.64 | 4.85 | 8.49 |
| Water service, copper, type K, 3/4" | 1.000 | L.F. | .083 | 2.23 | 3.26 | 5.49 |
| 1" diameter | 1.000 | L.F. | .093 | 2.76 | 3.66 | 6.42 |
| PVC, 3/4" | 1.000 | L.F. | .121 | 1.90 | 4.74 | 6.64 |
| 1" diameter | 1.000 | L.F. | .134 | 2.06 | 5.25 | 7.31 |
| Backfill, bottom of trench 2' wide no compact, 2' deep, pipe, 2" diameter | .226 | L.F. | .053 | | 1.38 | 1.38 |
| 4" diameter | .212 | L.F. | .050 | | 1.29 | 1.29 |
| 6" diameter | .185 | L.F. | .044 | | 1.13 | 1.13 |
| 4' deep, pipe, 2" diameter | .819 | C.Y. | .193 | | 5.00 | 5.00 |
| 4" diameter | .805 | C.Y. | .189 | | 4.91 | 4.91 |
| 6" diameter | .778 | C.Y. | .183 | | 4.75 | 4.75 |
| 6' deep, pipe, 2" diameter | 1.700 | C.Y. | .400 | | 10.35 | 10.35 |
| 4" diameter | 1.690 | C.Y. | .398 | | 10.30 | 10.30 |
| 6" diameter | 1.660 | C.Y. | .391 | | 10.15 | 10.15 |
| 8' deep, pipe, 2" diameter | 2.890 | C.Y. | .680 | | 17.65 | 17.65 |
| 4" diameter | 2.870 | C.Y. | .675 | | 17.50 | 17.50 |
| 6" diameter | 2.850 | C.Y. | .671 | | 17.40 | 17.40 |

**SITE WORK**

**1**

89

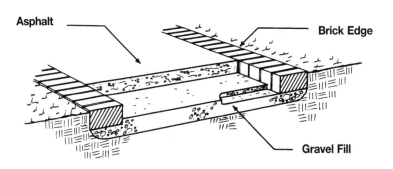

Asphalt — Brick Edge — Gravel Fill

**SITE WORK 1**

| System Description | QUAN. | UNIT | LABOR HOURS | COST PER S.F. | | |
|---|---|---|---|---|---|---|
| | | | | MAT. | INST. | TOTAL |
| **ASPHALT SIDEWALK SYSTEM, 3′ WIDE WALK** | | | | | | |
| Gravel fill, 4″ deep | 1.000 | S.F. | .001 | .59 | .03 | .62 |
| Compact fill | .012 | C.Y. | .001 | | .01 | .01 |
| Handgrade | 1.000 | S.F. | .004 | | .11 | .11 |
| Walking surface, bituminous paving, 2″ thick | 1.000 | S.F. | .007 | .45 | .23 | .68 |
| Edging, brick, laid on edge | .670 | L.F. | .079 | 1.33 | 2.51 | 3.84 |
| | | | | | | |
| TOTAL | | | .092 | 2.37 | 2.89 | 5.26 |
| **CONCRETE SIDEWALK SYSTEM, 3′ WIDE WALK** | | | | | | |
| Gravel fill, 4″ deep | 1.000 | S.F. | .001 | .59 | .03 | .62 |
| Compact fill | .012 | C.Y. | .001 | | .01 | .01 |
| Handgrade | 1.000 | S.F. | .004 | | .11 | .11 |
| Walking surface, concrete, 4″ thick | 1.000 | S.F. | .040 | 1.27 | 1.25 | 2.52 |
| Edging, brick, laid on edge | .670 | L.F. | .079 | 1.33 | 2.51 | 3.84 |
| | | | | | | |
| TOTAL | | | .125 | 3.19 | 3.91 | 7.10 |
| **PAVERS, BRICK SIDEWALK SYSTEM, 3′ WIDE WALK** | | | | | | |
| Sand base fill, 4″ deep | 1.000 | S.F. | .001 | .16 | .06 | .22 |
| Compact fill | .012 | C.Y. | .001 | | .01 | .01 |
| Handgrade | 1.000 | S.F. | .004 | | .11 | .11 |
| Walking surface, brick pavers | 1.000 | S.F. | .160 | 2.36 | 5.05 | 7.41 |
| Edging, redwood, untreated, 1″ x 4″ | .670 | L.F. | .032 | 1.95 | 1.15 | 3.10 |
| | | | | | | |
| TOTAL | | | .198 | 4.47 | 6.38 | 10.85 |

The costs in this system are based on a cost per square foot of sidewalk area. Concrete used is 3000 p.s.i.

| Description | QUAN. | UNIT | LABOR HOURS | COST PER S.F. | | |
|---|---|---|---|---|---|---|
| | | | | MAT. | INST. | TOTAL |
| | | | | | | |
| | | | | | | |
| | | | | | | |
| | | | | | | |
| | | | | | | |

**Important: See the Reference Section for critical supporting data - Reference Nos., Crews & Location Factors**

| Sidewalk Price Sheet | QUAN. | UNIT | LABOR HOURS | COST PER S.F. MAT. | COST PER S.F. INST. | COST PER S.F. TOTAL |
|---|---|---|---|---|---|---|
| Base, crushed stone, 3" deep | 1.000 | S.F. | .001 | .43 | .06 | .49 |
| 6" deep | 1.000 | S.F. | .001 | .87 | .06 | .93 |
| 9" deep | 1.000 | S.F. | .002 | 1.30 | .09 | 1.39 |
| 12" deep | 1.000 | S.F. | .002 | 1.78 | .11 | 1.89 |
| Bank run gravel, 6" deep | 1.000 | S.F. | .001 | .88 | .05 | .93 |
| 9" deep | 1.000 | S.F. | .001 | 1.32 | .07 | 1.39 |
| 12" deep | 1.000 | S.F. | .001 | 1.76 | .08 | 1.84 |
| Compact base, 3" deep | .009 | C.Y. | .001 | | .01 | .01 |
| 6" deep | .019 | C.Y. | .001 | | .03 | .03 |
| 9" deep | .028 | C.Y. | .001 | | .04 | .04 |
| Handgrade | 1.000 | S.F. | .004 | | .11 | .11 |
| Surface, brick, pavers dry joints, laid flat, running bond | 1.000 | S.F. | .160 | 2.36 | 5.05 | 7.41 |
| Basket weave | 1.000 | S.F. | .168 | 3.04 | 5.35 | 8.39 |
| Herringbone | 1.000 | S.F. | .174 | 3.04 | 5.50 | 8.54 |
| Laid on edge, running bond | 1.000 | S.F. | .229 | 2.35 | 7.25 | 9.60 |
| Mortar jts. laid flat, running bond | 1.000 | S.F. | .192 | 2.83 | 6.05 | 8.88 |
| Basket weave | 1.000 | S.F. | .202 | 3.65 | 6.40 | 10.05 |
| Herringbone | 1.000 | S.F. | .209 | 3.65 | 6.60 | 10.25 |
| Laid on edge, running bond | 1.000 | S.F. | .274 | 2.82 | 8.70 | 11.52 |
| Bituminous paving, 1-1/2" thick | 1.000 | S.F. | .006 | .34 | .16 | .50 |
| 2" thick | 1.000 | S.F. | .007 | .45 | .23 | .68 |
| 2-1/2" thick | 1.000 | S.F. | .008 | .57 | .24 | .81 |
| Sand finish, 3/4" thick | 1.000 | S.F. | .001 | .20 | .07 | .27 |
| 1" thick | 1.000 | S.F. | .001 | .24 | .09 | .33 |
| Concrete, reinforced, broom finish, 4" thick | 1.000 | S.F. | .040 | 1.27 | 1.25 | 2.52 |
| 5" thick | 1.000 | S.F. | .044 | 1.69 | 1.38 | 3.07 |
| 6" thick | 1.000 | S.F. | .047 | 1.97 | 1.48 | 3.45 |
| Crushed stone, white marble, 3" thick | 1.000 | S.F. | .009 | .24 | .24 | .48 |
| Bluestone, 3" thick | 1.000 | S.F. | .009 | .22 | .24 | .46 |
| Flagging, bluestone, 1" | 1.000 | S.F. | .198 | 4.62 | 6.25 | 10.87 |
| 1-1/2" | 1.000 | S.F. | .188 | 8.40 | 5.95 | 14.35 |
| Slate, natural cleft, 3/4" | 1.000 | S.F. | .174 | 4.96 | 5.50 | 10.46 |
| Random rect., 1/2" | 1.000 | S.F. | .152 | 10.75 | 4.82 | 15.57 |
| Granite blocks | 1.000 | S.F. | .174 | 6.80 | 5.50 | 12.30 |
| Edging, corrugated aluminum, 4", 3' wide walk | .666 | L.F. | .008 | .22 | .29 | .51 |
| 4' wide walk | .500 | L.F. | .006 | .17 | .22 | .39 |
| 6", 3' wide walk | .666 | L.F. | .010 | .27 | .35 | .62 |
| 4' wide walk | .500 | L.F. | .007 | .21 | .26 | .47 |
| Redwood-cedar-cypress, 1" x 4", 3' wide walk | .666 | L.F. | .021 | .98 | .76 | 1.74 |
| 4' wide walk | .500 | L.F. | .016 | .74 | .57 | 1.31 |
| 2" x 4", 3' wide walk | .666 | L.F. | .032 | 1.95 | 1.15 | 3.10 |
| 4' wide walk | .500 | L.F. | .024 | 1.47 | .87 | 2.34 |
| Brick, dry joints, 3' wide walk | .666 | L.F. | .079 | 1.33 | 2.51 | 3.84 |
| 4' wide walk | .500 | L.F. | .059 | .99 | 1.88 | 2.87 |
| Mortar joints, 3' wide walk | .666 | L.F. | .095 | 1.59 | 3.02 | 4.61 |
| 4' wide walk | .500 | L.F. | .071 | 1.19 | 2.25 | 3.44 |
| | | | | | | |
| | | | | | | |
| | | | | | | |
| | | | | | | |
| | | | | | | |

SITE WORK

1

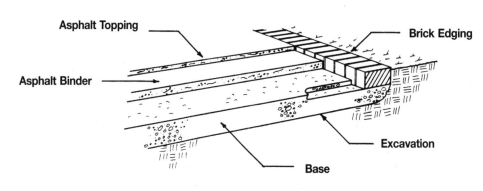

Asphalt Topping — Asphalt Binder — Brick Edging — Excavation — Base

**SITE WORK 1**

| System Description | QUAN. | UNIT | LABOR HOURS | COST PER S.F. | | |
|---|---|---|---|---|---|---|
| | | | | MAT. | INST. | TOTAL |
| **ASPHALT DRIVEWAY TO 10' WIDE** | | | | | | |
| Excavation, driveway to 10' wide, 6" deep | .019 | C.Y. | .001 | | .03 | .03 |
| Base, 6" crushed stone | 1.000 | S.F. | .001 | .87 | .06 | .93 |
| Handgrade base | 1.000 | S.F. | .004 | | .11 | .11 |
| 2" thick base | 1.000 | S.F. | .002 | .45 | .12 | .57 |
| 1" topping | 1.000 | S.F. | .001 | .24 | .09 | .33 |
| Edging, brick pavers | .200 | L.F. | .024 | .40 | .75 | 1.15 |
| | | | | | | |
| TOTAL | | | .033 | 1.96 | 1.16 | 3.12 |
| **CONCRETE DRIVEWAY TO 10' WIDE** | | | | | | |
| Excavation, driveway to 10' wide, 6" deep | .019 | C.Y. | .001 | | .03 | .03 |
| Base, 6" crushed stone | 1.000 | S.F. | .001 | .87 | .06 | .93 |
| Handgrade base | 1.000 | S.F. | .004 | | .11 | .11 |
| Surface, concrete, 4" thick | 1.000 | S.F. | .040 | 1.27 | 1.25 | 2.52 |
| Edging, brick pavers | .200 | L.F. | .024 | .40 | .75 | 1.15 |
| | | | | | | |
| TOTAL | | | .070 | 2.54 | 2.20 | 4.74 |
| **PAVERS, BRICK DRIVEWAY TO 10' WIDE** | | | | | | |
| Excavation, driveway to 10' wide, 6" deep | .019 | C.Y. | .001 | | .03 | .03 |
| Base, 6" sand | 1.000 | S.F. | .001 | .25 | .09 | .34 |
| Handgrade base | 1.000 | S.F. | .004 | | .11 | .11 |
| Surface, pavers, brick laid flat, running bond | 1.000 | S.F. | .160 | 2.36 | 5.05 | 7.41 |
| Edging, redwood, untreated, 2" x 4" | .200 | L.F. | .010 | .59 | .35 | .94 |
| | | | | | | |
| TOTAL | | | .176 | 3.20 | 5.63 | 8.83 |

| Description | QUAN. | UNIT | LABOR HOURS | COST PER S.F. | | |
|---|---|---|---|---|---|---|
| | | | | MAT. | INST. | TOTAL |
| | | | | | | |
| | | | | | | |
| | | | | | | |
| | | | | | | |

| Driveway Price Sheet | QUAN. | UNIT | LABOR HOURS | COST PER S.F. | | |
|---|---|---|---|---|---|---|
| | | | | MAT. | INST. | TOTAL |
| Excavation, by machine, 10' wide, 6" deep | .019 | C.Y. | .001 | | .03 | .03 |
| 12" deep | .037 | C.Y. | .001 | | .05 | .05 |
| 18" deep | .055 | C.Y. | .001 | | .08 | .08 |
| 20' wide, 6" deep | .019 | C.Y. | .001 | | .03 | .03 |
| 12" deep | .037 | C.Y. | .001 | | .05 | .05 |
| 18" deep | .055 | C.Y. | .001 | | .08 | .08 |
| Base, crushed stone, 10' wide, 3" deep | 1.000 | S.F. | .001 | .44 | .04 | .48 |
| 6" deep | 1.000 | S.F. | .001 | .87 | .06 | .93 |
| 9" deep | 1.000 | S.F. | .002 | 1.30 | .09 | 1.39 |
| 20' wide, 3" deep | 1.000 | S.F. | .001 | .44 | .04 | .48 |
| 6" deep | 1.000 | S.F. | .001 | .87 | .06 | .93 |
| 9" deep | 1.000 | S.F. | .002 | 1.30 | .09 | 1.39 |
| Bank run gravel, 10' wide, 3" deep | 1.000 | S.F. | .001 | .44 | .03 | .47 |
| 6" deep | 1.000 | S.F. | .001 | .88 | .05 | .93 |
| 9" deep | 1.000 | S.F. | .001 | 1.32 | .07 | 1.39 |
| 20' wide, 3" deep | 1.000 | S.F. | .001 | .44 | .03 | .47 |
| 6" deep | 1.000 | S.F. | .001 | .88 | .05 | .93 |
| 9" deep | 1.000 | S.F. | .001 | 1.32 | .07 | 1.39 |
| Handgrade, 10' wide | 1.000 | S.F. | .004 | | .11 | .11 |
| 20' wide | 1.000 | S.F. | .004 | | .11 | .11 |
| Surface, asphalt, 10' wide, 3/4" topping, 1" base | 1.000 | S.F. | .002 | .55 | .16 | .71 |
| 2" base | 1.000 | S.F. | .003 | .65 | .19 | .84 |
| 1" topping, 1" base | 1.000 | S.F. | .002 | .59 | .18 | .77 |
| 2" base | 1.000 | S.F. | .003 | .69 | .21 | .90 |
| 20' wide, 3/4" topping, 1" base | 1.000 | S.F. | .002 | .55 | .16 | .71 |
| 2" base | 1.000 | S.F. | .003 | .65 | .19 | .84 |
| 1" topping, 1" base | 1.000 | S.F. | .002 | .59 | .18 | .77 |
| 2" base | 1.000 | S.F. | .003 | .69 | .21 | .90 |
| Concrete, 10' wide, 4" thick | 1.000 | S.F. | .040 | 1.27 | 1.25 | 2.52 |
| 6" thick | 1.000 | S.F. | .047 | 1.97 | 1.48 | 3.45 |
| 20' wide, 4" thick | 1.000 | S.F. | .040 | 1.27 | 1.25 | 2.52 |
| 6" thick | 1.000 | S.F. | .047 | 1.97 | 1.48 | 3.45 |
| Paver, brick 10' wide dry joints, running bond, laid flat | 1.000 | S.F. | .160 | 2.36 | 5.05 | 7.41 |
| Laid on edge | 1.000 | S.F. | .229 | 2.35 | 7.25 | 9.60 |
| Mortar joints, laid flat | 1.000 | S.F. | .192 | 2.83 | 6.05 | 8.88 |
| Laid on edge | 1.000 | S.F. | .274 | 2.82 | 8.70 | 11.52 |
| 20' wide, running bond, dry jts., laid flat | 1.000 | S.F. | .160 | 2.36 | 5.05 | 7.41 |
| Laid on edge | 1.000 | S.F. | .229 | 2.35 | 7.25 | 9.60 |
| Mortar joints, laid flat | 1.000 | S.F. | .192 | 2.83 | 6.05 | 8.88 |
| Laid on edge | 1.000 | S.F. | .274 | 2.82 | 8.70 | 11.52 |
| Crushed stone, 10' wide, white marble, 3" | 1.000 | S.F. | .009 | .24 | .24 | .48 |
| Bluestone, 3" | 1.000 | S.F. | .009 | .22 | .24 | .46 |
| 20' wide, white marble, 3" | 1.000 | S.F. | .009 | .24 | .24 | .48 |
| Bluestone, 3" | 1.000 | S.F. | .009 | .22 | .24 | .46 |
| Soil cement, 10' wide | 1.000 | S.F. | .007 | .22 | .63 | .85 |
| 20' wide | 1.000 | S.F. | .007 | .22 | .63 | .85 |
| Granite blocks, 10' wide | 1.000 | S.F. | .174 | 6.80 | 5.50 | 12.30 |
| 20' wide | 1.000 | S.F. | .174 | 6.80 | 5.50 | 12.30 |
| Asphalt block, solid 1-1/4" thick | 1.000 | S.F. | .119 | 4.21 | 3.75 | 7.96 |
| Solid 3" thick | 1.000 | S.F. | .123 | 5.90 | 3.90 | 9.80 |
| Edging, brick, 10' wide | .200 | L.F. | .024 | .40 | .75 | 1.15 |
| 20' wide | .100 | L.F. | .012 | .20 | .38 | .58 |
| Redwood, untreated 2" x 4", 10' wide | .200 | L.F. | .010 | .59 | .35 | .94 |
| 20' wide | .100 | L.F. | .005 | .29 | .17 | .46 |
| Granite, 4 1/2" x 12" straight, 10' wide | .200 | L.F. | .032 | 1.02 | 1.44 | 2.46 |
| 20' wide | .100 | L.F. | .016 | .51 | .72 | 1.23 |
| Finishes, asphalt sealer, 10' wide | 1.000 | S.F. | .023 | .50 | .62 | 1.12 |
| 20' wide | 1.000 | S.F. | .023 | .50 | .62 | 1.12 |
| Concrete, exposed aggregate 10' wide | 1.000 | S.F. | .013 | 4.22 | .42 | 4.64 |
| 20' wide | 1.000 | S.F. | .013 | 4.22 | .42 | 4.64 |

**SITE WORK**

**1**

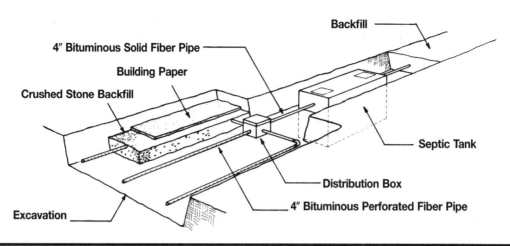

SITE WORK

| System Description | QUAN. | UNIT | LABOR HOURS | COST EACH | | |
|---|---|---|---|---|---|---|
| | | | | MAT. | INST. | TOTAL |
| **SEPTIC SYSTEM WITH 1000 S.F. LEACHING FIELD, 1000 GALLON TANK** | | | | | | |
| Tank, 1000 gallon, concrete | 1.000 | Ea. | 3.500 | 570.00 | 121.00 | 691.00 |
| Distribution box, concrete | 1.000 | Ea. | 1.000 | 104.00 | 26.00 | 130.00 |
| 4" PVC pipe | 25.000 | L.F. | 1.600 | 39.25 | 43.25 | 82.50 |
| Tank and field excavation | 119.000 | C.Y. | 6.565 | | 654.50 | 654.50 |
| Crushed stone backfill | 76.000 | C.Y. | 12.160 | 1710.00 | 446.88 | 2156.88 |
| Backfill with excavated material | 36.000 | C.Y. | .240 | | 36.00 | 36.00 |
| Building paper | 125.000 | S.Y. | 2.430 | 22.50 | 90.00 | 112.50 |
| 4" PVC perforated pipe | 145.000 | L.F. | 9.280 | 227.65 | 250.85 | 478.50 |
| 4" pipe fittings | 2.000 | Ea. | 1.939 | 14.00 | 68.00 | 82.00 |
| | | | | | | |
| TOTAL | | | 38.714 | 2687.40 | 1736.48 | 4423.88 |
| **SEPTIC SYSTEM WITH 2 LEACHING PITS, 1000 GALLON TANK** | | | | | | |
| Tank, 1000 gallon, concrete | 1.000 | Ea. | 3.500 | 570.00 | 121.00 | 691.00 |
| Distribution box, concrete | 1.000 | Ea. | 1.000 | 104.00 | 26.00 | 130.00 |
| 4" PVC pipe | 75.000 | L.F. | 4.800 | 117.75 | 129.75 | 247.50 |
| Excavation for tank only | 20.000 | C.Y. | 1.103 | | 110.00 | 110.00 |
| Crushed stone backfill | 10.000 | C.Y. | 1.600 | 225.00 | 58.80 | 283.80 |
| Backfill with excavated material | 55.000 | C.Y. | .367 | | 55.00 | 55.00 |
| Pits, 6' diameter, including excavation and stone backfill | 2.000 | Ea. | | 1200.00 | | 1200.00 |
| | | | | | | |
| TOTAL | | | 12.370 | 2216.75 | 500.55 | 2717.30 |

The costs in this system include all necessary piping and excavation.

| Description | QUAN. | UNIT | LABOR HOURS | COST EACH | | |
|---|---|---|---|---|---|---|
| | | | | MAT. | INST. | TOTAL |
| | | | | | | |
| | | | | | | |
| | | | | | | |
| | | | | | | |
| | | | | | | |

**Important: See the Reference Section for critical supporting data - Reference Nos., Crews & Location Factors**

| Septic Systems Price Sheet | QUAN. | UNIT | LABOR HOURS | MAT. | INST. | TOTAL |
|---|---|---|---|---|---|---|
| | | | | COST EACH | | |
| Tank, precast concrete, 1000 gallon | 1.000 | Ea. | 3.500 | 570.00 | 121.00 | 691.00 |
| 2000 gallon | 1.000 | Ea. | 5.600 | 1125.00 | 194.00 | 1319.00 |
| Distribution box, concrete, 5 outlets | 1.000 | Ea. | 1.000 | 104.00 | 26.00 | 130.00 |
| 12 outlets | 1.000 | Ea. | 2.000 | 270.00 | 51.50 | 321.50 |
| 4" pipe, PVC, solid | 25.000 | L.F. | 1.600 | 39.50 | 43.50 | 83.00 |
| | | | | | | |
| Tank and field excavation, 1000 S.F. field | 119.000 | C.Y. | 6.565 | | 655.00 | 655.00 |
| 2000 S.F. field | 190.000 | C.Y. | 10.482 | | 1050.00 | 1050.00 |
| Tank excavation only, 1000 gallon tank | 20.000 | C.Y. | 1.103 | | 110.00 | 110.00 |
| 2000 gallon tank | 32.000 | C.Y. | 1.765 | | 176.00 | 176.00 |
| Backfill, crushed stone 1000 S.F. field | 76.000 | C.Y. | 12.160 | 1700.00 | 445.00 | 2145.00 |
| 2000 S.F. field | 140.000 | C.Y. | 22.400 | 3150.00 | 825.00 | 3975.00 |
| Backfill with excavated material, 1000 S.F. field | 36.000 | C.Y. | .240 | | 36.00 | 36.00 |
| 2000 S.F. field | 60.000 | C.Y. | .400 | | 60.00 | 60.00 |
| 6' diameter pits | 55.000 | C.Y. | .367 | | 55.00 | 55.00 |
| 3' diameter pits | 42.000 | C.Y. | .280 | | 42.00 | 42.00 |
| Building paper, 1000 S.F. field | 125.000 | S.Y. | 2.376 | 22.00 | 88.00 | 110.00 |
| 2000 S.F. field | 250.000 | S.Y. | 4.860 | 45.00 | 180.00 | 225.00 |
| | | | | | | |
| 4" pipe, PVC, perforated, 1000 S.F. field | 145.000 | L.F. | 9.280 | 228.00 | 251.00 | 479.00 |
| 2000 S.F. field | 265.000 | L.F. | 16.960 | 415.00 | 460.00 | 875.00 |
| Pipe fittings, bituminous fiber, 1000 S.F. field | 2.000 | Ea. | 1.939 | 14.00 | 68.00 | 82.00 |
| 2000 S.F. field | 4.000 | Ea. | 3.879 | 28.00 | 136.00 | 164.00 |
| Leaching pit, including excavation and stone backfill, 3' diameter | 1.000 | Ea. | | 450.00 | | 450.00 |
| 6' diameter | 1.000 | Ea. | | 600.00 | | 600.00 |

| System Description | QUAN. | UNIT | LABOR HOURS | COST PER UNIT | | |
|---|---|---|---|---|---|---|
| | | | | MAT. | INST. | TOTAL |
| Chain link fence | | | | | | |
|    Galv.9ga. wire, 1-5/8"post 10'O.C., 1-3/8"top rail, 2"corner post, 3'hi | 1.000 | L.F. | .130 | 4.59 | 3.50 | 8.09 |
|    4' high | 1.000 | L.F. | .141 | 7.00 | 3.81 | 10.81 |
|    6' high | 1.000 | L.F. | .209 | 7.90 | 5.65 | 13.55 |
|    Add for gate 3' wide 1-3/8" frame 3' high | 1.000 | Ea. | 2.000 | 41.00 | 54.00 | 95.00 |
|    4' high | 1.000 | Ea. | 2.400 | 51.00 | 65.00 | 116.00 |
|    6' high | 1.000 | Ea. | 2.400 | 92.00 | 65.00 | 157.00 |
|    Add for gate 4' wide 1-3/8" frame 3' high | 1.000 | Ea. | 2.667 | 48.00 | 72.00 | 120.00 |
|    4' high | 1.000 | Ea. | 2.667 | 63.00 | 72.00 | 135.00 |
|    6' high | 1.000 | Ea. | 3.000 | 116.00 | 81.00 | 197.00 |
|    Alum.9ga. wire, 1-5/8"post, 10'O.C., 1-3/8"top rail, 2"corner post,3'hi | 1.000 | L.F. | .130 | 5.50 | 3.50 | 9.00 |
|    4' high | 1.000 | L.F. | .141 | 6.30 | 3.81 | 10.11 |
|    6' high | 1.000 | L.F. | .209 | 8.10 | 5.65 | 13.75 |
|    Add for gate 3' wide 1-3/8" frame 3' high | 1.000 | Ea. | 2.000 | 54.00 | 54.00 | 108.00 |
|    4' high | 1.000 | Ea. | 2.400 | 73.50 | 65.00 | 138.50 |
|    6' high | 1.000 | Ea. | 2.400 | 110.00 | 65.00 | 175.00 |
|    Add for gate 4' wide 1-3/8" frame 3' high | 1.000 | Ea. | 2.400 | 73.50 | 65.00 | 138.50 |
|    4' high | 1.000 | Ea. | 2.667 | 98.00 | 72.00 | 170.00 |
|    6' high | 1.000 | Ea. | 3.000 | 153.00 | 81.00 | 234.00 |
|    Vinyl 9ga. wire, 1-5/8"post 10'O.C., 1-3/8"top rail, 2"corner post,3'hi | 1.000 | L.F. | .130 | 4.90 | 3.50 | 8.40 |
|    4' high | 1.000 | L.F. | .141 | 8.20 | 3.81 | 12.01 |
|    6' high | 1.000 | L.F. | .209 | 9.40 | 5.65 | 15.05 |
|    Add for gate 3' wide 1-3/8" frame 3' high | 1.000 | Ea. | 2.000 | 61.50 | 54.00 | 115.50 |
|    4' high | 1.000 | Ea. | 2.400 | 79.50 | 65.00 | 144.50 |
|    6' high | 1.000 | Ea. | 2.400 | 123.00 | 65.00 | 188.00 |
|    Add for gate 4' wide 1-3/8" frame 3' high | 1.000 | Ea. | 2.400 | 83.50 | 65.00 | 148.50 |
|    4' high | 1.000 | Ea. | 2.667 | 110.00 | 72.00 | 182.00 |
|    6' high | 1.000 | Ea. | 3.000 | 159.00 | 81.00 | 240.00 |
| Tennis court, chain link fence, 10' high | | | | | | |
|    Galv.11ga.wire, 2"post 10'O.C., 1-3/8"top rail, 2-1/2"corner post | 1.000 | L.F. | .253 | 12.25 | 6.80 | 19.05 |
|    Add for gate 3' wide 1-3/8" frame | 1.000 | Ea. | 2.400 | 153.00 | 65.00 | 218.00 |
|    Alum.11ga.wire, 2"post 10'O.C., 1-3/8"top rail, 2-1/2"corner post | 1.000 | L.F. | .253 | 17.15 | 6.80 | 23.95 |
|    Add for gate 3' wide 1-3/8" frame | 1.000 | Ea. | 2.400 | 196.00 | 65.00 | 261.00 |
|    Vinyl 11ga.wire,2"post 10' O.C.,1-3/8"top rail,2-1/2"corner post | 1.000 | L.F. | .253 | 14.70 | 6.80 | 21.50 |
|    Add for gate 3' wide 1-3/8" frame | 1.000 | Ea. | 2.400 | 221.00 | 65.00 | 286.00 |
| Railings, commercial | | | | | | |
|    Aluminum balcony rail, 1-1/2" posts with pickets | 1.000 | L.F. | .164 | 42.50 | 8.00 | 50.50 |
|    With expanded metal panels | 1.000 | L.F. | .164 | 54.00 | 8.00 | 62.00 |
|    With porcelain enamel panel inserts | 1.000 | L.F. | .164 | 46.00 | 8.00 | 54.00 |
|    Mild steel, ornamental rounded top rail | 1.000 | L.F. | .164 | 40.50 | 8.00 | 48.50 |
|    As above, but pitch down stairs | 1.000 | L.F. | .183 | 44.00 | 8.90 | 52.90 |
|    Steel pipe, welded, 1-1/2" round, painted | 1.000 | L.F. | .160 | 11.90 | 7.80 | 19.70 |
|    Galvanized | 1.000 | L.F. | .160 | 16.80 | 7.80 | 24.60 |
|    Residential, stock units, mild steel, deluxe | 1.000 | L.F. | .102 | 9.25 | 4.94 | 14.19 |
|    Economy | 1.000 | L.F. | .102 | 6.95 | 4.94 | 11.89 |

SITE WORK 1

| System Description | QUAN. | UNIT | LABOR HOURS | COST PER UNIT | | |
|---|---|---|---|---|---|---|
| | | | | MAT. | INST. | TOTAL |
| Basketweave, 3/8"x4" boards, 2"x4" stringers on spreaders, 4"x4" posts | | | | | | |
| No. 1 cedar, 6' high | 1.000 | L.F. | .150 | 8.10 | 4.05 | 12.15 |
| Treated pine, 6' high | 1.000 | L.F. | .160 | 9.85 | 4.32 | 14.17 |
| | | | | | | |
| Board fence, 1"x4" boards, 2"x4" rails, 4"x4" posts | | | | | | |
| Preservative treated, 2 rail, 3' high | 1.000 | L.F. | .166 | 6.00 | 4.47 | 10.47 |
| 4' high | 1.000 | L.F. | .178 | 6.60 | 4.80 | 11.40 |
| 3 rail, 5' high | 1.000 | L.F. | .185 | 7.45 | 4.98 | 12.43 |
| 6' high | 1.000 | L.F. | .192 | 8.55 | 5.20 | 13.75 |
| Western cedar, No. 1, 2 rail, 3' high | 1.000 | L.F. | .166 | 6.55 | 4.47 | 11.02 |
| 3 rail, 4' high | 1.000 | L.F. | .178 | 7.75 | 4.80 | 12.55 |
| 5' high | 1.000 | L.F. | .185 | 8.95 | 4.98 | 13.93 |
| 6' high | 1.000 | L.F. | .192 | 9.80 | 5.20 | 15.00 |
| No. 1 cedar, 2 rail, 3' high | 1.000 | L.F. | .166 | 9.85 | 4.47 | 14.32 |
| 4' high | 1.000 | L.F. | .178 | 11.20 | 4.80 | 16.00 |
| 3 rail, 5' high | 1.000 | L.F. | .185 | 12.95 | 4.98 | 17.93 |
| 6' high | 1.000 | L.F. | .192 | 14.45 | 5.20 | 19.65 |
| | | | | | | |
| Shadow box, 1"x6" boards, 2"x4" rails, 4"x4" posts | | | | | | |
| Fir, pine or spruce, treated, 3 rail, 6' high | 1.000 | L.F. | .160 | 11.05 | 4.32 | 15.37 |
| No. 1 cedar, 3 rail, 4' high | 1.000 | L.F. | .185 | 13.55 | 4.98 | 18.53 |
| 6' high | 1.000 | L.F. | .192 | 16.75 | 5.20 | 21.95 |
| Open rail, split rails, No. 1 cedar, 2 rail, 3' high | 1.000 | L.F. | .150 | 5.45 | 4.05 | 9.50 |
| 3 rail, 4' high | 1.000 | L.F. | .160 | 7.35 | 4.32 | 11.67 |
| No. 2 cedar, 2 rail, 3' high | 1.000 | L.F. | .150 | 4.24 | 4.05 | 8.29 |
| 3 rail, 4' high | 1.000 | L.F. | .160 | 4.83 | 4.32 | 9.15 |
| Open rail, rustic rails, No. 1 cedar, 2 rail, 3' high | 1.000 | L.F. | .150 | 3.40 | 4.05 | 7.45 |
| 3 rail, 4' high | 1.000 | L.F. | .160 | 4.55 | 4.32 | 8.87 |
| No. 2 cedar, 2 rail, 3' high | 1.000 | L.F. | .150 | 3.26 | 4.05 | 7.31 |
| 3 rail, 4' high | 1.000 | L.F. | .160 | 3.44 | 4.32 | 7.76 |
| Rustic picket, molded pine pickets, 2 rail, 3' high | 1.000 | L.F. | .171 | 4.81 | 4.63 | 9.44 |
| 3 rail, 4' high | 1.000 | L.F. | .197 | 5.55 | 5.30 | 10.85 |
| No. 1 cedar, 2 rail, 3' high | 1.000 | L.F. | .171 | 6.55 | 4.63 | 11.18 |
| 3 rail, 4' high | 1.000 | L.F. | .197 | 7.55 | 5.30 | 12.85 |
| Picket fence, fir, pine or spruce, preserved, treated | | | | | | |
| 2 rail, 3' high | 1.000 | L.F. | .171 | 4.18 | 4.63 | 8.81 |
| 3 rail, 4' high | 1.000 | L.F. | .185 | 4.94 | 4.98 | 9.92 |
| Western cedar, 2 rail, 3' high | 1.000 | L.F. | .171 | 5.25 | 4.63 | 9.88 |
| 3 rail, 4' high | 1.000 | L.F. | .185 | 5.35 | 4.98 | 10.33 |
| No. 1 cedar, 2 rail, 3' high | 1.000 | L.F. | .171 | 10.45 | 4.63 | 15.08 |
| 3 rail, 4' high | 1.000 | L.F. | .185 | 12.20 | 4.98 | 17.18 |
| | | | | | | |
| Stockade, No. 1 cedar, 3-1/4" rails, 6' high | 1.000 | L.F. | .150 | 9.85 | 4.05 | 13.90 |
| 8' high | 1.000 | L.F. | .155 | 12.80 | 4.18 | 16.98 |
| No. 2 cedar, treated rails, 6' high | 1.000 | L.F. | .150 | 9.85 | 4.05 | 13.90 |
| Treated pine, treated rails, 6' high | 1.000 | L.F. | .150 | 9.65 | 4.05 | 13.70 |
| Gates, No. 2 cedar, picket, 3'-6" wide 4' high | 1.000 | Ea. | 2.667 | 52.50 | 72.00 | 124.50 |
| No. 2 cedar, rustic round, 3' wide, 3' high | 1.000 | Ea. | 2.667 | 66.50 | 72.00 | 138.50 |
| No. 2 cedar, stockade screen, 3'-6" wide, 6' high | 1.000 | Ea. | 3.000 | 58.00 | 81.00 | 139.00 |
| General, wood, 3'-6" wide, 4' high | 1.000 | Ea. | 2.400 | 51.00 | 65.00 | 116.00 |
| 6' high | 1.000 | Ea. | 3.000 | 63.50 | 81.00 | 144.50 |
| | | | | | | |
| | | | | | | |
| | | | | | | |

**SITE WORK** **1**

**For information about Means Estimating Seminars, see yellow pages 11 and 12 in back of book**

# Division 2
# Foundations

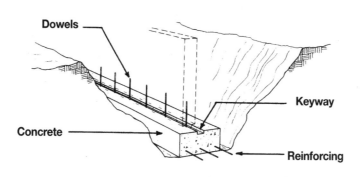

Dowels

Keyway

Concrete

Reinforcing

| System Description | QUAN. | UNIT | LABOR HOURS | COST PER L.F. | | |
|---|---|---|---|---|---|---|
| | | | | MAT. | INST. | TOTAL |
| **8″ THICK BY 18″ WIDE FOOTING** | | | | | | |
| Concrete, 3000 psi | .040 | C.Y. | | 3.04 | | 3.04 |
| Place concrete, direct chute | .040 | C.Y. | .016 | | .47 | .47 |
| Forms, footing, 4 uses | 1.330 | SFCA | .103 | .81 | 3.18 | 3.99 |
| Reinforcing, 1/2″ diameter bars, 2 each | 1.380 | Lb. | .011 | .44 | .44 | .88 |
| Keyway, 2″ x 4″, beveled, 4 uses | 1.000 | L.F. | .015 | .21 | .54 | .75 |
| Dowels, 1/2″ diameter bars, 2′ long, 6′ O.C. | .166 | Ea. | .006 | .07 | .23 | .30 |
| TOTAL | | | .151 | 4.57 | 4.86 | 9.43 |
| **12″ THICK BY 24″ WIDE FOOTING** | | | | | | |
| Concrete, 3000 psi | .070 | C.Y. | | 5.32 | | 5.32 |
| Place concrete, direct chute | .070 | C.Y. | .028 | | .82 | .82 |
| Forms, footing, 4 uses | 2.000 | SFCA | .155 | 1.22 | 4.78 | 6.00 |
| Reinforcing, 1/2″ diameter bars, 2 each | 1.380 | Lb. | .011 | .44 | .44 | .88 |
| Keyway, 2″ x 4″, beveled, 4 uses | 1.000 | L.F. | .015 | .21 | .54 | .75 |
| Dowels, 1/2″ diameter bars, 2′ long, 6′ O.C. | .166 | Ea. | .006 | .07 | .23 | .30 |
| TOTAL | | | .215 | 7.26 | 6.81 | 14.07 |
| **12″ THICK BY 36″ WIDE FOOTING** | | | | | | |
| Concrete, 3000 psi | .110 | C.Y. | | 8.36 | | 8.36 |
| Place concrete, direct chute | .110 | C.Y. | .044 | | 1.28 | 1.28 |
| Forms, footing, 4 uses | 2.000 | SFCA | .155 | 1.22 | 4.78 | 6.00 |
| Reinforcing, 1/2″ diameter bars, 2 each | 1.380 | Lb. | .011 | .44 | .44 | .88 |
| Keyway, 2″ x 4″, beveled, 4 uses | 1.000 | L.F. | .015 | .21 | .54 | .75 |
| Dowels, 1/2″ diameter bars, 2′ long, 6′ O.C. | .166 | Ea. | .006 | .07 | .23 | .30 |
| TOTAL | | | .231 | 10.30 | 7.27 | 17.57 |

The footing costs in this system are on a cost per linear foot basis.

| Description | QUAN. | UNIT | LABOR HOURS | COST PER S.F. | | |
|---|---|---|---|---|---|---|
| | | | | MAT. | INST. | TOTAL |
| | | | | | | |
| | | | | | | |
| | | | | | | |
| | | | | | | |
| | | | | | | |

**Important: See the Reference Section for critical supporting data - Reference Nos., Crews & Location Factors**

| Footing Price Sheet | QUAN. | UNIT | LABOR HOURS | COST PER L.F. MAT. | INST. | TOTAL |
|---|---|---|---|---|---|---|
| Concrete, 8" thick by 18" wide footing | | | | | | |
| 2000 psi concrete | .040 | C.Y. | | 2.86 | | 2.86 |
| 2500 psi concrete | .040 | C.Y. | | 2.56 | | 2.56 |
| 3000 psi concrete | .040 | C.Y. | | 3.04 | | 3.04 |
| 3500 psi concrete | .040 | C.Y. | | 3.12 | | 3.12 |
| 4000 psi concrete | .040 | C.Y. | | 3.26 | | 3.26 |
| 12" thick by 24" wide footing | | | | | | |
| 2000 psi concrete | .070 | C.Y. | | 5.00 | | 5.00 |
| 2500 psi concrete | .070 | C.Y. | | 4.48 | | 4.48 |
| 3000 psi concrete | .070 | C.Y. | | 5.30 | | 5.30 |
| 3500 psi concrete | .070 | C.Y. | | 5.45 | | 5.45 |
| 4000 psi concrete | .070 | C.Y. | | 5.70 | | 5.70 |
| 12" thick by 36" wide footing | | | | | | |
| 2000 psi concrete | .110 | C.Y. | | 7.85 | | 7.85 |
| 2500 psi concrete | .110 | C.Y. | | 7.05 | | 7.05 |
| 3000 psi concrete | .110 | C.Y. | | 8.35 | | 8.35 |
| 3500 psi concrete | .110 | C.Y. | | 8.60 | | 8.60 |
| 4000 psi concrete | .110 | C.Y. | | 8.95 | | 8.95 |
| Place concrete, 8" thick by 18" wide footing, direct chute | .040 | C.Y. | .016 | | .47 | .47 |
| Pumped concrete | .040 | C.Y. | .017 | | .70 | .70 |
| Crane & bucket | .040 | C.Y. | .032 | | 1.44 | 1.44 |
| 12" thick by 24" wide footing, direct chute | .070 | C.Y. | .028 | | .82 | .82 |
| Pumped concrete | .070 | C.Y. | .030 | | 1.24 | 1.24 |
| Crane & bucket | .070 | C.Y. | .056 | | 2.52 | 2.52 |
| 12" thick by 36" wide footing, direct chute | .110 | C.Y. | .044 | | 1.28 | 1.28 |
| Pumped concrete | .110 | C.Y. | .047 | | 1.94 | 1.94 |
| Crane & bucket | .110 | C.Y. | .088 | | 3.97 | 3.97 |
| Forms, 8" thick footing, 1 use | 1.330 | SFCA | .140 | .21 | 4.35 | 4.56 |
| 4 uses | 1.330 | SFCA | .103 | .81 | 3.18 | 3.99 |
| 12" thick footing, 1 use | 2.000 | SFCA | .211 | .32 | 6.55 | 6.87 |
| 4 uses | 2.000 | SFCA | .155 | 1.22 | 4.78 | 6.00 |
| Reinforcing, 3/8" diameter bar, 1 each | .400 | Lb. | .003 | .13 | .13 | .26 |
| 2 each | .800 | Lb. | .006 | .26 | .26 | .52 |
| 3 each | 1.200 | Lb. | .009 | .38 | .38 | .76 |
| 1/2" diameter bar, 1 each | .700 | Lb. | .005 | .22 | .22 | .44 |
| 2 each | 1.380 | Lb. | .011 | .44 | .44 | .88 |
| 3 each | 2.100 | Lb. | .016 | .67 | .67 | 1.34 |
| 5/8" diameter bar, 1 each | 1.040 | Lb. | .008 | .33 | .33 | .66 |
| 2 each | 2.080 | Lb. | .016 | .67 | .67 | 1.34 |
| Keyway, beveled, 2" x 4", 1 use | 1.000 | L.F. | .030 | .42 | 1.08 | 1.50 |
| 2 uses | 1.000 | L.F. | .023 | .32 | .81 | 1.13 |
| 2" x 6", 1 use | 1.000 | L.F. | .032 | .62 | 1.14 | 1.76 |
| 2 uses | 1.000 | L.F. | .024 | .47 | .86 | 1.33 |
| Dowels, 2 feet long, 6' O.C., 3/8" bar | .166 | Ea. | .005 | .04 | .21 | .25 |
| 1/2" bar | .166 | Ea. | .006 | .07 | .23 | .30 |
| 5/8" bar | .166 | Ea. | .006 | .12 | .26 | .38 |
| 3/4" bar | .166 | Ea. | .006 | .12 | .26 | .38 |

**2 FOUNDATIONS**

101

| System Description | QUAN. | UNIT | LABOR HOURS | COST PER S.F. | | |
|---|---|---|---|---|---|---|
| | | | | MAT. | INST. | TOTAL |
| **8" WALL, GROUTED, FULL HEIGHT** | | | | | | |
| Concrete block, 8" x 16" x 8" | 1.000 | S.F. | .094 | 1.85 | 3.06 | 4.91 |
| Masonry reinforcing, every second course | .750 | L.F. | .002 | .09 | .08 | .17 |
| Parging, plastering with portland cement plaster, 1 coat | 1.000 | S.F. | .014 | .23 | .47 | .70 |
| Dampproofing, bituminous coating, 1 coat | 1.000 | S.F. | .012 | .07 | .41 | .48 |
| Insulation, 1" rigid polystyrene | 1.000 | S.F. | .010 | .37 | .36 | .73 |
| Grout, solid, pumped | 1.000 | S.F. | .059 | .94 | 1.93 | 2.87 |
| Anchor bolts, 1/2" diameter, 8" long, 4' O.C. | .060 | Ea. | .002 | .04 | .09 | .13 |
| Sill plate, 2" x 4", treated | .250 | L.F. | .007 | .14 | .26 | .40 |
| TOTAL | | | .200 | 3.73 | 6.66 | 10.39 |
| **12" WALL, GROUTED, FULL HEIGHT** | | | | | | |
| Concrete block, 8" x 16" x 12" | 1.000 | S.F. | .160 | 2.60 | 5.05 | 7.65 |
| Masonry reinforcing, every second course | .750 | L.F. | .003 | .11 | .11 | .22 |
| Parging, plastering with portland cement plaster, 1 coat | 1.000 | S.F. | .014 | .23 | .47 | .70 |
| Dampproofing, bituminous coating, 1 coat | 1.000 | S.F. | .012 | .07 | .41 | .48 |
| Insulation, 1" rigid polystyrene | 1.000 | S.F. | .010 | .37 | .36 | .73 |
| Grout, solid, pumped | 1.000 | S.F. | .063 | 1.53 | 2.05 | 3.58 |
| Anchor bolts, 1/2" diameter, 8" long, 4' O.C. | .060 | Ea. | .002 | .04 | .09 | .13 |
| Sill plate, 2" x 4", treated | .250 | L.F. | .007 | .14 | .26 | .40 |
| TOTAL | | | .271 | 5.09 | 8.80 | 13.89 |

The costs in this system are based on a square foot of wall. Do not subtract for window or door openings.

| Description | QUAN. | UNIT | LABOR HOURS | COST PER S.F. | | |
|---|---|---|---|---|---|---|
| | | | | MAT. | INST. | TOTAL |
| | | | | | | |
| | | | | | | |
| | | | | | | |
| | | | | | | |
| | | | | | | |

**Important: See the Reference Section for critical supporting data - Reference Nos., Crews & Location Factors**

| Block Wall Systems | QUAN. | UNIT | LABOR HOURS | COST PER S.F. | | |
|---|---|---|---|---|---|---|
| | | | | MAT. | INST. | TOTAL |
| Concrete, block, 8" x 16" x, 6" thick | 1.000 | S.F. | .089 | 1.72 | 2.85 | 4.57 |
| 8" thick | 1.000 | S.F. | .093 | 1.85 | 3.06 | 4.91 |
| 10" thick | 1.000 | S.F. | .095 | 2.54 | 3.71 | 6.25 |
| 12" thick | 1.000 | S.F. | .122 | 2.60 | 5.05 | 7.65 |
| Solid block, 8" x 16" x, 6" thick | 1.000 | S.F. | .091 | 1.86 | 2.95 | 4.81 |
| 8" thick | 1.000 | S.F. | .096 | 2.70 | 3.13 | 5.83 |
| 10" thick | 1.000 | S.F. | .096 | 2.70 | 3.13 | 5.83 |
| 12" thick | 1.000 | S.F. | .126 | 3.93 | 4.34 | 8.27 |
| Masonry reinforcing, wire strips, to 8" wide, every course | 1.500 | L.F. | .004 | .18 | .15 | .33 |
| Every 2nd course | .750 | L.F. | .002 | .09 | .08 | .17 |
| Every 3rd course | .500 | L.F. | .001 | .06 | .05 | .11 |
| Every 4th course | .400 | L.F. | .001 | .05 | .04 | .09 |
| Wire strips to 12" wide, every course | 1.500 | L.F. | .006 | .21 | .21 | .42 |
| Every 2nd course | .750 | L.F. | .003 | .11 | .11 | .22 |
| Every 3rd course | .500 | L.F. | .002 | .07 | .07 | .14 |
| Every 4th course | .400 | L.F. | .002 | .06 | .06 | .12 |
| Parging, plastering with portland cement plaster, 1 coat | 1.000 | S.F. | .014 | .23 | .47 | .70 |
| 2 coats | 1.000 | S.F. | .022 | .36 | .72 | 1.08 |
| Dampproofing, bituminous, brushed on, 1 coat | 1.000 | S.F. | .012 | .07 | .41 | .48 |
| 2 coats | 1.000 | S.F. | .016 | .11 | .54 | .65 |
| Sprayed on, 1 coat | 1.000 | S.F. | .010 | .08 | .33 | .41 |
| 2 coats | 1.000 | S.F. | .016 | .16 | .54 | .70 |
| Troweled on, 1/16" thick | 1.000 | S.F. | .016 | .17 | .54 | .71 |
| 1/8" thick | 1.000 | S.F. | .020 | .32 | .68 | 1.00 |
| 1/2" thick | 1.000 | S.F. | .023 | 1.03 | .77 | 1.80 |
| Insulation, rigid, fiberglass, 1.5#/C.F., unfaced | | | | | | |
| 1-1/2" thick R 6.2 | 1.000 | S.F. | .008 | .45 | .28 | .73 |
| 2" thick R 8.5 | 1.000 | S.F. | .008 | .50 | .28 | .78 |
| 3" thick R 13 | 1.000 | S.F. | .010 | .63 | .36 | .99 |
| Foamglass, 1-1/2" thick R 2.64 | 1.000 | S.F. | .010 | 1.76 | .36 | 2.12 |
| 2" thick R 5.26 | 1.000 | S.F. | .011 | 3.07 | .39 | 3.46 |
| Perlite, 1" thick R 2.77 | 1.000 | S.F. | .010 | .29 | .36 | .65 |
| 2" thick R 5.55 | 1.000 | S.F. | .011 | .55 | .39 | .94 |
| Polystyrene, extruded, 1" thick R 5.4 | 1.000 | S.F. | .010 | .37 | .36 | .73 |
| 2" thick R 10.8 | 1.000 | S.F. | .011 | 1.06 | .39 | 1.45 |
| Molded 1" thick R 3.85 | 1.000 | S.F. | .010 | .15 | .36 | .51 |
| 2" thick R 7.7 | 1.000 | S.F. | .011 | .57 | .39 | .96 |
| Grout, concrete block cores, 6" thick | 1.000 | S.F. | .044 | .71 | 1.45 | 2.16 |
| 8" thick | 1.000 | S.F. | .059 | .94 | 1.93 | 2.87 |
| 10" thick | 1.000 | S.F. | .061 | 1.24 | 1.99 | 3.23 |
| 12" thick | 1.000 | S.F. | .063 | 1.53 | 2.05 | 3.58 |
| Anchor bolts, 2' on center, 1/2" diameter, 8" long | .120 | Ea. | .005 | .08 | .17 | .25 |
| 12" long | .120 | Ea. | .005 | .15 | .18 | .33 |
| 3/4" diameter, 8" long | .120 | Ea. | .006 | .20 | .21 | .41 |
| 12" long | .120 | Ea. | .006 | .25 | .23 | .48 |
| 4' on center, 1/2" diameter, 8" long | .060 | Ea. | .002 | .04 | .09 | .13 |
| 12" long | .060 | Ea. | .003 | .08 | .09 | .17 |
| 3/4" diameter, 8" long | .060 | Ea. | .003 | .10 | .11 | .21 |
| 12" long | .060 | Ea. | .003 | .12 | .11 | .23 |
| Sill plates, treated, 2" x 4" | .250 | L.F. | .007 | .14 | .26 | .40 |
| 4" x 4" | .250 | L.F. | .007 | .40 | .24 | .64 |

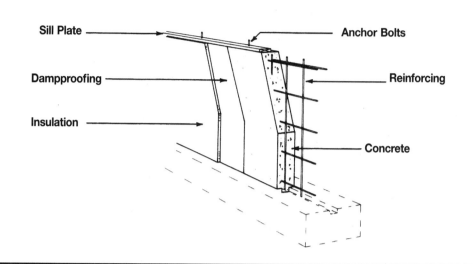

Sill Plate — Anchor Bolts

Dampproofing — Reinforcing

Insulation — Concrete

**FOUNDATIONS 2**

| System Description | QUAN. | UNIT | LABOR HOURS | COST PER S.F. | | |
|---|---|---|---|---|---|---|
| | | | | MAT. | INST. | TOTAL |
| **8″ THICK, POURED CONCRETE WALL** | | | | | | |
| Concrete, 8″ thick , 3000 psi | .025 | C.Y. | | 1.90 | | 1.90 |
| Forms, prefabricated plywood, 4 uses per month | 2.000 | SFCA | .076 | 1.22 | 2.40 | 3.62 |
| Reinforcing, light | .670 | Lb. | .004 | .21 | .15 | .36 |
| Placing concrete, direct chute | .025 | C.Y. | .013 | | .39 | .39 |
| Dampproofing, brushed on, 2 coats | 1.000 | S.F. | .016 | .11 | .54 | .65 |
| Rigid insulation, 1″ polystyrene | 1.000 | S.F. | .010 | .37 | .36 | .73 |
| Anchor bolts, 1/2″ diameter, 12″ long, 4′ O.C. | .060 | Ea. | .003 | .08 | .09 | .17 |
| Sill plates, 2″ x 4″, treated | .250 | L.F. | .007 | .14 | .26 | .40 |
| TOTAL | | | .129 | 4.03 | 4.19 | 8.22 |
| **12″ THICK, POURED CONCRETE WALL** | | | | | | |
| Concrete, 12″ thick, 3000 psi | .040 | C.Y. | | 3.04 | | 3.04 |
| Forms, prefabricated plywood, 4 uses per month | 2.000 | SFCA | .076 | 1.22 | 2.40 | 3.62 |
| Reinforcing, light | 1.000 | Lb. | .005 | .32 | .22 | .54 |
| Placing concrete, direct chute | .040 | C.Y. | .019 | | .56 | .56 |
| Dampproofing, brushed on, 2 coats | 1.000 | S.F. | .016 | .11 | .54 | .65 |
| Rigid insulation, 1″ polystyrene | 1.000 | S.F. | .010 | .37 | .36 | .73 |
| Anchor bolts, 1/2″ diameter, 12″ long, 4′ O.C. | .060 | Ea. | .003 | .08 | .09 | .17 |
| Sill plates, 2″ x 4″ treated | .250 | L.F. | .007 | .14 | .26 | .40 |
| TOTAL | | | .136 | 5.28 | 4.43 | 9.71 |

The costs in this system are based on sq. ft. of wall. Do not subtract
for window and door openings. The costs assume a 4′ high wall.

| Description | QUAN. | UNIT | LABOR HOURS | COST PER S.F. | | |
|---|---|---|---|---|---|---|
| | | | | MAT. | INST. | TOTAL |
| | | | | | | |
| | | | | | | |
| | | | | | | |
| | | | | | | |
| | | | | | | |

**Important: See the Reference Section for critical supporting data - Reference Nos., Crews & Location Factors**

| Concrete Wall Price Sheet | QUAN. | UNIT | LABOR HOURS | COST PER S.F. | | |
|---|---|---|---|---|---|---|
| | | | | MAT. | INST. | TOTAL |
| Formwork, prefabricated plywood, 1 use per month | 2.000 | SFCA | .081 | 3.70 | 2.56 | 6.26 |
| 4 uses per month | 2.000 | SFCA | .076 | 1.22 | 2.40 | 3.62 |
| Job built forms, 1 use per month | 2.000 | SFCA | .320 | 4.02 | 10.10 | 14.12 |
| 4 uses per month | 2.000 | SFCA | .221 | 1.58 | 6.95 | 8.53 |
| Reinforcing, 8" wall, light reinforcing | .670 | Lb. | .004 | .21 | .15 | .36 |
| Heavy reinforcing | 1.500 | Lb. | .008 | .48 | .33 | .81 |
| 10" wall, light reinforcing | .850 | Lb. | .005 | .27 | .19 | .46 |
| Heavy reinforcing | 2.000 | Lb. | .011 | .64 | .44 | 1.08 |
| 12" wall light reinforcing | 1.000 | Lb. | .005 | .32 | .22 | .54 |
| Heavy reinforcing | 2.250 | Lb. | .012 | .72 | .50 | 1.22 |
| Placing concrete, 8" wall, direct chute | .025 | C.Y. | .013 | | .39 | .39 |
| Pumped concrete | .025 | C.Y. | .016 | | .66 | .66 |
| Crane & bucket | .025 | C.Y. | .023 | | 1.01 | 1.01 |
| 10" wall, direct chute | .030 | C.Y. | .016 | | .47 | .47 |
| Pumped concrete | .030 | C.Y. | .019 | | .79 | .79 |
| Crane & bucket | .030 | C.Y. | .027 | | 1.21 | 1.21 |
| 12" wall, direct chute | .040 | C.Y. | .019 | | .56 | .56 |
| Pumped concrete | .040 | C.Y. | .023 | | .96 | .96 |
| Crane & bucket | .040 | C.Y. | .032 | | 1.44 | 1.44 |
| Dampproofing, bituminous, brushed on, 1 coat | 1.000 | S.F. | .012 | .07 | .41 | .48 |
| 2 coats | 1.000 | S.F. | .016 | .11 | .54 | .65 |
| Sprayed on, 1 coat | 1.000 | S.F. | .010 | .08 | .33 | .41 |
| 2 coats | 1.000 | S.F. | .016 | .16 | .54 | .70 |
| Troweled on, 1/16" thick | 1.000 | S.F. | .016 | .17 | .54 | .71 |
| 1/8" thick | 1.000 | S.F. | .020 | .32 | .68 | 1.00 |
| 1/2" thick | 1.000 | S.F. | .023 | 1.03 | .77 | 1.80 |
| Insulation rigid, fiberglass, 1.5#/C.F., unfaced | | | | | | |
| 1-1/2" thick, R 6.2 | 1.000 | S.F. | .008 | .45 | .28 | .73 |
| 2" thick, R 8.3 | 1.000 | S.F. | .008 | .50 | .28 | .78 |
| 3" thick, R 12.4 | 1.000 | S.F. | .010 | .63 | .36 | .99 |
| Foamglass, 1-1/2" thick R 2.64 | 1.000 | S.F. | .010 | 1.76 | .36 | 2.12 |
| 2" thick R 5.26 | 1.000 | S.F. | .011 | 3.07 | .39 | 3.46 |
| Perlite, 1" thick R 2.77 | 1.000 | S.F. | .010 | .29 | .36 | .65 |
| 2" thick R 5.55 | 1.000 | S.F. | .011 | .55 | .39 | .94 |
| Polystyrene, extruded, 1" thick R 5.40 | 1.000 | S.F. | .010 | .37 | .36 | .73 |
| 2" thick R 10.8 | 1.000 | S.F. | .011 | 1.06 | .39 | 1.45 |
| Molded, 1" thick R 3.85 | 1.000 | S.F. | .010 | .15 | .36 | .51 |
| 2" thick R 7.70 | 1.000 | S.F. | .011 | .57 | .39 | .96 |
| Anchor bolts, 2' on center, 1/2" diameter, 8" long | .120 | Ea. | .005 | .08 | .17 | .25 |
| 12" long | .120 | Ea. | .005 | .15 | .18 | .33 |
| 3/4" diameter, 8" long | .120 | Ea. | .006 | .20 | .21 | .41 |
| 12" long | .120 | Ea. | .006 | .25 | .23 | .48 |
| Sill plates, treated lumber, 2" x 4" | .250 | L.F. | .007 | .14 | .26 | .40 |
| 4" x 4" | .250 | L.F. | .007 | .40 | .24 | .64 |

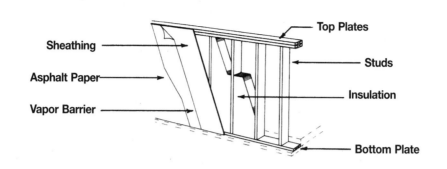

Sheathing — Top Plates

Asphalt Paper — Studs

Vapor Barrier — Insulation

— Bottom Plate

| System Description | QUAN. | UNIT | LABOR HOURS | COST PER S.F. | | |
|---|---|---|---|---|---|---|
| | | | | MAT. | INST. | TOTAL |
| **2″ X 4″ STUDS, 16″ O.C., WALL** | | | | | | |
| Studs, 2″ x 4″, 16″ O.C., treated | 1.000 | L.F. | .015 | .55 | .52 | 1.07 |
| Plates, double top plate, single bottom plate, treated, 2″ x 4″ | .750 | L.F. | .011 | .41 | .39 | .80 |
| Sheathing, 1/2″, exterior grade, CDX, treated | 1.000 | S.F. | .014 | .73 | .51 | 1.24 |
| Asphalt paper, 15# roll | 1.100 | S.F. | .002 | .02 | .09 | .11 |
| Vapor barrier, 4 mil polyethylene | 1.000 | S.F. | .002 | .03 | .08 | .11 |
| Insulation, batts, fiberglass, 3-1/2″ thick, R 11 | 1.000 | S.F. | .005 | .25 | .18 | .43 |
| TOTAL | | | .049 | 1.99 | 1.77 | 3.76 |
| **2″ X 6″ STUDS, 16″ O.C., WALL** | | | | | | |
| Studs, 2″ x 6″, 16″ O.C., treated | 1.000 | L.F. | .016 | 1.23 | .57 | 1.80 |
| Plates, double top plate, single bottom plate, treated, 2″ x 6″ | .750 | L.F. | .012 | .92 | .43 | 1.35 |
| Sheathing, 5/8″ exterior grade, CDX, treated | 1.000 | S.F. | .015 | .85 | .54 | 1.39 |
| Asphalt paper, 15# roll | 1.100 | S.F. | .002 | .02 | .09 | .11 |
| Vapor barrier, 4 mil polyethylene | 1.000 | S.F. | .002 | .03 | .08 | .11 |
| Insulation, batts, fiberglass, 6″ thick, R 19 | 1.000 | S.F. | .006 | .36 | .21 | .57 |
| TOTAL | | | .053 | 3.41 | 1.92 | 5.33 |
| **2″ X 8″ STUDS, 16″ O.C., WALL** | | | | | | |
| Studs, 2″ x 8″, 16″ O.C. treated | 1.000 | L.F. | .018 | 1.60 | .63 | 2.23 |
| Plates, double top plate, single bottom plate, treated, 2″ x 8″ | .750 | L.F. | .013 | 1.20 | .47 | 1.67 |
| Sheathing, 3/4″ exterior grade, CDX, treated | 1.000 | S.F. | .016 | 1.06 | .58 | 1.64 |
| Asphalt paper, 15# roll | 1.100 | S.F. | .002 | .02 | .09 | .11 |
| Vapor barrier, 4 mil polyethylene | 1.000 | S.F. | .002 | .03 | .08 | .11 |
| Insulation, batts, fiberglass, 9″ thick, R 30 | 1.000 | S.F. | .006 | .66 | .21 | .87 |
| TOTAL | | | .057 | 4.57 | 2.06 | 6.63 |

The costs in this system are based on a sq. ft. of wall area. Do not
subtract for window or door openings. The costs assume a 4′ high wall.

| Description | QUAN. | UNIT | LABOR HOURS | COST PER S.F. | | |
|---|---|---|---|---|---|---|
| | | | | MAT. | INST. | TOTAL |
| | | | | | | |
| | | | | | | |
| | | | | | | |
| | | | | | | |
| | | | | | | |

**Important: See the Reference Section for critical supporting data - Reference Nos., Crews & Location Factors**

# Wood Wall Foundation Price Sheet

| Wood Wall Foundation Price Sheet | QUAN. | UNIT | LABOR HOURS | COST PER S.F. | | |
|---|---|---|---|---|---|---|
| | | | | MAT. | INST. | TOTAL |
| Studs, treated, 2" x 4", 12" O.C. | 1.250 | L.F. | .018 | .69 | .65 | 1.34 |
| 16" O.C. | 1.000 | L.F. | .015 | .55 | .52 | 1.07 |
| 2" x 6", 12" O.C. | 1.250 | L.F. | .020 | 1.54 | .71 | 2.25 |
| 16" O.C. | 1.000 | L.F. | .016 | 1.23 | .57 | 1.80 |
| 2" x 8", 12" O.C. | 1.250 | L.F. | .022 | 2.00 | .79 | 2.79 |
| 16" O.C. | 1.000 | L.F. | .018 | 1.60 | .63 | 2.23 |
| Plates, treated double top single bottom, 2" x 4" | .750 | L.F. | .011 | .41 | .39 | .80 |
| 2" x 6" | .750 | L.F. | .012 | .92 | .43 | 1.35 |
| 2" x 8" | .750 | L.F. | .013 | 1.20 | .47 | 1.67 |
| Sheathing, treated exterior grade CDX, 1/2" thick | 1.000 | S.F. | .014 | .73 | .51 | 1.24 |
| 5/8" thick | 1.000 | S.F. | .015 | .85 | .54 | 1.39 |
| 3/4" thick | 1.000 | S.F. | .016 | 1.06 | .58 | 1.64 |
| Asphalt paper, 15# roll | 1.100 | S.F. | .002 | .02 | .09 | .11 |
| Vapor barrier, polyethylene, 4 mil | 1.000 | S.F. | .002 | .02 | .08 | .10 |
| 10 mil | 1.000 | S.F. | .002 | .06 | .08 | .14 |
| Insulation, rigid, fiberglass, 1.5#/C.F., unfaced | 1.000 | S.F. | .008 | .32 | .28 | .60 |
| 1-1/2" thick, R 6.2 | 1.000 | S.F. | .008 | .45 | .28 | .73 |
| 2" thick, R 8.3 | 1.000 | S.F. | .008 | .50 | .28 | .78 |
| 3" thick, R 12.4 | 1.000 | S.F. | .010 | .64 | .37 | 1.01 |
| Foamglass 1 1/2" thick, R 2.64 | 1.000 | S.F. | .010 | 1.76 | .36 | 2.12 |
| 2" thick, R 5.26 | 1.000 | S.F. | .011 | 3.07 | .39 | 3.46 |
| Perlite 1" thick, R 2.77 | 1.000 | S.F. | .010 | .29 | .36 | .65 |
| 2" thick, R 5.55 | 1.000 | S.F. | .011 | .55 | .39 | .94 |
| Polystyrene, extruded, 1" thick, R 5.40 | 1.000 | S.F. | .010 | .37 | .36 | .73 |
| 2" thick, R 10.8 | 1.000 | S.F. | .011 | 1.06 | .39 | 1.45 |
| Molded 1" thick, R 3.85 | 1.000 | S.F. | .010 | .15 | .36 | .51 |
| 2" thick, R 7.7 | 1.000 | S.F. | .011 | .57 | .39 | .96 |
| Non rigid, batts, fiberglass, paper backed, 3-1/2" thick roll, R 11 | 1.000 | S.F. | .005 | .25 | .18 | .43 |
| 6", R 19 | 1.000 | S.F. | .006 | .36 | .21 | .57 |
| 9", R 30 | 1.000 | S.F. | .006 | .66 | .21 | .87 |
| 12", R 38 | 1.000 | S.F. | .006 | .84 | .21 | 1.05 |
| Mineral fiber, paper backed, 3-1/2", R 13 | 1.000 | S.F. | .005 | .29 | .18 | .47 |
| 6", R 19 | 1.000 | S.F. | .005 | .43 | .18 | .61 |
| 10", R 30 | 1.000 | S.F. | .006 | .68 | .21 | .89 |

FOUNDATIONS

2

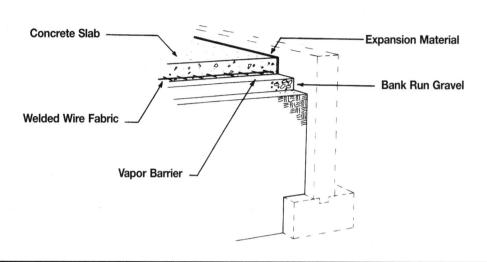

Concrete Slab — Expansion Material — Bank Run Gravel — Welded Wire Fabric — Vapor Barrier

| System Description | QUAN. | UNIT | LABOR HOURS | COST PER S.F. | | |
|---|---|---|---|---|---|---|
| | | | | MAT. | INST. | TOTAL |
| **4″ THICK SLAB** | | | | | | |
| Concrete, 4″ thick, 3000 psi concrete | .012 | C.Y. | | .91 | | .91 |
| Place concrete, direct chute | .012 | C.Y. | .005 | | .15 | .15 |
| Bank run gravel, 4″ deep | 1.000 | S.F. | .001 | .66 | .04 | .70 |
| Polyethylene vapor barrier, .006″ thick | 1.000 | S.F. | .002 | .03 | .08 | .11 |
| Edge forms, expansion material | .100 | L.F. | .005 | .05 | .17 | .22 |
| Welded wire fabric, 6 x 6, 10/10 (W1.4/W1.4) | 1.100 | S.F. | .005 | .09 | .21 | .30 |
| Steel trowel finish | 1.000 | S.F. | .015 | | .47 | .47 |
| TOTAL | | | .033 | 1.74 | 1.12 | 2.86 |
| **6″ THICK SLAB** | | | | | | |
| Concrete, 6″ thick, 3000 psi concrete | .019 | C.Y. | | 1.44 | | 1.44 |
| Place concrete, direct chute | .019 | C.Y. | .008 | | .24 | .24 |
| Bank run gravel, 4″ deep | 1.000 | S.F. | .001 | .66 | .04 | .70 |
| Polyethylene vapor barrier, .006″ thick | 1.000 | S.F. | .002 | .03 | .08 | .11 |
| Edge forms, expansion material | .100 | L.F. | .005 | .05 | .17 | .22 |
| Welded wire fabric, 6 x 6, 10/10 (W1.4/W1.4) | 1.100 | S.F. | .005 | .09 | .21 | .30 |
| Steel trowel finish | 1.000 | S.F. | .015 | | .47 | .47 |
| TOTAL | | | .036 | 2.27 | 1.21 | 3.48 |

The slab costs in this section are based on a cost per square foot of floor area.

| Description | QUAN. | UNIT | LABOR HOURS | COST PER S.F. | | |
|---|---|---|---|---|---|---|
| | | | | MAT. | INST. | TOTAL |
| | | | | | | |
| | | | | | | |
| | | | | | | |
| | | | | | | |
| | | | | | | |

   **Important: See the Reference Section for critical supporting data - Reference Nos., Crews & Location Factors**

| Floor Slab Price Sheet | QUAN. | UNIT | LABOR HOURS | COST PER S.F. | | |
|---|---|---|---|---|---|---|
| | | | | MAT. | INST. | TOTAL |
| Concrete, 4" thick slab, 2000 psi concrete | .012 | C.Y. | | .86 | | .86 |
| 2500 psi concrete | .012 | C.Y. | | .77 | | .77 |
| 3000 psi concrete | .012 | C.Y. | | .91 | | .91 |
| 3500 psi concrete | .012 | C.Y. | | .94 | | .94 |
| 4000 psi concrete | .012 | C.Y. | | .98 | | .98 |
| 4500 psi concrete | .012 | C.Y. | | 1.00 | | 1.00 |
| 5" thick slab, 2000 psi concrete | .015 | C.Y. | | 1.07 | | 1.07 |
| 2500 psi concrete | .015 | C.Y. | | .96 | | .96 |
| 3000 psi concrete | .015 | C.Y. | | 1.14 | | 1.14 |
| 3500 psi concrete | .015 | C.Y. | | 1.17 | | 1.17 |
| 4000 psi concrete | .015 | C.Y. | | 1.22 | | 1.22 |
| 4500 psi concrete | .015 | C.Y. | | 1.25 | | 1.25 |
| 6" thick slab, 2000 psi concrete | .019 | C.Y. | | 1.36 | | 1.36 |
| 2500 psi concrete | .019 | C.Y. | | 1.22 | | 1.22 |
| 3000 psi concrete | .019 | C.Y. | | 1.44 | | 1.44 |
| 3500 psi concrete | .019 | C.Y. | | 1.48 | | 1.48 |
| 4000 psi concrete | .019 | C.Y. | | 1.55 | | 1.55 |
| 4500 psi concrete | .019 | C.Y. | | 1.58 | | 1.58 |
| Place concrete, 4" slab, direct chute | .012 | C.Y. | .005 | | .15 | .15 |
| Pumped concrete | .012 | C.Y. | .006 | | .25 | .25 |
| Crane & bucket | .012 | C.Y. | .008 | | .35 | .35 |
| 5" slab, direct chute | .015 | C.Y. | .007 | | .19 | .19 |
| Pumped concrete | .015 | C.Y. | .007 | | .31 | .31 |
| Crane & bucket | .015 | C.Y. | .010 | | .44 | .44 |
| 6" slab, direct chute | .019 | C.Y. | .008 | | .24 | .24 |
| Pumped concrete | .019 | C.Y. | .009 | | .39 | .39 |
| Crane & bucket | .019 | C.Y. | .012 | | .56 | .56 |
| Gravel, bank run, 4" deep | 1.000 | S.F. | .001 | .66 | .04 | .70 |
| 6" deep | 1.000 | S.F. | .001 | .88 | .05 | .93 |
| 9" deep | 1.000 | S.F. | .001 | 1.32 | .07 | 1.39 |
| 12" deep | 1.000 | S.F. | .001 | 1.76 | .08 | 1.84 |
| 3/4" crushed stone, 3" deep | 1.000 | S.F. | .001 | .44 | .04 | .48 |
| 6" deep | 1.000 | S.F. | .001 | .87 | .06 | .93 |
| 9" deep | 1.000 | S.F. | .002 | 1.30 | .09 | 1.39 |
| 12" deep | 1.000 | S.F. | .002 | 1.78 | .11 | 1.89 |
| Vapor barrier polyethylene, .004" thick | 1.000 | S.F. | .002 | .02 | .08 | .10 |
| .006" thick | 1.000 | S.F. | .002 | .03 | .08 | .11 |
| Edge forms, expansion material, 4" thick slab | .100 | L.F. | .004 | .03 | .11 | .14 |
| 6" thick slab | .100 | L.F. | .005 | .05 | .17 | .22 |
| Welded wire fabric 6 x 6, 10/10 (W1.4/W1.4) | 1.100 | S.F. | .005 | .09 | .21 | .30 |
| 6 x 6, 6/6 (W2.9/W2.9) | 1.100 | S.F. | .006 | .14 | .25 | .39 |
| 4 x 4, 10/10 (W1.4/W1.4) | 1.100 | S.F. | .006 | .13 | .24 | .37 |
| Finish concrete, screed finish | 1.000 | S.F. | .009 | | .29 | .29 |
| Float finish | 1.000 | S.F. | .011 | | .36 | .36 |
| Steel trowel, for resilient floor | 1.000 | S.F. | .013 | | .43 | .43 |
| For finished floor | 1.000 | S.F. | .015 | | .47 | .47 |
| | | | | | | |
| | | | | | | |
| | | | | | | |

**2 FOUNDATIONS**

**For information about Means Estimating Seminars, see yellow pages 11 and 12 in back of book**

# Division 3
# Framing

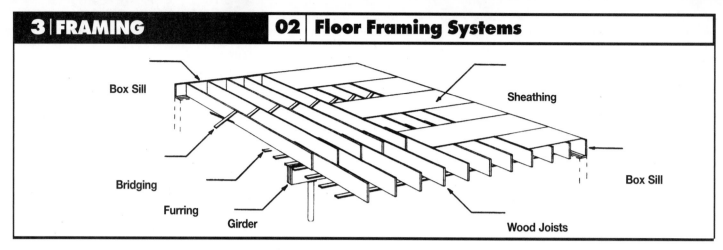

| System Description | QUAN. | UNIT | LABOR HOURS | COST PER S.F. | | |
|---|---|---|---|---|---|---|
| | | | | MAT. | INST. | TOTAL |
| **2" X 8", 16" O.C.** | | | | | | |
| Wood joists, 2" x 8", 16" O.C. | 1.000 | L.F. | .015 | .95 | .52 | 1.47 |
| Bridging, 1" x 3", 6' O.C. | .080 | Pr. | .005 | .02 | .18 | .20 |
| Box sills, 2" x 8" | .150 | L.F. | .002 | .14 | .08 | .22 |
| Concrete filled steel column, 4" diameter | .125 | L.F. | .002 | .09 | .09 | .18 |
| Girder, built up from three 2" x 8" | .125 | L.F. | .013 | .36 | .47 | .83 |
| Sheathing, plywood, subfloor, 5/8" CDX | 1.000 | S.F. | .012 | .62 | .42 | 1.04 |
| Furring, 1" x 3", 16" O.C. | 1.000 | L.F. | .023 | .21 | .81 | 1.02 |
| TOTAL | | | .072 | 2.39 | 2.57 | 4.96 |
| **2" X 10", 16" O.C.** | | | | | | |
| Wood joists, 2" x 10", 16" O.C. | 1.000 | L.F. | .018 | 1.37 | .63 | 2.00 |
| Bridging, 1" x 3", 6' O.C. | .080 | Pr. | .005 | .02 | .18 | .20 |
| Box sills, 2" x 10" | .150 | L.F. | .003 | .21 | .09 | .30 |
| Concrete filled steel column, 4" diameter | .125 | L.F. | .002 | .09 | .09 | .18 |
| Girder, built up from three 2" x 10" | .125 | L.F. | .014 | .51 | .50 | 1.01 |
| Sheathing, plywood, subfloor, 5/8" CDX | 1.000 | S.F. | .012 | .62 | .42 | 1.04 |
| Furring, 1" x 3",16" O.C. | 1.000 | L.F. | .023 | .21 | .81 | 1.02 |
| TOTAL | | | .077 | 3.03 | 2.72 | 5.75 |
| **2" X 12", 16" O.C.** | | | | | | |
| Wood joists, 2" x 12", 16" O.C. | 1.000 | L.F. | .018 | 1.87 | .65 | 2.52 |
| Bridging, 1" x 3", 6' O.C. | .080 | Pr. | .005 | .02 | .18 | .20 |
| Box sills, 2" x 12" | .150 | L.F. | .003 | .28 | .10 | .38 |
| Concrete filled steel column, 4" diameter | .125 | L.F. | .002 | .09 | .09 | .18 |
| Girder, built up from three 2" x 12" | .125 | L.F. | .015 | .70 | .53 | 1.23 |
| Sheathing, plywood, subfloor, 5/8" CDX | 1.000 | S.F. | .012 | .62 | .42 | 1.04 |
| Furring, 1" x 3", 16" O.C. | 1.000 | L.F. | .023 | .21 | .81 | 1.02 |
| TOTAL | | | .078 | 3.79 | 2.78 | 6.57 |

Floor costs on this page are given on a cost per square foot basis.

| Description | QUAN. | UNIT | LABOR HOURS | COST PER S.F. | | |
|---|---|---|---|---|---|---|
| | | | | MAT. | INST. | TOTAL |
| | | | | | | |
| | | | | | | |
| | | | | | | |
| | | | | | | |

# Floor Framing Price Sheet (Wood)

| | QUAN. | UNIT | LABOR HOURS | COST PER S.F. MAT. | COST PER S.F. INST. | COST PER S.F. TOTAL |
|---|---|---|---|---|---|---|
| Joists, #2 or better, pine, 2" x 4", 12" O.C. | 1.250 | L.F. | .016 | .49 | .58 | 1.07 |
| 16" O.C. | 1.000 | L.F. | .013 | .39 | .46 | .85 |
| 2" x 6", 12" O.C. | 1.250 | L.F. | .016 | .78 | .58 | 1.36 |
| 16" O.C. | 1.000 | L.F. | .013 | .62 | .46 | 1.08 |
| 2" x 8", 12" O.C. | 1.250 | L.F. | .018 | 1.19 | .65 | 1.84 |
| 16" O.C. | 1.000 | L.F. | .015 | .95 | .52 | 1.47 |
| 2" x 10", 12" O.C. | 1.250 | L.F. | .022 | 1.71 | .79 | 2.50 |
| 16" O.C. | 1.000 | L.F. | .018 | 1.37 | .63 | 2.00 |
| 2" x 12", 12" O.C. | 1.250 | L.F. | .023 | 2.34 | .81 | 3.15 |
| 16" O.C. | 1.000 | L.F. | .018 | 1.87 | .65 | 2.52 |
| Bridging, wood 1" x 3", joists 12" O.C. | .100 | Pr. | .006 | .03 | .22 | .25 |
| 16" O.C. | .080 | Pr. | .005 | .02 | .18 | .20 |
| Metal, galvanized, joists 12" O.C. | .100 | Pr. | .006 | .06 | .22 | .28 |
| 16" O.C. | .080 | Pr. | .005 | .05 | .18 | .23 |
| Compression type, joists 12" O.C. | .100 | Pr. | .004 | .14 | .14 | .28 |
| 16" O.C. | .080 | Pr. | .003 | .11 | .11 | .22 |
| Box sills, #2 or better pine, 2" x 4" | .150 | L.F. | .002 | .06 | .07 | .13 |
| 2" x 6" | .150 | L.F. | .002 | .09 | .07 | .16 |
| 2" x 8" | .150 | L.F. | .002 | .14 | .08 | .22 |
| 2" x 10" | .150 | L.F. | .003 | .21 | .09 | .30 |
| 2" x 12" | .150 | L.F. | .003 | .28 | .10 | .38 |
| Girders, including lally columns, 3 pieces spiked together, 2" x 8" | .125 | L.F. | .015 | .45 | .56 | 1.01 |
| 2" x 10" | .125 | L.F. | .016 | .60 | .59 | 1.19 |
| 2" x 12" | .125 | L.F. | .017 | .79 | .62 | 1.41 |
| Solid girders, 3" x 8" | .040 | L.F. | .004 | .19 | .15 | .34 |
| 3" x 10" | .040 | L.F. | .004 | .22 | .16 | .38 |
| 3" x 12" | .040 | L.F. | .004 | .25 | .16 | .41 |
| 4" x 8" | .040 | L.F. | .004 | .25 | .16 | .41 |
| 4" x 10" | .040 | L.F. | .004 | .28 | .17 | .45 |
| 4" x 12" | .040 | L.F. | .004 | .32 | .18 | .50 |
| Steel girders, bolted & including fabrication, wide flange shapes | | | | | | |
| 12" deep, 14#/l.f. | .040 | L.F. | .003 | .49 | .21 | .70 |
| 10" deep, 15#/l.f. | .040 | L.F. | .003 | .49 | .21 | .70 |
| 8" deep, 10#/l.f. | .040 | L.F. | .003 | .32 | .21 | .53 |
| 6" deep, 9#/l.f. | .040 | L.F. | .003 | .29 | .21 | .50 |
| 5" deep, 16#/l.f. | .040 | L.F. | .003 | .49 | .21 | .70 |
| Sheathing, plywood exterior grade CDX, 1/2" thick | 1.000 | S.F. | .011 | .54 | .41 | .95 |
| 5/8" thick | 1.000 | S.F. | .012 | .62 | .42 | 1.04 |
| 3/4" thick | 1.000 | S.F. | .013 | .73 | .46 | 1.19 |
| Boards, 1" x 8" laid regular | 1.000 | S.F. | .016 | 1.00 | .57 | 1.57 |
| Laid diagonal | 1.000 | S.F. | .019 | 1.00 | .67 | 1.67 |
| 1" x 10" laid regular | 1.000 | S.F. | .015 | 1.11 | .52 | 1.63 |
| Laid diagonal | 1.000 | S.F. | .018 | 1.12 | .63 | 1.75 |
| Furring, 1" x 3", 12" O.C. | 1.250 | L.F. | .029 | .26 | 1.01 | 1.27 |
| 16" O.C. | 1.000 | L.F. | .023 | .21 | .81 | 1.02 |
| 24" O.C. | .750 | L.F. | .017 | .16 | .61 | .77 |

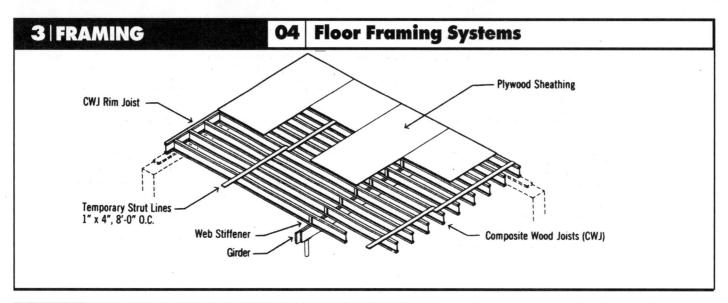

CWJ Rim Joist

Plywood Sheathing

Temporary Strut Lines 1" x 4", 8'-0" O.C.

Web Stiffener

Girder

Composite Wood Joists (CWJ)

**FRAMING 3**

| System Description | QUAN. | UNIT | LABOR HOURS | COST PER S.F. | | |
|---|---|---|---|---|---|---|
| | | | | MAT. | INST. | TOTAL |
| **9-1/2" COMPOSITE WOOD JOISTS, 16" O.C.** | | | | | | |
| CWJ, 9-1/2", 16" O.C., 15' span | 1.000 | L.F. | .018 | 1.55 | .64 | 2.19 |
| Temp. strut line, 1" x 4", 8' O.C. | .160 | L.F. | .003 | .06 | .11 | .17 |
| CWJ rim joist, 9-1/2" | .150 | L.F. | .003 | .23 | .10 | .33 |
| Concrete filled steel column, 4" diameter | .125 | L.F. | .002 | .09 | .09 | .18 |
| Girder, built up from three 2" x 8" | .125 | L.F. | .013 | .36 | .47 | .83 |
| Sheathing, plywood, subfloor, 5/8" CDX | 1.000 | S.F. | .012 | .62 | .42 | 1.04 |
| TOTAL | | | .051 | 2.91 | 1.83 | 4.74 |
| **11-1/2" COMPOSITE WOOD JOISTS, 16" O.C.** | | | | | | |
| CWJ, 11-1/2", 16" O.C., 18' span | 1.000 | L.F. | .018 | 1.68 | .65 | 2.33 |
| Temp. strut line, 1" x 4", 8' O.C. | .160 | L.F. | .003 | .06 | .11 | .17 |
| CWJ rim joist, 11-1/2" | .150 | L.F. | .003 | .25 | .10 | .35 |
| Concrete filled steel column, 4" diameter | .125 | L.F. | .002 | .09 | .09 | .18 |
| Girder, built up from three 2" x 10" | .125 | L.F. | .014 | .51 | .50 | 1.01 |
| Sheathing, plywood, subfloor, 5/8" CDX | 1.000 | S.F. | .012 | .62 | .42 | 1.04 |
| TOTAL | | | .052 | 3.21 | 1.87 | 5.08 |
| **14" COMPOSITE WOOD JOISTS, 16" O.C.** | | | | | | |
| CWJ, 14", 16" O.C., 22' span | 1.000 | L.F. | .020 | 1.83 | .70 | 2.53 |
| Temp. strut line, 1" x 4", 8' O.C. | .160 | L.F. | .003 | .06 | .11 | .17 |
| CWJ rim joist, 14" | .150 | L.F. | .003 | .27 | .10 | .37 |
| Concrete filled steel column, 4" diameter | .600 | L.F. | .002 | .09 | .09 | .18 |
| Girder, built up from three 2" x 12" | .600 | L.F. | .015 | .70 | .53 | 1.23 |
| Sheathing, plywood, subfloor, 5/8" CDX | 1.000 | S.F. | .012 | .62 | .42 | 1.04 |
| TOTAL | | | .055 | 3.57 | 1.95 | 5.52 |

Floor costs on this page are given on a cost per square foot basis.

| Description | QUAN. | UNIT | LABOR HOURS | COST PER S.F. | | |
|---|---|---|---|---|---|---|
| | | | | MAT. | INST. | TOTAL |
| | | | | | | |
| | | | | | | |
| | | | | | | |
| | | | | | | |

**Important: See the Reference Section for critical supporting data - Reference Nos., Crews & Location Factors**

# Floor Framing Price Sheet (Wood)

| Floor Framing Price Sheet (Wood) | QUAN. | UNIT | LABOR HOURS | COST PER S.F. | | |
|---|---|---|---|---|---|---|
| | | | | MAT. | INST. | TOTAL |
| Composite wood joist 9-1/2" deep, 12" O.C. | 1.250 | L.F. | .022 | 1.94 | .79 | 2.73 |
| 16" O.C. | 1.000 | L.F. | .018 | 1.55 | .64 | 2.19 |
| 11-1/2" deep, 12" O.C. | 1.250 | L.F. | .023 | 2.09 | .81 | 2.90 |
| 16" O.C. | 1.000 | L.F. | .018 | 1.68 | .65 | 2.33 |
| 14" deep, 12" O.C. | 1.250 | L.F. | .024 | 2.28 | .87 | 3.15 |
| 16" O.C. | 1.000 | L.F. | .020 | 1.83 | .70 | 2.53 |
| 16" deep, 12" O.C. | 1.250 | L.F. | .026 | 2.84 | .91 | 3.75 |
| 16" O.C. | 1.000 | L.F. | .021 | 2.28 | .73 | 3.01 |
| CWJ rim joist, 9-1/2" | .150 | L.F. | .003 | .23 | .10 | .33 |
| 11-1/2" | .150 | L.F. | .003 | .25 | .10 | .35 |
| 14" | .150 | L.F. | .003 | .27 | .10 | .37 |
| 16" | .150 | L.F. | .003 | .34 | .11 | .45 |
| Girders, including lally columns, 3 pieces spiked together, 2" x 8" | .125 | L.F. | .015 | .45 | .56 | 1.01 |
| 2" x 10" | .125 | L.F. | .016 | .60 | .59 | 1.19 |
| 2" x 12" | .125 | L.F. | .017 | .79 | .62 | 1.41 |
| Solid girders, 3" x 8" | .040 | L.F. | .004 | .19 | .15 | .34 |
| 3" x 10" | .040 | L.F. | .004 | .22 | .16 | .38 |
| 3" x 12" | .040 | L.F. | .004 | .25 | .16 | .41 |
| 4" x 8" | .040 | L.F. | .004 | .25 | .16 | .41 |
| 4" x 10" | .040 | L.F. | .004 | .28 | .17 | .45 |
| 4" x 12" | .040 | L.F. | .004 | .32 | .18 | .50 |
| Steel girders, bolted & including fabrication, wide flange shapes | | | | | | |
| 12" deep, 14#/l.f. | 1.000 | L.F. | .061 | 11.10 | 4.67 | 15.77 |
| 10" deep, 15#/l.f. | 1.000 | L.F. | .067 | 12.15 | 5.10 | 17.25 |
| 8" deep, 10#/l.f. | 1.000 | L.F. | .067 | 8.10 | 5.10 | 13.20 |
| 6" deep, 9#/l.f. | 1.000 | L.F. | .067 | 7.30 | 5.10 | 12.40 |
| 5" deep, 16#/l.f. | 1.000 | L.F. | .064 | 11.75 | 4.93 | 16.68 |
| Sheathing, plywood exterior grade CDX, 1/2" thick | 1.000 | S.F. | .011 | .54 | .41 | .95 |
| 5/8" thick | 1.000 | S.F. | .012 | .62 | .42 | 1.04 |
| 3/4" thick | 1.000 | S.F. | .013 | .73 | .46 | 1.19 |
| Boards, 1" x 8" laid regular | 1.000 | S.F. | .016 | 1.00 | .57 | 1.57 |
| Laid diagonal | 1.000 | S.F. | .019 | 1.00 | .67 | 1.67 |
| 1" x 10" laid regular | 1.000 | S.F. | .015 | 1.11 | .52 | 1.63 |
| Laid diagonal | 1.000 | S.F. | .018 | 1.12 | .63 | 1.75 |
| Furring, 1" x 3", 12" O.C. | 1.250 | L.F. | .029 | .26 | 1.01 | 1.27 |
| 16" O.C. | 1.000 | L.F. | .023 | .21 | .81 | 1.02 |
| 24" O.C. | .750 | L.F. | .017 | .16 | .61 | .77 |

**3 FRAMING**

115

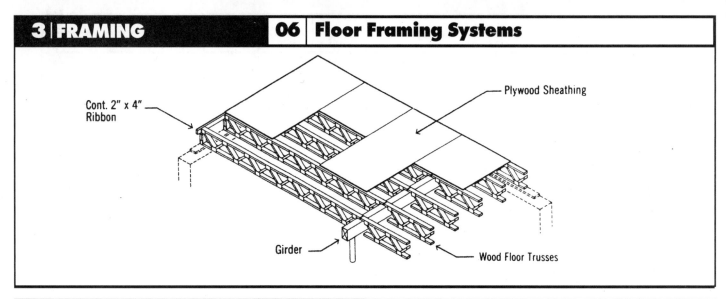

Cont. 2" x 4" Ribbon

Plywood Sheathing

Girder

Wood Floor Trusses

| System Description | QUAN. | UNIT | LABOR HOURS | COST PER S.F. | | |
|---|---|---|---|---|---|---|
| | | | | MAT. | INST. | TOTAL |
| **12" OPEN WEB JOISTS, 16" O.C.** | | | | | | |
| OWJ 12", 16" O.C., 21' span | 1.000 | L.F. | .018 | 1.65 | .65 | 2.30 |
| Continuous ribbing, 2" x 4" | .150 | L.F. | .002 | .06 | .07 | .13 |
| Concrete filled steel column, 4" diameter | .125 | L.F. | .002 | .09 | .09 | .18 |
| Girder, built up from three 2" x 8" | .125 | L.F. | .013 | .36 | .47 | .83 |
| Sheathing, plywood, subfloor, 5/8" CDX | 1.000 | S.F. | .012 | .62 | .42 | 1.04 |
| Furring, 1" x 3", 16" O.C. | 1.000 | L.F. | .023 | .21 | .81 | 1.02 |
| TOTAL | | | .070 | 2.99 | 2.51 | 5.50 |
| **14" OPEN WEB WOOD JOISTS, 16" O.C.** | | | | | | |
| OWJ 14", 16" O.C., 22' span | 1.000 | L.F. | .020 | 1.93 | .70 | 2.63 |
| Continuous ribbing, 2" x 4" | .150 | L.F. | .002 | .06 | .07 | .13 |
| Concrete filled steel column, 4" diameter | .125 | L.F. | .002 | .09 | .09 | .18 |
| Girder, built up from three 2" x 10" | .125 | L.F. | .014 | .51 | .50 | 1.01 |
| Sheathing, plywood, subfloor, 5/8" CDX | 1.000 | S.F. | .012 | .62 | .42 | 1.04 |
| Furring, 1" x 3",16" O.C. | 1.000 | L.F. | .023 | .21 | .81 | 1.02 |
| TOTAL | | | .073 | 3.42 | 2.59 | 6.01 |
| **16" OPEN WEB WOOD JOISTS, 16" O.C.** | | | | | | |
| OWJ 16", 16" O.C., 24' span | 1.000 | L.F. | .021 | 2.00 | .73 | 2.73 |
| Continuous ribbing, 2" x 4" | .150 | L.F. | .002 | .06 | .07 | .13 |
| Concrete filled steel column, 4" diameter | .125 | L.F. | .002 | .09 | .09 | .18 |
| Girder, built up from three 2" x 12" | .125 | L.F. | .015 | .70 | .53 | 1.23 |
| Sheathing, plywood, subfloor, 5/8" CDX | 1.000 | S.F. | .012 | .62 | .42 | 1.04 |
| Furring, 1" x 3", 16" O.C. | 1.000 | L.F. | .023 | .21 | .81 | 1.02 |
| TOTAL | | | .075 | 3.68 | 2.65 | 6.33 |

Floor costs on this page are given on a cost per square foot basis.

| Description | QUAN. | UNIT | LABOR HOURS | COST PER S.F. | | |
|---|---|---|---|---|---|---|
| | | | | MAT. | INST. | TOTAL |
| | | | | | | |
| | | | | | | |
| | | | | | | |
| | | | | | | |

**Important: See the Reference Section for critical supporting data - Reference Nos., Crews & Location Factors**

FRAMING 3

# Floor Framing Price Sheet (Wood)

| | QUAN. | UNIT | LABOR HOURS | COST PER S.F. MAT. | INST. | TOTAL |
|---|---|---|---|---|---|---|
| Open web joists, 12" deep, 12" O.C. | 1.250 | L.F. | .023 | 2.06 | .81 | 2.87 |
| 16" O.C. | 1.000 | L.F. | .018 | 1.65 | .65 | 2.30 |
| 14" deep, 12" O.C. | 1.250 | L.F. | .024 | 2.41 | .87 | 3.28 |
| 16" O.C. | 1.000 | L.F. | .020 | 1.93 | .70 | 2.63 |
| 16 " deep, 12" O.C. | 1.250 | L.F. | .026 | 2.50 | .91 | 3.41 |
| 16" O.C. | 1.000 | L.F. | .021 | 2.00 | .73 | 2.73 |
| 18" deep, 12" O.C. | 1.250 | L.F. | .027 | 2.59 | .96 | 3.55 |
| 16" O.C. | 1.000 | L.F. | .022 | 2.08 | .77 | 2.85 |
| Continuous ribbing, 2" x 4" | .150 | L.F. | .002 | .06 | .07 | .13 |
| 2" x 6" | .150 | L.F. | .002 | .09 | .07 | .16 |
| 2" x 8" | .150 | L.F. | .002 | .14 | .08 | .22 |
| 2" x 10" | .150 | L.F. | .003 | .21 | .09 | .30 |
| 2" x 12" | .150 | L.F. | .003 | .28 | .10 | .38 |
| Girders, including lally columns, 3 pieces spiked together, 2" x 8" | .125 | L.F. | .015 | .45 | .56 | 1.01 |
| 2" x 10" | .125 | L.F. | .016 | .60 | .59 | 1.19 |
| 2" x 12" | .125 | L.F. | .017 | .79 | .62 | 1.41 |
| Solid girders, 3" x 8" | .040 | L.F. | .004 | .19 | .15 | .34 |
| 3" x 10" | .040 | L.F. | .004 | .22 | .16 | .38 |
| 3" x 12" | .040 | L.F. | .004 | .25 | .16 | .41 |
| 4" x 8" | .040 | L.F. | .004 | .25 | .16 | .41 |
| 4" x 10" | .040 | L.F. | .004 | .28 | .17 | .45 |
| 4" x 12" | .040 | L.F. | .004 | .32 | .18 | .50 |
| Steel girders, bolted & including fabrication, wide flange shapes | | | | | | |
| 12" deep, 14#/l.f. | 1.000 | L.F. | .061 | 11.10 | 4.67 | 15.77 |
| 10" deep, 15#/l.f. | 1.000 | L.F. | .067 | 12.15 | 5.10 | 17.25 |
| 8" deep, 10#/l.f. | 1.000 | L.F. | .067 | 8.10 | 5.10 | 13.20 |
| 6" deep, 9#/l.f. | 1.000 | L.F. | .067 | 7.30 | 5.10 | 12.40 |
| 5" deep, 16#/l.f. | 1.000 | L.F. | .064 | 11.75 | 4.93 | 16.68 |
| Sheathing, plywood exterior grade CDX, 1/2" thick | 1.000 | S.F. | .011 | .54 | .41 | .95 |
| 5/8" thick | 1.000 | S.F. | .012 | .62 | .42 | 1.04 |
| 3/4" thick | 1.000 | S.F. | .013 | .73 | .46 | 1.19 |
| Boards, 1" x 8" laid regular | 1.000 | S.F. | .016 | 1.00 | .57 | 1.57 |
| Laid diagonal | 1.000 | S.F. | .019 | 1.00 | .67 | 1.67 |
| 1" x 10" laid regular | 1.000 | S.F. | .015 | 1.11 | .52 | 1.63 |
| Laid diagonal | 1.000 | S.F. | .018 | 1.12 | .63 | 1.75 |
| Furring, 1" x 3", 12" O.C. | 1.250 | L.F. | .029 | .26 | 1.01 | 1.27 |
| 16" O.C. | 1.000 | L.F. | .023 | .21 | .81 | 1.02 |
| 24" O.C. | .750 | L.F. | .017 | .16 | .61 | .77 |

**3 FRAMING**

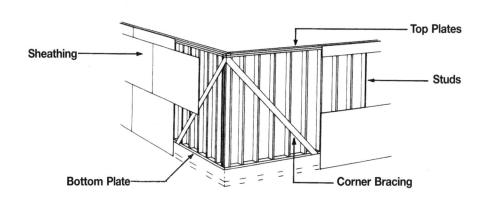

| System Description | QUAN. | UNIT | LABOR HOURS | COST PER S.F. | | |
|---|---|---|---|---|---|---|
| | | | | MAT. | INST. | TOTAL |
| **2″ X 4″, 16″ O.C.** | | | | | | |
| 2″ x 4″ studs, 16″ O.C. | 1.000 | L.F. | .015 | .39 | .52 | .91 |
| Plates, 2″ x 4″, double top, single bottom | .375 | L.F. | .005 | .15 | .20 | .35 |
| Corner bracing, let-in, 1″ x 6″ | .063 | L.F. | .003 | .04 | .12 | .16 |
| Sheathing, 1/2″ plywood, CDX | 1.000 | S.F. | .011 | .54 | .41 | .95 |
| TOTAL | | | .034 | 1.12 | 1.25 | 2.37 |
| **2″ X 4″, 24″ O.C.** | | | | | | |
| 2″ x 4″ studs, 24″ O.C. | .750 | L.F. | .011 | .29 | .39 | .68 |
| Plates, 2″ x 4″, double top, single bottom | .375 | L.F. | .005 | .15 | .20 | .35 |
| Corner bracing, let-in, 1″ x 6″ | .063 | L.F. | .002 | .04 | .08 | .12 |
| Sheathing, 1/2″ plywood, CDX | 1.000 | S.F. | .011 | .54 | .41 | .95 |
| TOTAL | | | .029 | 1.02 | 1.08 | 2.10 |
| **2″ X 6″, 16″ O.C.** | | | | | | |
| 2″ x 6″ studs, 16″ O.C. | 1.000 | L.F. | .016 | .62 | .57 | 1.19 |
| Plates, 2″ x 6″, double top, single bottom | .375 | L.F. | .006 | .23 | .21 | .44 |
| Corner bracing, let-in, 1″ x 6″ | .063 | L.F. | .003 | .04 | .12 | .16 |
| Sheathing, 1/2″ plywood, CDX | 1.000 | S.F. | .011 | .54 | .41 | .95 |
| TOTAL | | | .036 | 1.43 | 1.31 | 2.74 |
| **2″ X 6″, 24″ O.C.** | | | | | | |
| 2″ x 6″ studs, 24″ O.C. | .750 | L.F. | .012 | .47 | .43 | .90 |
| Plates, 2″ x 6″, double top, single bottom | .375 | L.F. | .006 | .23 | .21 | .44 |
| Corner bracing, let-in, 1″ x 6″ | .063 | L.F. | .002 | .04 | .08 | .12 |
| Sheathing, 1/2″ plywood, CDX | 1.000 | S.F. | .011 | .54 | .41 | .95 |
| TOTAL | | | .031 | 1.28 | 1.13 | 2.41 |

The wall costs on this page are given in cost per square foot of wall.
For window and door openings see below.

| Description | QUAN. | UNIT | LABOR HOURS | COST PER S.F. | | |
|---|---|---|---|---|---|---|
| | | | | MAT. | INST. | TOTAL |
| | | | | | | |
| | | | | | | |
| | | | | | | |

**Important: See the Reference Section for critical supporting data - Reference Nos., Crews & Location Factors**

| Exterior Wall Framing Price Sheet | QUAN. | UNIT | LABOR HOURS | COST PER S.F. | | |
|---|---|---|---|---|---|---|
| | | | | MAT. | INST. | TOTAL |
| Studs, #2 or better, 2" x 4", 12" O.C. | 1.250 | L.F. | .018 | .49 | .65 | 1.14 |
| 16" O.C. | 1.000 | L.F. | .015 | .39 | .52 | .91 |
| 24" O.C. | .750 | L.F. | .011 | .29 | .39 | .68 |
| 32" O.C. | .600 | L.F. | .009 | .23 | .31 | .54 |
| 2" x 6", 12" O.C. | 1.250 | L.F. | .020 | .78 | .71 | 1.49 |
| 16" O.C. | 1.000 | L.F. | .016 | .62 | .57 | 1.19 |
| 24" O.C. | .750 | L.F. | .012 | .47 | .43 | .90 |
| 32" O.C. | .600 | L.F. | .010 | .37 | .34 | .71 |
| 2" x 8", 12" O.C. | 1.250 | L.F. | .025 | 1.55 | .89 | 2.44 |
| 16" O.C. | 1.000 | L.F. | .020 | 1.24 | .71 | 1.95 |
| 24" O.C. | .750 | L.F. | .015 | .93 | .53 | 1.46 |
| 32" O.C. | .600 | L.F. | .012 | .74 | .43 | 1.17 |
| Plates, #2 or better, double top, single bottom, 2" x 4" | .375 | L.F. | .005 | .15 | .20 | .35 |
| 2" x 6" | .375 | L.F. | .006 | .23 | .21 | .44 |
| 2" x 8" | .375 | L.F. | .008 | .47 | .27 | .74 |
| Corner bracing, let-in 1" x 6" boards, studs, 12" O.C. | .070 | L.F. | .004 | .04 | .13 | .17 |
| 16" O.C. | .063 | L.F. | .003 | .04 | .12 | .16 |
| 24" O.C. | .063 | L.F. | .002 | .04 | .08 | .12 |
| 32" O.C. | .057 | L.F. | .002 | .03 | .07 | .10 |
| Let-in steel ("T" shape), studs, 12" O.C. | .070 | L.F. | .001 | .03 | .03 | .06 |
| 16" O.C. | .063 | L.F. | .001 | .03 | .03 | .06 |
| 24" O.C. | .063 | L.F. | .001 | .03 | .03 | .06 |
| 32" O.C. | .057 | L.F. | .001 | .03 | .03 | .06 |
| Sheathing, plywood CDX, 3/8" thick | 1.000 | S.F. | .010 | .48 | .37 | .85 |
| 1/2" thick | 1.000 | S.F. | .011 | .54 | .41 | .95 |
| 5/8" thick | 1.000 | S.F. | .012 | .62 | .44 | 1.06 |
| 3/4" thick | 1.000 | S.F. | .013 | .73 | .47 | 1.20 |
| Boards, 1" x 6", laid regular | 1.000 | S.F. | .025 | 1.43 | .88 | 2.31 |
| Laid diagonal | 1.000 | S.F. | .027 | 1.43 | .97 | 2.40 |
| 1" x 8", laid regular | 1.000 | S.F. | .021 | 1.43 | .74 | 2.17 |
| Laid diagonal | 1.000 | S.F. | .025 | 1.43 | .88 | 2.31 |
| Wood fiber, regular, no vapor barrier, 1/2" thick | 1.000 | S.F. | .013 | .58 | .47 | 1.05 |
| 5/8" thick | 1.000 | S.F. | .013 | .78 | .47 | 1.25 |
| Asphalt impregnated 25/32" thick | 1.000 | S.F. | .013 | .39 | .47 | .86 |
| 1/2" thick | 1.000 | S.F. | .013 | .32 | .47 | .79 |
| Polystyrene, regular, 3/4" thick | 1.000 | S.F. | .010 | .37 | .36 | .73 |
| 2" thick | 1.000 | S.F. | .011 | 1.06 | .39 | 1.45 |
| Fiberglass, foil faced, 1" thick | 1.000 | S.F. | .008 | .88 | .28 | 1.16 |
| 2" thick | 1.000 | S.F. | .009 | 1.49 | .32 | 1.81 |
| | | | | | | |

| Window & Door Openings | QUAN. | UNIT | LABOR HOURS | COST EACH | | |
|---|---|---|---|---|---|---|
| | | | | MAT. | INST. | TOTAL |
| The following costs are to be added to the total costs of the wall for each opening. Do not subtract the area of the openings. | | | | | | |
| Headers, 2" x 6" double, 2' long | 4.000 | L.F. | .178 | 2.48 | 6.30 | 8.78 |
| 3' long | 6.000 | L.F. | .267 | 3.72 | 9.50 | 13.22 |
| 4' long | 8.000 | L.F. | .356 | 4.96 | 12.65 | 17.61 |
| 5' long | 10.000 | L.F. | .444 | 6.20 | 15.80 | 22.00 |
| 2" x 8" double, 4' long | 8.000 | L.F. | .376 | 7.60 | 13.45 | 21.05 |
| 5' long | 10.000 | L.F. | .471 | 9.50 | 16.80 | 26.30 |
| 6' long | 12.000 | L.F. | .565 | 11.40 | 20.00 | 31.40 |
| 8' long | 16.000 | L.F. | .753 | 15.20 | 27.00 | 42.20 |
| 2" x 10" double, 4' long | 8.000 | L.F. | .400 | 10.95 | 14.25 | 25.20 |
| 6' long | 12.000 | L.F. | .600 | 16.45 | 21.50 | 37.95 |
| 8' long | 16.000 | L.F. | .800 | 22.00 | 28.50 | 50.50 |
| 10' long | 20.000 | L.F. | 1.000 | 27.50 | 35.50 | 63.00 |
| 2" x 12" double, 8' long | 16.000 | L.F. | .853 | 30.00 | 30.50 | 60.50 |
| 12' long | 24.000 | L.F. | 1.280 | 45.00 | 45.50 | 90.50 |

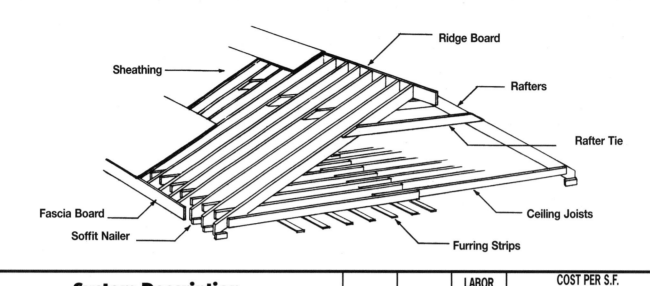

| System Description | QUAN. | UNIT | LABOR HOURS | COST PER S.F. | | |
|---|---|---|---|---|---|---|
| | | | | MAT. | INST. | TOTAL |
| **2" X 6" RAFTERS, 16" O.C., 4/12 PITCH** | | | | | | |
| Rafters, 2" x 6", 16" O.C., 4/12 pitch | 1.170 | L.F. | .019 | .73 | .67 | 1.40 |
| Ceiling joists, 2" x 4", 16" O.C. | 1.000 | L.F. | .013 | .39 | .46 | .85 |
| Ridge board, 2" x 6" | .050 | L.F. | .002 | .03 | .06 | .09 |
| Fascia board, 2" x 6" | .100 | L.F. | .005 | .07 | .19 | .26 |
| Rafter tie, 1" x 4", 4' O.C. | .060 | L.F. | .001 | .02 | .04 | .06 |
| Soffit nailer (outrigger), 2" x 4", 24" O.C. | .170 | L.F. | .004 | .07 | .16 | .23 |
| Sheathing, exterior, plywood, CDX, 1/2" thick | 1.170 | S.F. | .013 | .63 | .48 | 1.11 |
| Furring strips, 1" x 3", 16" O.C. | 1.000 | L.F. | .023 | .21 | .81 | 1.02 |
| TOTAL | | | .080 | 2.15 | 2.87 | 5.02 |
| **2" X 8" RAFTERS, 16" O.C., 4/12 PITCH** | | | | | | |
| Rafters, 2" x 8", 16" O.C., 4/12 pitch | 1.170 | L.F. | .020 | 1.11 | .70 | 1.81 |
| Ceiling joists, 2" x 6", 16" O.C. | 1.000 | L.F. | .013 | .62 | .46 | 1.08 |
| Ridge board, 2" x 8" | .050 | L.F. | .002 | .05 | .06 | .11 |
| Fascia board, 2" x 8" | .100 | L.F. | .007 | .10 | .25 | .35 |
| Rafter tie, 1" x 4", 4' O.C. | .060 | L.F. | .001 | .02 | .04 | .06 |
| Soffit nailer (outrigger), 2" x 4", 24" O.C. | .170 | L.F. | .004 | .07 | .16 | .23 |
| Sheathing, exterior, plywood, CDX, 1/2" thick | 1.170 | S.F. | .013 | .63 | .48 | 1.11 |
| Furring strips, 1" x 3", 16" O.C. | 1.000 | L.F. | .023 | .21 | .81 | 1.02 |
| TOTAL | | | .083 | 2.81 | 2.96 | 5.77 |

The cost of this system is based on the square foot of plan area.
All quantities have been adjusted accordingly.

| Description | QUAN. | UNIT | LABOR HOURS | COST PER S.F. | | |
|---|---|---|---|---|---|---|
| | | | | MAT. | INST. | TOTAL |
| | | | | | | |
| | | | | | | |
| | | | | | | |
| | | | | | | |
| | | | | | | |

| Gable End Roof Framing Price Sheet | QUAN. | UNIT | LABOR HOURS | COST PER S.F. | | |
|---|---|---|---|---|---|---|
| | | | | MAT. | INST. | TOTAL |
| Rafters, #2 or better, 16" O.C., 2" x 6", 4/12 pitch | 1.170 | L.F. | .019 | .73 | .67 | 1.40 |
| 8/12 pitch | 1.330 | L.F. | .027 | .82 | .94 | 1.76 |
| 2" x 8", 4/12 pitch | 1.170 | L.F. | .020 | 1.11 | .70 | 1.81 |
| 8/12 pitch | 1.330 | L.F. | .028 | 1.26 | 1.01 | 2.27 |
| 2" x 10", 4/12 pitch | 1.170 | L.F. | .030 | 1.60 | 1.05 | 2.65 |
| 8/12 pitch | 1.330 | L.F. | .043 | 1.82 | 1.53 | 3.35 |
| 24" O.C., 2" x 6", 4/12 pitch | .940 | L.F. | .015 | .58 | .54 | 1.12 |
| 8/12 pitch | 1.060 | L.F. | .021 | .66 | .75 | 1.41 |
| 2" x 8", 4/12 pitch | .940 | L.F. | .016 | .89 | .56 | 1.45 |
| 8/12 pitch | 1.060 | L.F. | .023 | 1.01 | .81 | 1.82 |
| 2" x 10", 4/12 pitch | .940 | L.F. | .024 | 1.29 | .85 | 2.14 |
| 8/12 pitch | 1.060 | L.F. | .034 | 1.45 | 1.22 | 2.67 |
| Ceiling joist, #2 or better, 2" x 4", 16" O.C. | 1.000 | L.F. | .013 | .39 | .46 | .85 |
| 24" O.C. | .750 | L.F. | .010 | .29 | .35 | .64 |
| 2" x 6", 16" O.C. | 1.000 | L.F. | .013 | .62 | .46 | 1.08 |
| 24" O.C. | .750 | L.F. | .010 | .47 | .35 | .82 |
| 2" x 8", 16" O.C. | 1.000 | L.F. | .015 | .95 | .52 | 1.47 |
| 24" O.C. | .750 | L.F. | .011 | .71 | .39 | 1.10 |
| 2" x 10", 16" O.C. | 1.000 | L.F. | .018 | 1.37 | .63 | 2.00 |
| 24" O.C. | .750 | L.F. | .013 | 1.03 | .47 | 1.50 |
| Ridge board, #2 or better, 1" x 6" | .050 | L.F. | .001 | .05 | .05 | .10 |
| 1" x 8" | .050 | L.F. | .001 | .07 | .05 | .12 |
| 1" x 10" | .050 | L.F. | .002 | .08 | .06 | .14 |
| 2" x 6" | .050 | L.F. | .002 | .03 | .06 | .09 |
| 2" x 8" | .050 | L.F. | .002 | .05 | .06 | .11 |
| 2" x 10" | .050 | L.F. | .002 | .07 | .07 | .14 |
| Fascia board, #2 or better, 1" x 6" | .100 | L.F. | .004 | .05 | .14 | .19 |
| 1" x 8" | .100 | L.F. | .005 | .06 | .16 | .22 |
| 1" x 10" | .100 | L.F. | .005 | .07 | .18 | .25 |
| 2" x 6" | .100 | L.F. | .006 | .08 | .20 | .28 |
| 2" x 8" | .100 | L.F. | .007 | .10 | .25 | .35 |
| 2" x 10" | .100 | L.F. | .004 | .27 | .13 | .40 |
| Rafter tie, #2 or better, 4' O.C., 1" x 4" | .060 | L.F. | .001 | .02 | .04 | .06 |
| 1" x 6" | .060 | L.F. | .001 | .03 | .05 | .08 |
| 2" x 4" | .060 | L.F. | .002 | .03 | .06 | .09 |
| 2" x 6" | .060 | L.F. | .002 | .04 | .07 | .11 |
| Soffit nailer (outrigger), 2" x 4", 16" O.C. | .220 | L.F. | .006 | .09 | .20 | .29 |
| 24" O.C. | .170 | L.F. | .004 | .07 | .16 | .23 |
| 2" x 6", 16" O.C. | .220 | L.F. | .006 | .10 | .23 | .33 |
| 24" O.C. | .170 | L.F. | .005 | .08 | .18 | .26 |
| Sheathing, plywood CDX, 4/12 pitch, 3/8" thick. | 1.170 | S.F. | .012 | .56 | .43 | .99 |
| 1/2" thick | 1.170 | S.F. | .013 | .63 | .48 | 1.11 |
| 5/8" thick | 1.170 | S.F. | .014 | .73 | .51 | 1.24 |
| 8/12 pitch, 3/8" | 1.330 | S.F. | .014 | .64 | .49 | 1.13 |
| 1/2" thick | 1.330 | S.F. | .015 | .72 | .55 | 1.27 |
| 5/8" thick | 1.330 | S.F. | .016 | .82 | .59 | 1.41 |
| Boards, 4/12 pitch roof, 1" x 6" | 1.170 | S.F. | .026 | 1.67 | .92 | 2.59 |
| 1" x 8" | 1.170 | S.F. | .021 | 1.67 | .76 | 2.43 |
| 8/12 pitch roof, 1" x 6" | 1.330 | S.F. | .029 | 1.90 | 1.05 | 2.95 |
| 1" x 8" | 1.330 | S.F. | .024 | 1.90 | .86 | 2.76 |
| Furring, 1" x 3", 12" O.C. | 1.200 | L.F. | .027 | .25 | .97 | 1.22 |
| 16" O.C. | 1.000 | L.F. | .023 | .21 | .81 | 1.02 |
| 24" O.C. | .800 | L.F. | .018 | .17 | .65 | .82 |
| | | | | | | |
| | | | | | | |
| | | | | | | |
| | | | | | | |

3 FRAMING

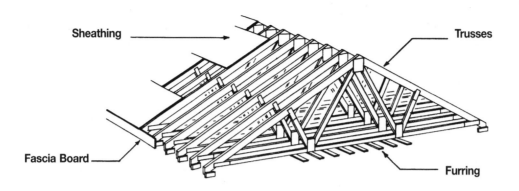

Sheathing → Trusses

Fascia Board → Furring

| System Description | QUAN. | UNIT | LABOR HOURS | COST PER S.F. | | |
|---|---|---|---|---|---|---|
| | | | | MAT. | INST. | TOTAL |
| **TRUSS, 16″ O.C., 4/12 PITCH, 1′ OVERHANG, 26′ SPAN** | | | | | | |
| Truss, 40# loading, 16″ O.C., 4/12 pitch, 26′ span | .030 | Ea. | .021 | 2.04 | 1.03 | 3.07 |
| Fascia board, 2″ x 6″ | .100 | L.F. | .005 | .07 | .19 | .26 |
| Sheathing, exterior, plywood, CDX, 1/2″ thick | 1.170 | S.F. | .013 | .63 | .48 | 1.11 |
| Furring, 1″ x 3″, 16″ O.C. | 1.000 | L.F. | .023 | .21 | .81 | 1.02 |
| TOTAL | | | .062 | 2.95 | 2.51 | 5.46 |
| **TRUSS, 16″ O.C., 8/12 PITCH, 1′ OVERHANG, 26′ SPAN** | | | | | | |
| Truss, 40# loading, 16″ O.C., 8/12 pitch, 26′ span | .030 | Ea. | .023 | 2.42 | 1.13 | 3.55 |
| Fascia board, 2″ x 6″ | .100 | L.F. | .005 | .07 | .19 | .26 |
| Sheathing, exterior, plywood, CDX, 1/2″ thick | 1.330 | S.F. | .015 | .72 | .55 | 1.27 |
| Furring, 1″ x 3″, 16″ O.C. | 1.000 | L.F. | .023 | .21 | .81 | 1.02 |
| TOTAL | | | .066 | 3.42 | 2.68 | 6.10 |
| **TRUSS, 24″ O.C., 4/12 PITCH, 1′ OVERHANG, 26′ SPAN** | | | | | | |
| Truss, 40# loading, 24″ O.C., 4/12 pitch, 26′ span | .020 | Ea. | .014 | 1.36 | .69 | 2.05 |
| Fascia board, 2″ x 6″ | .100 | L.F. | .005 | .07 | .19 | .26 |
| Sheathing, exterior, plywood, CDX, 1/2″ thick | 1.170 | S.F. | .013 | .63 | .48 | 1.11 |
| Furring, 1″ x 3″, 16″ O.C. | 1.000 | L.F. | .023 | .21 | .81 | 1.02 |
| TOTAL | | | .055 | 2.27 | 2.17 | 4.44 |
| **TRUSS, 24″ O.C., 8/12 PITCH, 1′ OVERHANG, 26′ SPAN** | | | | | | |
| Truss, 40# loading, 24″ O.C., 8/12 pitch, 26′ span | .020 | Ea. | .015 | 1.61 | .75 | 2.36 |
| Fascia board, 2″ x 6″ | .100 | L.F. | .005 | .07 | .19 | .26 |
| Sheathing, exterior, plywood, CDX, 1/2″ thick | 1.330 | S.F. | .015 | .72 | .55 | 1.27 |
| Furring, 1″ x 3″, 16″ O.C. | 1.000 | L.F. | .023 | .21 | .81 | 1.02 |
| TOTAL | | | .058 | 2.61 | 2.30 | 4.91 |

The cost of this system is based on the square foot of plan area.
A one foot overhang is included.

| Description | QUAN. | UNIT | LABOR HOURS | COST PER S.F. | | |
|---|---|---|---|---|---|---|
| | | | | MAT. | INST. | TOTAL |
| | | | | | | |
| | | | | | | |
| | | | | | | |

**Important: See the Reference Section for critical supporting data - Reference Nos., Crews & Location Factors**

# Truss Roof Framing Price Sheet

| Truss Roof Framing Price Sheet | QUAN. | UNIT | LABOR HOURS | COST PER S.F. | | |
|---|---|---|---|---|---|---|
| | | | | MAT. | INST. | TOTAL |
| Truss, 40# loading, including 1' overhang, 4/12 pitch, 24' span, 16" O.C. | .033 | Ea. | .022 | 1.60 | 1.08 | 2.68 |
| 24" O.C. | .022 | Ea. | .015 | 1.07 | .72 | 1.79 |
| 26' span, 16" O.C. | .030 | Ea. | .021 | 2.04 | 1.03 | 3.07 |
| 24" O.C. | .020 | Ea. | .014 | 1.36 | .69 | 2.05 |
| 28' span, 16" O.C. | .027 | Ea. | .020 | 1.59 | .99 | 2.58 |
| 24" O.C. | .019 | Ea. | .014 | 1.12 | .70 | 1.82 |
| 32' span, 16" O.C. | .024 | Ea. | .019 | 1.99 | .93 | 2.92 |
| 24" O.C. | .016 | Ea. | .013 | 1.33 | .62 | 1.95 |
| 36' span, 16" O.C. | .022 | Ea. | .019 | 2.29 | .94 | 3.23 |
| 24" O.C. | .015 | Ea. | .013 | 1.56 | .64 | 2.20 |
| 8/12 pitch, 24' span, 16" O.C. | .033 | Ea. | .024 | 2.44 | 1.16 | 3.60 |
| 24" O.C. | .022 | Ea. | .016 | 1.63 | .78 | 2.41 |
| 26' span, 16" O.C. | .030 | Ea. | .023 | 2.42 | 1.13 | 3.55 |
| 24" O.C. | .020 | Ea. | .015 | 1.61 | .75 | 2.36 |
| 28' span, 16" O.C. | .027 | Ea. | .022 | 2.34 | 1.07 | 3.41 |
| 24" O.C. | .019 | Ea. | .016 | 1.64 | .75 | 2.39 |
| 32' span, 16" O.C. | .024 | Ea. | .021 | 2.45 | 1.04 | 3.49 |
| 24" O.C. | .016 | Ea. | .014 | 1.63 | .70 | 2.33 |
| 36' span, 16" O.C. | .022 | Ea. | .021 | 2.66 | 1.04 | 3.70 |
| 24" O.C. | .015 | Ea. | .015 | 1.82 | .71 | 2.53 |
| Fascia board, #2 or better, 1" x 6" | .100 | L.F. | .004 | .05 | .14 | .19 |
| 1" x 8" | .100 | L.F. | .005 | .06 | .16 | .22 |
| 1" x 10" | .100 | L.F. | .005 | .07 | .18 | .25 |
| 2" x 6" | .100 | L.F. | .006 | .08 | .20 | .28 |
| 2" x 8" | .100 | L.F. | .007 | .10 | .25 | .35 |
| 2" x 10" | .100 | L.F. | .009 | .14 | .32 | .46 |
| Sheathing, plywood CDX, 4/12 pitch, 3/8" thick | 1.170 | S.F. | .012 | .56 | .43 | .99 |
| 1/2" thick | 1.170 | S.F. | .013 | .63 | .48 | 1.11 |
| 5/8" thick | 1.170 | S.F. | .014 | .73 | .51 | 1.24 |
| 8/12 pitch, 3/8" thick | 1.330 | S.F. | .014 | .64 | .49 | 1.13 |
| 1/2" thick | 1.330 | S.F. | .015 | .72 | .55 | 1.27 |
| 5/8" thick | 1.330 | S.F. | .016 | .82 | .59 | 1.41 |
| Boards, 4/12 pitch, 1" x 6" | 1.170 | S.F. | .026 | 1.67 | .92 | 2.59 |
| 1" x 8" | 1.170 | S.F. | .021 | 1.67 | .76 | 2.43 |
| 8/12 pitch, 1" x 6" | 1.330 | S.F. | .029 | 1.90 | 1.05 | 2.95 |
| 1" x 8" | 1.330 | S.F. | .024 | 1.90 | .86 | 2.76 |
| Furring, 1" x 3", 12" O.C. | 1.200 | L.F. | .027 | .25 | .97 | 1.22 |
| 16" O.C. | 1.000 | L.F. | .023 | .21 | .81 | 1.02 |
| 24" O.C. | .800 | L.F. | .018 | .17 | .65 | .82 |

**3 FRAMING**

Ceiling Joists

Sheathing

Fascia Board

Jack Rafters

Hip Rafter

| System Description | QUAN. | UNIT | LABOR HOURS | COST PER S.F. | | |
|---|---|---|---|---|---|---|
| | | | | MAT. | INST. | TOTAL |
| **2″ X 6″, 16″ O.C., 4/12 PITCH** | | | | | | |
| Hip and valley rafters, 2″ x 6″, ordinary | .160 | L.F. | .003 | .10 | .12 | .22 |
| Jack rafters, 2″ x 8″, 16″ O.C., 4/12 pitch | 1.430 | L.F. | .038 | .89 | 1.36 | 2.25 |
| Ceiling joists, 2″ x 6″, 16″ O.C. | 1.000 | L.F. | .013 | .39 | .46 | .85 |
| Fascia board, 2″ x 8″ | .220 | L.F. | .016 | .21 | .56 | .77 |
| Soffit nailer (outrigger), 2″ x 4″, 24″ O.C. | .220 | L.F. | .006 | .09 | .20 | .29 |
| Sheathing, 1/2″ exterior plywood, CDX | 1.570 | S.F. | .018 | .85 | .64 | 1.49 |
| Furring strips, 1″ x 3″, 16″ O.C. | 1.000 | L.F. | .023 | .21 | .81 | 1.02 |
| TOTAL | | | .117 | 2.74 | 4.15 | 6.89 |
| **2″ X 8″, 16″ O.C., 4/12 PITCH** | | | | | | |
| Hip rafters, 2″ x 6″, 4/12 pitch | .160 | L.F. | .004 | .15 | .13 | .28 |
| Jack rafters, 2″ x 6″, 16″ O.C., 4/12 pitch | 1.430 | L.F. | .047 | 1.36 | 1.66 | 3.02 |
| Ceiling joists, 2″ x 4″, 16″ O.C. | 1.000 | L.F. | .013 | .62 | .46 | 1.08 |
| Fascia board, 2″ x 6″ | .220 | L.F. | .012 | .16 | .43 | .59 |
| Soffit nailer (outrigger), 2″ x 4″, 24″ O.C. | .220 | L.F. | .006 | .09 | .20 | .29 |
| Sheathing, 1/2″ exterior plywood, CDX | 1.570 | S.F. | .018 | .85 | .64 | 1.49 |
| Furring strips, 1″ x 3″, 16″ O.C. | 1.000 | L.F. | .023 | .21 | .81 | 1.02 |
| TOTAL | | | .123 | 3.44 | 4.33 | 7.77 |

The cost of this system is based on S.F. of plan area. Measurement is area under the hip roof only. See gable roof system for added costs.

| Description | QUAN. | UNIT | LABOR HOURS | COST PER S.F. | | |
|---|---|---|---|---|---|---|
| | | | | MAT. | INST. | TOTAL |
| | | | | | | |
| | | | | | | |
| | | | | | | |
| | | | | | | |
| | | | | | | |

**Important: See the Reference Section for critical supporting data - Reference Nos., Crews & Location Factors**

# Hip Roof Framing Price Sheet

| | QUAN. | UNIT | LABOR HOURS | COST PER S.F. | | |
|---|---|---|---|---|---|---|
| | | | | MAT. | INST. | TOTAL |
| Hip rafters, #2 or better, 2" x 6", 4/12 pitch | .160 | L.F. | .003 | .10 | .12 | .22 |
| 8/12 pitch | .210 | L.F. | .006 | .13 | .20 | .33 |
| 2" x 8", 4/12 pitch | .160 | L.F. | .004 | .15 | .13 | .28 |
| 8/12 pitch | .210 | L.F. | .006 | .20 | .22 | .42 |
| 2" x 10", 4/12 pitch | .160 | L.F. | .004 | .22 | .16 | .38 |
| 8/12 pitch roof | .210 | L.F. | .008 | .29 | .27 | .56 |
| Jack rafters, #2 or better, 16" O.C., 2" x 6", 4/12 pitch | 1.430 | L.F. | .038 | .89 | 1.36 | 2.25 |
| 8/12 pitch | 1.800 | L.F. | .061 | 1.12 | 2.16 | 3.28 |
| 2" x 8", 4/12 pitch | 1.430 | L.F. | .047 | 1.36 | 1.66 | 3.02 |
| 8/12 pitch | 1.800 | L.F. | .075 | 1.71 | 2.66 | 4.37 |
| 2" x 10", 4/12 pitch | 1.430 | L.F. | .051 | 1.96 | 1.82 | 3.78 |
| 8/12 pitch | 1.800 | L.F. | .082 | 2.47 | 2.93 | 5.40 |
| 24" O.C., 2" x 6", 4/12 pitch | 1.150 | L.F. | .031 | .71 | 1.09 | 1.80 |
| 8/12 pitch | 1.440 | L.F. | .048 | .89 | 1.73 | 2.62 |
| 2" x 8", 4/12 pitch | 1.150 | L.F. | .038 | 1.09 | 1.33 | 2.42 |
| 8/12 pitch | 1.440 | L.F. | .060 | 1.37 | 2.13 | 3.50 |
| 2" x 10", 4/12 pitch | 1.150 | L.F. | .041 | 1.58 | 1.46 | 3.04 |
| 8/12 pitch | 1.440 | L.F. | .066 | 1.97 | 2.35 | 4.32 |
| Ceiling joists, #2 or better, 2" x 4", 16" O.C. | 1.000 | L.F. | .013 | .39 | .46 | .85 |
| 24" O.C. | .750 | L.F. | .010 | .29 | .35 | .64 |
| 2" x 6", 16" O.C. | 1.000 | L.F. | .013 | .62 | .46 | 1.08 |
| 24" O.C. | .750 | L.F. | .010 | .47 | .35 | .82 |
| 2" x 8", 16" O.C. | 1.000 | L.F. | .015 | .95 | .52 | 1.47 |
| 24" O.C. | .750 | L.F. | .011 | .71 | .39 | 1.10 |
| 2" x 10", 16" O.C. | 1.000 | L.F. | .018 | 1.37 | .63 | 2.00 |
| 24" O.C. | .750 | L.F. | .013 | 1.03 | .47 | 1.50 |
| Fascia board, #2 or better, 1" x 6" | .220 | L.F. | .009 | .11 | .31 | .42 |
| 1" x 8" | .220 | L.F. | .010 | .13 | .36 | .49 |
| 1" x 10" | .220 | L.F. | .011 | .15 | .40 | .55 |
| 2" x 6" | .220 | L.F. | .013 | .17 | .45 | .62 |
| 2" x 8" | .220 | L.F. | .016 | .21 | .56 | .77 |
| 2" x 10" | .220 | L.F. | .020 | .30 | .70 | 1.00 |
| Soffit nailer (outrigger), 2" x 4", 16" O.C. | .280 | L.F. | .007 | .11 | .26 | .37 |
| 24" O.C. | .220 | L.F. | .006 | .09 | .20 | .29 |
| 2" x 8", 16" O.C. | .280 | L.F. | .007 | .20 | .24 | .44 |
| 24" O.C. | .220 | L.F. | .005 | .16 | .19 | .35 |
| Sheathing, plywood CDX, 4/12 pitch, 3/8" thick | 1.570 | S.F. | .016 | .75 | .58 | 1.33 |
| 1/2" thick | 1.570 | S.F. | .018 | .85 | .64 | 1.49 |
| 5/8" thick | 1.570 | S.F. | .019 | .97 | .69 | 1.66 |
| 8/12 pitch, 3/8" thick | 1.900 | S.F. | .020 | .91 | .70 | 1.61 |
| 1/2" thick | 1.900 | S.F. | .022 | 1.03 | .78 | 1.81 |
| 5/8" thick | 1.900 | S.F. | .023 | 1.18 | .84 | 2.02 |
| Boards, 4/12 pitch, 1" x 6" boards | 1.450 | S.F. | .032 | 2.07 | 1.15 | 3.22 |
| 1" x 8" boards | 1.450 | S.F. | .027 | 2.07 | .94 | 3.01 |
| 8/12 pitch, 1" x 6" boards | 1.750 | S.F. | .039 | 2.50 | 1.38 | 3.88 |
| 1" x 8" boards | 1.750 | S.F. | .032 | 2.50 | 1.14 | 3.64 |
| Furring, 1" x 3", 12" O.C. | 1.200 | L.F. | .027 | .25 | .97 | 1.22 |
| 16" O.C. | 1.000 | L.F. | .023 | .21 | .81 | 1.02 |
| 24" O.C. | .800 | L.F. | .018 | .17 | .65 | .82 |

**3 FRAMING**

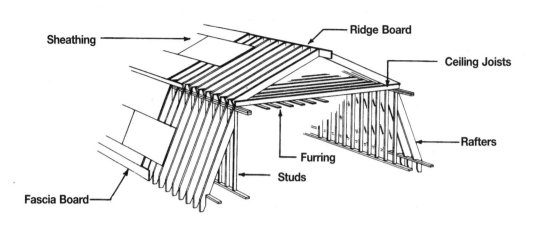

Sheathing — Ridge Board — Ceiling Joists — Rafters — Furring — Studs — Fascia Board

| System Description | QUAN. | UNIT | LABOR HOURS | COST PER S.F. | | |
|---|---|---|---|---|---|---|
| | | | | MAT. | INST. | TOTAL |
| **2″ X 6″ RAFTERS, 16″ O.C.** | | | | | | |
| Roof rafters, 2″ x 6″, 16″ O.C. | 1.430 | L.F. | .029 | .89 | 1.02 | 1.91 |
| Ceiling joists, 2″ x 6″, 16″ O.C. | .710 | L.F. | .009 | .44 | .33 | .77 |
| Stud wall, 2″ x 4″, 16″ O.C., including plates | .790 | L.F. | .012 | .31 | .44 | .75 |
| Furring strips, 1″ x 3″, 16″ O.C. | .710 | L.F. | .016 | .15 | .58 | .73 |
| Ridge board, 2″ x 8″ | .050 | L.F. | .002 | .05 | .06 | .11 |
| Fascia board, 2″ x 6″ | .100 | L.F. | .006 | .08 | .20 | .28 |
| Sheathing, exterior grade plywood, 1/2″ thick | 1.450 | S.F. | .017 | .78 | .59 | 1.37 |
| | | | | | | |
| TOTAL | | | .091 | 2.70 | 3.22 | 5.92 |
| **2″ X 8″ RAFTERS, 16″ O.C.** | | | | | | |
| Roof rafters, 2″ x 8″, 16″ O.C. | 1.430 | L.F. | .031 | 1.36 | 1.09 | 2.45 |
| Ceiling joists, 2″ x 6″, 16″ O.C. | .710 | L.F. | .009 | .44 | .33 | .77 |
| Stud wall, 2″ x 4″, 16″ O.C., including plates | .790 | L.F. | .012 | .31 | .44 | .75 |
| Furring strips, 1″ x 3″, 16″ O.C. | .710 | L.F. | .016 | .15 | .58 | .73 |
| Ridge board, 2″ x 8″ | .050 | L.F. | .002 | .05 | .06 | .11 |
| Fascia board, 2″ x 8″ | .100 | L.F. | .007 | .10 | .25 | .35 |
| Sheathing, exterior grade plywood, 1/2″ thick | 1.450 | S.F. | .017 | .78 | .59 | 1.37 |
| | | | | | | |
| TOTAL | | | .094 | 3.19 | 3.34 | 6.53 |

The cost of this system is based on the square foot of plan area on the first floor.

| Description | QUAN. | UNIT | LABOR HOURS | COST PER S.F. | | |
|---|---|---|---|---|---|---|
| | | | | MAT. | INST. | TOTAL |
| | | | | | | |
| | | | | | | |
| | | | | | | |
| | | | | | | |
| | | | | | | |
| | | | | | | |

**Important: See the Reference Section for critical supporting data - Reference Nos., Crews & Location Factors**

# Gambrel Roof Framing Price Sheet

| | QUAN. | UNIT | LABOR HOURS | COST PER S.F. MAT. | COST PER S.F. INST. | COST PER S.F. TOTAL |
|---|---|---|---|---|---|---|
| Roof rafters, #2 or better, 2" x 6", 16" O.C. | 1.430 | L.F. | .029 | .89 | 1.02 | 1.91 |
| 24" O.C. | 1.140 | L.F. | .023 | .71 | .81 | 1.52 |
| 2" x 8", 16" O.C. | 1.430 | L.F. | .031 | 1.36 | 1.09 | 2.45 |
| 24" O.C. | 1.140 | L.F. | .024 | 1.08 | .87 | 1.95 |
| 2" x 10", 16" O.C. | 1.430 | L.F. | .046 | 1.96 | 1.64 | 3.60 |
| 24" O.C. | 1.140 | L.F. | .037 | 1.56 | 1.31 | 2.87 |
| Ceiling joist, #2 or better, 2" x 4", 16" O.C. | .710 | L.F. | .009 | .28 | .33 | .61 |
| 24" O.C. | .570 | L.F. | .007 | .22 | .26 | .48 |
| 2" x 6", 16" O.C. | .710 | L.F. | .009 | .44 | .33 | .77 |
| 24" O.C. | .570 | L.F. | .007 | .35 | .26 | .61 |
| 2" x 8", 16" O.C. | .710 | L.F. | .010 | .67 | .37 | 1.04 |
| 24" O.C. | .570 | L.F. | .008 | .54 | .30 | .84 |
| Stud wall, #2 or better, 2" x 4", 16" O.C. | .790 | L.F. | .012 | .31 | .44 | .75 |
| 24" O.C. | .630 | L.F. | .010 | .25 | .35 | .60 |
| 2" x 6", 16" O.C. | .790 | L.F. | .014 | .49 | .51 | 1.00 |
| 24" O.C. | .630 | L.F. | .011 | .39 | .40 | .79 |
| Furring, 1" x 3", 16" O.C. | .710 | L.F. | .016 | .15 | .58 | .73 |
| 24" O.C. | .590 | L.F. | .013 | .12 | .48 | .60 |
| Ridge board, #2 or better, 1" x 6" | .050 | L.F. | .001 | .05 | .05 | .10 |
| 1" x 8" | .050 | L.F. | .001 | .07 | .05 | .12 |
| 1" x 10" | .050 | L.F. | .002 | .08 | .06 | .14 |
| 2" x 6" | .050 | L.F. | .002 | .03 | .06 | .09 |
| 2" x 8" | .050 | L.F. | .002 | .05 | .06 | .11 |
| 2" x 10" | .050 | L.F. | .002 | .07 | .07 | .14 |
| Fascia board, #2 or better, 1" x 6" | .100 | L.F. | .004 | .05 | .14 | .19 |
| 1" x 8" | .100 | L.F. | .005 | .06 | .16 | .22 |
| 1" x 10" | .100 | L.F. | .005 | .07 | .18 | .25 |
| 2" x 6" | .100 | L.F. | .006 | .08 | .20 | .28 |
| 2" x 8" | .100 | L.F. | .007 | .10 | .25 | .35 |
| 2" x 10" | .100 | L.F. | .009 | .14 | .32 | .46 |
| Sheathing, plywood, exterior grade CDX, 3/8" thick | 1.450 | S.F. | .015 | .70 | .54 | 1.24 |
| 1/2" thick | 1.450 | S.F. | .017 | .78 | .59 | 1.37 |
| 5/8" thick | 1.450 | S.F. | .018 | .90 | .64 | 1.54 |
| 3/4" thick | 1.450 | S.F. | .019 | 1.06 | .68 | 1.74 |
| Boards, 1" x 6", laid regular | 1.450 | S.F. | .032 | 2.07 | 1.15 | 3.22 |
| Laid diagonal | 1.450 | S.F. | .036 | 2.07 | 1.28 | 3.35 |
| 1" x 8", laid regular | 1.450 | S.F. | .027 | 2.07 | .94 | 3.01 |
| Laid diagonal | 1.450 | S.F. | .032 | 2.07 | 1.15 | 3.22 |

**3 FRAMING**

127

| System Description | QUAN. | UNIT | LABOR HOURS | COST PER S.F. | | |
|---|---|---|---|---|---|---|
| | | | | MAT. | INST. | TOTAL |
| **2″ X 6″ RAFTERS, 16″ O.C.** | | | | | | |
| Roof rafters, 2″ x 6″, 16″ O.C. | 1.210 | L.F. | .033 | .75 | 1.17 | 1.92 |
| Rafter plates, 2″ x 6″, double top, single bottom | .364 | L.F. | .010 | .23 | .35 | .58 |
| Ceiling joists, 2″ x 4″, 16″ O.C. | .920 | L.F. | .012 | .36 | .42 | .78 |
| Hip rafter, 2″ x 6″ | .070 | L.F. | .002 | .04 | .08 | .12 |
| Jack rafter, 2″ x 6″, 16″ O.C. | 1.000 | L.F. | .039 | .62 | 1.39 | 2.01 |
| Ridge board, 2″ x 6″ | .018 | L.F. | .001 | .01 | .02 | .03 |
| Sheathing, exterior grade plywood, 1/2″ thick | 2.210 | S.F. | .025 | 1.19 | .91 | 2.10 |
| Furring strips, 1″ x 3″, 16″ O.C. | .920 | L.F. | .021 | .19 | .75 | .94 |
| | | | | | | |
| TOTAL | | | .143 | 3.39 | 5.09 | 8.48 |
| **2″ X 8″ RAFTERS, 16″ O.C.** | | | | | | |
| Roof rafters, 2″ x 8″, 16″ O.C. | 1.210 | L.F. | .036 | 1.15 | 1.27 | 2.42 |
| Rafter plates, 2″ x 8″, double top, single bottom | .364 | L.F. | .011 | .35 | .38 | .73 |
| Ceiling joists, 2″ x 6″, 16″ O.C. | .920 | L.F. | .012 | .57 | .42 | .99 |
| Hip rafter, 2″ x 8″ | .070 | L.F. | .002 | .07 | .08 | .15 |
| Jack rafter, 2″ x 8″, 16″ O.C. | 1.000 | L.F. | .048 | .95 | 1.70 | 2.65 |
| Ridge board, 2″ x 8″ | .018 | L.F. | .001 | .02 | .02 | .04 |
| Sheathing, exterior grade plywood, 1/2″ thick | 2.210 | S.F. | .025 | 1.19 | .91 | 2.10 |
| Furring strips, 1″ x 3″, 16″ O.C. | .920 | L.F. | .021 | .19 | .75 | .94 |
| | | | | | | |
| TOTAL | | | .156 | 4.49 | 5.53 | 10.02 |

The cost of this system is based on the square foot of plan area.

| Description | QUAN. | UNIT | LABOR HOURS | COST PER S.F. | | |
|---|---|---|---|---|---|---|
| | | | | MAT. | INST. | TOTAL |
| | | | | | | |
| | | | | | | |
| | | | | | | |
| | | | | | | |
| | | | | | | |
| | | | | | | |

**Important: See the Reference Section for critical supporting data - Reference Nos., Crews & Location Factors**

| Mansard Roof Framing Price Sheet | QUAN. | UNIT | LABOR HOURS | COST PER S.F. | | |
|---|---|---|---|---|---|---|
| | | | | MAT. | INST. | TOTAL |
| Roof rafters, #2 or better, 2" x 6", 16" O.C. | 1.210 | L.F. | .033 | .75 | 1.17 | 1.92 |
| 24" O.C. | .970 | L.F. | .026 | .60 | .94 | 1.54 |
| 2" x 8", 16" O.C. | 1.210 | L.F. | .036 | 1.15 | 1.27 | 2.42 |
| 24" O.C. | .970 | L.F. | .029 | .92 | 1.02 | 1.94 |
| 2" x 10", 16" O.C. | 1.210 | L.F. | .046 | 1.66 | 1.62 | 3.28 |
| 24" O.C. | .970 | L.F. | .037 | 1.33 | 1.30 | 2.63 |
| Rafter plates, #2 or better double top single bottom, 2" x 6" | .364 | L.F. | .010 | .23 | .35 | .58 |
| 2" x 8" | .364 | L.F. | .011 | .35 | .38 | .73 |
| 2" x 10" | .364 | L.F. | .014 | .50 | .49 | .99 |
| Ceiling joist, #2 or better, 2" x 4", 16" O.C. | .920 | L.F. | .012 | .36 | .42 | .78 |
| 24" O.C. | .740 | L.F. | .009 | .29 | .34 | .63 |
| 2" x 6", 16" O.C. | .920 | L.F. | .012 | .57 | .42 | .99 |
| 24" O.C. | .740 | L.F. | .009 | .46 | .34 | .80 |
| 2" x 8", 16" O.C. | .920 | L.F. | .013 | .87 | .48 | 1.35 |
| 24" O.C. | .740 | L.F. | .011 | .70 | .38 | 1.08 |
| Hip rafter, #2 or better, 2" x 6" | .070 | L.F. | .002 | .04 | .08 | .12 |
| 2" x 8" | .070 | L.F. | .002 | .07 | .08 | .15 |
| 2" x 10" | .070 | L.F. | .003 | .10 | .11 | .21 |
| Jack rafter, #2 or better, 2" x 6", 16" O.C. | 1.000 | L.F. | .039 | .62 | 1.39 | 2.01 |
| 24" O.C. | .800 | L.F. | .031 | .50 | 1.11 | 1.61 |
| 2" x 8", 16" O.C. | 1.000 | L.F. | .048 | .95 | 1.70 | 2.65 |
| 24" O.C. | .800 | L.F. | .038 | .76 | 1.36 | 2.12 |
| Ridge board, #2 or better, 1" x 6" | .018 | L.F. | .001 | .02 | .02 | .04 |
| 1" x 8" | .018 | L.F. | .001 | .02 | .02 | .04 |
| 1" x 10" | .018 | L.F. | .001 | .03 | .02 | .05 |
| 2" x 6" | .018 | L.F. | .001 | .01 | .02 | .03 |
| 2" x 8" | .018 | L.F. | .001 | .02 | .02 | .04 |
| 2" x 10" | .018 | L.F. | .001 | .02 | .03 | .05 |
| Sheathing, plywood exterior grade CDX, 3/8" thick | 2.210 | S.F. | .023 | 1.06 | .82 | 1.88 |
| 1/2" thick | 2.210 | S.F. | .025 | 1.19 | .91 | 2.10 |
| 5/8" thick | 2.210 | S.F. | .027 | 1.37 | .97 | 2.34 |
| 3/4" thick | 2.210 | S.F. | .029 | 1.61 | 1.04 | 2.65 |
| Boards, 1" x 6", laid regular | 2.210 | S.F. | .049 | 3.16 | 1.75 | 4.91 |
| Laid diagonal | 2.210 | S.F. | .054 | 3.16 | 1.94 | 5.10 |
| 1" x 8", laid regular | 2.210 | S.F. | .040 | 3.16 | 1.44 | 4.60 |
| Laid diagonal | 2.210 | S.F. | .049 | 3.16 | 1.75 | 4.91 |
| Furring, 1" x 3", 12" O.C. | 1.150 | L.F. | .026 | .24 | .93 | 1.17 |
| 24" O.C. | .740 | L.F. | .017 | .16 | .60 | .76 |

**3  FRAMING**

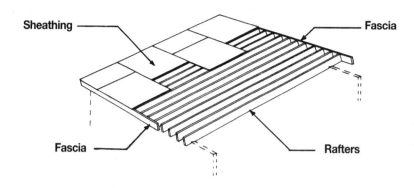

| System Description | QUAN. | UNIT | LABOR HOURS | COST PER S.F. | | |
|---|---|---|---|---|---|---|
| | | | | MAT. | INST. | TOTAL |
| **2″ X 6″,16″ O.C., 4/12 PITCH** | | | | | | |
| Rafters, 2″ x 6″, 16″ O.C., 4/12 pitch | 1.170 | L.F. | .019 | .73 | .67 | 1.40 |
| Fascia, 2″ x 6″ | .100 | L.F. | .006 | .08 | .20 | .28 |
| Bridging, 1″ x 3″, 6′ O.C. | .080 | Pr. | .005 | .02 | .18 | .20 |
| Sheathing, exterior grade plywood, 1/2″ thick | 1.230 | S.F. | .014 | .66 | .50 | 1.16 |
| TOTAL | | | .044 | 1.49 | 1.55 | 3.04 |
| **2″ X 6″, 24″ O.C., 4/12 PITCH** | | | | | | |
| Rafters, 2″ x 6″, 24″ O.C., 4/12 pitch | .940 | L.F. | .015 | .58 | .54 | 1.12 |
| Fascia, 2″ x 6″ | .100 | L.F. | .006 | .08 | .20 | .28 |
| Bridging, 1″ x 3″, 6′ O.C. | .060 | Pr. | .004 | .02 | .13 | .15 |
| Sheathing, exterior grade plywood, 1/2″ thick | 1.230 | S.F. | .014 | .66 | .50 | 1.16 |
| TOTAL | | | .039 | 1.34 | 1.37 | 2.71 |
| **2″ X 8″, 16″ O.C., 4/12 PITCH** | | | | | | |
| Rafters, 2″ x 8″, 16″ O.C., 4/12 pitch | 1.170 | L.F. | .020 | 1.11 | .70 | 1.81 |
| Fascia, 2″ x 8″ | .100 | L.F. | .007 | .10 | .25 | .35 |
| Bridging, 1″ x 3″, 6′ O.C. | .080 | Pr. | .005 | .02 | .18 | .20 |
| Sheathing, exterior grade plywood, 1/2″ thick | 1.230 | S.F. | .014 | .66 | .50 | 1.16 |
| TOTAL | | | .046 | 1.89 | 1.63 | 3.52 |
| **2″ X 8″, 24″ O.C., 4/12 PITCH** | | | | | | |
| Rafters, 2″ x 8″, 24″ O.C., 4/12 pitch | .940 | L.F. | .016 | .89 | .56 | 1.45 |
| Fascia, 2″ x 8″ | .100 | L.F. | .007 | .10 | .25 | .35 |
| Bridging, 1″ x 3″, 6′ O.C. | .060 | Pr. | .004 | .02 | .13 | .15 |
| Sheathing, exterior grade plywood, 1/2″ thick | 1.230 | S.F. | .014 | .66 | .50 | 1.16 |
| TOTAL | | | .041 | 1.67 | 1.44 | 3.11 |

The cost of this system is based on the square foot of plan area.
A 1′ overhang is assumed. No ceiling joists or furring are included.

| Description | QUAN. | UNIT | LABOR HOURS | COST PER S.F. | | |
|---|---|---|---|---|---|---|
| | | | | MAT. | INST. | TOTAL |
| | | | | | | |
| | | | | | | |
| | | | | | | |

**Important: See the Reference Section for critical supporting data - Reference Nos., Crews & Location Factors**

| Shed/Flat Roof Framing Price Sheet | QUAN. | UNIT | LABOR HOURS | COST PER S.F. | | |
|---|---|---|---|---|---|---|
| | | | | MAT. | INST. | TOTAL |
| Rafters, #2 or better, 16″ O.C., 2″ x 4″, 0 - 4/12 pitch | 1.170 | L.F. | .014 | .54 | .50 | 1.04 |
| 5/12 - 8/12 pitch | 1.330 | L.F. | .020 | .62 | .71 | 1.33 |
| 2″ x 6″, 0 - 4/12 pitch | 1.170 | L.F. | .019 | .73 | .67 | 1.40 |
| 5/12 - 8/12 pitch | 1.330 | L.F. | .027 | .82 | .94 | 1.76 |
| 2″ x 8″, 0 - 4/12 pitch | 1.170 | L.F. | .020 | 1.11 | .70 | 1.81 |
| 5/12 - 8/12 pitch | 1.330 | L.F. | .028 | 1.26 | 1.01 | 2.27 |
| 2″ x 10″, 0 - 4/12 pitch | 1.170 | L.F. | .030 | 1.60 | 1.05 | 2.65 |
| 5/12 - 8/12 pitch | 1.330 | L.F. | .043 | 1.82 | 1.53 | 3.35 |
| 24″ O.C., 2″ x 4″, 0 - 4/12 pitch | .940 | L.F. | .011 | .44 | .40 | .84 |
| 5/12 - 8/12 pitch | 1.060 | L.F. | .021 | .66 | .75 | 1.41 |
| 2″ x 6″, 0 - 4/12 pitch | .940 | L.F. | .015 | .58 | .54 | 1.12 |
| 5/12 - 8/12 pitch | 1.060 | L.F. | .021 | .66 | .75 | 1.41 |
| 2″ x 8″, 0 - 4/12 pitch | .940 | L.F. | .016 | .89 | .56 | 1.45 |
| 5/12 - 8/12 pitch | 1.060 | L.F. | .023 | 1.01 | .81 | 1.82 |
| 2″ x 10″, 0 - 4/12 pitch | .940 | L.F. | .024 | 1.29 | .85 | 2.14 |
| 5/12 - 8/12 pitch | 1.060 | L.F. | .034 | 1.45 | 1.22 | 2.67 |
| Fascia, #2 or better,, 1″ x 4″ | .100 | L.F. | .003 | .04 | .10 | .14 |
| 1″ x 6″ | .100 | L.F. | .004 | .05 | .14 | .19 |
| 1″ x 8″ | .100 | L.F. | .005 | .06 | .16 | .22 |
| 1″ x 10″ | .100 | L.F. | .005 | .07 | .18 | .25 |
| 2″ x 4″ | .100 | L.F. | .005 | .06 | .17 | .23 |
| 2″ x 6″ | .100 | L.F. | .006 | .08 | .20 | .28 |
| 2″ x 8″ | .100 | L.F. | .007 | .10 | .25 | .35 |
| 2″ x 10″ | .100 | L.F. | .009 | .14 | .32 | .46 |
| Bridging, wood 6′ O.C., 1″ x 3″, rafters, 16″ O.C. | .080 | Pr. | .005 | .02 | .18 | .20 |
| 24″ O.C. | .060 | Pr. | .004 | .02 | .13 | .15 |
| Metal, galvanized, rafters, 16″ O.C. | .080 | Pr. | .005 | .05 | .18 | .23 |
| 24″ O.C. | .060 | Pr. | .003 | .15 | .12 | .27 |
| Compression type, rafters, 16″ O.C. | .080 | Pr. | .003 | .11 | .11 | .22 |
| 24″ O.C. | .060 | Pr. | .002 | .08 | .09 | .17 |
| Sheathing, plywood, exterior grade, 3/8″ thick, flat 0 - 4/12 pitch | 1.230 | S.F. | .013 | .59 | .46 | 1.05 |
| 5/12 - 8/12 pitch | 1.330 | S.F. | .014 | .64 | .49 | 1.13 |
| 1/2″ thick, flat 0 - 4/12 pitch | 1.230 | S.F. | .014 | .66 | .50 | 1.16 |
| 5/12 - 8/12 pitch | 1.330 | S.F. | .015 | .72 | .55 | 1.27 |
| 5/8″ thick, flat 0 - 4/12 pitch | 1.230 | S.F. | .015 | .76 | .54 | 1.30 |
| 5/12 - 8/12 pitch | 1.330 | S.F. | .016 | .82 | .59 | 1.41 |
| 3/4″ thick, flat 0 - 4/12 pitch | 1.230 | S.F. | .016 | .90 | .58 | 1.48 |
| 5/12 - 8/12 pitch | 1.330 | S.F. | .018 | .97 | .63 | 1.60 |
| Boards, 1″ x 6″, laid regular, flat 0 - 4/12 pitch | 1.230 | S.F. | .027 | 1.76 | .97 | 2.73 |
| 5/12 - 8/12 pitch | 1.330 | S.F. | .041 | 1.90 | 1.46 | 3.36 |
| Laid diagonal, flat 0 - 4/12 pitch | 1.230 | S.F. | .030 | 1.76 | 1.08 | 2.84 |
| 5/12 - 8/12 pitch | 1.330 | S.F. | .044 | 1.90 | 1.58 | 3.48 |
| 1″ x 8″, laid regular, flat 0 - 4/12 pitch | 1.230 | S.F. | .022 | 1.76 | .80 | 2.56 |
| 5/12 - 8/12 pitch | 1.330 | S.F. | .034 | 1.90 | 1.20 | 3.10 |
| Laid diagonal, flat 0 - 4/12 pitch | 1.230 | S.F. | .027 | 1.76 | .97 | 2.73 |
| 5/12 - 8/12 pitch | 1.330 | S.F. | .044 | 1.90 | 1.58 | 3.48 |

**FRAMING**

**3**

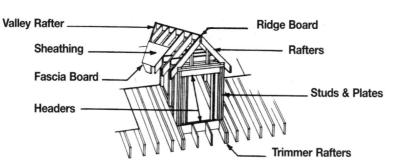

Valley Rafter — Ridge Board — Sheathing — Rafters — Fascia Board — Headers — Studs & Plates — Trimmer Rafters

| System Description | QUAN. | UNIT | LABOR HOURS | COST PER S.F. | | |
|---|---|---|---|---|---|---|
| | | | | MAT. | INST. | TOTAL |
| **2″ X 6″, 16″ O.C.** | | | | | | |
| Dormer rafter, 2″ x 6″, 16″ O.C. | 1.330 | L.F. | .036 | .82 | 1.29 | 2.11 |
| Ridge board, 2″ x 6″ | .280 | L.F. | .009 | .17 | .32 | .49 |
| Trimmer rafters, 2″ x 6″ | .880 | L.F. | .014 | .55 | .50 | 1.05 |
| Wall studs & plates, 2″ x 4″, 16″ O.C. | 3.160 | L.F. | .056 | 1.23 | 1.99 | 3.22 |
| Fascia, 2″ x 6″ | .220 | L.F. | .012 | .16 | .43 | .59 |
| Valley rafter, 2″ x 6″, 16″ O.C. | .280 | L.F. | .009 | .17 | .31 | .48 |
| Cripple rafter, 2″ x 6″, 16″ O.C. | .560 | L.F. | .022 | .35 | .78 | 1.13 |
| Headers, 2″ x 6″, doubled | .670 | L.F. | .030 | .42 | 1.06 | 1.48 |
| Ceiling joist, 2″ x 4″, 16″ O.C. | 1.000 | L.F. | .013 | .39 | .46 | .85 |
| Sheathing, exterior grade plywood, 1/2″ thick | 3.610 | S.F. | .041 | 1.95 | 1.48 | 3.43 |
| TOTAL | | | .242 | 6.21 | 8.62 | 14.83 |
| **2″ X 8″, 16″ O.C.** | | | | | | |
| Dormer rafter, 2″ x 8″, 16″ O.C. | 1.330 | L.F. | .039 | 1.26 | 1.40 | 2.66 |
| Ridge board, 2″ x 8″ | .280 | L.F. | .010 | .27 | .36 | .63 |
| Trimmer rafter, 2″ x 8″ | .880 | L.F. | .015 | .84 | .53 | 1.37 |
| Wall studs & plates, 2″ x 4″, 16″ O.C. | 3.160 | L.F. | .056 | 1.23 | 1.99 | 3.22 |
| Fascia, 2″ x 8″ | .220 | L.F. | .016 | .21 | .56 | .77 |
| Valley rafter, 2″ x 8″, 16″ O.C. | .280 | L.F. | .010 | .27 | .34 | .61 |
| Cripple rafter, 2″ x 8″, 16″ O.C. | .560 | L.F. | .027 | .53 | .95 | 1.48 |
| Headers, 2″ x 8″, doubled | .670 | L.F. | .032 | .64 | 1.13 | 1.77 |
| Ceiling joist, 2″ x 4″, 16″ O.C. | 1.000 | L.F. | .013 | .39 | .46 | .85 |
| Sheathing,, exterior grade plywood, 1/2″ thick | 3.610 | S.F. | .041 | 1.95 | 1.48 | 3.43 |
| TOTAL | | | .259 | 7.59 | 9.20 | 16.79 |

The cost in this system is based on the square foot of plan area.
The measurement being the plan area of the dormer only.

| Description | QUAN. | UNIT | LABOR HOURS | COST PER S.F. | | |
|---|---|---|---|---|---|---|
| | | | | MAT. | INST. | TOTAL |
| | | | | | | |
| | | | | | | |
| | | | | | | |
| | | | | | | |
| | | | | | | |
| | | | | | | |

**Important: See the Reference Section for critical supporting data - Reference Nos., Crews & Location Factors**

| Gable Dormer Framing Price Sheet | QUAN. | UNIT | LABOR HOURS | COST PER S.F. | | |
|---|---|---|---|---|---|---|
| | | | | MAT. | INST. | TOTAL |
| Dormer rafters, #2 or better, 2" x 4", 16" O.C. | 1.330 | L.F. | .029 | .66 | 1.03 | 1.69 |
| 24" O.C. | 1.060 | L.F. | .023 | .53 | .82 | 1.35 |
| 2" x 6", 16" O.C. | 1.330 | L.F. | .036 | .82 | 1.29 | 2.11 |
| 24" O.C. | 1.060 | L.F. | .029 | .66 | 1.03 | 1.69 |
| 2" x 8", 16" O.C. | 1.330 | L.F. | .039 | 1.26 | 1.40 | 2.66 |
| 24" O.C. | 1.060 | L.F. | .031 | 1.01 | 1.11 | 2.12 |
| Ridge board, #2 or better, 1" x 4" | .280 | L.F. | .006 | .22 | .21 | .43 |
| 1" x 6" | .280 | L.F. | .007 | .28 | .27 | .55 |
| 1" x 8" | .280 | L.F. | .008 | .37 | .29 | .66 |
| 2" x 4" | .280 | L.F. | .007 | .14 | .26 | .40 |
| 2" x 6" | .280 | L.F. | .009 | .17 | .32 | .49 |
| 2" x 8" | .280 | L.F. | .010 | .27 | .36 | .63 |
| Trimmer rafters, #2 or better, 2" x 4" | .880 | L.F. | .011 | .44 | .40 | .84 |
| 2" x 6" | .880 | L.F. | .014 | .55 | .50 | 1.05 |
| 2" x 8" | .880 | L.F. | .015 | .84 | .53 | 1.37 |
| 2" x 10" | .880 | L.F. | .022 | 1.21 | .79 | 2.00 |
| Wall studs & plates, #2 or better, 2" x 4" studs, 16" O.C. | 3.160 | L.F. | .056 | 1.23 | 1.99 | 3.22 |
| 24" O.C. | 2.800 | L.F. | .050 | 1.09 | 1.76 | 2.85 |
| 2" x 6" studs, 16" O.C. | 3.160 | L.F. | .063 | 1.96 | 2.24 | 4.20 |
| 24" O.C. | 2.800 | L.F. | .056 | 1.74 | 1.99 | 3.73 |
| Fascia, #2 or better, 1" x 4" | .220 | L.F. | .006 | .09 | .23 | .32 |
| 1" x 6" | .220 | L.F. | .008 | .10 | .28 | .38 |
| 1" x 8" | .220 | L.F. | .009 | .12 | .32 | .44 |
| 2" x 4" | .220 | L.F. | .011 | .14 | .38 | .52 |
| 2" x 6" | .220 | L.F. | .014 | .18 | .48 | .66 |
| 2" x 8" | .220 | L.F. | .016 | .21 | .56 | .77 |
| Valley rafter, #2 or better, 2" x 4" | .280 | L.F. | .007 | .14 | .25 | .39 |
| 2" x 6" | .280 | L.F. | .009 | .17 | .31 | .48 |
| 2" x 8" | .280 | L.F. | .010 | .27 | .34 | .61 |
| 2" x 10" | .280 | L.F. | .012 | .38 | .42 | .80 |
| Cripple rafter, #2 or better, 2" x 4", 16" O.C. | .560 | L.F. | .018 | .28 | .63 | .91 |
| 24" O.C. | .450 | L.F. | .014 | .22 | .50 | .72 |
| 2" x 6", 16" O.C. | .560 | L.F. | .022 | .35 | .78 | 1.13 |
| 24" O.C. | .450 | L.F. | .018 | .28 | .63 | .91 |
| 2" x 8", 16" O.C. | .560 | L.F. | .027 | .53 | .95 | 1.48 |
| 24" O.C. | .450 | L.F. | .021 | .43 | .77 | 1.20 |
| Headers, #2 or better double header, 2" x 4" | .670 | L.F. | .024 | .33 | .85 | 1.18 |
| 2" x 6" | .670 | L.F. | .030 | .42 | 1.06 | 1.48 |
| 2" x 8" | .670 | L.F. | .032 | .64 | 1.13 | 1.77 |
| 2" x 10" | .670 | L.F. | .034 | .92 | 1.19 | 2.11 |
| Ceiling joist, #2 or better, 2" x 4", 16" O.C. | 1.000 | L.F. | .013 | .39 | .46 | .85 |
| 24" O.C. | .800 | L.F. | .010 | .31 | .37 | .68 |
| 2" x 6", 16" O.C. | 1.000 | L.F. | .013 | .62 | .46 | 1.08 |
| 24" O.C. | .800 | L.F. | .010 | .50 | .37 | .87 |
| Sheathing, plywood exterior grade, 3/8" thick | 3.610 | S.F. | .038 | 1.73 | 1.34 | 3.07 |
| 1/2" thick | 3.610 | S.F. | .041 | 1.95 | 1.48 | 3.43 |
| 5/8" thick | 3.610 | S.F. | .044 | 2.24 | 1.59 | 3.83 |
| 3/4" thick | 3.610 | S.F. | .048 | 2.64 | 1.70 | 4.34 |
| Boards, 1" x 6", laid regular | 3.610 | S.F. | .089 | 5.15 | 3.18 | 8.33 |
| Laid diagonal | 3.610 | S.F. | .099 | 5.15 | 3.50 | 8.65 |
| 1" x 8", laid regular | 3.610 | S.F. | .076 | 5.15 | 2.67 | 7.82 |
| Laid diagonal | 3.610 | S.F. | .089 | 5.15 | 3.18 | 8.33 |
| | | | | | | |
| | | | | | | |
| | | | | | | |
| | | | | | | |

**3 FRAMING**

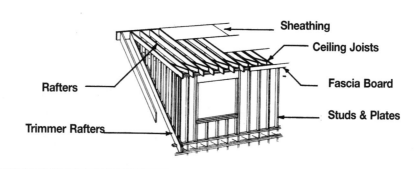

Sheathing
Ceiling Joists
Fascia Board
Studs & Plates
Rafters
Trimmer Rafters

| System Description | QUAN. | UNIT | LABOR HOURS | COST PER S.F. | | |
|---|---|---|---|---|---|---|
| | | | | MAT. | INST. | TOTAL |
| **2″ X 6″ RAFTERS, 16″ O.C.** | | | | | | |
| Dormer rafter, 2″ x 6″, 16″ O.C. | 1.080 | L.F. | .029 | .67 | 1.05 | 1.72 |
| Trimmer rafter, 2″ x 6″ | .400 | L.F. | .006 | .25 | .23 | .48 |
| Studs & plates, 2″ x 4″, 16″ O.C. | 2.750 | L.F. | .049 | 1.07 | 1.73 | 2.80 |
| Fascia, 2″ x 6″ | .250 | L.F. | .014 | .18 | .48 | .66 |
| Ceiling joist, 2″ x 4″, 16″ O.C. | 1.000 | L.F. | .013 | .39 | .46 | .85 |
| Sheathing, exterior grade plywood, CDX, 1/2″ thick | 2.940 | S.F. | .034 | 1.59 | 1.21 | 2.80 |
| TOTAL | | | .145 | 4.15 | 5.16 | 9.31 |
| **2″ X 8″ RAFTERS, 16″ O.C.** | | | | | | |
| Dormer rafter, 2″ x 8″, 16″ O.C. | 1.080 | L.F. | .032 | 1.03 | 1.13 | 2.16 |
| Trimmer rafter, 2″ x 8″ | .400 | L.F. | .007 | .38 | .24 | .62 |
| Studs & plates, 2″ x 4″, 16″ O.C. | 2.750 | L.F. | .049 | 1.07 | 1.73 | 2.80 |
| Fascia, 2″ x 8″ | .250 | L.F. | .018 | .24 | .63 | .87 |
| Ceiling joist, 2″ x 6″, 16″ O.C. | 1.000 | L.F. | .013 | .62 | .46 | 1.08 |
| Sheathing, exterior grade plywood, CDX, 1/2″ thick | 2.940 | S.F. | .034 | 1.59 | 1.21 | 2.80 |
| TOTAL | | | .153 | 4.93 | 5.40 | 10.33 |
| **2″ X 10″ RAFTERS, 16″ O.C.** | | | | | | |
| Dormer rafter, 2″ x 10″, 16″ O.C. | 1.080 | L.F. | .041 | 1.48 | 1.45 | 2.93 |
| Trimmer rafter, 2″ x 10″ | .400 | L.F. | .010 | .55 | .36 | .91 |
| Studs & plates, 2″ x 4″, 16″ O.C. | 2.750 | L.F. | .049 | 1.07 | 1.73 | 2.80 |
| Fascia, 2″ x 10″ | .250 | L.F. | .022 | .34 | .79 | 1.13 |
| Ceiling joist, 2″ x 6″, 16″ O.C. | 1.000 | L.F. | .013 | .62 | .46 | 1.08 |
| Sheathing, exterior grade plywood, CDX, 1/2″ thick | 2.940 | S.F. | .034 | 1.59 | 1.21 | 2.80 |
| TOTAL | | | .169 | 5.65 | 6.00 | 11.65 |

The cost in this system is based on the square foot of plan area.
The measurement is the plan area of the dormer only.

| Description | QUAN. | UNIT | LABOR HOURS | COST PER S.F. | | |
|---|---|---|---|---|---|---|
| | | | | MAT. | INST. | TOTAL |
| | | | | | | |
| | | | | | | |
| | | | | | | |
| | | | | | | |
| | | | | | | |

**Important: See the Reference Section for critical supporting data - Reference Nos., Crews & Location Factors**

## Shed Dormer Framing Price Sheet

| Shed Dormer Framing Price Sheet | QUAN. | UNIT | LABOR HOURS | COST PER S.F. | | |
|---|---|---|---|---|---|---|
| | | | | MAT. | INST. | TOTAL |
| Dormer rafters, #2 or better, 2" x 4", 16" O.C. | 1.080 | L.F. | .023 | .54 | .84 | 1.38 |
| 24" O.C. | .860 | L.F. | .019 | .43 | .67 | 1.10 |
| 2" x 6", 16" O.C. | 1.080 | L.F. | .029 | .67 | 1.05 | 1.72 |
| 24" O.C. | .860 | L.F. | .023 | .53 | .83 | 1.36 |
| 2" x 8", 16" O.C. | 1.080 | L.F. | .032 | 1.03 | 1.13 | 2.16 |
| 24" O.C. | .860 | L.F. | .025 | .82 | .90 | 1.72 |
| 2" x 10", 16" O.C. | 1.080 | L.F. | .041 | 1.48 | 1.45 | 2.93 |
| 24" O.C. | .860 | L.F. | .032 | 1.18 | 1.15 | 2.33 |
| Trimmer rafter, #2 or better, 2" x 4" | .400 | L.F. | .005 | .20 | .18 | .38 |
| 2" x 6" | .400 | L.F. | .006 | .25 | .23 | .48 |
| 2" x 8" | .400 | L.F. | .007 | .38 | .24 | .62 |
| 2" x 10" | .400 | L.F. | .010 | .55 | .36 | .91 |
| Studs & plates, #2 or better, 2" x 4", 16" O.C. | 2.750 | L.F. | .049 | 1.07 | 1.73 | 2.80 |
| 24" O.C. | 2.200 | L.F. | .039 | .86 | 1.39 | 2.25 |
| 2" x 6", 16" O.C. | 2.750 | L.F. | .055 | 1.71 | 1.95 | 3.66 |
| 24" O.C. | 2.200 | L.F. | .044 | 1.36 | 1.56 | 2.92 |
| Fascia, #2 or better, 1" x 4" | .250 | L.F. | .006 | .09 | .23 | .32 |
| 1" x 6" | .250 | L.F. | .008 | .10 | .28 | .38 |
| 1" x 8" | .250 | L.F. | .009 | .12 | .32 | .44 |
| 2" x 4" | .250 | L.F. | .011 | .14 | .38 | .52 |
| 2" x 6" | .250 | L.F. | .014 | .18 | .48 | .66 |
| 2" x 8" | .250 | L.F. | .018 | .24 | .63 | .87 |
| Ceiling joist, #2 or better, 2" x 4", 16" O.C. | 1.000 | L.F. | .013 | .39 | .46 | .85 |
| 24" O.C. | .800 | L.F. | .010 | .31 | .37 | .68 |
| 2" x 6", 16" O.C. | 1.000 | L.F. | .013 | .62 | .46 | 1.08 |
| 24" O.C. | .800 | L.F. | .010 | .50 | .37 | .87 |
| 2" x 8", 16" O.C. | 1.000 | L.F. | .015 | .95 | .52 | 1.47 |
| 24" O.C. | .800 | L.F. | .012 | .76 | .42 | 1.18 |
| Sheathing, plywood exterior grade, 3/8" thick | 2.940 | S.F. | .031 | 1.41 | 1.09 | 2.50 |
| 1/2" thick | 2.940 | S.F. | .034 | 1.59 | 1.21 | 2.80 |
| 5/8" thick | 2.940 | S.F. | .036 | 1.82 | 1.29 | 3.11 |
| 3/4" thick | 2.940 | S.F. | .039 | 2.15 | 1.38 | 3.53 |
| Boards, 1" x 6", laid regular | 2.940 | S.F. | .072 | 4.20 | 2.59 | 6.79 |
| Laid diagonal | 2.940 | S.F. | .080 | 4.20 | 2.85 | 7.05 |
| 1" x 8", laid regular | 2.940 | S.F. | .062 | 4.20 | 2.18 | 6.38 |
| Laid diagonal | 2.940 | S.F. | .072 | 4.20 | 2.59 | 6.79 |
| | | | | | | |
| | | | | | | |
| | | | | | | |
| | | | | | | |

## Window Openings

| Window Openings | QUAN. | UNIT | LABOR HOURS | COST EACH | | |
|---|---|---|---|---|---|---|
| | | | | MAT. | INST. | TOTAL |
| The following are to be added to the total cost of the dormers for window openings. Do not subtract window area from the stud wall quantities. | | | | | | |
| Headers, 2" x 6" doubled, 2' long | 4.000 | L.F. | .178 | 2.48 | 6.30 | 8.78 |
| 3' long | 6.000 | L.F. | .267 | 3.72 | 9.50 | 13.22 |
| 4' long | 8.000 | L.F. | .356 | 4.96 | 12.65 | 17.61 |
| 5' long | 10.000 | L.F. | .444 | 6.20 | 15.80 | 22.00 |
| 2" x 8" doubled, 4' long | 8.000 | L.F. | .376 | 7.60 | 13.45 | 21.05 |
| 5' long | 10.000 | L.F. | .471 | 9.50 | 16.80 | 26.30 |
| 6' long | 12.000 | L.F. | .565 | 11.40 | 20.00 | 31.40 |
| 8' long | 16.000 | L.F. | .753 | 15.20 | 27.00 | 42.20 |
| 2" x 10" doubled, 4' long | 8.000 | L.F. | .400 | 10.95 | 14.25 | 25.20 |
| 6' long | 12.000 | L.F. | .600 | 16.45 | 21.50 | 37.95 |
| 8' long | 16.000 | L.F. | .800 | 22.00 | 28.50 | 50.50 |
| 10' long | 20.000 | L.F. | 1.000 | 27.50 | 35.50 | 63.00 |

**3 FRAMING**

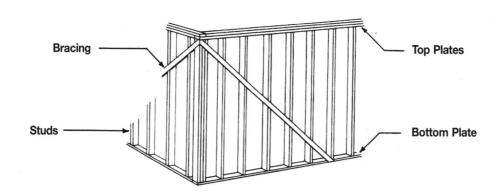

Bracing — Top Plates

Studs — Bottom Plate

| System Description | QUAN. | UNIT | LABOR HOURS | COST PER S.F. | | |
|---|---|---|---|---|---|---|
| | | | | MAT. | INST. | TOTAL |
| **2" X 4", 16" O.C.** | | | | | | |
| 2" x 4" studs, #2 or better, 16" O.C. | 1.000 | L.F. | .015 | .39 | .52 | .91 |
| Plates, double top, single bottom | .375 | L.F. | .005 | .15 | .20 | .35 |
| Cross bracing, let-in, 1" x 6" | .080 | L.F. | .004 | .04 | .15 | .19 |
| TOTAL | | | .024 | .58 | .87 | 1.45 |
| **2" X 4", 24" O.C.** | | | | | | |
| 2" x 4" studs, #2 or better, 24" O.C. | .800 | L.F. | .012 | .31 | .42 | .73 |
| Plates, double top, single bottom | .375 | L.F. | .005 | .15 | .20 | .35 |
| Cross bracing, let-in, 1" x 6" | .080 | L.F. | .003 | .04 | .10 | .14 |
| TOTAL | | | .020 | .50 | .72 | 1.22 |
| **2" X 6", 16" O.C.** | | | | | | |
| 2" x 6" studs, #2 or better, 16" O.C. | 1.000 | L.F. | .016 | .62 | .57 | 1.19 |
| Plates, double top, single bottom | .375 | L.F. | .006 | .23 | .21 | .44 |
| Cross bracing, let-in, 1" x 6" | .080 | L.F. | .004 | .04 | .15 | .19 |
| TOTAL | | | .026 | .89 | .93 | 1.82 |
| **2" X 6", 24" O.C.** | | | | | | |
| 2" x 6" studs, #2 or better, 24" O.C. | .800 | L.F. | .013 | .50 | .46 | .96 |
| Plates, double top, single bottom | .375 | L.F. | .006 | .23 | .21 | .44 |
| Cross bracing, let-in, 1" x 6" | .080 | L.F. | .003 | .04 | .10 | .14 |
| TOTAL | | | .022 | .77 | .77 | 1.54 |

The costs in this system are based on a square foot of wall area. Do not subtract for door or window openings.

| Description | QUAN. | UNIT | LABOR HOURS | COST PER S.F. | | |
|---|---|---|---|---|---|---|
| | | | | MAT. | INST. | TOTAL |
| | | | | | | |
| | | | | | | |
| | | | | | | |

**Important: See the Reference Section for critical supporting data - Reference Nos., Crews & Location Factors**

| Partition Framing Price Sheet | QUAN. | UNIT | LABOR HOURS | COST PER S.F. | | |
|---|---|---|---|---|---|---|
| | | | | MAT. | INST. | TOTAL |
| Wood studs, #2 or better, 2" x 4", 12" O.C. | 1.250 | L.F. | .018 | .49 | .65 | 1.14 |
| 16" O.C. | 1.000 | L.F. | .015 | .39 | .52 | .91 |
| 24" O.C. | .800 | L.F. | .012 | .31 | .42 | .73 |
| 32" O.C. | .650 | L.F. | .009 | .25 | .34 | .59 |
| 2" x 6", 12" O.C. | 1.250 | L.F. | .020 | .78 | .71 | 1.49 |
| 16" O.C. | 1.000 | L.F. | .016 | .62 | .57 | 1.19 |
| 24" O.C. | .800 | L.F. | .013 | .50 | .46 | .96 |
| 32" O.C. | .650 | L.F. | .010 | .40 | .37 | .77 |
| Plates, #2 or better double top single bottom, 2" x 4" | .375 | L.F. | .005 | .15 | .20 | .35 |
| 2" x 6" | .375 | L.F. | .006 | .23 | .21 | .44 |
| 2" x 8" | .375 | L.F. | .005 | .36 | .20 | .56 |
| Cross bracing, let-in, 1" x 6" boards studs, 12" O.C. | .080 | L.F. | .005 | .06 | .19 | .25 |
| 16" O.C. | .080 | L.F. | .004 | .04 | .15 | .19 |
| 24" O.C. | .080 | L.F. | .003 | .04 | .10 | .14 |
| 32" O.C. | .080 | L.F. | .002 | .04 | .08 | .12 |
| Let-in steel (T shaped) studs, 12" O.C. | .080 | L.F. | .001 | .04 | .05 | .09 |
| 16" O.C. | .080 | L.F. | .001 | .04 | .04 | .08 |
| 24" O.C. | .080 | L.F. | .001 | .04 | .04 | .08 |
| 32" O.C. | .080 | L.F. | .001 | .03 | .03 | .06 |
| Steel straps studs, 12" O.C. | .080 | L.F. | .001 | .06 | .04 | .10 |
| 16" O.C. | .080 | L.F. | .001 | .05 | .04 | .09 |
| 24" O.C. | .080 | L.F. | .001 | .05 | .04 | .09 |
| 32" O.C. | .080 | L.F. | .001 | .05 | .03 | .08 |
| Metal studs, load bearing 24" O.C., 20 ga. galv., 2-1/2" wide | 1.000 | S.F. | .015 | .36 | .53 | .89 |
| 3-5/8" wide | 1.000 | S.F. | .015 | .43 | .54 | .97 |
| 4" wide | 1.000 | S.F. | .016 | .45 | .56 | 1.01 |
| 6" wide | 1.000 | S.F. | .016 | .58 | .57 | 1.15 |
| 16 ga., 2-1/2" wide | 1.000 | S.F. | .017 | .42 | .61 | 1.03 |
| 3-5/8" wide | 1.000 | S.F. | .017 | .50 | .62 | 1.12 |
| 4" wide | 1.000 | S.F. | .018 | .53 | .64 | 1.17 |
| 6" wide | 1.000 | S.F. | .018 | .67 | .65 | 1.32 |
| Non-load bearing 24" O.C., 25 ga. galv., 1-5/8" wide | 1.000 | S.F. | .011 | .13 | .37 | .50 |
| 2-1/2" wide | 1.000 | S.F. | .011 | .14 | .38 | .52 |
| 3-5/8" wide | 1.000 | S.F. | .011 | .16 | .38 | .54 |
| 4" wide | 1.000 | S.F. | .011 | .17 | .38 | .55 |
| 6" wide | 1.000 | S.F. | .011 | .26 | .39 | .65 |
| 20 ga., 2-1/2" wide | 1.000 | S.F. | .013 | .22 | .47 | .69 |
| 3-5/8" wide | 1.000 | S.F. | .014 | .25 | .48 | .73 |
| 4" wide | 1.000 | S.F. | .014 | .27 | .48 | .75 |
| 6" wide | 1.000 | S.F. | .014 | .33 | .49 | .82 |

| Window & Door Openings | QUAN. | UNIT | LABOR HOURS | COST EACH | | |
|---|---|---|---|---|---|---|
| | | | | MAT. | INST. | TOTAL |
| The following costs are to be added to the total costs of the walls. Do not subtract openings from total wall area. | | | | | | |
| Headers, 2" x 6" double, 2' long | 4.000 | L.F. | .178 | 2.48 | 6.30 | 8.78 |
| 3' long | 6.000 | L.F. | .267 | 3.72 | 9.50 | 13.22 |
| 4' long | 8.000 | L.F. | .356 | 4.96 | 12.65 | 17.61 |
| 5' long | 10.000 | L.F. | .444 | 6.20 | 15.80 | 22.00 |
| 2" x 8" double, 4' long | 8.000 | L.F. | .376 | 7.60 | 13.45 | 21.05 |
| 5' long | 10.000 | L.F. | .471 | 9.50 | 16.80 | 26.30 |
| 6' long | 12.000 | L.F. | .565 | 11.40 | 20.00 | 31.40 |
| 8' long | 16.000 | L.F. | .753 | 15.20 | 27.00 | 42.20 |
| 2" x 10" double, 4' long | 8.000 | L.F. | .400 | 10.95 | 14.25 | 25.20 |
| 6' long | 12.000 | L.F. | .600 | 16.45 | 21.50 | 37.95 |
| 8' long | 16.000 | L.F. | .800 | 22.00 | 28.50 | 50.50 |
| 10' long | 20.000 | L.F. | 1.000 | 27.50 | 35.50 | 63.00 |
| 2" x 12" double, 8' long | 16.000 | L.F. | .853 | 30.00 | 30.50 | 60.50 |
| 12' long | 24.000 | L.F. | 1.280 | 45.00 | 45.50 | 90.50 |

# Division 4
# Exterior Walls

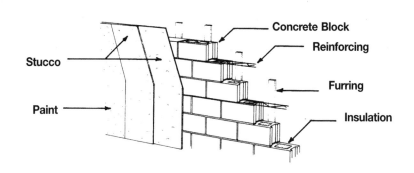

| System Description | QUAN. | UNIT | LABOR HOURS | COST PER S.F. | | |
|---|---|---|---|---|---|---|
| | | | | MAT. | INST. | TOTAL |
| **6″ THICK CONCRETE BLOCK WALL** | | | | | | |
| 6″ thick concrete block, 6″ x 8″ x 16″ | 1.000 | S.F. | .100 | 1.46 | 3.25 | 4.71 |
| Masonry reinforcing, truss strips every other course | .625 | L.F. | .002 | .08 | .06 | .14 |
| Furring, 1″ x 3″, 16″ O.C. | 1.000 | L.F. | .016 | .21 | .58 | .79 |
| Masonry insulation, poured vermiculite | 1.000 | S.F. | .013 | .60 | .47 | 1.07 |
| Stucco, 2 coats | 1.000 | S.F. | .069 | .19 | 2.26 | 2.45 |
| Masonry paint, 2 coats | 1.000 | S.F. | .016 | .18 | .51 | .69 |
| TOTAL | | | .216 | 2.72 | 7.13 | 9.85 |
| **8″ THICK CONCRETE BLOCK WALL** | | | | | | |
| 8″ thick concrete block, 8″ x 8″ x 16″ | 1.000 | S.F. | .107 | 1.59 | 3.46 | 5.05 |
| Masonry reinforcing, truss strips every other course | .625 | L.F. | .002 | .08 | .06 | .14 |
| Furring, 1″ x 3″, 16″ O.C. | 1.000 | L.F. | .016 | .21 | .58 | .79 |
| Masonry insulation, poured vermiculite | 1.000 | S.F. | .018 | .79 | .62 | 1.41 |
| Stucco, 2 coats | 1.000 | S.F. | .069 | .19 | 2.26 | 2.45 |
| Masonry paint, 2 coats | 1.000 | S.F. | .016 | .18 | .51 | .69 |
| TOTAL | | | .228 | 3.04 | 7.49 | 10.53 |
| **12″ THICK CONCRETE BLOCK WALL** | | | | | | |
| 12″ thick concrete block, 12″ x 8″ x 16″ | 1.000 | S.F. | .141 | 2.32 | 4.47 | 6.79 |
| Masonry reinforcing, truss strips every other course | .625 | L.F. | .003 | .09 | .09 | .18 |
| Furring, 1″ x 3″, 16″ O.C. | 1.000 | L.F. | .016 | .21 | .58 | .79 |
| Masonry insulation, poured vermiculite | 1.000 | S.F. | .026 | 1.17 | .92 | 2.09 |
| Stucco, 2 coats | 1.000 | S.F. | .069 | .19 | 2.26 | 2.45 |
| Masonry paint, 2 coats | 1.000 | S.F. | .016 | .18 | .51 | .69 |
| TOTAL | | | .271 | 4.16 | 8.83 | 12.99 |

Costs for this system are based on a square foot of wall area. Do not subtract for window openings.

| Description | QUAN. | UNIT | LABOR HOURS | COST PER S.F. | | |
|---|---|---|---|---|---|---|
| | | | | MAT. | INST. | TOTAL |
| | | | | | | |
| | | | | | | |
| | | | | | | |
| | | | | | | |
| | | | | | | |

**Important: See the Reference Section for critical supporting data - Reference Nos., Crews & Location Factors**

| Masonry Block Price Sheet | QUAN. | UNIT | LABOR HOURS | COST PER S.F. | | |
|---|---|---|---|---|---|---|
| | | | | MAT. | INST. | TOTAL |
| Block concrete, 8" x 16" regular, 4" thick | 1.000 | S.F. | .093 | .99 | 3.02 | 4.01 |
| 6" thick | 1.000 | S.F. | .100 | 1.46 | 3.25 | 4.71 |
| 8" thick | 1.000 | S.F. | .107 | 1.59 | 3.46 | 5.05 |
| 10" thick | 1.000 | S.F. | .111 | 2.26 | 3.61 | 5.87 |
| 12" thick | 1.000 | S.F. | .141 | 2.32 | 4.47 | 6.79 |
| | | | | | | |
| Solid block, 4" thick | 1.000 | S.F. | .096 | 1.37 | 3.13 | 4.50 |
| 6" thick | 1.000 | S.F. | .104 | 1.60 | 3.37 | 4.97 |
| 8" thick | 1.000 | S.F. | .111 | 2.43 | 3.61 | 6.04 |
| 10" thick | 1.000 | S.F. | .133 | 3.29 | 4.20 | 7.49 |
| 12" thick | 1.000 | S.F. | .148 | 3.65 | 4.67 | 8.32 |
| | | | | | | |
| Lightweight, 4" thick | 1.000 | S.F. | .093 | .99 | 3.02 | 4.01 |
| 6" thick | 1.000 | S.F. | .100 | 1.46 | 3.25 | 4.71 |
| 8" thick | 1.000 | S.F. | .107 | 1.59 | 3.46 | 5.05 |
| 10" thick | 1.000 | S.F. | .111 | 2.26 | 3.61 | 5.87 |
| 12" thick | 1.000 | S.F. | .141 | 2.32 | 4.47 | 6.79 |
| | | | | | | |
| Split rib profile, 4" thick | 1.000 | S.F. | .116 | 2.11 | 3.76 | 5.87 |
| 6" thick | 1.000 | S.F. | .123 | 2.67 | 4.00 | 6.67 |
| 8" thick | 1.000 | S.F. | .131 | 3.22 | 4.33 | 7.55 |
| 10" thick | 1.000 | S.F. | .157 | 3.67 | 4.95 | 8.62 |
| 12" thick | 1.000 | S.F. | .175 | 4.08 | 5.50 | 9.58 |
| | | | | | | |
| Masonry reinforcing, wire truss strips, every course, 8" block | 1.375 | L.F. | .004 | .17 | .14 | .31 |
| 12" block | 1.375 | L.F. | .006 | .19 | .19 | .38 |
| Every other course, 8" block | .625 | L.F. | .002 | .08 | .06 | .14 |
| 12" block | .625 | L.F. | .003 | .09 | .09 | .18 |
| Furring, wood, 1" x 3", 12" O.C. | 1.250 | L.F. | .020 | .26 | .73 | .99 |
| 16" O.C. | 1.000 | L.F. | .016 | .21 | .58 | .79 |
| 24" O.C. | .800 | L.F. | .013 | .17 | .46 | .63 |
| 32" O.C. | .640 | L.F. | .010 | .13 | .37 | .50 |
| Steel, 3/4" channels, 12" O.C. | 1.250 | L.F. | .034 | .18 | 1.11 | 1.29 |
| 16" O.C. | 1.000 | L.F. | .030 | .16 | .99 | 1.15 |
| 24" O.C. | .800 | L.F. | .023 | .11 | .75 | .86 |
| 32" O.C. | .640 | L.F. | .018 | .09 | .60 | .69 |
| Masonry insulation, vermiculite or perlite poured 4" thick | 1.000 | S.F. | .009 | .39 | .31 | .70 |
| 6" thick | 1.000 | S.F. | .013 | .59 | .47 | 1.06 |
| 8" thick | 1.000 | S.F. | .018 | .79 | .62 | 1.41 |
| 10" thick | 1.000 | S.F. | .021 | .96 | .76 | 1.72 |
| 12" thick | 1.000 | S.F. | .026 | 1.17 | .92 | 2.09 |
| | | | | | | |
| Block inserts polystyrene, 6" thick | 1.000 | S.F. | | .90 | | .90 |
| 8" thick | 1.000 | S.F. | | .90 | | .90 |
| 10" thick | 1.000 | S.F. | | 1.07 | | 1.07 |
| 12" thick | 1.000 | S.F. | | 1.12 | | 1.12 |
| Stucco, 1 coat | 1.000 | S.F. | .057 | .15 | 1.85 | 2.00 |
| 2 coats | 1.000 | S.F. | .069 | .19 | 2.26 | 2.45 |
| 3 coats | 1.000 | S.F. | .081 | .22 | 2.65 | 2.87 |
| | | | | | | |
| Painting, 1 coat | 1.000 | S.F. | .011 | .12 | .35 | .47 |
| 2 coats | 1.000 | S.F. | .016 | .18 | .51 | .69 |
| Primer & 1 coat | 1.000 | S.F. | .013 | .19 | .42 | .61 |
| 2 coats | 1.000 | S.F. | .018 | .25 | .58 | .83 |
| Lath, metal lath expanded 2.5 lb/S.Y., painted | 1.000 | S.F. | .010 | .17 | .34 | .51 |
| Galvanized | 1.000 | S.F. | .012 | .19 | .37 | .56 |
| | | | | | | |
| | | | | | | |

**4 EXTERIOR WALLS**

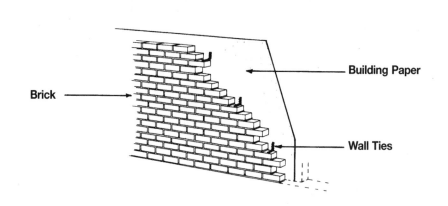

Brick

Building Paper

Wall Ties

| System Description | QUAN. | UNIT | LABOR HOURS | COST PER S.F. | | |
|---|---|---|---|---|---|---|
| | | | | MAT. | INST. | TOTAL |
| **SELECT COMMON BRICK** | | | | | | |
| Brick, select common, running bond | 1.000 | S.F. | .174 | 2.96 | 5.65 | 8.61 |
| Wall ties, 7/8" x 7", 22 gauge | 1.000 | Ea. | .008 | .05 | .27 | .32 |
| Building paper, #15 asphalt | 1.100 | S.F. | .002 | .02 | .09 | .11 |
| Trim, pine, painted | .125 | L.F. | .004 | .08 | .14 | .22 |
| TOTAL | | | .188 | 3.11 | 6.15 | 9.26 |
| **RED FACED COMMON BRICK** | | | | | | |
| Brick, common, red faced, running bond | 1.000 | S.F. | .182 | 2.96 | 5.90 | 8.86 |
| Wall ties, 7/8" x 7", 22 gauge | 1.000 | Ea. | .008 | .05 | .27 | .32 |
| Building paper, #15 asphalt | 1.100 | S.F. | .002 | .02 | .09 | .11 |
| Trim, pine, painted | .125 | L.F. | .004 | .08 | .14 | .22 |
| TOTAL | | | .196 | 3.11 | 6.40 | 9.51 |
| **BUFF OR GREY FACE BRICK** | | | | | | |
| Brick, buff or grey | 1.000 | S.F. | .182 | 3.14 | 5.90 | 9.04 |
| Wall ties, 7/8" x 7", 22 gauge | 1.000 | Ea. | .008 | .05 | .27 | .32 |
| Building paper, #15 asphalt | 1.100 | S.F. | .002 | .02 | .09 | .11 |
| Trim, pine, painted | .125 | L.F. | .004 | .08 | .14 | .22 |
| TOTAL | | | .196 | 3.29 | 6.40 | 9.69 |
| **STONE WORK, ROUGH STONE, AVERAGE** | | | | | | |
| Stone work, rough stone, average | 1.000 | S.F. | .179 | 7.44 | 5.66 | 13.10 |
| Wall ties, 7/8" x 7", 22 gauge | 1.000 | Ea. | .008 | .05 | .27 | .32 |
| Building paper, #15 asphalt | 1.000 | S.F. | .002 | .02 | .09 | .11 |
| Trim, pine, painted | .125 | L.F. | .004 | .08 | .14 | .22 |
| TOTAL | | | .193 | 7.59 | 6.16 | 13.75 |

The costs in this system are based on a square foot of wall area. Do not subtract area for window & door openings.

| Description | QUAN. | UNIT | LABOR HOURS | COST PER S.F. | | |
|---|---|---|---|---|---|---|
| | | | | MAT. | INST. | TOTAL |
| | | | | | | |
| | | | | | | |
| | | | | | | |

**Important: See the Reference Section for critical supporting data - Reference Nos., Crews & Location Factors**

EXTERIOR WALLS 4

| Brick/Stone Veneer Price Sheet | QUAN. | UNIT | LABOR HOURS | COST PER S.F. | | |
|---|---|---|---|---|---|---|
| | | | | MAT. | INST. | TOTAL |
| Brick | | | | | | |
| Select common, running bond | 1.000 | S.F. | .174 | 2.96 | 5.65 | 8.61 |
| Red faced, running bond | 1.000 | S.F. | .182 | 2.96 | 5.90 | 8.86 |
| Buff or grey faced, running bond | 1.000 | S.F. | .182 | 3.14 | 5.90 | 9.04 |
| Header every 6th course | 1.000 | S.F. | .216 | 3.45 | 7.00 | 10.45 |
| English bond | 1.000 | S.F. | .286 | 4.42 | 9.30 | 13.72 |
| Flemish bond | 1.000 | S.F. | .195 | 3.12 | 6.35 | 9.47 |
| Common bond | 1.000 | S.F. | .267 | 3.93 | 8.65 | 12.58 |
| Stack bond | 1.000 | S.F. | .182 | 3.14 | 5.90 | 9.04 |
| Jumbo, running bond | 1.000 | S.F. | .092 | 3.88 | 2.99 | 6.87 |
| Norman, running bond | 1.000 | S.F. | .125 | 3.85 | 4.06 | 7.91 |
| Norwegian, running bond | 1.000 | S.F. | .107 | 2.32 | 3.46 | 5.78 |
| Economy, running bond | 1.000 | S.F. | .129 | 2.75 | 4.19 | 6.94 |
| Engineer, running bond | 1.000 | S.F. | .154 | 2.25 | 5.00 | 7.25 |
| Roman, running bond | 1.000 | S.F. | .160 | 4.97 | 5.20 | 10.17 |
| Utility, running bond | 1.000 | S.F. | .089 | 3.52 | 2.89 | 6.41 |
| Glazed, running bond | 1.000 | S.F. | .190 | 10.90 | 6.20 | 17.10 |
| Stone work, rough stone, average | 1.000 | S.F. | .179 | 7.45 | 5.65 | 13.10 |
| Maximum | 1.000 | S.F. | .267 | 11.10 | 8.45 | 19.55 |
| Wall ties, galvanized, corrugated 7/8" x 7", 22 gauge | 1.000 | Ea. | .008 | .05 | .27 | .32 |
| 16 gauge | 1.000 | Ea. | .008 | .15 | .27 | .42 |
| Cavity wall, every 3rd course 6" long Z type, 1/4" diameter | 1.330 | L.F. | .010 | .29 | .35 | .64 |
| 3/16" diameter | 1.330 | L.F. | .010 | .20 | .35 | .55 |
| 8" long, Z type, 1/4" diameter | 1.330 | L.F. | .010 | .34 | .35 | .69 |
| 3/16" diameter | 1.330 | L.F. | .010 | .16 | .35 | .51 |
| Building paper, aluminum and kraft laminated foil, 1 side | 1.000 | S.F. | .002 | .04 | .08 | .12 |
| 2 sides | 1.000 | S.F. | .002 | .06 | .08 | .14 |
| #15 asphalt paper | 1.100 | S.F. | .002 | .02 | .09 | .11 |
| Polyethylene, .002" thick | 1.000 | S.F. | .002 | .01 | .08 | .09 |
| .004" thick | 1.000 | S.F. | .002 | .02 | .08 | .10 |
| .006" thick | 1.000 | S.F. | .002 | .03 | .08 | .11 |
| .010" thick | 1.000 | S.F. | .002 | .06 | .08 | .14 |
| Trim, 1" x 4", cedar | .125 | L.F. | .005 | .18 | .18 | .36 |
| Fir | .125 | L.F. | .005 | .07 | .18 | .25 |
| Redwood | .125 | L.F. | .005 | .18 | .18 | .36 |
| White pine | .125 | L.F. | .005 | .07 | .18 | .25 |

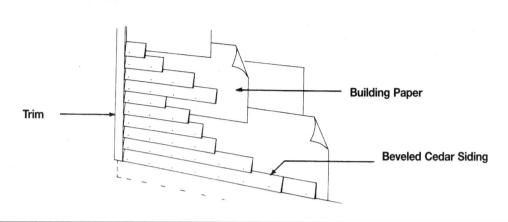

Trim

Building Paper

Beveled Cedar Siding

| System Description | QUAN. | UNIT | LABOR HOURS | COST PER S.F. | | |
|---|---|---|---|---|---|---|
| | | | | MAT. | INST. | TOTAL |
| **1/2" X 6" BEVELED CEDAR SIDING, "A" GRADE** | | | | | | |
| 1/2" x 6" beveled cedar siding | 1.000 | S.F. | .032 | 1.82 | 1.14 | 2.96 |
| #15 asphalt felt paper | 1.100 | S.F. | .002 | .02 | .09 | .11 |
| Trim, cedar | .125 | L.F. | .005 | .18 | .18 | .36 |
| Paint, primer & 2 coats | 1.000 | S.F. | .017 | .17 | .55 | .72 |
| TOTAL | | | .056 | 2.19 | 1.96 | 4.15 |
| **1/2" X 8" BEVELED CEDAR SIDING, "A" GRADE** | | | | | | |
| 1/2" x 8" beveled cedar siding | 1.000 | S.F. | .029 | 2.09 | 1.04 | 3.13 |
| #15 asphalt felt paper | 1.100 | S.F. | .002 | .02 | .09 | .11 |
| Trim, cedar | .125 | L.F. | .005 | .18 | .18 | .36 |
| Paint, primer & 2 coats | 1.000 | S.F. | .017 | .17 | .55 | .72 |
| TOTAL | | | .053 | 2.46 | 1.86 | 4.32 |
| **1" X 4" TONGUE & GROOVE, REDWOOD, VERTICAL GRAIN** | | | | | | |
| Redwood, clear, vertical grain, 1" x 10" | 1.000 | S.F. | .018 | 3.62 | .65 | 4.27 |
| #15 asphalt felt paper | 1.100 | S.F. | .002 | .02 | .09 | .11 |
| Trim, redwood | .125 | L.F. | .005 | .18 | .18 | .36 |
| Sealer, 1 coat, stain, 1 coat | 1.000 | S.F. | .013 | .11 | .43 | .54 |
| TOTAL | | | .038 | 3.93 | 1.35 | 5.28 |
| **1" X 6" TONGUE & GROOVE, REDWOOD, VERTICAL GRAIN** | | | | | | |
| Redwood, clear, vertical grain, 1" x 10" | 1.000 | S.F. | .019 | 3.73 | .67 | 4.40 |
| #15 asphalt felt paper | 1.100 | S.F. | .002 | .02 | .09 | .11 |
| Trim, redwood | .125 | L.F. | .005 | .18 | .18 | .36 |
| Sealer, 1 coat, stain, 1 coat | 1.000 | S.F. | .013 | .11 | .43 | .54 |
| TOTAL | | | .039 | 4.04 | 1.37 | 5.41 |

The costs in this system are based on a square foot of wall area.
Do not subtract area for door or window openings.

| Description | QUAN. | UNIT | LABOR HOURS | COST PER S.F. | | |
|---|---|---|---|---|---|---|
| | | | | MAT. | INST. | TOTAL |
| | | | | | | |
| | | | | | | |
| | | | | | | |

**Important: See the Reference Section for critical supporting data - Reference Nos., Crews & Location Factors**

| Wood Siding Price Sheet | QUAN. | UNIT | LABOR HOURS | COST PER S.F. | | |
|---|---|---|---|---|---|---|
| | | | | MAT. | INST. | TOTAL |
| Siding, beveled cedar, "A" grade, 1/2" x 6" | 1.000 | S.F. | .028 | 1.82 | 1.14 | 2.96 |
| 1/2" x 8" | 1.000 | S.F. | .023 | 2.09 | 1.04 | 3.13 |
| "B" grade, 1/2" x 6" | 1.000 | S.F. | .032 | 2.02 | 1.27 | 3.29 |
| 1/2" x 8" | 1.000 | S.F. | .029 | 2.32 | 1.16 | 3.48 |
| Clear grade, 1/2" x 6" | 1.000 | S.F. | .028 | 2.28 | 1.43 | 3.71 |
| 1/2" x 8" | 1.000 | S.F. | .023 | 2.61 | 1.30 | 3.91 |
| Redwood, clear vertical grain, 1/2" x 6" | 1.000 | S.F. | .036 | 2.88 | 1.27 | 4.15 |
| 1/2" x 8" | 1.000 | S.F. | .032 | 2.33 | 1.14 | 3.47 |
| Clear all heart vertical grain, 1/2" x 6" | 1.000 | S.F. | .028 | 3.20 | 1.41 | 4.61 |
| 1/2" x 8" | 1.000 | S.F. | .023 | 2.59 | 1.27 | 3.86 |
| Siding board & batten, cedar, "B" grade, 1" x 10" | 1.000 | S.F. | .031 | 2.22 | 1.10 | 3.32 |
| 1" x 12" | 1.000 | S.F. | .031 | 2.22 | 1.10 | 3.32 |
| Redwood, clear vertical grain, 1" x 6" | 1.000 | S.F. | .043 | 2.76 | 1.73 | 4.49 |
| 1" x 8" | 1.000 | S.F. | .018 | 2.52 | 1.52 | 4.04 |
| White pine, #2 & better, 1" x 10" | 1.000 | S.F. | .029 | .74 | 1.04 | 1.78 |
| 1" x 12" | 1.000 | S.F. | .029 | .74 | 1.04 | 1.78 |
| Siding vertical, tongue & groove, cedar "B" grade, 1" x 4" | 1.000 | S.F. | .033 | 2.00 | .65 | 2.65 |
| 1" x 6" | 1.000 | S.F. | .024 | 2.06 | .67 | 2.73 |
| 1" x 8" | 1.000 | S.F. | .024 | 2.12 | .69 | 2.81 |
| 1" x 10" | 1.000 | S.F. | .021 | 2.18 | .71 | 2.89 |
| "A" grade, 1" x 4" | 1.000 | S.F. | .033 | 1.83 | .60 | 2.43 |
| 1" x 6" | 1.000 | S.F. | .024 | 1.88 | .61 | 2.49 |
| 1" x 8" | 1.000 | S.F. | .024 | 1.93 | .63 | 2.56 |
| 1" x 10" | 1.000 | S.F. | .021 | 1.98 | .65 | 2.63 |
| Clear vertical grain, 1" x 4" | 1.000 | S.F. | .033 | 1.69 | .55 | 2.24 |
| 1" x 6" | 1.000 | S.F. | .024 | 1.73 | .56 | 2.29 |
| 1" x 8" | 1.000 | S.F. | .024 | 1.77 | .58 | 2.35 |
| 1" x 10" | 1.000 | S.F. | .021 | 1.82 | .59 | 2.41 |
| Redwood, clear vertical grain, 1" x 4" | 1.000 | S.F. | .033 | 3.62 | .65 | 4.27 |
| 1" x 6" | 1.000 | S.F. | .024 | 3.73 | .67 | 4.40 |
| 1" x 8" | 1.000 | S.F. | .024 | 3.83 | .69 | 4.52 |
| 1" x 10" | 1.000 | S.F. | .021 | 3.95 | .71 | 4.66 |
| Clear all heart vertical grain, 1" x 4" | 1.000 | S.F. | .033 | 3.32 | .60 | 3.92 |
| 1" x 6" | 1.000 | S.F. | .024 | 3.41 | .61 | 4.02 |
| 1" x 8" | 1.000 | S.F. | .024 | 3.50 | .63 | 4.13 |
| 1" x 10" | 1.000 | S.F. | .021 | 3.59 | .65 | 4.24 |
| White pine, 1" x 10" | 1.000 | S.F. | .024 | .77 | .71 | 1.48 |
| Siding plywood, texture 1-11 cedar, 3/8" thick | 1.000 | S.F. | .024 | 1.19 | .84 | 2.03 |
| 5/8" thick | 1.000 | S.F. | .024 | 2.35 | .84 | 3.19 |
| Redwood, 3/8" thick | 1.000 | S.F. | .024 | 1.19 | .84 | 2.03 |
| 5/8" thick | 1.000 | S.F. | .024 | 1.98 | .84 | 2.82 |
| Fir, 3/8" thick | 1.000 | S.F. | .024 | .64 | .84 | 1.48 |
| 5/8" thick | 1.000 | S.F. | .024 | .95 | .84 | 1.79 |
| Southern yellow pine, 3/8" thick | 1.000 | S.F. | .024 | .64 | .84 | 1.48 |
| 5/8" thick | 1.000 | S.F. | .024 | .89 | .84 | 1.73 |
| Hard board, 7/16" thick primed, plain finish | 1.000 | S.F. | .025 | 1.14 | .88 | 2.02 |
| Board finish | 1.000 | S.F. | .023 | .80 | .81 | 1.61 |
| Polyvinyl coated, 3/8" thick | 1.000 | S.F. | .021 | .97 | .76 | 1.73 |
| 5/8" thick | 1.000 | S.F. | .024 | .89 | .84 | 1.73 |
| Paper, #15 asphalt felt | 1.100 | S.F. | .002 | .02 | .09 | .11 |
| Trim, cedar | .125 | L.F. | .005 | .18 | .18 | .36 |
| Fir | .125 | L.F. | .005 | .07 | .18 | .25 |
| Redwood | .125 | L.F. | .005 | .18 | .18 | .36 |
| White pine | .125 | L.F. | .005 | .07 | .18 | .25 |
| Painting, primer, & 1 coat | 1.000 | S.F. | .013 | .11 | .43 | .54 |
| 2 coats | 1.000 | S.F. | .017 | .17 | .55 | .72 |
| Stain, sealer, & 1 coat | 1.000 | S.F. | .017 | .09 | .55 | .64 |
| 2 coats | 1.000 | S.F. | .019 | .14 | .60 | .74 |

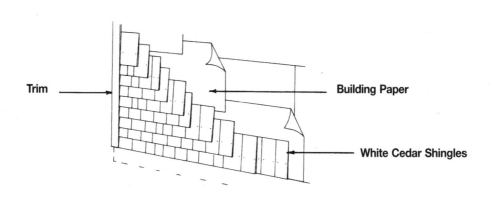

| System Description | QUAN. | UNIT | LABOR HOURS | COST PER S.F. | | |
|---|---|---|---|---|---|---|
| | | | | MAT. | INST. | TOTAL |
| **WHITE CEDAR SHINGLES, 5″ EXPOSURE** | | | | | | |
| White cedar shingles, 16″ long, grade "A", 5″ exposure | 1.000 | S.F. | .033 | 1.18 | 1.19 | 2.37 |
| #15 asphalt felt paper | 1.100 | S.F. | .002 | .02 | .09 | .11 |
| Trim, cedar | .125 | S.F. | .005 | .18 | .18 | .36 |
| Paint, primer & 1 coat | 1.000 | S.F. | .017 | .09 | .55 | .64 |
| TOTAL | | | .057 | 1.47 | 2.01 | 3.48 |
| **NO. 1 PERFECTIONS, 5-1/2″ EXPOSURE** | | | | | | |
| No. 1 perfections, red cedar, 5-1/2″ exposure | 1.000 | S.F. | .029 | 1.76 | 1.04 | 2.80 |
| #15 asphalt felt paper | 1.100 | S.F. | .002 | .02 | .09 | .11 |
| Trim, cedar | .125 | S.F. | .005 | .18 | .18 | .36 |
| Stain, sealer & 1 coat | 1.000 | S.F. | .017 | .09 | .55 | .64 |
| TOTAL | | | .053 | 2.05 | 1.86 | 3.91 |
| **RESQUARED & REBUTTED PERFECTIONS, 5-1/2″ EXPOSURE** | | | | | | |
| Resquared & rebutted perfections, 5-1/2″ exposure | 1.000 | S.F. | .027 | 2.19 | .95 | 3.14 |
| #15 asphalt felt paper | 1.100 | S.F. | .002 | .02 | .09 | .11 |
| Trim, cedar | .125 | S.F. | .005 | .18 | .18 | .36 |
| Stain, sealer & 1 coat | 1.000 | S.F. | .017 | .09 | .55 | .64 |
| TOTAL | | | .051 | 2.48 | 1.77 | 4.25 |
| **HAND-SPLIT SHAKES, 8-1/2″ EXPOSURE** | | | | | | |
| Hand-split red cedar shakes, 18″ long, 8-1/2″ exposure | 1.000 | S.F. | .040 | 1.07 | 1.42 | 2.49 |
| #15 asphalt felt paper | 1.100 | S.F. | .002 | .02 | .09 | .11 |
| Trim, cedar | .125 | S.F. | .005 | .18 | .18 | .36 |
| Stain, sealer & 1 coat | 1.000 | S.F. | .017 | .09 | .55 | .64 |
| TOTAL | | | .064 | 1.36 | 2.24 | 3.60 |

The costs in this system are based on a square foot of wall area.
Do not subtract area for door or window openings.

| Description | QUAN. | UNIT | LABOR HOURS | COST PER S.F. | | |
|---|---|---|---|---|---|---|
| | | | | MAT. | INST. | TOTAL |
| | | | | | | |
| | | | | | | |
| | | | | | | |

EXTERIOR WALLS   4

| Shingle Siding Price Sheet | QUAN. | UNIT | LABOR HOURS | COST PER S.F. | | |
|---|---|---|---|---|---|---|
| | | | | MAT. | INST. | TOTAL |
| Shingles wood, white cedar 16" long, "A" grade, 5" exposure | 1.000 | S.F. | .033 | 1.18 | 1.19 | 2.37 |
| 7" exposure | 1.000 | S.F. | .030 | 1.06 | 1.07 | 2.13 |
| 8-1/2" exposure | 1.000 | S.F. | .032 | .68 | 1.14 | 1.82 |
| 10" exposure | 1.000 | S.F. | .028 | .59 | .99 | 1.58 |
| "B" grade, 5" exposure | 1.000 | S.F. | .040 | 1.21 | 1.42 | 2.63 |
| 7" exposure | 1.000 | S.F. | .028 | .85 | .99 | 1.84 |
| 8-1/2" exposure | 1.000 | S.F. | .024 | .73 | .85 | 1.58 |
| 10" exposure | 1.000 | S.F. | .020 | .61 | .71 | 1.32 |
| Fire retardant, "A" grade, 5" exposure | 1.000 | S.F. | .033 | 1.51 | 1.19 | 2.70 |
| 7" exposure | 1.000 | S.F. | .030 | 1.36 | 1.07 | 2.43 |
| 8-1/2" exposure | 1.000 | S.F. | .027 | 1.20 | .95 | 2.15 |
| 10" exposure | 1.000 | S.F. | .023 | 1.06 | .83 | 1.89 |
| "B" grade, 5" exposure | 1.000 | S.F. | .040 | 1.54 | 1.42 | 2.96 |
| 7" exposure | 1.000 | S.F. | .028 | 1.08 | .99 | 2.07 |
| 8-1/2" exposure | 1.000 | S.F. | .024 | .93 | .85 | 1.78 |
| 10" exposure | 1.000 | S.F. | .020 | .78 | .71 | 1.49 |
| No. 1 perfections red cedar, 18" long, 5-1/2" exposure | 1.000 | S.F. | .029 | 1.76 | 1.04 | 2.80 |
| 7" exposure | 1.000 | S.F. | .036 | 1.29 | 1.27 | 2.56 |
| 8-1/2" exposure | 1.000 | S.F. | .032 | 1.16 | 1.14 | 2.30 |
| 10" exposure | 1.000 | S.F. | .025 | .90 | .89 | 1.79 |
| Fire retardant, 5" exposure | 1.000 | S.F. | .029 | 2.08 | 1.04 | 3.12 |
| 7" exposure | 1.000 | S.F. | .036 | 1.61 | 1.27 | 2.88 |
| 8-1/2" exposure | 1.000 | S.F. | .032 | 1.44 | 1.14 | 2.58 |
| 10" exposure | 1.000 | S.F. | .025 | 1.12 | .89 | 2.01 |
| Resquared & rebutted, 5-1/2" exposure | 1.000 | S.F. | .027 | 2.19 | .95 | 3.14 |
| 7" exposure | 1.000 | S.F. | .024 | 1.97 | .86 | 2.83 |
| 8-1/2" exposure | 1.000 | S.F. | .021 | 1.75 | .76 | 2.51 |
| 10" exposure | 1.000 | S.F. | .019 | 1.53 | .67 | 2.20 |
| Fire retardant, 5" exposure | 1.000 | S.F. | .027 | 2.51 | .95 | 3.46 |
| 7" exposure | 1.000 | S.F. | .024 | 2.25 | .86 | 3.11 |
| 8-1/2" exposure | 1.000 | S.F. | .021 | 2.00 | .76 | 2.76 |
| 10" exposure | 1.000 | S.F. | .023 | 1.35 | .81 | 2.16 |
| Hand-split, red cedar, 24" long, 7" exposure | 1.000 | S.F. | .045 | 2.13 | 1.60 | 3.73 |
| 8-1/2" exposure | 1.000 | S.F. | .038 | 1.82 | 1.37 | 3.19 |
| 10" exposure | 1.000 | S.F. | .032 | 1.52 | 1.14 | 2.66 |
| 12" exposure | 1.000 | S.F. | .026 | 1.22 | .91 | 2.13 |
| Fire retardant, 7" exposure | 1.000 | S.F. | .045 | 2.59 | 1.60 | 4.19 |
| 8-1/2" exposure | 1.000 | S.F. | .038 | 2.22 | 1.37 | 3.59 |
| 10" exposure | 1.000 | S.F. | .032 | 1.85 | 1.14 | 2.99 |
| 12" exposure | 1.000 | S.F. | .026 | 1.48 | .91 | 2.39 |
| 18" long, 5" exposure | 1.000 | S.F. | .068 | 1.82 | 2.41 | 4.23 |
| 7" exposure | 1.000 | S.F. | .048 | 1.28 | 1.70 | 2.98 |
| 8-1/2" exposure | 1.000 | S.F. | .040 | 1.07 | 1.42 | 2.49 |
| 10" exposure | 1.000 | S.F. | .036 | .96 | 1.28 | 2.24 |
| Fire retardant, 5" exposure | 1.000 | S.F. | .068 | 2.38 | 2.41 | 4.79 |
| 7" exposure | 1.000 | S.F. | .048 | 1.68 | 1.70 | 3.38 |
| 8-1/2" exposure | 1.000 | S.F. | .040 | 1.40 | 1.42 | 2.82 |
| 10" exposure | 1.000 | S.F. | .036 | 1.26 | 1.28 | 2.54 |
| Paper, #15 asphalt felt | 1.100 | S.F. | .002 | .02 | .08 | .10 |
| Trim, cedar | .125 | S.F. | .005 | .18 | .18 | .36 |
| Fir | .125 | S.F. | .005 | .07 | .18 | .25 |
| Redwood | .125 | S.F. | .005 | .18 | .18 | .36 |
| White pine | .125 | S.F. | .005 | .07 | .18 | .25 |
| Painting, primer, & 1 coat | 1.000 | S.F. | .013 | .11 | .43 | .54 |
| 2 coats | 1.000 | S.F. | .017 | .17 | .55 | .72 |
| Staining, sealer, & 1 coat | 1.000 | S.F. | .017 | .09 | .55 | .64 |
| 2 coats | 1.000 | S.F. | .019 | .14 | .60 | .74 |

**EXTERIOR WALLS**

**4**

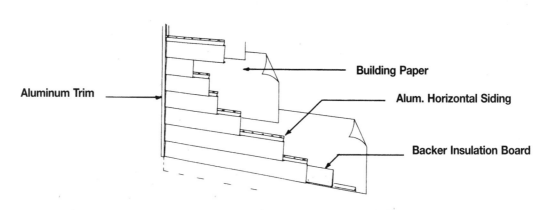

Aluminum Trim

Building Paper

Alum. Horizontal Siding

Backer Insulation Board

| System Description | QUAN. | UNIT | LABOR HOURS | COST PER S.F. | | |
|---|---|---|---|---|---|---|
| | | | | MAT. | INST. | TOTAL |
| **ALUMINUM CLAPBOARD SIDING, 8″ WIDE, WHITE** | | | | | | |
| Aluminum horizontal siding, 8″ clapboard | 1.000 | S.F. | .031 | 1.38 | 1.11 | 2.49 |
| Backer, insulation board | 1.000 | S.F. | .008 | .45 | .28 | .73 |
| Trim, aluminum | .600 | L.F. | .016 | .62 | .56 | 1.18 |
| Paper, #15 asphalt felt | 1.100 | S.F. | .002 | .02 | .09 | .11 |
| TOTAL | | | .057 | 2.47 | 2.04 | 4.51 |
| **ALUMINUM VERTICAL BOARD & BATTEN, WHITE** | | | | | | |
| Aluminum vertical board & batten | 1.000 | S.F. | .027 | 1.54 | .97 | 2.51 |
| Backer insulation board | 1.000 | S.F. | .008 | .45 | .28 | .73 |
| Trim, aluminum | .600 | L.F. | .016 | .62 | .56 | 1.18 |
| Paper, #15 asphalt felt | 1.100 | S.F. | .002 | .02 | .09 | .11 |
| TOTAL | | | .053 | 2.63 | 1.90 | 4.53 |
| **VINYL CLAPBOARD SIDING, 8″ WIDE, WHITE** | | | | | | |
| PVC vinyl horizontal siding, 8″ clapboard | 1.000 | S.F. | .032 | .65 | 1.15 | 1.80 |
| Backer, insulation board | 1.000 | S.F. | .008 | .45 | .28 | .73 |
| Trim, vinyl | .600 | L.F. | .014 | .39 | .50 | .89 |
| Paper, #15 asphalt felt | 1.100 | S.F. | .002 | .02 | .09 | .11 |
| TOTAL | | | .056 | 1.51 | 2.02 | 3.53 |
| **VINYL VERTICAL BOARD & BATTEN, WHITE** | | | | | | |
| PVC vinyl vertical board & batten | 1.000 | S.F. | .029 | 1.51 | 1.04 | 2.55 |
| Backer, insulation board | 1.000 | S.F. | .008 | .45 | .28 | .73 |
| Trim, vinyl | .600 | L.F. | .014 | .39 | .50 | .89 |
| Paper, #15 asphalt felt | 1.100 | S.F. | .002 | .02 | .09 | .11 |
| TOTAL | | | .053 | 2.37 | 1.91 | 4.28 |

The costs in this system are on a square foot of wall basis.
Do not subtract openings from wall area.

| Description | QUAN. | UNIT | LABOR HOURS | COST PER S.F. | | |
|---|---|---|---|---|---|---|
| | | | | MAT. | INST. | TOTAL |
| | | | | | | |
| | | | | | | |
| | | | | | | |

**Important: See the Reference Section for critical supporting data - Reference Nos., Crews & Location Factors**

EXTERIOR WALLS 4

# Metal & Plastic Siding Price Sheet

| | QUAN. | UNIT | LABOR HOURS | COST PER S.F. MAT. | COST PER S.F. INST. | COST PER S.F. TOTAL |
|---|---|---|---|---|---|---|
| Siding, aluminum, .024" thick, smooth, 8" wide, white | 1.000 | S.F. | .031 | 1.38 | 1.11 | 2.49 |
| Color | 1.000 | S.F. | .031 | 1.47 | 1.11 | 2.58 |
| Double 4" pattern, 8" wide, white | 1.000 | S.F. | .031 | 1.31 | 1.11 | 2.42 |
| Color | 1.000 | S.F. | .031 | 1.40 | 1.11 | 2.51 |
| Double 5" pattern, 10" wide, white | 1.000 | S.F. | .029 | 1.31 | 1.04 | 2.35 |
| Color | 1.000 | S.F. | .029 | 1.40 | 1.04 | 2.44 |
| Embossed, single, 8" wide, white | 1.000 | S.F. | .031 | 1.63 | 1.11 | 2.74 |
| Color | 1.000 | S.F. | .031 | 1.72 | 1.11 | 2.83 |
| Double 4" pattern, 8" wide, white | 1.000 | S.F. | .031 | 1.49 | 1.11 | 2.60 |
| Color | 1.000 | S.F. | .031 | 1.58 | 1.11 | 2.69 |
| Double 5" pattern, 10" wide, white | 1.000 | S.F. | .029 | 1.49 | 1.04 | 2.53 |
| Color | 1.000 | S.F. | .029 | 1.58 | 1.04 | 2.62 |
| Alum siding with insulation board, smooth, 8" wide, white | 1.000 | S.F. | .031 | 1.32 | 1.11 | 2.43 |
| Color | 1.000 | S.F. | .031 | 1.41 | 1.11 | 2.52 |
| Double 4" pattern, 8" wide, white | 1.000 | S.F. | .031 | 1.30 | 1.11 | 2.41 |
| Color | 1.000 | S.F. | .031 | 1.39 | 1.11 | 2.50 |
| Double 5" pattern, 10" wide, white | 1.000 | S.F. | .029 | 1.30 | 1.04 | 2.34 |
| Color | 1.000 | S.F. | .029 | 1.39 | 1.04 | 2.43 |
| Embossed, single, 8" wide, white | 1.000 | S.F. | .031 | 1.52 | 1.11 | 2.63 |
| Color | 1.000 | S.F. | .031 | 1.61 | 1.11 | 2.72 |
| Double 4" pattern, 8" wide, white | 1.000 | S.F. | .031 | 1.54 | 1.11 | 2.65 |
| Color | 1.000 | S.F. | .031 | 1.63 | 1.11 | 2.74 |
| Double 5" pattern, 10" wide, white | 1.000 | S.F. | .029 | 1.54 | 1.04 | 2.58 |
| Color | 1.000 | S.F. | .029 | 1.63 | 1.04 | 2.67 |
| Aluminum, shake finish, 10" wide, white | 1.000 | S.F. | .029 | 1.63 | 1.04 | 2.67 |
| Color | 1.000 | S.F. | .029 | 1.72 | 1.04 | 2.76 |
| Aluminum, vertical, 12" wide, white | 1.000 | S.F. | .027 | 1.54 | .97 | 2.51 |
| Color | 1.000 | S.F. | .027 | 1.63 | .97 | 2.60 |
| Vinyl siding, 8" wide, smooth, white | 1.000 | S.F. | .032 | .65 | 1.15 | 1.80 |
| Color | 1.000 | S.F. | .032 | .74 | 1.15 | 1.89 |
| 10" wide, Dutch lap, smooth, white | 1.000 | S.F. | .029 | .70 | 1.04 | 1.74 |
| Color | 1.000 | S.F. | .029 | .79 | 1.04 | 1.83 |
| Double 4" pattern, 8" wide, white | 1.000 | S.F. | .032 | .61 | 1.15 | 1.76 |
| Color | 1.000 | S.F. | .032 | .70 | 1.15 | 1.85 |
| Double 5" pattern, 10" wide, white | 1.000 | S.F. | .029 | .57 | 1.04 | 1.61 |
| Color | 1.000 | S.F. | .029 | .66 | 1.04 | 1.70 |
| Embossed, single, 8" wide, white | 1.000 | S.F. | .032 | .73 | 1.15 | 1.88 |
| Color | 1.000 | S.F. | .032 | .82 | 1.15 | 1.97 |
| 10" wide, white | 1.000 | S.F. | .029 | .74 | 1.04 | 1.78 |
| Color | 1.000 | S.F. | .029 | .83 | 1.04 | 1.87 |
| Double 4" pattern, 8" wide, white | 1.000 | S.F. | .032 | .63 | 1.15 | 1.78 |
| Color | 1.000 | S.F. | .032 | .72 | 1.15 | 1.87 |
| Double 5" pattern, 10" wide, white | 1.000 | S.F. | .029 | .65 | 1.04 | 1.69 |
| Color | 1.000 | S.F. | .029 | .74 | 1.04 | 1.78 |
| Vinyl, shake finish, 10" wide, white | 1.000 | S.F. | .029 | 2.09 | 1.04 | 3.13 |
| Color | 1.000 | S.F. | .029 | 2.18 | 1.04 | 3.22 |
| Vinyl, vertical, double 5" pattern, 10" wide, white | 1.000 | S.F. | .029 | 1.51 | 1.04 | 2.55 |
| Color | 1.000 | S.F. | .029 | 1.60 | 1.04 | 2.64 |
| Backer board, installed in siding panels 8" or 10" wide | 1.000 | S.F. | .008 | .45 | .28 | .73 |
| 4' x 8' sheets, polystyrene, 3/4" thick | 1.000 | S.F. | .010 | .37 | .36 | .73 |
| 4' x 8' fiberboard, plain | 1.000 | S.F. | .008 | .45 | .28 | .73 |
| Trim, aluminum, white | .600 | L.F. | .016 | .62 | .56 | 1.18 |
| Color | .600 | L.F. | .016 | .67 | .56 | 1.23 |
| Vinyl, white | .600 | L.F. | .014 | .39 | .50 | .89 |
| Color | .600 | L.F. | .014 | .44 | .50 | .94 |
| Paper, #15 asphalt felt | 1.100 | S.F. | .002 | .02 | .09 | .11 |
| Kraft paper, plain | 1.100 | S.F. | .002 | .04 | .09 | .13 |
| Foil backed | 1.100 | S.F. | .002 | .07 | .09 | .16 |

| Description | QUAN. | UNIT | LABOR HOURS | COST PER S.F. | | |
|---|---|---|---|---|---|---|
| | | | | MAT. | INST. | TOTAL |
| Poured insulation, cellulose fiber, R3.8 per inch (1" thick) | 1.000 | S.F. | .003 | .04 | .12 | .16 |
| Fiberglass , R4.0 per inch (1" thick) | 1.000 | S.F. | .003 | .03 | .12 | .15 |
| Mineral wool, R3.0 per inch (1" thick) | 1.000 | S.F. | .003 | .03 | .12 | .15 |
| Polystyrene, R4.0 per inch (1" thick) | 1.000 | S.F. | .003 | .20 | .12 | .32 |
| Vermiculite, R2.7 per inch (1" thick) | 1.000 | S.F. | .003 | .15 | .12 | .27 |
| Perlite, R2.7 per inch (1" thick) | 1.000 | S.F. | .003 | .15 | .12 | .27 |
| Reflective insulation, aluminum foil reinforced with scrim | 1.000 | S.F. | .004 | .15 | .15 | .30 |
| Reinforced with woven polyolefin | 1.000 | S.F. | .004 | .19 | .15 | .34 |
| With single bubble air space, R8.8 | 1.000 | S.F. | .005 | .30 | .19 | .49 |
| With double bubble air space, R9.8 | 1.000 | S.F. | .005 | .32 | .19 | .51 |
| Rigid insulation, fiberglass, unfaced, | | | | | | |
| 1-1/2" thick, R6.2 | 1.000 | S.F. | .008 | .45 | .28 | .73 |
| 2" thick, R8.3 | 1.000 | S.F. | .008 | .50 | .28 | .78 |
| 2-1/2" thick, R10.3 | 1.000 | S.F. | .010 | .63 | .36 | .99 |
| 3" thick, R12.4 | 1.000 | S.F. | .010 | .63 | .36 | .99 |
| Foil faced, 1" thick, R4.3 | 1.000 | S.F. | .008 | .88 | .28 | 1.16 |
| 1-1/2" thick, R6.2 | 1.000 | S.F. | .008 | 1.19 | .28 | 1.47 |
| 2" thick, R8.7 | 1.000 | S.F. | .009 | 1.49 | .32 | 1.81 |
| 2-1/2" thick, R10.9 | 1.000 | S.F. | .010 | 1.76 | .36 | 2.12 |
| 3" thick, R13.0 | 1.000 | S.F. | .010 | 1.91 | .36 | 2.27 |
| Foam glass, 1-1/2" thick R2.64 | 1.000 | S.F. | .010 | 1.76 | .36 | 2.12 |
| 2" thick R5.26 | 1.000 | S.F. | .011 | 3.07 | .39 | 3.46 |
| Perlite, 1" thick R2.77 | 1.000 | S.F. | .010 | .29 | .36 | .65 |
| 2" thick R5.55 | 1.000 | S.F. | .011 | .55 | .39 | .94 |
| Polystyrene, extruded, blue, 2.2#/C.F., 3/4" thick R4 | 1.000 | S.F. | .010 | .37 | .36 | .73 |
| 1-1/2" thick R8.1 | 1.000 | S.F. | .011 | .74 | .39 | 1.13 |
| 2" thick R10.8 | 1.000 | S.F. | .011 | 1.06 | .39 | 1.45 |
| Molded bead board, white, 1" thick R3.85 | 1.000 | S.F. | .010 | .15 | .36 | .51 |
| 1-1/2" thick, R5.6 | 1.000 | S.F. | .011 | .42 | .39 | .81 |
| 2" thick, R7.7 | 1.000 | S.F. | .011 | .57 | .39 | .96 |
| Non-rigid insulation, batts | | | | | | |
| Fiberglass, kraft faced, 3-1/2" thick, R11, 11" wide | 1.000 | S.F. | .005 | .25 | .18 | .43 |
| 15" wide | 1.000 | S.F. | .005 | .25 | .18 | .43 |
| 23" wide | 1.000 | S.F. | .005 | .25 | .18 | .43 |
| 6" thick, R19, 11" wide | 1.000 | S.F. | .006 | .36 | .21 | .57 |
| 15" wide | 1.000 | S.F. | .006 | .36 | .21 | .57 |
| 23" wide | 1.000 | S.F. | .006 | .36 | .21 | .57 |
| 9" thick, R30, 15" wide | 1.000 | S.F. | .006 | .66 | .21 | .87 |
| 23" wide | 1.000 | S.F. | .006 | .66 | .21 | .87 |
| 12" thick, R38, 15" wide | 1.000 | S.F. | .006 | .84 | .21 | 1.05 |
| 23" wide | 1.000 | S.F. | .006 | .84 | .21 | 1.05 |
| Fiberglass, foil faced, 3-1/2" thick, R11, 15" wide | 1.000 | S.F. | .005 | .37 | .18 | .55 |
| 23" wide | 1.000 | S.F. | .005 | .37 | .18 | .55 |
| 6" thick, R19, 15" thick | 1.000 | S.F. | .005 | .45 | .18 | .63 |
| 23" wide | 1.000 | S.F. | .005 | .45 | .18 | .63 |
| 9" thick, R30, 15" wide | 1.000 | S.F. | .006 | .78 | .21 | .99 |
| 23" wide | 1.000 | S.F. | .006 | .78 | .21 | .99 |
| | | | | | | |
| | | | | | | |
| | | | | | | |
| | | | | | | |

| Insulation Systems | QUAN. | UNIT | LABOR HOURS | COST PER S.F. | | |
|---|---|---|---|---|---|---|
| | | | | MAT. | INST. | TOTAL |
| Non-rigid insulation batts | | | | | | |
| Fiberglass unfaced, 3-1/2" thick, R11, 15" wide | 1.000 | S.F. | .005 | .23 | .18 | .41 |
| 23" wide | 1.000 | S.F. | .005 | .23 | .18 | .41 |
| 6" thick, R19, 15" wide | 1.000 | S.F. | .006 | .37 | .21 | .58 |
| 23" wide | 1.000 | S.F. | .006 | .37 | .21 | .58 |
| 9" thick, R19, 15" wide | 1.000 | S.F. | .007 | .66 | .25 | .91 |
| 23" wide | 1.000 | S.F. | .007 | .66 | .25 | .91 |
| 12" thick, R38, 15" wide | 1.000 | S.F. | .007 | .84 | .25 | 1.09 |
| 23" wide | 1.000 | S.F. | .007 | .84 | .25 | 1.09 |
| | | | | | | |
| Mineral fiber batts, 3" thick, R11 | 1.000 | S.F. | .005 | .29 | .18 | .47 |
| 3-1/2" thick, R13 | 1.000 | S.F. | .005 | .29 | .18 | .47 |
| 6" thick, R19 | 1.000 | S.F. | .005 | .43 | .18 | .61 |
| 6-1/2" thick, R22 | 1.000 | S.F. | .005 | .43 | .18 | .61 |
| 10" thick, R30 | 1.000 | S.F. | .006 | .68 | .21 | .89 |

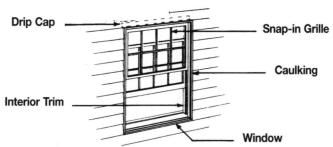

| System Description | QUAN. | UNIT | LABOR HOURS | COST EACH | | |
|---|---|---|---|---|---|---|
| | | | | MAT. | INST. | TOTAL |
| **BUILDER'S QUALITY WOOD WINDOW 2' X 3', DOUBLE HUNG** | | | | | | |
| Window, primed, builder's quality, 2' x 3', insulating glass | 1.000 | Ea. | .800 | 243.00 | 28.50 | 271.50 |
| Trim, interior casing | 11.000 | L.F. | .367 | 8.25 | 13.09 | 21.34 |
| Paint, interior & exterior, primer & 2 coats | 2.000 | Face | 1.778 | 2.02 | 57.00 | 59.02 |
| Caulking | 10.000 | L.F. | .323 | 1.60 | 11.50 | 13.10 |
| Snap-in grille | 1.000 | Set | .333 | 43.00 | 11.85 | 54.85 |
| Drip cap, metal | 2.000 | L.F. | .040 | .44 | 1.42 | 1.86 |
| TOTAL | | | 3.641 | 298.31 | 123.36 | 421.67 |
| **PLASTIC CLAD WOOD WINDOW 3' X 4', DOUBLE HUNG** | | | | | | |
| Window, plastic clad, premium, 3' x 4', insulating glass | 1.000 | Ea. | .889 | 271.00 | 31.50 | 302.50 |
| Trim, interior casing | 15.000 | L.F. | .500 | 11.25 | 17.85 | 29.10 |
| Paint, interior, primer & 2 coats | 1.000 | Face | .889 | 1.01 | 28.50 | 29.51 |
| Caulking | 14.000 | L.F. | .452 | 2.24 | 16.10 | 18.34 |
| Snap-in grille | 1.000 | Set | .333 | 43.00 | 11.85 | 54.85 |
| TOTAL | | | 3.063 | 328.50 | 105.80 | 434.30 |
| **METAL CLAD WOOD WINDOW, 3' X 5', DOUBLE HUNG** | | | | | | |
| Window, metal clad, deluxe, 3' x 5', insulating glass | 1.000 | Ea. | 1.000 | 289.00 | 35.50 | 324.50 |
| Trim, interior casing | 17.000 | L.F. | .567 | 12.75 | 20.23 | 32.98 |
| Paint, interior, primer & 2 coats | 1.000 | Face | .889 | 1.01 | 28.50 | 29.51 |
| Caulking | 16.000 | L.F. | .516 | 2.56 | 18.40 | 20.96 |
| Snap-in grille | 1.000 | Set | .235 | 119.00 | 8.40 | 127.40 |
| Drip cap, metal | 3.000 | L.F. | .060 | .66 | 2.13 | 2.79 |
| TOTAL | | | 3.267 | 424.98 | 113.16 | 538.14 |

The cost of this system is on a cost per each window basis.

| Description | QUAN. | UNIT | LABOR HOURS | COST EACH | | |
|---|---|---|---|---|---|---|
| | | | | MAT. | INST. | TOTAL |
| | | | | | | |
| | | | | | | |
| | | | | | | |
| | | | | | | |
| | | | | | | |

**Important: See the Reference Section for critical supporting data - Reference Nos., Crews & Location Factors**

| Double Hung Window Price Sheet | QUAN. | UNIT | LABOR HOURS | COST EACH | | |
|---|---|---|---|---|---|---|
| | | | | MAT. | INST. | TOTAL |
| Windows, double-hung, builder's quality, 2' x 3', single glass | 1.000 | Ea. | .800 | 232.00 | 28.50 | 260.50 |
| Insulating glass | 1.000 | Ea. | .800 | 243.00 | 28.50 | 271.50 |
| 3' x 4', single glass | 1.000 | Ea. | .889 | 285.00 | 31.50 | 316.50 |
| Insulating glass | 1.000 | Ea. | .889 | 296.00 | 31.50 | 327.50 |
| 4' x 4'-6", single glass | 1.000 | Ea. | 1.000 | 355.00 | 35.50 | 390.50 |
| Insulating glass | 1.000 | Ea. | 1.000 | 370.00 | 35.50 | 405.50 |
| Plastic clad premium insulating glass, 2'-6" x 3' | 1.000 | Ea. | .800 | 241.00 | 28.50 | 269.50 |
| 3' x 3'-6" | 1.000 | Ea. | .800 | 289.00 | 28.50 | 317.50 |
| 3' x 4' | 1.000 | Ea. | .889 | 271.00 | 31.50 | 302.50 |
| 3' x 4'-6" | 1.000 | Ea. | .889 | 335.00 | 31.50 | 366.50 |
| 3' x 5' | 1.000 | Ea. | 1.000 | 345.00 | 35.50 | 380.50 |
| 3'-6" x 6' | 1.000 | Ea. | 1.000 | 425.00 | 35.50 | 460.50 |
| Metal clad deluxe insulating glass, 2'-6" x 3' | 1.000 | Ea. | .800 | 198.00 | 28.50 | 226.50 |
| 3' x 3'-6" | 1.000 | Ea. | .800 | 235.00 | 28.50 | 263.50 |
| 3' x 4' | 1.000 | Ea. | .889 | 250.00 | 31.50 | 281.50 |
| 3' x 4'-6" | 1.000 | Ea. | .889 | 270.00 | 31.50 | 301.50 |
| 3' x 5' | 1.000 | Ea. | 1.000 | 289.00 | 35.50 | 324.50 |
| 3'-6" x 6' | 1.000 | Ea. | 1.000 | 350.00 | 35.50 | 385.50 |
| Trim, interior casing, window 2' x 3' | 11.000 | L.F. | .367 | 8.25 | 13.10 | 21.35 |
| 2'-6" x 3' | 12.000 | L.F. | .400 | 9.00 | 14.30 | 23.30 |
| 3' x 3'-6" | 14.000 | L.F. | .467 | 10.50 | 16.65 | 27.15 |
| 3' x 4' | 15.000 | L.F. | .500 | 11.25 | 17.85 | 29.10 |
| 3' x 4'-6" | 16.000 | L.F. | .533 | 12.00 | 19.05 | 31.05 |
| 3' x 5' | 17.000 | L.F. | .567 | 12.75 | 20.00 | 32.75 |
| 3'-6" x 6' | 20.000 | L.F. | .667 | 15.00 | 24.00 | 39.00 |
| 4' x 4'-6" | 18.000 | L.F. | .600 | 13.50 | 21.50 | 35.00 |
| Paint or stain, interior or exterior, 2' x 3' window, 1 coat | 1.000 | Face | .444 | .37 | 14.20 | 14.57 |
| 2 coats | 1.000 | Face | .727 | .74 | 23.00 | 23.74 |
| Primer & 1 coat | 1.000 | Face | .727 | .68 | 23.00 | 23.68 |
| Primer & 2 coats | 1.000 | Face | .889 | 1.01 | 28.50 | 29.51 |
| 3' x 4' window, 1 coat | 1.000 | Face | .667 | .76 | 21.50 | 22.26 |
| 2 coats | 1.000 | Face | .667 | .85 | 21.50 | 22.35 |
| Primer & 1 coat | 1.000 | Face | .727 | 1.03 | 23.00 | 24.03 |
| Primer & 2 coats | 1.000 | Face | .889 | 1.01 | 28.50 | 29.51 |
| 4' x 4'-6" window, 1 coat | 1.000 | Face | .667 | .76 | 21.50 | 22.26 |
| 2 coats | 1.000 | Face | .667 | .85 | 21.50 | 22.35 |
| Primer & 1 coat | 1.000 | Face | .727 | 1.03 | 23.00 | 24.03 |
| Primer & 2 coats | 1.000 | Face | .889 | 1.01 | 28.50 | 29.51 |
| Caulking, window, 2' x 3' | 10.000 | L.F. | .323 | 1.60 | 11.50 | 13.10 |
| 2'-6" x 3' | 11.000 | L.F. | .355 | 1.76 | 12.65 | 14.41 |
| 3' x 3'-6" | 13.000 | L.F. | .419 | 2.08 | 14.95 | 17.03 |
| 3' x 4' | 14.000 | L.F. | .452 | 2.24 | 16.10 | 18.34 |
| 3' x 4'-6" | 15.000 | L.F. | .484 | 2.40 | 17.25 | 19.65 |
| 3' x 5' | 16.000 | L.F. | .516 | 2.56 | 18.40 | 20.96 |
| 3'-6" x 6' | 19.000 | L.F. | .613 | 3.04 | 22.00 | 25.04 |
| 4' x 4'-6" | 17.000 | L.F. | .548 | 2.72 | 19.55 | 22.27 |
| Grilles, glass size to, 16" x 24" per sash | 1.000 | Set | .333 | 43.00 | 11.85 | 54.85 |
| 32" x 32" per sash | 1.000 | Set | .235 | 119.00 | 8.40 | 127.40 |
| Drip cap, aluminum, 2' long | 2.000 | L.F. | .040 | .44 | 1.42 | 1.86 |
| 3' long | 3.000 | L.F. | .060 | .66 | 2.13 | 2.79 |
| 4' long | 4.000 | L.F. | .080 | .88 | 2.84 | 3.72 |
| Wood, 2' long | 2.000 | L.F. | .067 | 1.50 | 2.38 | 3.88 |
| 3' long | 3.000 | L.F. | .100 | 2.25 | 3.57 | 5.82 |
| 4' long | 4.000 | L.F. | .133 | 3.00 | 4.76 | 7.76 |
| | | | | | | |
| | | | | | | |
| | | | | | | |

**EXTERIOR WALLS**

**4**

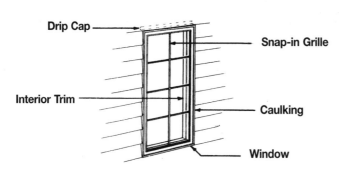

- Drip Cap
- Snap-in Grille
- Interior Trim
- Caulking
- Window

| System Description | QUAN. | UNIT | LABOR HOURS | COST EACH | | |
|---|---|---|---|---|---|---|
| | | | | MAT. | INST. | TOTAL |
| **BUILDER'S QUALITY WINDOW, WOOD, 2' BY 3', CASEMENT** | | | | | | |
| Window, primed, builder's quality, 2' x 3', insulating glass | 1.000 | Ea. | .800 | 224.00 | 28.50 | 252.50 |
| Trim, interior casing | 11.000 | L.F. | .367 | 8.25 | 13.09 | 21.34 |
| Paint, interior & exterior, primer & 2 coats | 2.000 | Face | 1.778 | 2.02 | 57.00 | 59.02 |
| Caulking | 10.000 | L.F. | .323 | 1.60 | 11.50 | 13.10 |
| Snap-in grille | 1.000 | Ea. | .267 | 24.00 | 9.50 | 33.50 |
| Drip cap, metal | 2.000 | L.F. | .040 | .44 | 1.42 | 1.86 |
| TOTAL | | | 3.575 | 260.31 | 121.01 | 381.32 |
| **PLASTIC CLAD WOOD WINDOW, 2' X 4', CASEMENT** | | | | | | |
| Window, plastic clad, premium, 2' x 4', insulating glass | 1.000 | Ea. | .889 | 287.00 | 31.50 | 318.50 |
| Trim, interior casing | 13.000 | L.F. | .433 | 9.75 | 15.47 | 25.22 |
| Paint, interior, primer & 2 coats | 1.000 | Ea. | .889 | 1.01 | 28.50 | 29.51 |
| Caulking | 12.000 | L.F. | .387 | 1.92 | 13.80 | 15.72 |
| Snap-in grille | 1.000 | Ea. | .267 | 24.00 | 9.50 | 33.50 |
| TOTAL | | | 2.865 | 323.68 | 98.77 | 422.45 |
| **METAL CLAD WOOD WINDOW, 2' X 5', CASEMENT** | | | | | | |
| Window, metal clad, deluxe, 2' x 5', insulating glass | 1.000 | Ea. | 1.000 | 256.00 | 35.50 | 291.50 |
| Trim, interior casing | 15.000 | L.F. | .500 | 11.25 | 17.85 | 29.10 |
| Paint, interior, primer & 2 coats | 1.000 | Ea. | .889 | 1.01 | 28.50 | 29.51 |
| Caulking | 14.000 | L.F. | .452 | 2.24 | 16.10 | 18.34 |
| Snap-in grille | 1.000 | Ea. | .250 | 35.00 | 8.90 | 43.90 |
| Drip cap, metal | 12.000 | L.F. | .040 | .44 | 1.42 | 1.86 |
| TOTAL | | | 3.131 | 305.94 | 108.27 | 414.21 |

The cost of this system is on a cost per each window basis.

| Description | QUAN. | UNIT | LABOR HOURS | COST EACH | | |
|---|---|---|---|---|---|---|
| | | | | MAT. | INST. | TOTAL |
| | | | | | | |
| | | | | | | |
| | | | | | | |
| | | | | | | |
| | | | | | | |

**Important: See the Reference Section for critical supporting data - Reference Nos., Crews & Location Factors**

# Casement Window Price Sheet

| | QUAN. | UNIT | LABOR HOURS | COST EACH | | |
|---|---|---|---|---|---|---|
| | | | | MAT. | INST. | TOTAL |
| Window, casement, builders quality, 2' x 3', single glass | 1.000 | Ea. | .800 | 184.00 | 28.50 | 212.50 |
| Insulating glass | 1.000 | Ea. | .800 | 224.00 | 28.50 | 252.50 |
| 2' x 4'-6", single glass | 1.000 | Ea. | .727 | 705.00 | 26.00 | 731.00 |
| Insulating glass | 1.000 | Ea. | .727 | 595.00 | 26.00 | 621.00 |
| 2' x 6', single glass | 1.000 | Ea. | .889 | 920.00 | 31.50 | 951.50 |
| Insulating glass | 1.000 | Ea. | .889 | 900.00 | 31.50 | 931.50 |
| Plastic clad premium insulating glass, 2' x 3' | 1.000 | Ea. | .800 | 187.00 | 28.50 | 215.50 |
| 2' x 4' | 1.000 | Ea. | .889 | 198.00 | 31.50 | 229.50 |
| 2' x 5' | 1.000 | Ea. | 1.000 | 270.00 | 35.50 | 305.50 |
| 2' x 6' | 1.000 | Ea. | 1.000 | 340.00 | 35.50 | 375.50 |
| Metal clad deluxe insulating glass, 2' x 3' | 1.000 | Ea. | .800 | 187.00 | 28.50 | 215.50 |
| 2' x 4' | 1.000 | Ea. | .889 | 225.00 | 31.50 | 256.50 |
| 2' x 5' | 1.000 | Ea. | 1.000 | 256.00 | 35.50 | 291.50 |
| 2' x 6' | 1.000 | Ea. | 1.000 | 294.00 | 35.50 | 329.50 |
| Trim, interior casing, window 2' x 3' | 11.000 | L.F. | .367 | 8.25 | 13.10 | 21.35 |
| 2' x 4' | 13.000 | L.F. | .433 | 9.75 | 15.45 | 25.20 |
| 2' x 4'-6" | 14.000 | L.F. | .467 | 10.50 | 16.65 | 27.15 |
| 2' x 5' | 15.000 | L.F. | .500 | 11.25 | 17.85 | 29.10 |
| 2' x 6' | 17.000 | L.F. | .567 | 12.75 | 20.00 | 32.75 |
| Paint or stain, interior or exterior, 2' x 3' window, 1 coat | 1.000 | Face | .444 | .37 | 14.20 | 14.57 |
| 2 coats | 1.000 | Face | .727 | .74 | 23.00 | 23.74 |
| Primer & 1 coat | 1.000 | Face | .727 | .68 | 23.00 | 23.68 |
| Primer & 2 coats | 1.000 | Face | .889 | 1.01 | 28.50 | 29.51 |
| 2' x 4' window, 1 coat | 1.000 | Face | .444 | .37 | 14.20 | 14.57 |
| 2 coats | 1.000 | Face | .727 | .74 | 23.00 | 23.74 |
| Primer & 1 coat | 1.000 | Face | .727 | .68 | 23.00 | 23.68 |
| Primer & 2 coats | 1.000 | Face | .889 | 1.01 | 28.50 | 29.51 |
| 2' x 6' window, 1 coat | 1.000 | Face | .667 | .76 | 21.50 | 22.26 |
| 2 coats | 1.000 | Face | .667 | .85 | 21.50 | 22.35 |
| Primer & 1 coat | 1.000 | Face | .727 | 1.03 | 23.00 | 24.03 |
| Primer & 2 coats | 1.000 | Face | .889 | 1.01 | 28.50 | 29.51 |
| Caulking, window, 2' x 3' | 10.000 | L.F. | .323 | 1.60 | 11.50 | 13.10 |
| 2' x 4' | 12.000 | L.F. | .387 | 1.92 | 13.80 | 15.72 |
| 2' x 4'-6" | 13.000 | L.F. | .419 | 2.08 | 14.95 | 17.03 |
| 2' x 5' | 14.000 | L.F. | .452 | 2.24 | 16.10 | 18.34 |
| 2' x 6' | 16.000 | L.F. | .516 | 2.56 | 18.40 | 20.96 |
| Grilles, glass size, to 20" x 36" | 1.000 | Ea. | .267 | 24.00 | 9.50 | 33.50 |
| To 20" x 56" | 1.000 | Ea. | .250 | 35.00 | 8.90 | 43.90 |
| Drip cap, metal, 2' long | 2.000 | L.F. | .040 | .44 | 1.42 | 1.86 |
| Wood, 2' long | 2.000 | L.F. | .067 | 1.50 | 2.38 | 3.88 |

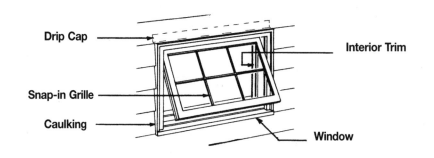

| System Description | QUAN. | UNIT | LABOR HOURS | COST EACH | | |
|---|---|---|---|---|---|---|
| | | | | MAT. | INST. | TOTAL |
| **BUILDER'S QUALITY WINDOW, WOOD, 34″ X 22″, AWNING** | | | | | | |
| Window, builder quality, 34″ x 22″, insulating glass | 1.000 | Ea. | .800 | 238.00 | 28.50 | 266.50 |
| Trim, interior casing | 10.500 | L.F. | .350 | 7.88 | 12.50 | 20.38 |
| Paint, interior & exterior, primer & 2 coats | 2.000 | Face | 1.778 | 2.02 | 57.00 | 59.02 |
| Caulking | 9.500 | L.F. | .306 | 1.52 | 10.93 | 12.45 |
| Snap-in grille | 1.000 | Ea. | .267 | 19.60 | 9.50 | 29.10 |
| Drip cap, metal | 3.000 | L.F. | .060 | .66 | 2.13 | 2.79 |
| TOTAL | | | 3.561 | 269.68 | 120.56 | 390.24 |
| **PLASTIC CLAD WOOD WINDOW, 40″ X 28″, AWNING** | | | | | | |
| Window, plastic clad, premium, 40″ x 28″, insulating glass | 1.000 | Ea. | .889 | 299.00 | 31.50 | 330.50 |
| Trim interior casing | 13.500 | L.F. | .450 | 10.13 | 16.07 | 26.20 |
| Paint, interior, primer & 2 coats | 1.000 | Face | .889 | 1.01 | 28.50 | 29.51 |
| Caulking | 12.500 | L.F. | .403 | 2.00 | 14.38 | 16.38 |
| Snap-in grille | 1.000 | Ea. | .267 | 19.60 | 9.50 | 29.10 |
| TOTAL | | | 2.898 | 331.74 | 99.95 | 431.69 |
| **METAL CLAD WOOD WINDOW, 48″ X 36″, AWNING** | | | | | | |
| Window, metal clad, deluxe, 48″ x 36″, insulating glass | 1.000 | Ea. | 1.000 | 335.00 | 35.50 | 370.50 |
| Trim, interior casing | 15.000 | L.F. | .500 | 11.25 | 17.85 | 29.10 |
| Paint, interior, primer & 2 coats | 1.000 | Face | .889 | 1.01 | 28.50 | 29.51 |
| Caulking | 14.000 | L.F. | .452 | 2.24 | 16.10 | 18.34 |
| Snap-in grille | 1.000 | Ea. | .250 | 28.50 | 8.90 | 37.40 |
| Drip cap, metal | 4.000 | L.F. | .080 | .88 | 2.84 | 3.72 |
| TOTAL | | | 3.171 | 378.88 | 109.69 | 488.57 |

The cost of this system is on a cost per each window basis.

| Description | QUAN. | UNIT | LABOR HOURS | COST EACH | | |
|---|---|---|---|---|---|---|
| | | | | MAT. | INST. | TOTAL |
| | | | | | | |
| | | | | | | |
| | | | | | | |
| | | | | | | |
| | | | | | | |

**Important: See the Reference Section for critical supporting data - Reference Nos., Crews & Location Factors**

| Awning Window Price Sheet | QUAN. | UNIT | LABOR HOURS | COST EACH | | |
|---|---|---|---|---|---|---|
| | | | | MAT. | INST. | TOTAL |
| Windows, awning, builder's quality, 34" x 22", insulated glass | 1.000 | Ea. | .800 | 225.00 | 28.50 | 253.50 |
| Low E glass | 1.000 | Ea. | .800 | 238.00 | 28.50 | 266.50 |
| 40" x 28", insulated glass | 1.000 | Ea. | .889 | 287.00 | 31.50 | 318.50 |
| Low E glass | 1.000 | Ea. | .889 | 300.00 | 31.50 | 331.50 |
| 48" x 36", insulated glass | 1.000 | Ea. | 1.000 | 435.00 | 35.50 | 470.50 |
| Low E glass | 1.000 | Ea. | 1.000 | 455.00 | 35.50 | 490.50 |
| Plastic clad premium insulating glass, 34" x 22" | 1.000 | Ea. | .800 | 234.00 | 28.50 | 262.50 |
| 40" x 22" | 1.000 | Ea. | .800 | 258.00 | 28.50 | 286.50 |
| 36" x 28" | 1.000 | Ea. | .889 | 275.00 | 31.50 | 306.50 |
| 36" x 36" | 1.000 | Ea. | .889 | 299.00 | 31.50 | 330.50 |
| 48" x 28" | 1.000 | Ea. | 1.000 | 325.00 | 35.50 | 360.50 |
| 60" x 36" | 1.000 | Ea. | 1.000 | 525.00 | 35.50 | 560.50 |
| Metal clad deluxe insulating glass, 34" x 22" | 1.000 | Ea. | .800 | 209.00 | 28.50 | 237.50 |
| 40" x 22" | 1.000 | Ea. | .800 | 259.00 | 28.50 | 287.50 |
| 36" x 25" | 1.000 | Ea. | .889 | 246.00 | 31.50 | 277.50 |
| 40" x 30" | 1.000 | Ea. | .889 | 305.00 | 31.50 | 336.50 |
| 48" x 28" | 1.000 | Ea. | 1.000 | 315.00 | 35.50 | 350.50 |
| 60" x 36" | 1.000 | Ea. | 1.000 | 335.00 | 35.50 | 370.50 |
| Trim, interior casing window, 34" x 22" | 10.500 | L.F. | .350 | 7.90 | 12.50 | 20.40 |
| 40" x 22" | 11.500 | L.F. | .383 | 8.65 | 13.70 | 22.35 |
| 36" x 28" | 12.500 | L.F. | .417 | 9.40 | 14.90 | 24.30 |
| 40" x 28" | 13.500 | L.F. | .450 | 10.15 | 16.05 | 26.20 |
| 48" x 28" | 14.500 | L.F. | .483 | 10.90 | 17.25 | 28.15 |
| 48" x 36" | 15.000 | L.F. | .500 | 11.25 | 17.85 | 29.10 |
| Paint or stain, interior or exterior, 34" x 22", 1 coat | 1.000 | Face | .444 | .37 | 14.20 | 14.57 |
| 2 coats | 1.000 | Face | .727 | .74 | 23.00 | 23.74 |
| Primer & 1 coat | 1.000 | Face | .727 | .68 | 23.00 | 23.68 |
| Primer & 2 coats | 1.000 | Face | .889 | 1.01 | 28.50 | 29.51 |
| 36" x 28", 1 coat | 1.000 | Face | .444 | .37 | 14.20 | 14.57 |
| 2 coats | 1.000 | Face | .727 | .74 | 23.00 | 23.74 |
| Primer & 1 coat | 1.000 | Face | .727 | .68 | 23.00 | 23.68 |
| Primer & 2 coats | 1.000 | Face | .889 | 1.01 | 28.50 | 29.51 |
| 48" x 36", 1 coat | 1.000 | Face | .667 | .76 | 21.50 | 22.26 |
| 2 coats | 1.000 | Face | .667 | .85 | 21.50 | 22.35 |
| Primer & 1 coat | 1.000 | Face | .727 | 1.03 | 23.00 | 24.03 |
| Primer & 2 coats | 1.000 | Face | .889 | 1.01 | 28.50 | 29.51 |
| Caulking, window, 34" x 22" | 9.500 | L.F. | .306 | 1.52 | 10.95 | 12.47 |
| 40" x 22" | 10.500 | L.F. | .339 | 1.68 | 12.10 | 13.78 |
| 36" x 28" | 11.500 | L.F. | .371 | 1.84 | 13.25 | 15.09 |
| 40" x 28" | 12.500 | L.F. | .403 | 2.00 | 14.40 | 16.40 |
| 48" x 28" | 13.500 | L.F. | .436 | 2.16 | 15.55 | 17.71 |
| 48" x 36" | 14.000 | L.F. | .452 | 2.24 | 16.10 | 18.34 |
| Grilles, glass size, to 28" by 16" | 1.000 | Ea. | .267 | 19.60 | 9.50 | 29.10 |
| To 44" by 24" | 1.000 | Ea. | .250 | 28.50 | 8.90 | 37.40 |
| Drip cap, aluminum, 3' long | 3.000 | L.F. | .060 | .66 | 2.13 | 2.79 |
| 3'-6" long | 3.500 | L.F. | .070 | .77 | 2.49 | 3.26 |
| 4' long | 4.000 | L.F. | .080 | .88 | 2.84 | 3.72 |
| Wood, 3' long | 3.000 | L.F. | .100 | 2.25 | 3.57 | 5.82 |
| 3'-6" long | 3.500 | L.F. | .117 | 2.63 | 4.17 | 6.80 |
| 4' long | 4.000 | L.F. | .133 | 3.00 | 4.76 | 7.76 |
| | | | | | | |
| | | | | | | |
| | | | | | | |
| | | | | | | |
| | | | | | | |

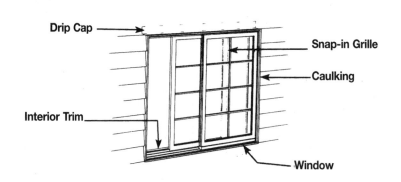

Drip Cap — Snap-in Grille — Caulking — Interior Trim — Window

| System Description | QUAN. | UNIT | LABOR HOURS | COST EACH | | |
|---|---|---|---|---|---|---|
| | | | | MAT. | INST. | TOTAL |
| **BUILDER'S QUALITY WOOD WINDOW, 3′ X 2′, SLIDING** | | | | | | |
| Window, primed, builder's quality, 3′ x 2′, insul. glass | 1.000 | Ea. | .800 | 187.00 | 28.50 | 215.50 |
| Trim, interior casing | 11.000 | L.F. | .367 | 8.25 | 13.09 | 21.34 |
| Paint, interior & exterior, primer & 2 coats | 2.000 | Face | 1.778 | 2.02 | 57.00 | 59.02 |
| Caulking | 10.000 | L.F. | .323 | 1.60 | 11.50 | 13.10 |
| Snap-in grille | 1.000 | Set | .333 | 23.00 | 11.85 | 34.85 |
| Drip cap, metal | 3.000 | L.F. | .060 | .66 | 2.13 | 2.79 |
| TOTAL | | | 3.661 | 222.53 | 124.07 | 346.60 |
| **PLASTIC CLAD WOOD WINDOW, 4′ X 3′-6″, SLIDING** | | | | | | |
| Window, plastic clad, premium, 4′ x 3′-6″, insulating glass | 1.000 | Ea. | .889 | 650.00 | 31.50 | 681.50 |
| Trim, interior casing | 16.000 | L.F. | .533 | 12.00 | 19.04 | 31.04 |
| Paint, interior, primer & 2 coats | 1.000 | Face | .889 | 1.01 | 28.50 | 29.51 |
| Caulking | 17.000 | L.F. | .548 | 2.72 | 19.55 | 22.27 |
| Snap-in grille | 1.000 | Set | .333 | 23.00 | 11.85 | 34.85 |
| TOTAL | | | 3.192 | 688.73 | 110.44 | 799.17 |
| **METAL CLAD WOOD WINDOW, 6′ X 5′, SLIDING** | | | | | | |
| Window, metal clad, deluxe, 6′ x 5′, insulating glass | 1.000 | Ea. | 1.000 | 640.00 | 35.50 | 675.50 |
| Trim, interior casing | 23.000 | L.F. | .767 | 17.25 | 27.37 | 44.62 |
| Paint, interior, primer & 2 coats | 1.000 | Face | .889 | 1.01 | 28.50 | 29.51 |
| Caulking | 22.000 | L.F. | .710 | 3.52 | 25.30 | 28.82 |
| Snap-in grille | 1.000 | Set | .364 | 48.00 | 12.95 | 60.95 |
| Drip cap, metal | 6.000 | L.F. | .120 | 1.32 | 4.26 | 5.58 |
| TOTAL | | | 3.850 | 711.10 | 133.88 | 844.98 |

The cost of this system is on a cost per each window basis.

| Description | QUAN. | UNIT | LABOR HOURS | COST EACH | | |
|---|---|---|---|---|---|---|
| | | | | MAT. | INST. | TOTAL |
| | | | | | | |
| | | | | | | |
| | | | | | | |
| | | | | | | |
| | | | | | | |

**Important: See the Reference Section for critical supporting data - Reference Nos., Crews & Location Factors**

| Sliding Window Price Sheet | QUAN. | UNIT | LABOR HOURS | COST EACH MAT. | INST. | TOTAL |
|---|---|---|---|---|---|---|
| Windows, sliding, builder's quality, 3' x 3', single glass | 1.000 | Ea. | .800 | 148.00 | 28.50 | 176.50 |
| Insulating glass | 1.000 | Ea. | .800 | 187.00 | 28.50 | 215.50 |
| 4' x 3'-6", single glass | 1.000 | Ea. | .889 | 176.00 | 31.50 | 207.50 |
| Insulating glass | 1.000 | Ea. | .889 | 220.00 | 31.50 | 251.50 |
| 6' x 5', single glass | 1.000 | Ea. | 1.000 | 325.00 | 35.50 | 360.50 |
| Insulating glass | 1.000 | Ea. | 1.000 | 385.00 | 35.50 | 420.50 |
| Plastic clad premium insulating glass, 3' x 3' | 1.000 | Ea. | .800 | 540.00 | 28.50 | 568.50 |
| 4' x 3'-6" | 1.000 | Ea. | .889 | 650.00 | 31.50 | 681.50 |
| 5' x 4' | 1.000 | Ea. | .889 | 750.00 | 31.50 | 781.50 |
| 6' x 5' | 1.000 | Ea. | 1.000 | 935.00 | 35.50 | 970.50 |
| Metal clad deluxe insulating glass, 3' x 3' | 1.000 | Ea. | .800 | 300.00 | 28.50 | 328.50 |
| 4' x 3'-6" | 1.000 | Ea. | .889 | 370.00 | 31.50 | 401.50 |
| 5' x 4' | 1.000 | Ea. | .889 | 445.00 | 31.50 | 476.50 |
| 6' x 5' | 1.000 | Ea. | 1.000 | 640.00 | 35.50 | 675.50 |
| Trim, interior casing, window 3' x 2' | 11.000 | L.F. | .367 | 8.25 | 13.10 | 21.35 |
| 3' x 3' | 13.000 | L.F. | .433 | 9.75 | 15.45 | 25.20 |
| 4' x 3'-6" | 16.000 | L.F. | .533 | 12.00 | 19.05 | 31.05 |
| 5' x 4' | 19.000 | L.F. | .633 | 14.25 | 22.50 | 36.75 |
| 6' x 5' | 23.000 | L.F. | .767 | 17.25 | 27.50 | 44.75 |
| Paint or stain, interior or exterior, 3' x 2' window, 1 coat | 1.000 | Face | .444 | .37 | 14.20 | 14.57 |
| 2 coats | 1.000 | Face | .727 | .74 | 23.00 | 23.74 |
| Primer & 1 coat | 1.000 | Face | .727 | .68 | 23.00 | 23.68 |
| Primer & 2 coats | 1.000 | Face | .889 | 1.01 | 28.50 | 29.51 |
| 4' x 3'-6" window, 1 coat | 1.000 | Face | .667 | .76 | 21.50 | 22.26 |
| 2 coats | 1.000 | Face | .667 | .85 | 21.50 | 22.35 |
| Primer & 1 coat | 1.000 | Face | .727 | 1.03 | 23.00 | 24.03 |
| Primer & 2 coats | 1.000 | Face | .889 | 1.01 | 28.50 | 29.51 |
| 6' x 5' window, 1 coat | 1.000 | Face | .889 | 2.10 | 28.50 | 30.60 |
| 2 coats | 1.000 | Face | 1.333 | 3.84 | 42.50 | 46.34 |
| Primer & 1 coat | 1.000 | Face | 1.333 | 3.62 | 42.50 | 46.12 |
| Primer & 2 coats | 1.000 | Face | 1.600 | 5.40 | 51.00 | 56.40 |
| Caulking, window, 3' x 2' | 10.000 | L.F. | .323 | 1.60 | 11.50 | 13.10 |
| 3' x 3' | 12.000 | L.F. | .387 | 1.92 | 13.80 | 15.72 |
| 4' x 3'-6" | 15.000 | L.F. | .484 | 2.40 | 17.25 | 19.65 |
| 5' x 4' | 18.000 | L.F. | .581 | 2.88 | 20.50 | 23.38 |
| 6' x 5' | 22.000 | L.F. | .710 | 3.52 | 25.50 | 29.02 |
| Grilles, glass size, to 14" x 36" | 1.000 | Set | .333 | 23.00 | 11.85 | 34.85 |
| To 36" x 36" | 1.000 | Set | .364 | 48.00 | 12.95 | 60.95 |
| Drip cap, aluminum, 3' long | 3.000 | L.F. | .060 | .66 | 2.13 | 2.79 |
| 4' long | 4.000 | L.F. | .080 | .88 | 2.84 | 3.72 |
| 5' long | 5.000 | L.F. | .100 | 1.10 | 3.55 | 4.65 |
| 6' long | 6.000 | L.F. | .120 | 1.32 | 4.26 | 5.58 |
| Wood, 3' long | 3.000 | L.F. | .100 | 2.25 | 3.57 | 5.82 |
| 4' long | 4.000 | L.F. | .133 | 3.00 | 4.76 | 7.76 |
| 5' long | 5.000 | L.F. | .167 | 3.75 | 5.95 | 9.70 |
| 6' long | 6.000 | L.F. | .200 | 4.50 | 7.15 | 11.65 |

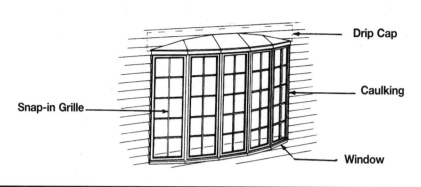

Drip Cap

Caulking

Snap-in Grille

Window

| System Description | QUAN. | UNIT | LABOR HOURS | COST EACH | | |
|---|---|---|---|---|---|---|
| | | | | MAT. | INST. | TOTAL |
| **AWNING TYPE BOW WINDOW, BUILDER'S QUALITY, 8' X 5'** | | | | | | |
| Window, primed, builder's quality, 8' x 5', insulating glass | 1.000 | Ea. | 1.600 | 1250.00 | 57.00 | 1307.00 |
| Trim, interior casing | 27.000 | L.F. | .900 | 20.25 | 32.13 | 52.38 |
| Paint, interior & exterior, primer & 1 coat | 2.000 | Face | 3.200 | 10.80 | 102.00 | 112.80 |
| Drip cap, vinyl | 1.000 | Ea. | .533 | 77.50 | 19.00 | 96.50 |
| Caulking | 26.000 | L.F. | .839 | 4.16 | 29.90 | 34.06 |
| Snap-in grilles | 1.000 | Set | 1.067 | 96.00 | 38.00 | 134.00 |
| | | | | | | |
| TOTAL | | | 8.139 | 1458.71 | 278.03 | 1736.74 |
| **CASEMENT TYPE BOW WINDOW, PLASTIC CLAD, 10' X 6'** | | | | | | |
| Window, plastic clad, premium, 10' x 6', insulating glass | 1.000 | Ea. | 2.286 | 1925.00 | 81.50 | 2006.50 |
| Trim, interior casing | 33.000 | L.F. | 1.100 | 24.75 | 39.27 | 64.02 |
| Paint, interior, primer & 1 coat | 1.000 | Face | 1.778 | 2.02 | 57.00 | 59.02 |
| Drip cap, vinyl | 1.000 | Ea. | .615 | 84.50 | 22.00 | 106.50 |
| Caulking | 32.000 | L.F. | 1.032 | 5.12 | 36.80 | 41.92 |
| Snap-in grilles | 1.000 | Set | 1.333 | 120.00 | 47.50 | 167.50 |
| | | | | | | |
| TOTAL | | | 8.144 | 2161.39 | 284.07 | 2445.46 |
| **DOUBLE HUNG TYPE, METAL CLAD, 9' X 5'** | | | | | | |
| Window, metal clad, deluxe, 9' x 5', insulating glass | 1.000 | Ea. | 2.667 | 1150.00 | 95.00 | 1245.00 |
| Trim, interior casing | 29.000 | L.F. | .967 | 21.75 | 34.51 | 56.26 |
| Paint, interior, primer & 1 coat | 1.000 | Face | 1.778 | 2.02 | 57.00 | 59.02 |
| Drip cap, vinyl | 1.000 | Set | .615 | 84.50 | 22.00 | 106.50 |
| Caulking | 28.000 | L.F. | .903 | 4.48 | 32.20 | 36.68 |
| Snap-in grilles | 1.000 | Set | 1.067 | 96.00 | 38.00 | 134.00 |
| | | | | | | |
| TOTAL | | | 7.997 | 1358.75 | 278.71 | 1637.46 |

The cost of this system is on a cost per each window basis.

| Description | QUAN. | UNIT | LABOR HOURS | COST EACH | | |
|---|---|---|---|---|---|---|
| | | | | MAT. | INST. | TOTAL |
| | | | | | | |
| | | | | | | |
| | | | | | | |
| | | | | | | |
| | | | | | | |

**Important: See the Reference Section for critical supporting data - Reference Nos., Crews & Location Factors**

| Bow/Bay Window Price Sheet | QUAN. | UNIT | LABOR HOURS | COST EACH | | |
|---|---|---|---|---|---|---|
| | | | | MAT. | INST. | TOTAL |
| Windows, bow awning type, builder's quality, 8' x 5', insulating glass | 1.000 | Ea. | 1.600 | 960.00 | 57.00 | 1017.00 |
| Low E glass | 1.000 | Ea. | 1.600 | 1250.00 | 57.00 | 1307.00 |
| 12' x 6', insulating glass | 1.000 | Ea. | 2.667 | 1300.00 | 95.00 | 1395.00 |
| Low E glass | 1.000 | Ea. | 2.667 | 1375.00 | 95.00 | 1470.00 |
| Plastic clad premium insulating glass, 6' x 4' | 1.000 | Ea. | 1.600 | 1050.00 | 57.00 | 1107.00 |
| 9' x 4' | 1.000 | Ea. | 2.000 | 1375.00 | 71.00 | 1446.00 |
| 10' x 5' | 1.000 | Ea. | 2.286 | 2350.00 | 81.50 | 2431.50 |
| 12' x 6' | 1.000 | Ea. | 2.667 | 2975.00 | 95.00 | 3070.00 |
| Metal clad deluxe insulating glass, 6' x 4' | 1.000 | Ea. | 1.600 | 890.00 | 57.00 | 947.00 |
| 9' x 4' | 1.000 | Ea. | 2.000 | 1250.00 | 71.00 | 1321.00 |
| 10' x 5' | 1.000 | Ea. | 2.286 | 1725.00 | 81.50 | 1806.50 |
| 12' x 6' | 1.000 | Ea. | 2.667 | 2400.00 | 95.00 | 2495.00 |
| Bow casement type, builder's quality, 8' x 5', single glass | 1.000 | Ea. | 1.600 | 1500.00 | 57.00 | 1557.00 |
| Insulating glass | 1.000 | Ea. | 1.600 | 1825.00 | 57.00 | 1882.00 |
| 12' x 6', single glass | 1.000 | Ea. | 2.667 | 1900.00 | 95.00 | 1995.00 |
| Insulating glass | 1.000 | Ea. | 2.667 | 1950.00 | 95.00 | 2045.00 |
| Plastic clad premium insulating glass, 8' x 5' | 1.000 | Ea. | 1.600 | 1300.00 | 57.00 | 1357.00 |
| 10' x 5' | 1.000 | Ea. | 2.000 | 1825.00 | 71.00 | 1896.00 |
| 10' x 6' | 1.000 | Ea. | 2.286 | 1925.00 | 81.50 | 2006.50 |
| 12' x 6' | 1.000 | Ea. | 2.667 | 2275.00 | 95.00 | 2370.00 |
| Metal clad deluxe insulating glass, 8' x 5' | 1.000 | Ea. | 1.600 | 1325.00 | 57.00 | 1382.00 |
| 10' x 5' | 1.000 | Ea. | 2.000 | 1450.00 | 71.00 | 1521.00 |
| 10' x 6' | 1.000 | Ea. | 2.286 | 1700.00 | 81.50 | 1781.50 |
| 12' x 6' | 1.000 | Ea. | 2.667 | 2350.00 | 95.00 | 2445.00 |
| Bow, double hung type, builder's quality, 8' x 4', single glass | 1.000 | Ea. | 1.600 | 1050.00 | 57.00 | 1107.00 |
| Insulating glass | 1.000 | Ea. | 1.600 | 1150.00 | 57.00 | 1207.00 |
| 9' x 5', single glass | 1.000 | Ea. | 2.667 | 1150.00 | 95.00 | 1245.00 |
| Insulating glass | 1.000 | Ea. | 2.667 | 1200.00 | 95.00 | 1295.00 |
| Plastic clad premium insulating glass, 7' x 4' | 1.000 | Ea. | 1.600 | 1100.00 | 57.00 | 1157.00 |
| 8' x 4' | 1.000 | Ea. | 2.000 | 1125.00 | 71.00 | 1196.00 |
| 8' x 5' | 1.000 | Ea. | 2.286 | 1175.00 | 81.50 | 1256.50 |
| 9' x 5' | 1.000 | Ea. | 2.667 | 1200.00 | 95.00 | 1295.00 |
| Metal clad deluxe insulating glass, 7' x 4' | 1.000 | Ea. | 1.600 | 1025.00 | 57.00 | 1082.00 |
| 8' x 4' | 1.000 | Ea. | 2.000 | 1050.00 | 71.00 | 1121.00 |
| 8' x 5' | 1.000 | Ea. | 2.286 | 1100.00 | 81.50 | 1181.50 |
| 9' x 5' | 1.000 | Ea. | 2.667 | 1150.00 | 95.00 | 1245.00 |
| Trim, interior casing, window 7' x 4' | 1.000 | Ea. | .767 | 17.25 | 27.50 | 44.75 |
| 8' x 5' | 1.000 | Ea. | .900 | 20.50 | 32.00 | 52.50 |
| 10' x 6' | 1.000 | Ea. | 1.100 | 25.00 | 39.50 | 64.50 |
| 12' x 6' | 1.000 | Ea. | 1.233 | 28.00 | 44.00 | 72.00 |
| Paint or stain, interior, or exterior, 7' x 4' window, 1 coat | 1.000 | Face | .889 | 2.10 | 28.50 | 30.60 |
| Primer & 1 coat | 1.000 | Face | 1.333 | 3.62 | 42.50 | 46.12 |
| 8' x 5' window, 1 coat | 1.000 | Face | .889 | 2.10 | 28.50 | 30.60 |
| Primer & 1 coat | 1.000 | Face | 1.333 | 3.62 | 42.50 | 46.12 |
| 10' x 6' window, 1 coat | 1.000 | Face | 1.333 | 1.52 | 43.00 | 44.52 |
| Primer & 1 coat | 1.000 | Face | 1.778 | 2.02 | 57.00 | 59.02 |
| 12' x 6' window, 1 coat | 1.000 | Face | 1.778 | 4.20 | 57.00 | 61.20 |
| Primer & 1 coat | 1.000 | Face | 2.667 | 7.25 | 85.00 | 92.25 |
| Drip cap, vinyl moulded window, 7' long | 1.000 | Ea. | .533 | 77.50 | 19.00 | 96.50 |
| 8' long | 1.000 | Ea. | .533 | 77.50 | 19.00 | 96.50 |
| 10' long | 1.000 | Ea. | .615 | 84.50 | 22.00 | 106.50 |
| 12' long | 1.000 | Ea. | .615 | 84.50 | 22.00 | 106.50 |
| Caulking, window, 7' x 4' | 1.000 | Ea. | .710 | 3.52 | 25.50 | 29.02 |
| 8' x 5' | 1.000 | Ea. | .839 | 4.16 | 30.00 | 34.16 |
| 10' x 6' | 1.000 | Ea. | 1.032 | 5.10 | 37.00 | 42.10 |
| 12' x 6' | 1.000 | Ea. | 1.161 | 5.75 | 41.50 | 47.25 |
| Grilles, window, 7' x 4' | 1.000 | Set | .800 | 72.00 | 28.50 | 100.50 |
| 8' x 5' | 1.000 | Set | 1.067 | 96.00 | 38.00 | 134.00 |
| 10' x 6' | 1.000 | Set | 1.333 | 120.00 | 47.50 | 167.50 |
| 12' x 6' | 1.000 | Set | 1.600 | 144.00 | 57.00 | 201.00 |

**4 EXTERIOR WALLS**

161

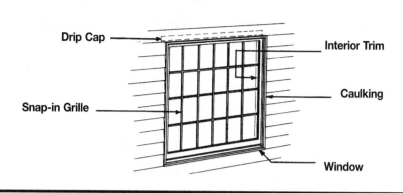

**EXTERIOR WALLS 4**

| System Description | QUAN. | UNIT | LABOR HOURS | COST EACH | | |
|---|---|---|---|---|---|---|
| | | | | MAT. | INST. | TOTAL |
| **BUILDER'S QUALITY PICTURE WINDOW, 4' X 4'** | | | | | | |
| Window, primed, builder's quality, 4' x 4', insulating glass | 1.000 | Ea. | 1.333 | 300.00 | 47.50 | 347.50 |
| Trim, interior casing | 17.000 | L.F. | .567 | 12.75 | 20.23 | 32.98 |
| Paint, interior & exterior, primer & 2 coats | 2.000 | Face | 1.778 | 2.02 | 57.00 | 59.02 |
| Caulking | 16.000 | L.F. | .516 | 2.56 | 18.40 | 20.96 |
| Snap-in grille | 1.000 | Ea. | .267 | 137.00 | 9.50 | 146.50 |
| Drip cap, metal | 4.000 | L.F. | .080 | .88 | 2.84 | 3.72 |
| TOTAL | | | 4.541 | 455.21 | 155.47 | 610.68 |
| **PLASTIC CLAD WOOD WINDOW, 4'-6" X 6'-6"** | | | | | | |
| Window, plastic clad, prem., 4'-6" x 6'-6", insul. glass | 1.000 | Ea. | 1.455 | 670.00 | 52.00 | 722.00 |
| Trim, interior casing | 23.000 | L.F. | .767 | 17.25 | 27.37 | 44.62 |
| Paint, interior, primer & 2 coats | 1.000 | Face | .889 | 1.01 | 28.50 | 29.51 |
| Caulking | 22.000 | L.F. | .710 | 3.52 | 25.30 | 28.82 |
| Snap-in grille | 1.000 | Ea. | .267 | 137.00 | 9.50 | 146.50 |
| TOTAL | | | 4.088 | 828.78 | 142.67 | 971.45 |
| **METAL CLAD WOOD WINDOW, 6'-6" X 6'-6"** | | | | | | |
| Window, metal clad, deluxe, 6'-6" x 6'-6", insulating glass | 1.000 | Ea. | 1.600 | 550.00 | 57.00 | 607.00 |
| Trim interior casing | 27.000 | L.F. | .900 | 20.25 | 32.13 | 52.38 |
| Paint, interior, primer & 2 coats | 1.000 | Face | 1.600 | 5.40 | 51.00 | 56.40 |
| Caulking | 26.000 | L.F. | .839 | 4.16 | 29.90 | 34.06 |
| Snap-in grille | 1.000 | Ea. | .267 | 137.00 | 9.50 | 146.50 |
| Drip cap, metal | 6.500 | L.F. | .130 | 1.43 | 4.62 | 6.05 |
| TOTAL | | | 5.336 | 718.24 | 184.15 | 902.39 |

The cost of this system is on a cost per each window basis.

| Description | QUAN. | UNIT | LABOR HOURS | COST EACH | | |
|---|---|---|---|---|---|---|
| | | | | MAT. | INST. | TOTAL |
| | | | | | | |
| | | | | | | |
| | | | | | | |
| | | | | | | |
| | | | | | | |

**Important: See the Reference Section for critical supporting data - Reference Nos., Crews & Location Factors**

# Fixed Window Price Sheet

| Fixed Window Price Sheet | QUAN. | UNIT | LABOR HOURS | COST EACH MAT. | COST EACH INST. | COST EACH TOTAL |
|---|---|---|---|---|---|---|
| Window-picture, builder's quality, 4' x 4', single glass | 1.000 | Ea. | 1.333 | 275.00 | 47.50 | 322.50 |
| Insulating glass | 1.000 | Ea. | 1.333 | 300.00 | 47.50 | 347.50 |
| 4' x 4'-6", single glass | 1.000 | Ea. | 1.455 | 295.00 | 52.00 | 347.00 |
| Insulating glass | 1.000 | Ea. | 1.455 | 325.00 | 52.00 | 377.00 |
| 5' x 4', single glass | 1.000 | Ea. | 1.455 | 365.00 | 52.00 | 417.00 |
| Insulating glass | 1.000 | Ea. | 1.455 | 410.00 | 52.00 | 462.00 |
| 6' x 4'-6", single glass | 1.000 | Ea. | 1.600 | 465.00 | 57.00 | 522.00 |
| Insulating glass | 1.000 | Ea. | 1.600 | 520.00 | 57.00 | 577.00 |
| Plastic clad premium insulating glass, 4' x 4' | 1.000 | Ea. | 1.333 | 405.00 | 47.50 | 452.50 |
| 4'-6" x 6'-6" | 1.000 | Ea. | 1.455 | 670.00 | 52.00 | 722.00 |
| 5'-6" x 6'-6" | 1.000 | Ea. | 1.600 | 870.00 | 57.00 | 927.00 |
| 6'-6" x 6'-6" | 1.000 | Ea. | 1.600 | 885.00 | 57.00 | 942.00 |
| Metal clad deluxe insulating glass, 4' x 4' | 1.000 | Ea. | 1.333 | 294.00 | 47.50 | 341.50 |
| 4'-6" x 6'-6" | 1.000 | Ea. | 1.455 | 435.00 | 52.00 | 487.00 |
| 5'-6" x 6'-6" | 1.000 | Ea. | 1.600 | 475.00 | 57.00 | 532.00 |
| 6'-6" x 6'-6" | 1.000 | Ea. | 1.600 | 550.00 | 57.00 | 607.00 |
| Trim, interior casing, window 4' x 4' | 17.000 | L.F. | .567 | 12.75 | 20.00 | 32.75 |
| 4'-6" x 4'-6" | 19.000 | L.F. | .633 | 14.25 | 22.50 | 36.75 |
| 5'-0" x 4'-0" | 19.000 | L.F. | .633 | 14.25 | 22.50 | 36.75 |
| 4'-6" x 6'-6" | 23.000 | L.F. | .767 | 17.25 | 27.50 | 44.75 |
| 5'-6" x 6'-6" | 25.000 | L.F. | .833 | 18.75 | 30.00 | 48.75 |
| 6'-6" x 6'-6" | 27.000 | L.F. | .900 | 20.50 | 32.00 | 52.50 |
| Paint or stain, interior or exterior, 4' x 4' window, 1 coat | 1.000 | Face | .667 | .76 | 21.50 | 22.26 |
| 2 coats | 1.000 | Face | .667 | .85 | 21.50 | 22.35 |
| Primer & 1 coat | 1.000 | Face | .727 | 1.03 | 23.00 | 24.03 |
| Primer & 2 coats | 1.000 | Face | .889 | 1.01 | 28.50 | 29.51 |
| 4'-6" x 6'-6" window, 1 coat | 1.000 | Face | .667 | .76 | 21.50 | 22.26 |
| 2 coats | 1.000 | Face | .667 | .85 | 21.50 | 22.35 |
| Primer & 1 coat | 1.000 | Face | .727 | 1.03 | 23.00 | 24.03 |
| Primer & 2 coats | 1.000 | Face | .889 | 1.01 | 28.50 | 29.51 |
| 6'-6" x 6'-6" window, 1 coat | 1.000 | Face | .889 | 2.10 | 28.50 | 30.60 |
| 2 coats | 1.000 | Face | 1.333 | 3.84 | 42.50 | 46.34 |
| Primer & 1 coat | 1.000 | Face | 1.333 | 3.62 | 42.50 | 46.12 |
| Primer & 2 coats | 1.000 | Face | 1.600 | 5.40 | 51.00 | 56.40 |
| Caulking, window, 4' x 4' | 1.000 | Ea. | .516 | 2.56 | 18.40 | 20.96 |
| 4'-6" x 4'-6" | 1.000 | Ea. | .581 | 2.88 | 20.50 | 23.38 |
| 5'-0" x 4'-0" | 1.000 | Ea. | .581 | 2.88 | 20.50 | 23.38 |
| 4'-6" x 6'-6" | 1.000 | Ea. | .710 | 3.52 | 25.50 | 29.02 |
| 5'-6" x 6'-6" | 1.000 | Ea. | .774 | 3.84 | 27.50 | 31.34 |
| 6'-6" x 6'-6" | 1.000 | Ea. | .839 | 4.16 | 30.00 | 34.16 |
| Grilles, glass size, to 48" x 48" | 1.000 | Ea. | .267 | 137.00 | 9.50 | 146.50 |
| To 60" x 68" | 1.000 | Ea. | .286 | 150.00 | 10.15 | 160.15 |
| Drip cap, aluminum, 4' long | 4.000 | L.F. | .080 | .88 | 2.84 | 3.72 |
| 4'-6" long | 4.500 | L.F. | .090 | .99 | 3.20 | 4.19 |
| 5' long | 5.000 | L.F. | .100 | 1.10 | 3.55 | 4.65 |
| 6' long | 6.000 | L.F. | .120 | 1.32 | 4.26 | 5.58 |
| Wood, 4' long | 4.000 | L.F. | .133 | 3.00 | 4.76 | 7.76 |
| 4'-6" long | 4.500 | L.F. | .150 | 3.38 | 5.35 | 8.73 |
| 5' long | 5.000 | L.F. | .167 | 3.75 | 5.95 | 9.70 |
| 6' long | 6.000 | L.F. | .200 | 4.50 | 7.15 | 11.65 |

**EXTERIOR WALLS**

**4**

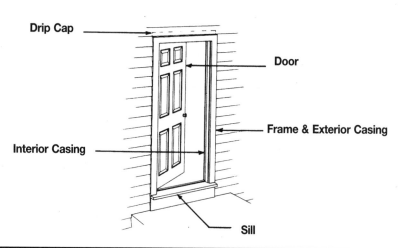

**Drip Cap**

**Door**

**Frame & Exterior Casing**

**Interior Casing**

**Sill**

| System Description | QUAN. | UNIT | LABOR HOURS | COST EACH | | |
|---|---|---|---|---|---|---|
| | | | | MAT. | INST. | TOTAL |
| **COLONIAL, 6 PANEL, 3' X 6'-8", WOOD** | | | | | | |
| Door, 3' x 6'-8" x 1-3/4" thick, pine, 6 panel colonial | 1.000 | Ea. | 1.067 | 385.00 | 38.00 | 423.00 |
| Frame, 5-13/16" deep, incl. exterior casing & drip cap | 17.000 | L.F. | .725 | 126.65 | 25.84 | 152.49 |
| Interior casing, 2-1/2" wide | 18.000 | L.F. | .600 | 13.50 | 21.42 | 34.92 |
| Sill, 8/4 x 8" deep | 3.000 | L.F. | .480 | 43.95 | 17.10 | 61.05 |
| Butt hinges, brass, 4-1/2" x 4-1/2" | 1.500 | Pr. | | 13.73 | | 13.73 |
| Lockset | 1.000 | Ea. | .571 | 31.00 | 20.50 | 51.50 |
| Weatherstripping, metal, spring type, bronze | 1.000 | Set | 1.053 | 17.20 | 37.50 | 54.70 |
| Paint, interior & exterior, primer & 2 coats. | 2.000 | Face | 1.778 | 10.70 | 57.00 | 67.70 |
| | | | | | | |
| TOTAL | | | 6.274 | 641.73 | 217.36 | 859.09 |
| **SOLID CORE BIRCH, FLUSH, 3' X 6'-8"** | | | | | | |
| Door, 3' x 6'-8", 1-3/4" thick, birch, flush solid core | 1.000 | Ea. | 1.067 | 90.50 | 38.00 | 128.50 |
| Frame, 5-13/16" deep, incl. exterior casing & drip cap | 17.000 | L.F. | .725 | 126.65 | 25.84 | 152.49 |
| Interior casing, 2-1/2" wide | 18.000 | L.F. | .600 | 13.50 | 21.42 | 34.92 |
| Sill, 8/4 x 8" deep | 3.000 | L.F. | .480 | 43.95 | 17.10 | 61.05 |
| Butt hinges, brass, 4-1/2" x 4-1/2" | 1.500 | Pr. | | 13.73 | | 13.73 |
| Lockset | 1.000 | Ea. | .571 | 31.00 | 20.50 | 51.50 |
| Weatherstripping, metal, spring type, bronze | 1.000 | Set | 1.053 | 17.20 | 37.50 | 54.70 |
| Paint, Interior & exterior, primer & 2 coats | 2.000 | Face | 1.778 | 9.98 | 57.00 | 66.98 |
| | | | | | | |
| TOTAL | | | 6.274 | 346.51 | 217.36 | 563.87 |

These systems are on a cost per each door basis.

| Description | QUAN. | UNIT | LABOR HOURS | COST EACH | | |
|---|---|---|---|---|---|---|
| | | | | MAT. | INST. | TOTAL |
| | | | | | | |
| | | | | | | |
| | | | | | | |
| | | | | | | |
| | | | | | | |

**EXTERIOR WALLS   4**

| Entrance Door Price Sheet | QUAN. | UNIT | LABOR HOURS | COST EACH | | |
|---|---|---|---|---|---|---|
| | | | | MAT. | INST. | TOTAL |
| Door exterior wood 1-3/4″ thick, pine, dutch door, 2′-8″ x 6′-8″ minimum | 1.000 | Ea. | 1.333 | 655.00 | 47.50 | 702.50 |
| Maximum | 1.000 | Ea. | 1.600 | 695.00 | 57.00 | 752.00 |
| 3′-0″ x 6′-8″, minimum | 1.000 | Ea. | 1.333 | 685.00 | 47.50 | 732.50 |
| Maximum | 1.000 | Ea. | 1.600 | 740.00 | 57.00 | 797.00 |
| Colonial, 6 panel, 2′-8″ x 6′-8″ | 1.000 | Ea. | 1.000 | 355.00 | 35.50 | 390.50 |
| 3′-0″ x 6′-8″ | 1.000 | Ea. | 1.067 | 385.00 | 38.00 | 423.00 |
| 8 panel, 2′-6″ x 6′-8″ | 1.000 | Ea. | 1.000 | 525.00 | 35.50 | 560.50 |
| 3′-0″ x 6′-8″ | 1.000 | Ea. | 1.067 | 575.00 | 38.00 | 613.00 |
| Flush, birch, solid core, 2′-8″ x 6′-8″ | 1.000 | Ea. | 1.000 | 82.00 | 35.50 | 117.50 |
| 3′-0″ x 6′-8″ | 1.000 | Ea. | 1.067 | 90.50 | 38.00 | 128.50 |
| Porch door, 2′-8″ x 6′-8″ | 1.000 | Ea. | 1.000 | 207.00 | 35.50 | 242.50 |
| 3′-0″ x 6′-8″ | 1.000 | Ea. | 1.067 | 238.00 | 38.00 | 276.00 |
| Hand carved mahogany, 2′-8″ x 6′-8″ | 1.000 | Ea. | 1.067 | 475.00 | 38.00 | 513.00 |
| 3′-0″ x 6′-8″ | 1.000 | Ea. | 1.067 | 510.00 | 38.00 | 548.00 |
| Rosewood, 2′-8″ x 6′-8″ | 1.000 | Ea. | 1.067 | 740.00 | 38.00 | 778.00 |
| 3′-0″ x 6-8″ | 1.000 | Ea. | 1.067 | 770.00 | 38.00 | 808.00 |
| Door, metal clad wood 1-3/8″ thick raised panel, 2′-8″ x 6′-8″ | 1.000 | Ea. | 1.067 | 264.00 | 38.00 | 302.00 |
| 3′-0″ x 6′-8″ | 1.000 | Ea. | 1.067 | 262.00 | 38.00 | 300.00 |
| Deluxe metal door, 2′-8″ x 6′-8″ | 1.000 | Ea. | 1.231 | 410.00 | 44.00 | 454.00 |
| 3′-0″ x 6′-8″ | 1.000 | Ea. | 1.231 | 405.00 | 44.00 | 449.00 |
| Frame, pine, including exterior trim & drip cap, 5/4, x 4-9/16″ deep | 17.000 | L.F. | .725 | 73.50 | 26.00 | 99.50 |
| 5-13/16″ deep | 17.000 | L.F. | .725 | 127.00 | 26.00 | 153.00 |
| 6-9/16″ deep | 17.000 | L.F. | .725 | 164.00 | 26.00 | 190.00 |
| Safety glass lites, add | 1.000 | Ea. | | 26.50 | | 26.50 |
| Interior casing, 2′-8″ x 6′-8″ door | 18.000 | L.F. | .600 | 13.50 | 21.50 | 35.00 |
| 3′-0″ x 6′-8″ door | 19.000 | L.F. | .633 | 14.25 | 22.50 | 36.75 |
| Sill, oak, 8/4 x 8″ deep | 3.000 | L.F. | .480 | 44.00 | 17.10 | 61.10 |
| 8/4 x 10″ deep | 3.000 | L.F. | .533 | 59.50 | 19.05 | 78.55 |
| Butt hinges, steel plated, 4-1/2″ x 4-1/2″, plain | 1.500 | Pr. | | 13.75 | | 13.75 |
| Ball bearing | 1.500 | Pr. | | 31.00 | | 31.00 |
| Bronze, 4-1/2″ x 4-1/2″, plain | 1.500 | Pr. | | 15.60 | | 15.60 |
| Ball bearing | 1.500 | Pr. | | 32.50 | | 32.50 |
| Lockset, minimum | 1.000 | Ea. | .571 | 31.00 | 20.50 | 51.50 |
| Maximum | 1.000 | Ea. | 1.000 | 130.00 | 35.50 | 165.50 |
| Weatherstripping, metal, interlocking, zinc | 1.000 | Set | 2.667 | 12.80 | 95.00 | 107.80 |
| Bronze | 1.000 | Set | 2.667 | 20.00 | 95.00 | 115.00 |
| Spring type, bronze | 1.000 | Set | 1.053 | 17.20 | 37.50 | 54.70 |
| Rubber, minimum | 1.000 | Set | 1.053 | 4.55 | 37.50 | 42.05 |
| Maximum | 1.000 | Set | 1.143 | 5.20 | 40.50 | 45.70 |
| Felt minimum | 1.000 | Set | .571 | 2.10 | 20.50 | 22.60 |
| Maximum | 1.000 | Set | .615 | 2.28 | 22.00 | 24.28 |
| Paint or stain, flush door, interior or exterior, 1 coat | 2.000 | Face | .941 | 3.64 | 30.00 | 33.64 |
| 2 coats | 2.000 | Face | 1.455 | 7.25 | 46.00 | 53.25 |
| Primer & 1 coat | 2.000 | Face | 1.455 | 6.55 | 46.00 | 52.55 |
| Primer & 2 coats | 2.000 | Face | 1.778 | 10.00 | 57.00 | 67.00 |
| Paneled door, interior & exterior, 1 coat | 2.000 | Face | 1.143 | 3.88 | 36.50 | 40.38 |
| 2 coats | 2.000 | Face | 2.000 | 7.75 | 64.00 | 71.75 |
| Primer & 1 coat | 2.000 | Face | 1.455 | 7.00 | 46.00 | 53.00 |
| Primer & 2 coats | 2.000 | Face | 1.778 | 10.70 | 57.00 | 67.70 |
| | | | | | | |
| | | | | | | |
| | | | | | | |
| | | | | | | |
| | | | | | | |

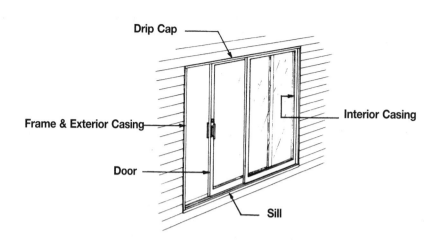

Drip Cap

Frame & Exterior Casing

Door

Interior Casing

Sill

| System Description | QUAN. | UNIT | LABOR HOURS | COST EACH | | |
|---|---|---|---|---|---|---|
| | | | | MAT. | INST. | TOTAL |
| **WOOD SLIDING DOOR, 8′ WIDE, PREMIUM** | | | | | | |
| Wood, 5/8″ thick tempered insul. glass, 8′ wide, premium | 1.000 | Ea. | 5.333 | 1225.00 | 190.00 | 1415.00 |
| Interior casing | 22.000 | L.F. | .733 | 16.50 | 26.18 | 42.68 |
| Exterior casing | 22.000 | L.F. | .733 | 16.50 | 26.18 | 42.68 |
| Sill, oak, 8/4 x 8″ deep | 8.000 | L.F. | 1.280 | 117.20 | 45.60 | 162.80 |
| Drip cap | 8.000 | L.F. | .160 | 1.76 | 5.68 | 7.44 |
| Paint, interior & exterior, primer & 2 coats | 2.000 | Face | 2.816 | 14.96 | 89.76 | 104.72 |
| TOTAL | | | 11.055 | 1391.92 | 383.40 | 1775.32 |
| **ALUMINUM SLIDING DOOR, 8′ WIDE, PREMIUM** | | | | | | |
| Aluminum, 5/8″ tempered insul. glass, 8′ wide, premium | 1.000 | Ea. | 5.333 | 1375.00 | 190.00 | 1565.00 |
| Interior casing | 22.000 | L.F. | .733 | 16.50 | 26.18 | 42.68 |
| Exterior casing | 22.000 | L.F. | .733 | 16.50 | 26.18 | 42.68 |
| Sill, oak, 8/4 x 8″ deep | 8.000 | L.F. | 1.280 | 117.20 | 45.60 | 162.80 |
| Drip cap | 8.000 | L.F. | .160 | 1.76 | 5.68 | 7.44 |
| Paint, interior & exterior, primer & 2 coats | 2.000 | Face | 2.816 | 14.96 | 89.76 | 104.72 |
| TOTAL | | | 11.055 | 1541.92 | 383.40 | 1925.32 |

The cost of this system is on a cost per each door basis.

| Description | QUAN. | UNIT | LABOR HOURS | COST EACH | | |
|---|---|---|---|---|---|---|
| | | | | MAT. | INST. | TOTAL |
| | | | | | | |
| | | | | | | |
| | | | | | | |
| | | | | | | |
| | | | | | | |
| | | | | | | |
| | | | | | | |
| | | | | | | |

**Important: See the Reference Section for critical supporting data - Reference Nos., Crews & Location Factors**

| Sliding Door Price Sheet | QUAN. | UNIT | LABOR HOURS | MAT. | INST. | TOTAL |
|---|---|---|---|---|---|---|
| Sliding door, wood, 5/8" thick, tempered insul. glass, 6' wide, premium | 1.000 | Ea. | 4.000 | 1050.00 | 142.00 | 1192.00 |
| Economy | 1.000 | Ea. | 4.000 | 760.00 | 142.00 | 902.00 |
| 8'wide, wood premium | 1.000 | Ea. | 5.333 | 1225.00 | 190.00 | 1415.00 |
| Economy | 1.000 | Ea. | 5.333 | 885.00 | 190.00 | 1075.00 |
| 12' wide, wood premium | 1.000 | Ea. | 6.400 | 1950.00 | 228.00 | 2178.00 |
| Economy | 1.000 | Ea. | 6.400 | 1275.00 | 228.00 | 1503.00 |
| Aluminum, 5/8" thick, tempered insul. glass, 6'wide, premium | 1.000 | Ea. | 4.000 | 1400.00 | 142.00 | 1542.00 |
| Economy | 1.000 | Ea. | 4.000 | 570.00 | 142.00 | 712.00 |
| 8'wide, premium | 1.000 | Ea. | 5.333 | 1375.00 | 190.00 | 1565.00 |
| Economy | 1.000 | Ea. | 5.333 | 1175.00 | 190.00 | 1365.00 |
| 12' wide, premium | 1.000 | Ea. | 6.400 | 2300.00 | 228.00 | 2528.00 |
| Economy | 1.000 | Ea. | 6.400 | 1425.00 | 228.00 | 1653.00 |
| Interior casing, 6' wide door | 20.000 | L.F. | .667 | 15.00 | 24.00 | 39.00 |
| 8' wide door | 22.000 | L.F. | .733 | 16.50 | 26.00 | 42.50 |
| 12' wide door | 26.000 | L.F. | .867 | 19.50 | 31.00 | 50.50 |
| Exterior casing, 6' wide door | 20.000 | L.F. | .667 | 15.00 | 24.00 | 39.00 |
| 8' wide door | 22.000 | L.F. | .733 | 16.50 | 26.00 | 42.50 |
| 12' wide door | 26.000 | L.F. | .867 | 19.50 | 31.00 | 50.50 |
| Sill, oak, 8/4 x 8" deep, 6' wide door | 6.000 | L.F. | .960 | 88.00 | 34.00 | 122.00 |
| 8' wide door | 8.000 | L.F. | 1.280 | 117.00 | 45.50 | 162.50 |
| 12' wide door | 12.000 | L.F. | 1.920 | 176.00 | 68.50 | 244.50 |
| 8/4 x 10" deep, 6' wide door | 6.000 | L.F. | 1.067 | 119.00 | 38.00 | 157.00 |
| 8' wide door | 8.000 | L.F. | 1.422 | 158.00 | 51.00 | 209.00 |
| 12' wide door | 12.000 | L.F. | 2.133 | 237.00 | 76.00 | 313.00 |
| Drip cap, 6' wide door | 6.000 | L.F. | .120 | 1.32 | 4.26 | 5.58 |
| 8' wide door | 8.000 | L.F. | .160 | 1.76 | 5.70 | 7.46 |
| 12' wide door | 12.000 | L.F. | .240 | 2.64 | 8.50 | 11.14 |
| Paint or stain, interior & exterior, 6' wide door, 1 coat | 2.000 | Face | 1.600 | 4.80 | 51.00 | 55.80 |
| 2 coats | 2.000 | Face | 1.600 | 4.80 | 51.00 | 55.80 |
| Primer & 1 coat | 2.000 | Face | 1.778 | 9.60 | 57.00 | 66.60 |
| Primer & 2 coats | 2.000 | Face | 2.560 | 13.60 | 81.50 | 95.10 |
| 8' wide door, 1 coat | 2.000 | Face | 1.760 | 5.30 | 56.50 | 61.80 |
| 2 coats | 2.000 | Face | 1.760 | 5.30 | 56.50 | 61.80 |
| Primer & 1 coat | 2.000 | Face | 1.955 | 10.55 | 62.50 | 73.05 |
| Primer & 2 coats | 2.000 | Face | 2.816 | 14.95 | 90.00 | 104.95 |
| 12' wide door, 1 coat | 2.000 | Face | 2.080 | 6.25 | 66.50 | 72.75 |
| 2 coats | 2.000 | Face | 2.080 | 6.25 | 66.50 | 72.75 |
| Primer & 1 coat | 2.000 | Face | 2.311 | 12.50 | 74.00 | 86.50 |
| Primer & 2 coats | 2.000 | Face | 3.328 | 17.70 | 106.00 | 123.70 |
| Aluminum door, trim only, interior & exterior, 6' door, 1 coat | 2.000 | Face | .800 | 2.40 | 25.50 | 27.90 |
| 2 coats | 2.000 | Face | .800 | 2.40 | 25.50 | 27.90 |
| Primer & 1 coat | 2.000 | Face | .889 | 4.80 | 28.50 | 33.30 |
| Primer & 2 coats | 2.000 | Face | 1.280 | 6.80 | 41.00 | 47.80 |
| 8' wide door, 1 coat | 2.000 | Face | .880 | 2.64 | 28.00 | 30.64 |
| 2 coats | 2.000 | Face | .880 | 2.64 | 28.00 | 30.64 |
| Primer & 1 coat | 2.000 | Face | .978 | 5.30 | 31.00 | 36.30 |
| Primer & 2 coats | 2.000 | Face | 1.408 | 7.50 | 45.00 | 52.50 |
| 12' wide door, 1 coat | 2.000 | Face | 1.040 | 3.12 | 33.50 | 36.62 |
| 2 coats | 2.000 | Face | 1.040 | 3.12 | 33.50 | 36.62 |
| Primer & 1 coat | 2.000 | Face | 1.155 | 6.25 | 37.00 | 43.25 |
| Primer & 2 coats | 2.000 | Face | 1.664 | 8.85 | 53.00 | 61.85 |

**EXTERIOR WALLS**

**4**

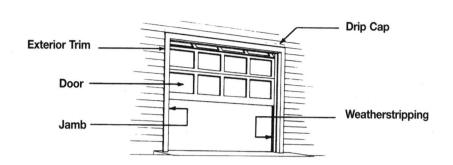

Exterior Trim

Drip Cap

Door

Jamb

Weatherstripping

| System Description | QUAN. | UNIT | LABOR HOURS | COST EACH | | |
|---|---|---|---|---|---|---|
| | | | | MAT. | INST. | TOTAL |
| **OVERHEAD, SECTIONAL GARAGE DOOR, 9' X 7'** | | | | | | |
| Wood, overhead sectional door, std., incl. hardware, 9' x 7' | 1.000 | Ea. | 2.000 | 415.00 | 71.00 | 486.00 |
| Jamb & header blocking, 2" x 6" | 25.000 | L.F. | .901 | 15.50 | 32.00 | 47.50 |
| Exterior trim | 25.000 | L.F. | .833 | 18.75 | 29.75 | 48.50 |
| Paint, interior & exterior, primer & 2 coats | 2.000 | Face | 3.556 | 21.40 | 114.00 | 135.40 |
| Weatherstripping, molding type | 1.000 | Set | .767 | 17.25 | 27.37 | 44.62 |
| Drip cap | 9.000 | L.F. | .180 | 1.98 | 6.39 | 8.37 |
| TOTAL | | | 8.237 | 489.88 | 280.51 | 770.39 |
| **OVERHEAD, SECTIONAL GARAGE DOOR, 16' X 7'** | | | | | | |
| Wood, overhead sectional, std., incl. hardware, 16' x 7' | 1.000 | Ea. | 2.667 | 830.00 | 95.00 | 925.00 |
| Jamb & header blocking, 2" x 6" | 30.000 | L.F. | 1.081 | 18.60 | 38.40 | 57.00 |
| Exterior trim | 30.000 | L.F. | 1.000 | 22.50 | 35.70 | 58.20 |
| Paint, interior & exterior, primer & 2 coats | 2.000 | Face | 5.333 | 32.10 | 171.00 | 203.10 |
| Weatherstripping, molding type | 1.000 | Set | 1.000 | 22.50 | 35.70 | 58.20 |
| Drip cap | 16.000 | L.F. | .320 | 3.52 | 11.36 | 14.88 |
| TOTAL | | | 11.401 | 929.22 | 387.16 | 1316.38 |
| **OVERHEAD, SWING-UP TYPE, GARAGE DOOR, 16' X 7'** | | | | | | |
| Wood, overhead, swing-up, std., incl. hardware, 16' x 7' | 1.000 | Ea. | 2.667 | 615.00 | 95.00 | 710.00 |
| Jamb & header blocking, 2" x 6" | 30.000 | L.F. | 1.081 | 18.60 | 38.40 | 57.00 |
| Exterior trim | 30.000 | L.F. | 1.000 | 22.50 | 35.70 | 58.20 |
| Paint, interior & exterior, primer & 2 coats | 2.000 | Face | 5.333 | 32.10 | 171.00 | 203.10 |
| Weatherstripping, molding type | 1.000 | Set | 1.000 | 22.50 | 35.70 | 58.20 |
| Drip cap | 16.000 | L.F. | .320 | 3.52 | 11.36 | 14.88 |
| TOTAL | | | 11.401 | 714.22 | 387.16 | 1101.38 |

This system is on a cost per each door basis.

| Description | QUAN. | UNIT | LABOR HOURS | COST EACH | | |
|---|---|---|---|---|---|---|
| | | | | MAT. | INST. | TOTAL |
| | | | | | | |
| | | | | | | |
| | | | | | | |
| | | | | | | |
| | | | | | | |

EXTERIOR WALLS 4

**Important: See the Reference Section for critical supporting data - Reference Nos., Crews & Location Factors**

| Resi Garage Door Price Sheet | QUAN. | UNIT | LABOR HOURS | COST EACH | | |
|---|---|---|---|---|---|---|
| | | | | MAT. | INST. | TOTAL |
| Overhead, sectional, including hardware, fiberglass, 9' x 7', standard | 1.000 | Ea. | 3.030 | 560.00 | 108.00 | 668.00 |
| Deluxe | 1.000 | Ea. | 3.030 | 710.00 | 108.00 | 818.00 |
| 16' x 7', standard | 1.000 | Ea. | 2.667 | 900.00 | 95.00 | 995.00 |
| Deluxe | 1.000 | Ea. | 2.667 | 1125.00 | 95.00 | 1220.00 |
| Hardboard, 9' x 7', standard | 1.000 | Ea. | 2.000 | 370.00 | 71.00 | 441.00 |
| Deluxe | 1.000 | Ea. | 2.000 | 445.00 | 71.00 | 516.00 |
| 16' x 7', standard | 1.000 | Ea. | 2.667 | 655.00 | 95.00 | 750.00 |
| Deluxe | 1.000 | Ea. | 2.667 | 760.00 | 95.00 | 855.00 |
| Metal, 9' x 7', standard | 1.000 | Ea. | 3.030 | 293.00 | 108.00 | 401.00 |
| Deluxe | 1.000 | Ea. | 2.000 | 585.00 | 71.00 | 656.00 |
| 16' x 7', standard | 1.000 | Ea. | 5.333 | 575.00 | 190.00 | 765.00 |
| Deluxe | 1.000 | Ea. | 2.667 | 800.00 | 95.00 | 895.00 |
| Wood, 9' x 7', standard | 1.000 | Ea. | 2.000 | 415.00 | 71.00 | 486.00 |
| Deluxe | 1.000 | Ea. | 2.000 | 1175.00 | 71.00 | 1246.00 |
| 16' x 7', standard | 1.000 | Ea. | 2.667 | 830.00 | 95.00 | 925.00 |
| Deluxe | 1.000 | Ea. | 2.667 | 1725.00 | 95.00 | 1820.00 |
| Overhead swing-up type including hardware, fiberglass, 9' x 7', standard | 1.000 | Ea. | 2.000 | 570.00 | 71.00 | 641.00 |
| Deluxe | 1.000 | Ea. | 2.000 | 675.00 | 71.00 | 746.00 |
| 16' x 7', standard | 1.000 | Ea. | 2.667 | 720.00 | 95.00 | 815.00 |
| Deluxe | 1.000 | Ea. | 2.667 | 835.00 | 95.00 | 930.00 |
| Hardboard, 9' x 7', standard | 1.000 | Ea. | 2.000 | 295.00 | 71.00 | 366.00 |
| Deluxe | 1.000 | Ea. | 2.000 | 390.00 | 71.00 | 461.00 |
| 16' x 7', standard | 1.000 | Ea. | 2.667 | 410.00 | 95.00 | 505.00 |
| Deluxe | 1.000 | Ea. | 2.667 | 610.00 | 95.00 | 705.00 |
| Metal, 9' x 7', standard | 1.000 | Ea. | 2.000 | 325.00 | 71.00 | 396.00 |
| Deluxe | 1.000 | Ea. | 2.000 | 565.00 | 71.00 | 636.00 |
| 16' x 7', standard | 1.000 | Ea. | 2.667 | 505.00 | 95.00 | 600.00 |
| Deluxe | 1.000 | Ea. | 2.667 | 815.00 | 95.00 | 910.00 |
| Wood, 9' x 7', standard | 1.000 | Ea. | 2.000 | 355.00 | 71.00 | 426.00 |
| Deluxe | 1.000 | Ea. | 2.000 | 615.00 | 71.00 | 686.00 |
| 16' x 7', standard | 1.000 | Ea. | 2.667 | 615.00 | 95.00 | 710.00 |
| Deluxe | 1.000 | Ea. | 2.667 | 865.00 | 95.00 | 960.00 |
| Jamb & header blocking, 2" x 6", 9' x 7' door | 25.000 | L.F. | .901 | 15.50 | 32.00 | 47.50 |
| 16' x 7' door | 30.000 | L.F. | 1.081 | 18.60 | 38.50 | 57.10 |
| 2" x 8", 9' x 7' door | 25.000 | L.F. | 1.000 | 24.00 | 35.50 | 59.50 |
| 16' x 7' door | 30.000 | L.F. | 1.200 | 28.50 | 42.50 | 71.00 |
| Exterior trim, 9' x 7' door | 25.000 | L.F. | .833 | 18.75 | 30.00 | 48.75 |
| 16' x 7' door | 30.000 | L.F. | 1.000 | 22.50 | 35.50 | 58.00 |
| Paint or stain, interior & exterior, 9' x 7' door, 1 coat | 1.000 | Face | 2.286 | 7.75 | 73.00 | 80.75 |
| 2 coats | 1.000 | Face | 4.000 | 15.50 | 128.00 | 143.50 |
| Primer & 1 coat | 1.000 | Face | 2.909 | 14.05 | 92.00 | 106.05 |
| Primer & 2 coats | 1.000 | Face | 3.556 | 21.50 | 114.00 | 135.50 |
| 16' x 7' door, 1 coat | 1.000 | Face | 3.429 | 11.65 | 110.00 | 121.65 |
| 2 coats | 1.000 | Face | 6.000 | 23.50 | 192.00 | 215.50 |
| Primer & 1 coat | 1.000 | Face | 4.364 | 21.00 | 138.00 | 159.00 |
| Primer & 2 coats | 1.000 | Face | 5.333 | 32.00 | 171.00 | 203.00 |
| Weatherstripping, molding type, 9' x 7' door | 1.000 | Set | .767 | 17.25 | 27.50 | 44.75 |
| 16' x 7' door | 1.000 | Set | 1.000 | 22.50 | 35.50 | 58.00 |
| Drip cap, 9' door | 9.000 | L.F. | .180 | 1.98 | 6.40 | 8.38 |
| 16' door | 16.000 | L.F. | .320 | 3.52 | 11.35 | 14.87 |
| Garage door opener, economy | 1.000 | Ea. | 1.000 | 237.00 | 35.50 | 272.50 |
| Deluxe, including remote control | 1.000 | Ea. | 1.000 | 340.00 | 35.50 | 375.50 |
| | | | | | | |
| | | | | | | |
| | | | | | | |
| | | | | | | |

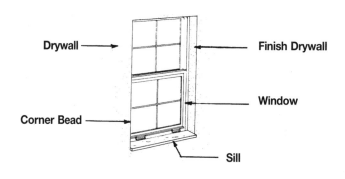

Drywall →    ← Finish Drywall

← Window

Corner Bead →    Sill

| System Description | QUAN. | UNIT | LABOR HOURS | COST EACH | | |
|---|---|---|---|---|---|---|
| | | | | MAT. | INST. | TOTAL |
| **SINGLE HUNG, 2' X 3' OPENING** | | | | | | |
| Window, 2' x 3' opening, enameled, insulating glass | 1.000 | Ea. | 1.600 | 173.00 | 71.50 | 244.50 |
| Blocking, 1" x 3" furring strip nailers | 10.000 | L.F. | .146 | 2.10 | 5.20 | 7.30 |
| Drywall, 1/2" thick, standard | 5.000 | S.F. | .040 | 1.15 | 1.40 | 2.55 |
| Corner bead, 1" x 1", galvanized steel | 8.000 | L.F. | .160 | .88 | 5.68 | 6.56 |
| Finish drywall, tape and finish corners inside and outside | 16.000 | L.F. | .233 | 1.12 | 8.32 | 9.44 |
| Sill, slate | 2.000 | L.F. | .400 | 16.30 | 12.50 | 28.80 |
| TOTAL | | | 2.579 | 194.55 | 104.60 | 299.15 |
| **SLIDING, 3' X 2' OPENING** | | | | | | |
| Window, 3' x 2' opening, enameled, insulating glass | 1.000 | Ea. | 1.600 | 178.00 | 71.50 | 249.50 |
| Blocking, 1" x 3" furring strip nailers | 10.000 | L.F. | .146 | 2.10 | 5.20 | 7.30 |
| Drywall, 1/2" thick, standard | 5.000 | S.F. | .040 | 1.15 | 1.40 | 2.55 |
| Corner bead, 1" x 1", galvanized steel | 7.000 | L.F. | .140 | .77 | 4.97 | 5.74 |
| Finish drywall, tape and finish corners inside and outside | 14.000 | L.F. | .204 | .98 | 7.28 | 8.26 |
| Sill, slate | 3.000 | L.F. | .600 | 24.45 | 18.75 | 43.20 |
| TOTAL | | | 2.730 | 207.45 | 109.10 | 316.55 |
| **AWNING, 3'-1" X 3'-2"** | | | | | | |
| Window, 3'-1" x 3'-2" opening, enameled, insul. glass | 1.000 | Ea. | 1.600 | 199.00 | 71.50 | 270.50 |
| Blocking, 1" x 3" furring strip, nailers | 12.500 | L.F. | .182 | 2.63 | 6.50 | 9.13 |
| Drywall, 1/2" thick, standard | 4.500 | S.F. | .036 | 1.04 | 1.26 | 2.30 |
| Corner bead, 1" x 1", galvanized steel | 9.250 | L.F. | .185 | 1.02 | 6.57 | 7.59 |
| Finish drywall, tape and finish corners, inside and outside | 18.500 | L.F. | .269 | 1.30 | 9.62 | 10.92 |
| Sill, slate | 3.250 | L.F. | .650 | 26.49 | 20.31 | 46.80 |
| TOTAL | | | 2.922 | 231.48 | 115.76 | 347.24 |

| Description | QUAN. | UNIT | LABOR HOURS | COST PER S.F. | | |
|---|---|---|---|---|---|---|
| | | | | MAT. | INST. | TOTAL |
| | | | | | | |
| | | | | | | |
| | | | | | | |
| | | | | | | |
| | | | | | | |

**EXTERIOR WALLS 4**

**Important: See the Reference Section for critical supporting data - Reference Nos., Crews & Location Factors**

# Aluminum Window Price Sheet

| Aluminum Window Price Sheet | QUAN. | UNIT | LABOR HOURS | COST EACH | | |
|---|---|---|---|---|---|---|
| | | | | MAT. | INST. | TOTAL |
| Window, aluminum, awning, 3'-1" x 3'-2", standard glass | 1.000 | Ea. | 1.600 | 213.00 | 71.50 | 284.50 |
| Insulating glass | 1.000 | Ea. | 1.600 | 199.00 | 71.50 | 270.50 |
| 4'-5" x 5'-3", standard glass | 1.000 | Ea. | 2.000 | 300.00 | 89.00 | 389.00 |
| Insulating glass | 1.000 | Ea. | 2.000 | 345.00 | 89.00 | 434.00 |
| Casement, 3'-1" x 3'-2", standard glass | 1.000 | Ea. | 1.600 | 298.00 | 71.50 | 369.50 |
| Insulating glass | 1.000 | Ea. | 1.600 | 298.00 | 71.50 | 369.50 |
| Single hung, 2' x 3', standard glass | 1.000 | Ea. | 1.600 | 143.00 | 71.50 | 214.50 |
| Insulating glass | 1.000 | Ea. | 1.600 | 173.00 | 71.50 | 244.50 |
| 2'-8" x 6'-8", standard glass | 1.000 | Ea. | 2.000 | 305.00 | 89.00 | 394.00 |
| Insulating glass | 1.000 | Ea. | 2.000 | 390.00 | 89.00 | 479.00 |
| 3'-4" x 5'-0", standard glass | 1.000 | Ea. | 1.778 | 196.00 | 79.50 | 275.50 |
| Insulating glass | 1.000 | Ea. | 1.778 | 275.00 | 79.50 | 354.50 |
| Sliding, 3' x 2', standard glass | 1.000 | Ea. | 1.600 | 161.00 | 71.50 | 232.50 |
| Insulating glass | 1.000 | Ea. | 1.600 | 178.00 | 71.50 | 249.50 |
| 5' x 3', standard glass | 1.000 | Ea. | 1.778 | 204.00 | 79.50 | 283.50 |
| Insulating glass | 1.000 | Ea. | 1.778 | 286.00 | 79.50 | 365.50 |
| 8' x 4', standard glass | 1.000 | Ea. | 2.667 | 294.00 | 119.00 | 413.00 |
| Insulating glass | 1.000 | Ea. | 2.667 | 475.00 | 119.00 | 594.00 |
| Blocking, 1" x 3" furring, opening 3' x 2' | 10.000 | L.F. | .146 | 2.10 | 5.20 | 7.30 |
| 3' x 3' | 12.500 | L.F. | .182 | 2.63 | 6.50 | 9.13 |
| 3' x 5' | 16.000 | L.F. | .233 | 3.36 | 8.30 | 11.66 |
| 4' x 4' | 16.000 | L.F. | .233 | 3.36 | 8.30 | 11.66 |
| 4' x 5' | 18.000 | L.F. | .262 | 3.78 | 9.35 | 13.13 |
| 4' x 6' | 20.000 | L.F. | .291 | 4.20 | 10.40 | 14.60 |
| 4' x 8' | 24.000 | L.F. | .349 | 5.05 | 12.50 | 17.55 |
| 6'-8" x 2'-8" | 19.000 | L.F. | .276 | 3.99 | 9.90 | 13.89 |
| Drywall, 1/2" thick, standard, opening 3' x 2' | 5.000 | S.F. | .040 | 1.15 | 1.40 | 2.55 |
| 3' x 3' | 6.000 | S.F. | .048 | 1.38 | 1.68 | 3.06 |
| 3' x 5' | 8.000 | S.F. | .064 | 1.84 | 2.24 | 4.08 |
| 4' x 4' | 8.000 | S.F. | .064 | 1.84 | 2.24 | 4.08 |
| 4' x 5' | 9.000 | S.F. | .072 | 2.07 | 2.52 | 4.59 |
| 4' x 6' | 10.000 | S.F. | .080 | 2.30 | 2.80 | 5.10 |
| 4' x 8' | 12.000 | S.F. | .096 | 2.76 | 3.36 | 6.12 |
| 6'-8" x 2' | 9.500 | S.F. | .076 | 2.19 | 2.66 | 4.85 |
| Corner bead, 1" x 1", galvanized steel, opening 3' x 2' | 7.000 | L.F. | .140 | .77 | 4.97 | 5.74 |
| 3' x 3' | 9.000 | L.F. | .180 | .99 | 6.40 | 7.39 |
| 3' x 5' | 11.000 | L.F. | .220 | 1.21 | 7.80 | 9.01 |
| 4' x 4' | 12.000 | L.F. | .240 | 1.32 | 8.50 | 9.82 |
| 4' x 5' | 13.000 | L.F. | .260 | 1.43 | 9.25 | 10.68 |
| 4' x 6' | 14.000 | L.F. | .280 | 1.54 | 9.95 | 11.49 |
| 4' x 8' | 16.000 | L.F. | .320 | 1.76 | 11.35 | 13.11 |
| 6'-8" x 2' | 15.000 | L.F. | .300 | 1.65 | 10.65 | 12.30 |
| Tape and finish corners, inside and outside, opening 3' x 2' | 14.000 | L.F. | .204 | .98 | 7.30 | 8.28 |
| 3' x 3' | 18.000 | L.F. | .262 | 1.26 | 9.35 | 10.61 |
| 3' x 5' | 22.000 | L.F. | .320 | 1.54 | 11.45 | 12.99 |
| 4' x 4' | 24.000 | L.F. | .349 | 1.68 | 12.50 | 14.18 |
| 4' x 5' | 26.000 | L.F. | .378 | 1.82 | 13.50 | 15.32 |
| 4' x 6' | 28.000 | L.F. | .407 | 1.96 | 14.55 | 16.51 |
| 4' x 8' | 32.000 | L.F. | .466 | 2.24 | 16.65 | 18.89 |
| 6'-8" x 2' | 30.000 | L.F. | .437 | 2.10 | 15.60 | 17.70 |
| Sill, slate, 2' long | 2.000 | L.F. | .400 | 16.30 | 12.50 | 28.80 |
| 3' long | 3.000 | L.F. | .600 | 24.50 | 18.75 | 43.25 |
| 4' long | 4.000 | L.F. | .800 | 32.50 | 25.00 | 57.50 |
| Wood, 1-5/8" x 5-1/8", 2' long | 2.000 | L.F. | .128 | 5.20 | 4.56 | 9.76 |
| 3' long | 3.000 | L.F. | .192 | 7.75 | 6.85 | 14.60 |
| 4' long | 4.000 | L.F. | .256 | 10.35 | 9.10 | 19.45 |

Aluminum Window → 

Aluminum Door ←

| System Description | QUAN. | UNIT | LABOR HOURS | COST EACH | | |
|---|---|---|---|---|---|---|
| | | | | MAT. | INST. | TOTAL |
| Storm door, aluminum, combination, storm & screen, anodized, 2'-6" x 6'-8" | 1.000 | Ea. | 1.067 | 154.00 | 38.00 | 192.00 |
| 2'-8" x 6'-8" | 1.000 | Ea. | 1.143 | 185.00 | 40.50 | 225.50 |
| 3'-0" x 6'-8" | 1.000 | Ea. | 1.143 | 185.00 | 40.50 | 225.50 |
| Mill finish, 2'-6" x 6'-8" | 1.000 | Ea. | 1.067 | 217.00 | 38.00 | 255.00 |
| 2'-8" x 6'-8" | 1.000 | Ea. | 1.143 | 217.00 | 40.50 | 257.50 |
| 3'-0" x 6'-8" | 1.000 | Ea. | 1.143 | 236.00 | 40.50 | 276.50 |
| Painted, 2'-6" x 6'-8" | 1.000 | Ea. | 1.067 | 212.00 | 38.00 | 250.00 |
| 2'-8" x 6'-8" | 1.000 | Ea. | 1.143 | 220.00 | 40.50 | 260.50 |
| 3'-0" x 6'-8" | 1.000 | Ea. | 1.143 | 227.00 | 40.50 | 267.50 |
| Wood, combination, storm & screen, crossbuck, 2'-6" x 6'-9" | 1.000 | Ea. | 1.455 | 242.00 | 52.00 | 294.00 |
| 2'-8" x 6'-9" | 1.000 | Ea. | 1.600 | 245.00 | 57.00 | 302.00 |
| 3'-0" x 6'-9" | 1.000 | Ea. | 1.778 | 254.00 | 63.50 | 317.50 |
| Full lite, 2'-6" x 6'-9" | 1.000 | Ea. | 1.455 | 245.00 | 52.00 | 297.00 |
| 2'-8" x 6'-9" | 1.000 | Ea. | 1.600 | 245.00 | 57.00 | 302.00 |
| 3'-0" x 6'-9" | 1.000 | Ea. | 1.778 | 254.00 | 63.50 | 317.50 |
| | | | | | | |
| Windows, aluminum, combination storm & screen, basement, 1'-10" x 1'-0" | 1.000 | Ea. | .533 | 28.50 | 19.00 | 47.50 |
| 2'-9" x 1'-6" | 1.000 | Ea. | .533 | 31.00 | 19.00 | 50.00 |
| 3'-4" x 2'-0" | 1.000 | Ea. | .533 | 37.50 | 19.00 | 56.50 |
| Double hung, anodized, 2'-0" x 3'-5" | 1.000 | Ea. | .533 | 74.00 | 19.00 | 93.00 |
| 2'-6" x 5'-0" | 1.000 | Ea. | .571 | 99.00 | 20.50 | 119.50 |
| 4'-0" x 6'-0" | 1.000 | Ea. | .640 | 210.00 | 23.00 | 233.00 |
| Painted, 2'-0" x 3'-5" | 1.000 | Ea. | .533 | 88.00 | 19.00 | 107.00 |
| 2'-6" x 5'-0" | 1.000 | Ea. | .571 | 142.00 | 20.50 | 162.50 |
| 4'-0" x 6'-0" | 1.000 | Ea. | .640 | 254.00 | 23.00 | 277.00 |
| | | | | | | |
| Fixed window, anodized, 4'-6" x 4'-6" | 1.000 | Ea. | .640 | 113.00 | 23.00 | 136.00 |
| 5'-8" x 4'-6" | 1.000 | Ea. | .800 | 128.00 | 28.50 | 156.50 |
| Painted, 4'-6" x 4'-6" | 1.000 | Ea. | .640 | 113.00 | 23.00 | 136.00 |
| 5'-8" x 4'-6" | 1.000 | Ea. | .800 | 128.00 | 28.50 | 156.50 |

EXTERIOR WALLS 4

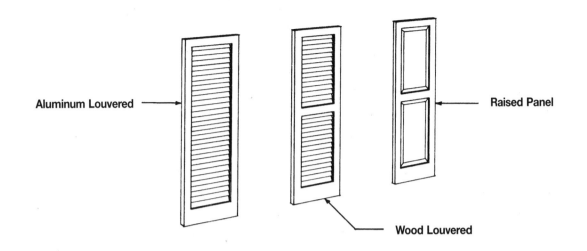

Aluminum Louvered →

← Raised Panel

Wood Louvered

| System Description | QUAN. | UNIT | LABOR HOURS | COST PER PAIR | | |
|---|---|---|---|---|---|---|
| | | | | MAT. | INST. | TOTAL |
| Shutters, exterior blinds, aluminum, louvered, 1'-4" wide, 3"-0" long | 1.000 | Set | .800 | 43.50 | 28.50 | 72.00 |
| 4'-0" long | 1.000 | Set | .800 | 52.00 | 28.50 | 80.50 |
| 5'-4" long | 1.000 | Set | .800 | 68.50 | 28.50 | 97.00 |
| 6'-8" long | 1.000 | Set | .889 | 87.00 | 31.50 | 118.50 |
| Wood, louvered, 1'-2" wide, 3'-3" long | 1.000 | Set | .800 | 82.50 | 28.50 | 111.00 |
| 4'-7" long | 1.000 | Set | .800 | 111.00 | 28.50 | 139.50 |
| 5'-3" long | 1.000 | Set | .800 | 126.00 | 28.50 | 154.50 |
| 1'-6" wide, 3'-3" long | 1.000 | Set | .800 | 87.50 | 28.50 | 116.00 |
| 4'-7" long | 1.000 | Set | .800 | 123.00 | 28.50 | 151.50 |
| Polystyrene, solid raised panel, 3'-3" wide, 3'-0" long | 1.000 | Set | .800 | 172.00 | 28.50 | 200.50 |
| 3'-11" long | 1.000 | Set | .800 | 183.00 | 28.50 | 211.50 |
| 5'-3" long | 1.000 | Set | .800 | 240.00 | 28.50 | 268.50 |
| 6'-8" long | 1.000 | Set | .889 | 270.00 | 31.50 | 301.50 |
| Polystyrene, louvered, 1'-2" wide, 3'-3" long | 1.000 | Set | .800 | 44.00 | 28.50 | 72.50 |
| 4'-7" long | 1.000 | Set | .800 | 54.50 | 28.50 | 83.00 |
| 5'-3" long | 1.000 | Set | .800 | 58.50 | 28.50 | 87.00 |
| 6'-8" long | 1.000 | Set | .889 | 95.50 | 31.50 | 127.00 |
| Vinyl, louvered, 1'-2" wide, 4'-7" long | 1.000 | Set | .720 | 51.50 | 25.50 | 77.00 |
| 1'-4" x 6'-8" long | 1.000 | Set | .889 | 87.00 | 31.50 | 118.50 |
| | | | | | | |
| | | | | | | |
| | | | | | | |
| | | | | | | |
| | | | | | | |
| | | | | | | |
| | | | | | | |
| | | | | | | |
| | | | | | | |
| | | | | | | |
| | | | | | | |
| | | | | | | |

**4 EXTERIOR WALLS**

**For information about Means Estimating Seminars, see yellow pages 11 and 12 in back of book**

# Division 5
# Roofing

Labels: Ridge Shingles, Building Paper, Shingles, Rake Board, Drip Edge, Gutter, Soffit & Fascia, Downspouts

| System Description | QUAN. | UNIT | LABOR HOURS | COST PER S.F. | | |
|---|---|---|---|---|---|---|
| | | | | MAT. | INST. | TOTAL |
| **ASPHALT, ROOF SHINGLES, CLASS A** | | | | | | |
| Shingles, inorganic class A, 210-235 lb./sq., 4/12 pitch | 1.160 | S.F. | .017 | .37 | .59 | .96 |
| Drip edge, metal, 5″ wide | .150 | L.F. | .003 | .04 | .11 | .15 |
| Building paper, #15 felt | 1.300 | S.F. | .002 | .03 | .06 | .09 |
| Ridge shingles, asphalt | .042 | L.F. | .001 | .03 | .03 | .06 |
| Soffit & fascia, white painted aluminum, 1′ overhang | .083 | L.F. | .012 | .20 | .43 | .63 |
| Rake trim, 1″ x 6″ | .040 | L.F. | .002 | .03 | .06 | .09 |
| Rake trim, prime and paint | .040 | L.F. | .002 | .01 | .06 | .07 |
| Gutter, seamless, aluminum painted | .083 | L.F. | .006 | .10 | .22 | .32 |
| Downspouts, aluminum painted | .035 | L.F. | .002 | .04 | .06 | .10 |
| TOTAL | | | .047 | .85 | 1.62 | 2.47 |
| **WOOD, CEDAR SHINGLES NO. 1 PERFECTIONS, 18″ LONG** | | | | | | |
| Shingles, wood, cedar, No. 1 perfections, 4/12 pitch | 1.160 | S.F. | .035 | 2.11 | 1.25 | 3.36 |
| Drip edge, metal, 5″ wide | .150 | L.F. | .003 | .04 | .11 | .15 |
| Building paper, #15 felt | 1.300 | S.F. | .002 | .03 | .06 | .09 |
| Ridge shingles, cedar | .042 | L.F. | .001 | .11 | .04 | .15 |
| Soffit & fascia, white painted aluminum, 1′ overhang | .083 | L.F. | .012 | .20 | .43 | .63 |
| Rake trim, 1″ x 6″ | .040 | L.F. | .002 | .03 | .06 | .09 |
| Rake trim, prime and paint | .040 | L.F. | .002 | .01 | .06 | .07 |
| Gutter, seamless, aluminum, painted | .083 | L.F. | .006 | .10 | .22 | .32 |
| Downspouts, aluminum, painted | .035 | L.F. | .002 | .04 | .06 | .10 |
| TOTAL | | | .065 | 2.67 | 2.29 | 4.96 |

The prices in these systems are based on a square foot of plan area.
All quantities have been adjusted accordingly.

| Description | QUAN. | UNIT | LABOR HOURS | COST PER S.F. | | |
|---|---|---|---|---|---|---|
| | | | | MAT. | INST. | TOTAL |
| | | | | | | |
| | | | | | | |
| | | | | | | |
| | | | | | | |
| | | | | | | |

ROOFING 5

| Gable End Roofing Price Sheet | QUAN. | UNIT | LABOR HOURS | COST PER S.F. | | |
|---|---|---|---|---|---|---|
| | | | | MAT. | INST. | TOTAL |
| Shingles, asphalt, inorganic, class A, 210-235 lb./sq., 4/12 pitch | 1.160 | S.F. | .017 | .37 | .59 | .96 |
| 8/12 pitch | 1.330 | S.F. | .019 | .40 | .64 | 1.04 |
| Laminated, multi-layered, 240-260 lb./sq., 4/12 pitch | 1.160 | S.F. | .021 | .52 | .72 | 1.24 |
| 8/12 pitch | 1.330 | S.F. | .023 | .57 | .78 | 1.35 |
| Premium laminated, multi-layered, 260-300 lb./sq., 4/12 pitch | 1.160 | S.F. | .027 | .65 | .93 | 1.58 |
| 8/12 pitch | 1.330 | S.F. | .030 | .71 | 1.01 | 1.72 |
| Clay tile, Spanish tile, red, 4/12 pitch | 1.160 | S.F. | .053 | 3.96 | 1.81 | 5.77 |
| 8/12 pitch | 1.330 | S.F. | .058 | 4.29 | 1.96 | 6.25 |
| Mission tile, red, 4/12 pitch | 1.160 | S.F. | .083 | 8.35 | 2.83 | 11.18 |
| 8/12 pitch | 1.330 | S.F. | .090 | 9.05 | 3.07 | 12.12 |
| French tile, red, 4/12 pitch | 1.160 | S.F. | .071 | 7.55 | 2.41 | 9.96 |
| 8/12 pitch | 1.330 | S.F. | .077 | 8.20 | 2.61 | 10.81 |
| Slate, Buckingham, Virginia, black, 4/12 pitch | 1.160 | S.F. | .055 | 7.15 | 1.86 | 9.01 |
| 8/12 pitch | 1.330 | S.F. | .059 | 7.75 | 2.02 | 9.77 |
| Vermont, black or grey, 4/12 pitch | 1.160 | S.F. | .055 | 4.56 | 1.86 | 6.42 |
| 8/12 pitch | 1.330 | S.F. | .059 | 4.94 | 2.02 | 6.96 |
| Wood, No. 1 red cedar, 5X, 16" long, 5" exposure, 4/12 pitch | 1.160 | S.F. | .038 | 1.99 | 1.37 | 3.36 |
| 8/12 pitch | 1.330 | S.F. | .042 | 2.16 | 1.48 | 3.64 |
| Fire retardant, 4/12 pitch | 1.160 | S.F. | .038 | 2.39 | 1.37 | 3.76 |
| 8/12 pitch | 1.330 | S.F. | .042 | 2.59 | 1.48 | 4.07 |
| 18" long, No.1 perfections, 5" exposure, 4/12 pitch | 1.160 | S.F. | .035 | 2.11 | 1.25 | 3.36 |
| 8/12 pitch | 1.330 | S.F. | .038 | 2.29 | 1.35 | 3.64 |
| Fire retardant, 4/12 pitch | 1.160 | S.F. | .035 | 2.49 | 1.25 | 3.74 |
| 8/12 pitch | 1.330 | S.F. | .038 | 2.70 | 1.35 | 4.05 |
| Resquared & rebutted, 18" long, 6" exposure, 4/12 pitch | 1.160 | S.F. | .032 | 2.63 | 1.14 | 3.77 |
| 8/12 pitch | 1.330 | S.F. | .035 | 2.85 | 1.24 | 4.09 |
| Fire retardant, 4/12 pitch | 1.160 | S.F. | .032 | 3.01 | 1.14 | 4.15 |
| 8/12 pitch | 1.330 | S.F. | .035 | 3.26 | 1.24 | 4.50 |
| Wood shakes hand split, 24" long, 10" exposure, 4/12 pitch | 1.160 | S.F. | .038 | 1.82 | 1.37 | 3.19 |
| 8/12 pitch | 1.330 | S.F. | .042 | 1.98 | 1.48 | 3.46 |
| Fire retardant, 4/12 pitch | 1.160 | S.F. | .038 | 2.22 | 1.37 | 3.59 |
| 8/12 pitch | 1.330 | S.F. | .042 | 2.41 | 1.48 | 3.89 |
| 18" long, 8" exposure, 4/12 pitch | 1.160 | S.F. | .048 | 1.28 | 1.70 | 2.98 |
| 8/12 pitch | 1.330 | S.F. | .052 | 1.39 | 1.85 | 3.24 |
| Fire retardant, 4/12 pitch | 1.160 | S.F. | .048 | 1.68 | 1.70 | 3.38 |
| 8/12 pitch | 1.330 | S.F. | .052 | 1.82 | 1.85 | 3.67 |
| Drip edge, metal, 5" wide | .150 | L.F. | .003 | .04 | .11 | .15 |
| 8" wide | .150 | L.F. | .003 | .05 | .11 | .16 |
| Building paper, #15 asphalt felt | 1.300 | S.F. | .002 | .03 | .06 | .09 |
| Ridge shingles, asphalt | .042 | L.F. | .001 | .03 | .03 | .06 |
| Clay | .042 | L.F. | .002 | .40 | .06 | .46 |
| Slate | .042 | L.F. | .002 | .39 | .06 | .45 |
| Wood, shingles | .042 | L.F. | .001 | .11 | .04 | .15 |
| Shakes | .042 | L.F. | .001 | .11 | .04 | .15 |
| Soffit & fascia, aluminum, vented, 1' overhang | .083 | L.F. | .012 | .20 | .43 | .63 |
| 2' overhang | .083 | L.F. | .013 | .29 | .47 | .76 |
| Vinyl, vented, 1' overhang | .083 | L.F. | .011 | .14 | .39 | .53 |
| 2' overhang | .083 | L.F. | .012 | .19 | .43 | .62 |
| Wood, board fascia, plywood soffit, 1' overhang | .083 | L.F. | .004 | .02 | .12 | .14 |
| 2' overhang | .083 | L.F. | .006 | .03 | .18 | .21 |
| Rake trim, painted, 1" x 6" | .040 | L.F. | .004 | .04 | .12 | .16 |
| 1" x 8" | .040 | L.F. | .004 | .07 | .13 | .20 |
| Gutter, 5" box, aluminum, seamless, painted | .083 | L.F. | .006 | .10 | .22 | .32 |
| Vinyl | .083 | L.F. | .006 | .09 | .21 | .30 |
| Downspout, 2" x 3", aluminum, one story house | .035 | L.F. | .001 | .04 | .06 | .10 |
| Two story house | .060 | L.F. | .003 | .06 | .10 | .16 |
| Vinyl, one story house | .035 | L.F. | .002 | .04 | .06 | .10 |
| Two story house | .060 | L.F. | .003 | .06 | .10 | .16 |

**5  ROOFING**

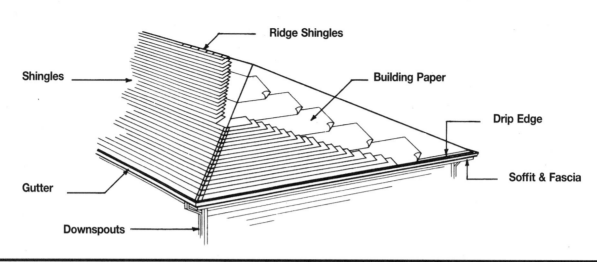

| System Description | QUAN. | UNIT | LABOR HOURS | COST PER S.F. | | |
|---|---|---|---|---|---|---|
| | | | | MAT. | INST. | TOTAL |
| **ASPHALT, ROOF SHINGLES, CLASS A** | | | | | | |
| Shingles, inorganic, class A, 210-235 lb./sq. 4/12 pitch | 1.570 | S.F. | .023 | .50 | .78 | 1.28 |
| Drip edge, metal, 5″ wide | .122 | L.F. | .002 | .03 | .09 | .12 |
| Building paper, #15 asphalt felt | 1.800 | S.F. | .002 | .04 | .08 | .12 |
| Ridge shingles, asphalt | .075 | L.F. | .002 | .06 | .06 | .12 |
| Soffit & fascia, white painted aluminum, 1′ overhang | .120 | L.F. | .017 | .29 | .62 | .91 |
| Gutter, seamless, aluminum, painted | .120 | L.F. | .008 | .14 | .31 | .45 |
| Downspouts, aluminum, painted | .035 | L.F. | .002 | .04 | .06 | .10 |
| | | | | | | |
| TOTAL | | | .056 | 1.10 | 2.00 | 3.10 |
| **WOOD, CEDAR SHINGLES, NO. 1 PERFECTIONS, 18″ LONG** | | | | | | |
| Shingles, red cedar, No. 1 perfections, 5″ exp., 4/12 pitch | 1.570 | S.F. | .047 | 2.82 | 1.66 | 4.48 |
| Drip edge, metal, 5″ wide | .122 | L.F. | .002 | .03 | .09 | .12 |
| Building paper, #15 asphalt felt | 1.800 | S.F. | .002 | .04 | .08 | .12 |
| Ridge shingles, wood, cedar | .075 | L.F. | .002 | .20 | .08 | .28 |
| Soffit & fascia, white painted aluminum, 1′ overhang | .120 | L.F. | .017 | .29 | .62 | .91 |
| Gutter, seamless, aluminum, painted | .120 | L.F. | .008 | .14 | .31 | .45 |
| Downspouts, aluminum, painted | .035 | L.F. | .002 | .04 | .06 | .10 |
| | | | | | | |
| TOTAL | | | .080 | 3.56 | 2.90 | 6.46 |

The prices in these systems are based on a square foot of plan area.
All quantities have been adjusted accordingly.

| Description | QUAN. | UNIT | LABOR HOURS | COST PER S.F. | | |
|---|---|---|---|---|---|---|
| | | | | MAT. | INST. | TOTAL |
| | | | | | | |
| | | | | | | |
| | | | | | | |
| | | | | | | |
| | | | | | | |

# Hip Roof - Roofing Price Sheet

| | QUAN. | UNIT | LABOR HOURS | COST PER S.F. MAT. | COST PER S.F. INST. | COST PER S.F. TOTAL |
|---|---|---|---|---|---|---|
| Shingles, asphalt, inorganic, class A, 210-235 lb./sq., 4/12 pitch | 1.570 | S.F. | .023 | .50 | .78 | 1.28 |
| 8/12 pitch | 1.850 | S.F. | .028 | .59 | .93 | 1.52 |
| Laminated, multi-layered, 240-260 lb./sq., 4/12 pitch | 1.570 | S.F. | .028 | .70 | .96 | 1.66 |
| 8/12 pitch | 1.850 | S.F. | .034 | .83 | 1.14 | 1.97 |
| Prem. laminated, multi-layered, 260-300 lb./sq., 4/12 pitch | 1.570 | S.F. | .037 | .87 | 1.24 | 2.11 |
| 8/12 pitch | 1.850 | S.F. | .043 | 1.04 | 1.47 | 2.51 |
| Clay tile, Spanish tile, red, 4/12 pitch | 1.570 | S.F. | .071 | 5.30 | 2.42 | 7.72 |
| 8/12 pitch | 1.850 | S.F. | .084 | 6.25 | 2.87 | 9.12 |
| Mission tile, red, 4/12 pitch | 1.570 | S.F. | .111 | 11.10 | 3.78 | 14.88 |
| 8/12 pitch | 1.850 | S.F. | .132 | 13.20 | 4.48 | 17.68 |
| French tile, red, 4/12 pitch | 1.570 | S.F. | .095 | 10.10 | 3.22 | 13.32 |
| 8/12 pitch | 1.850 | S.F. | .113 | 11.95 | 3.82 | 15.77 |
| Slate, Buckingham, Virginia, black, 4/12 pitch | 1.570 | S.F. | .073 | 9.50 | 2.48 | 11.98 |
| 8/12 pitch | 1.850 | S.F. | .087 | 11.30 | 2.95 | 14.25 |
| Vermont, black or grey, 4/12 pitch | 1.570 | S.F. | .073 | 6.10 | 2.48 | 8.58 |
| 8/12 pitch | 1.850 | S.F. | .087 | 7.20 | 2.95 | 10.15 |
| Wood, red cedar, No.1 5X, 16" long, 5" exposure, 4/12 pitch | 1.570 | S.F. | .051 | 2.66 | 1.82 | 4.48 |
| 8/12 pitch | 1.850 | S.F. | .061 | 3.15 | 2.17 | 5.32 |
| Fire retardant, 4/12 pitch | 1.570 | S.F. | .051 | 3.19 | 1.82 | 5.01 |
| 8/12 pitch | 1.850 | S.F. | .061 | 3.78 | 2.17 | 5.95 |
| 18" long, No.1 perfections, 5" exposure, 4/12 pitch | 1.570 | S.F. | .047 | 2.82 | 1.66 | 4.48 |
| 8/12 pitch | 1.850 | S.F. | .055 | 3.34 | 1.98 | 5.32 |
| Fire retardant, 4/12 pitch | 1.570 | S.F. | .047 | 3.32 | 1.66 | 4.98 |
| 8/12 pitch | 1.850 | S.F. | .055 | 3.94 | 1.98 | 5.92 |
| Resquared & rebutted, 18" long, 6" exposure, 4/12 pitch | 1.570 | S.F. | .043 | 3.50 | 1.52 | 5.02 |
| 8/12 pitch | 1.850 | S.F. | .051 | 4.16 | 1.81 | 5.97 |
| Fire retardant, 4/12 pitch | 1.570 | S.F. | .043 | 4.00 | 1.52 | 5.52 |
| 8/12 pitch | 1.850 | S.F. | .051 | 4.76 | 1.81 | 6.57 |
| Wood shakes hand split, 24" long, 10" exposure, 4/12 pitch | 1.570 | S.F. | .051 | 2.43 | 1.82 | 4.25 |
| 8/12 pitch | 1.850 | S.F. | .061 | 2.89 | 2.17 | 5.06 |
| Fire retardant, 4/12 pitch | 1.570 | S.F. | .051 | 2.96 | 1.82 | 4.78 |
| 8/12 pitch | 1.850 | S.F. | .061 | 3.52 | 2.17 | 5.69 |
| 18" long, 8" exposure, 4/12 pitch | 1.570 | S.F. | .064 | 1.71 | 2.27 | 3.98 |
| 8/12 pitch | 1.850 | S.F. | .076 | 2.03 | 2.70 | 4.73 |
| Fire retardant, 4/12 pitch | 1.570 | S.F. | .064 | 2.24 | 2.27 | 4.51 |
| 8/12 pitch | 1.850 | S.F. | .076 | 2.66 | 2.70 | 5.36 |
| Drip edge, metal, 5" wide | .122 | L.F. | .002 | .03 | .09 | .12 |
| 8" wide | .122 | L.F. | .002 | .04 | .09 | .13 |
| Building paper, #15 asphalt felt | 1.800 | S.F. | .002 | .04 | .08 | .12 |
| Ridge shingles, asphalt | .075 | L.F. | .002 | .06 | .06 | .12 |
| Clay | .075 | L.F. | .003 | .71 | .10 | .81 |
| Slate | .075 | L.F. | .003 | .70 | .10 | .80 |
| Wood, shingles | .075 | L.F. | .002 | .20 | .08 | .28 |
| Shakes | .075 | L.F. | .002 | .20 | .08 | .28 |
| Soffit & fascia, aluminum, vented, 1' overhang | .120 | L.F. | .017 | .29 | .62 | .91 |
| 2' overhang | .120 | L.F. | .019 | .42 | .68 | 1.10 |
| Vinyl, vented, 1' overhang | .120 | L.F. | .016 | .20 | .57 | .77 |
| 2' overhang | .120 | L.F. | .017 | .28 | .62 | .90 |
| Wood, board fascia, plywood soffit, 1' overhang | .120 | L.F. | .004 | .02 | .12 | .14 |
| 2' overhang | .120 | L.F. | .006 | .03 | .18 | .21 |
| Gutter, 5" box, aluminum, seamless, painted | .120 | L.F. | .008 | .14 | .31 | .45 |
| Vinyl | .120 | L.F. | .009 | .13 | .31 | .44 |
| Downspout, 2" x 3", aluminum, one story house | .035 | L.F. | .002 | .04 | .06 | .10 |
| Two story house | .060 | L.F. | .003 | .06 | .10 | .16 |
| Vinyl, one story house | .035 | L.F. | .001 | .04 | .06 | .10 |
| Two story house | .060 | L.F. | .003 | .06 | .10 | .16 |

**5 ROOFING**

Labeled diagram: Shingles, Ridge Shingles, Building Paper, Rake Boards, Soffit, Drip Edge

| System Description | QUAN. | UNIT | LABOR HOURS | COST PER S.F. | | |
|---|---|---|---|---|---|---|
| | | | | MAT. | INST. | TOTAL |
| **ASPHALT, ROOF SHINGLES, CLASS A** | | | | | | |
| Shingles, asphalt, inorganic, class A, 210-235 lb./sq. | 1.450 | S.F. | .022 | .47 | .74 | 1.21 |
| Drip edge, metal, 5" wide | .146 | L.F. | .003 | .04 | .10 | .14 |
| Building paper, #15 asphalt felt | 1.500 | S.F. | .002 | .04 | .07 | .11 |
| Ridge shingles, asphalt | .042 | L.F. | .001 | .03 | .03 | .06 |
| Soffit & fascia, painted aluminum, 1' overhang | .083 | L.F. | .012 | .20 | .43 | .63 |
| Rake trim, 1" x 6" | .063 | L.F. | .003 | .05 | .09 | .14 |
| Rake trim, prime and paint | .063 | L.F. | .003 | .02 | .09 | .11 |
| Gutter, seamless, aluminum, painted | .083 | L.F. | .006 | .10 | .22 | .32 |
| Downspouts, aluminum, painted | .042 | L.F. | .002 | .05 | .07 | .12 |
| TOTAL | | | .054 | 1.00 | 1.84 | 2.84 |
| **WOOD, CEDAR SHINGLES, NO. 1 PERFECTIONS, 18" LONG** | | | | | | |
| Shingles, wood, red cedar, No. 1 perfections, 5" exposure | 1.450 | S.F. | .044 | 2.64 | 1.56 | 4.20 |
| Drip edge, metal, 5" wide | .146 | L.F. | .003 | .04 | .10 | .14 |
| Building paper, #15 asphalt felt | 1.500 | S.F. | .002 | .04 | .07 | .11 |
| Ridge shingles, wood | .042 | L.F. | .001 | .11 | .04 | .15 |
| Soffit & fascia, white painted aluminum, 1' overhang | .083 | L.F. | .012 | .20 | .43 | .63 |
| Rake trim, 1" x 6" | .063 | L.F. | .003 | .05 | .09 | .14 |
| Rake trim, prime and paint | .063 | L.F. | .001 | .01 | .04 | .05 |
| Gutter, seamless, aluminum, painted | .083 | L.F. | .006 | .10 | .22 | .32 |
| Downspouts, aluminum, painted | .042 | L.F. | .002 | .05 | .07 | .12 |
| TOTAL | | | .074 | 3.24 | 2.62 | 5.86 |

The prices in this system are based on a square foot of plan area.
All quantities have been adjusted accordingly.

| Description | QUAN. | UNIT | LABOR HOURS | COST PER S.F. | | |
|---|---|---|---|---|---|---|
| | | | | MAT. | INST. | TOTAL |
| | | | | | | |
| | | | | | | |
| | | | | | | |
| | | | | | | |

ROOFING 5

| Gambrel Roofing Price Sheet | QUAN. | UNIT | LABOR HOURS | COST PER S.F. | | |
|---|---|---|---|---|---|---|
| | | | | MAT. | INST. | TOTAL |
| Shingles, asphalt, standard, inorganic, class A, 210-235 lb./sq. | 1.450 | S.F. | .022 | .47 | .74 | 1.21 |
| Laminated, multi-layered, 240-260 lb./sq. | 1.450 | S.F. | .027 | .65 | .90 | 1.55 |
| Premium laminated, multi-layered, 260-300 lb./sq. | 1.450 | S.F. | .034 | .82 | 1.16 | 1.98 |
| Slate, Buckingham, Virginia, black | 1.450 | S.F. | .069 | 8.95 | 2.33 | 11.28 |
| Vermont, black or grey | 1.450 | S.F. | .069 | 5.70 | 2.33 | 8.03 |
| Wood, red cedar, No.1 5X, 16" long, 5" exposure, plain | 1.450 | S.F. | .048 | 2.49 | 1.71 | 4.20 |
| Fire retardant | 1.450 | S.F. | .048 | 2.99 | 1.71 | 4.70 |
| 18" long, No.1 perfections, 6" exposure, plain | 1.450 | S.F. | .044 | 2.64 | 1.56 | 4.20 |
| Fire retardant | 1.450 | S.F. | .044 | 3.11 | 1.56 | 4.67 |
| Resquared & rebutted, 18" long, 6" exposure, plain | 1.450 | S.F. | .040 | 3.29 | 1.43 | 4.72 |
| Fire retardant | 1.450 | S.F. | .040 | 3.75 | 1.43 | 5.18 |
| Shakes, hand split, 24" long, 10" exposure, plain | 1.450 | S.F. | .048 | 2.28 | 1.71 | 3.99 |
| Fire retardant | 1.450 | S.F. | .048 | 2.78 | 1.71 | 4.49 |
| 18" long, 8" exposure, plain | 1.450 | S.F. | .060 | 1.61 | 2.13 | 3.74 |
| Fire retardant | 1.450 | S.F. | .060 | 2.11 | 2.13 | 4.24 |
| Drip edge, metal, 5" wide | .146 | L.F. | .003 | .04 | .10 | .14 |
| 8" wide | .146 | L.F. | .003 | .05 | .10 | .15 |
| Building paper, #15 asphalt felt | 1.500 | S.F. | .002 | .04 | .07 | .11 |
| Ridge shingles, asphalt | .042 | L.F. | .001 | .03 | .03 | .06 |
| Slate | .042 | L.F. | .002 | .39 | .06 | .45 |
| Wood, shingles | .042 | L.F. | .001 | .11 | .04 | .15 |
| Shakes | .042 | L.F. | .001 | .11 | .04 | .15 |
| Soffit & fascia, aluminum, vented, 1' overhang | .083 | L.F. | .012 | .20 | .43 | .63 |
| 2' overhang | .083 | L.F. | .013 | .29 | .47 | .76 |
| Vinyl vented, 1' overhang | .083 | L.F. | .011 | .14 | .39 | .53 |
| 2' overhang | .083 | L.F. | .012 | .19 | .43 | .62 |
| Wood board fascia, plywood soffit, 1' overhang | .083 | L.F. | .004 | .02 | .12 | .14 |
| 2' overhang | .083 | L.F. | .006 | .03 | .18 | .21 |
| Rake trim, painted, 1" x 6" | .063 | L.F. | .006 | .07 | .18 | .25 |
| 1" x 8" | .063 | L.F. | .007 | .09 | .24 | .33 |
| Gutter, 5" box, aluminum, seamless, painted | .083 | L.F. | .006 | .10 | .22 | .32 |
| Vinyl | .083 | L.F. | .006 | .09 | .21 | .30 |
| Downspout 2" x 3", aluminum, one story house | .042 | L.F. | .002 | .04 | .07 | .11 |
| Two story house | .070 | L.F. | .003 | .07 | .11 | .18 |
| Vinyl, one story house | .042 | L.F. | .002 | .04 | .07 | .11 |
| Two story house | .070 | L.F. | .003 | .07 | .11 | .18 |

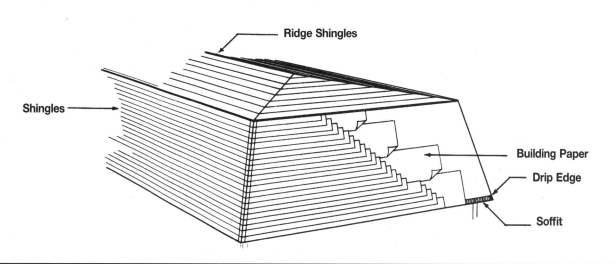

| System Description | QUAN. | UNIT | LABOR HOURS | COST PER S.F. | | |
|---|---|---|---|---|---|---|
| | | | | MAT. | INST. | TOTAL |
| **ASPHALT, ROOF SHINGLES, CLASS A** | | | | | | |
| Shingles, standard inorganic class A 210-235 lb./sq. | 2.210 | S.F. | .032 | .68 | 1.08 | 1.76 |
| Drip edge, metal, 5" wide | .122 | L.F. | .002 | .03 | .09 | .12 |
| Building paper, #15 asphalt felt | 2.300 | S.F. | .003 | .05 | .11 | .16 |
| Ridge shingles, asphalt | .090 | L.F. | .002 | .07 | .07 | .14 |
| Soffit & fascia, white painted aluminum, 1' overhang | .122 | L.F. | .018 | .29 | .63 | .92 |
| Gutter, seamless, aluminum, painted | .122 | L.F. | .008 | .15 | .32 | .47 |
| Downspouts, aluminum, painted | .042 | L.F. | .002 | .05 | .07 | .12 |
| | | | | | | |
| TOTAL | | | .067 | 1.32 | 2.37 | 3.69 |
| **WOOD, CEDAR SHINGLES, NO. 1 PERFECTIONS, 18" LONG** | | | | | | |
| Shingles, wood, red cedar, No. 1 perfections, 5" exposure | 2.210 | S.F. | .064 | 3.87 | 2.29 | 6.16 |
| Drip edge, metal, 5" wide | .122 | L.F. | .002 | .03 | .09 | .12 |
| Building paper, #15 asphalt felt | 2.300 | S.F. | .003 | .05 | .11 | .16 |
| Ridge shingles, wood | .090 | L.F. | .003 | .24 | .09 | .33 |
| Soffit & fascia, white painted aluminum, 1' overhang | .122 | L.F. | .018 | .29 | .63 | .92 |
| Gutter, seamless, aluminum, painted | .122 | L.F. | .008 | .15 | .32 | .47 |
| Downspouts, aluminum, painted | .042 | L.F. | .002 | .05 | .07 | .12 |
| | | | | | | |
| TOTAL | | | .100 | 4.68 | 3.60 | 8.28 |

The prices in these systems are based on a square foot of plan area.
All quantities have been adjusted accordingly.

| Description | QUAN. | UNIT | LABOR HOURS | COST PER S.F. | | |
|---|---|---|---|---|---|---|
| | | | | MAT. | INST. | TOTAL |
| | | | | | | |
| | | | | | | |
| | | | | | | |
| | | | | | | |
| | | | | | | |

**Important: See the Reference Section for critical supporting data - Reference Nos., Crews & Location Factors**

ROOFING 5

| Mansard Roofing Price Sheet | QUAN. | UNIT | LABOR HOURS | COST PER S.F. | | |
|---|---|---|---|---|---|---|
| | | | | MAT. | INST. | TOTAL |
| Shingles, asphalt, standard, inorganic, class A, 210-235 lb./sq. | 2.210 | S.F. | .032 | .68 | 1.08 | 1.76 |
| Laminated, multi-layered, 240-260 lb./sq. | 2.210 | S.F. | .039 | .96 | 1.32 | 2.28 |
| Premium laminated, multi-layered, 260-300 lb./sq. | 2.210 | S.F. | .050 | 1.20 | 1.71 | 2.91 |
| Slate Buckingham, Virginia, black | 2.210 | S.F. | .101 | 13.10 | 3.41 | 16.51 |
| Vermont, black or grey | 2.210 | S.F. | .101 | 8.35 | 3.41 | 11.76 |
| Wood, red cedar, No.1 5X, 16" long, 5" exposure, plain | 2.210 | S.F. | .070 | 3.65 | 2.51 | 6.16 |
| Fire retardant | 2.210 | S.F. | .070 | 4.38 | 2.51 | 6.89 |
| 18" long, No.1 perfections 6" exposure, plain | 2.210 | S.F. | .064 | 3.87 | 2.29 | 6.16 |
| Fire retardant | 2.210 | S.F. | .064 | 4.56 | 2.29 | 6.85 |
| Resquared & rebutted, 18" long, 6" exposure, plain | 2.210 | S.F. | .059 | 4.82 | 2.09 | 6.91 |
| Fire retardant | 2.210 | S.F. | .059 | 5.50 | 2.09 | 7.59 |
| Shakes, hand split, 24" long 10" exposure, plain | 2.210 | S.F. | .070 | 3.34 | 2.51 | 5.85 |
| Fire retardant | 2.210 | S.F. | .070 | 4.07 | 2.51 | 6.58 |
| 18" long, 8" exposure, plain | 2.210 | S.F. | .088 | 2.35 | 3.12 | 5.47 |
| Fire retardant | 2.210 | S.F. | .088 | 3.08 | 3.12 | 6.20 |
| Drip edge, metal, 5" wide | .122 | S.F. | .002 | .03 | .09 | .12 |
| 8" wide | .122 | S.F. | .002 | .04 | .09 | .13 |
| Building paper, #15 asphalt felt | 2.300 | S.F. | .003 | .05 | .11 | .16 |
| Ridge shingles, asphalt | .090 | L.F. | .002 | .07 | .07 | .14 |
| Slate | .090 | L.F. | .004 | .84 | .12 | .96 |
| Wood, shingles | .090 | L.F. | .003 | .24 | .09 | .33 |
| Shakes | .090 | L.F. | .003 | .24 | .09 | .33 |
| Soffit & fascia, aluminum vented, 1' overhang | .122 | L.F. | .018 | .29 | .63 | .92 |
| 2' overhang | .122 | L.F. | .020 | .43 | .70 | 1.13 |
| Vinyl vented, 1' overhang | .122 | L.F. | .016 | .20 | .58 | .78 |
| 2' overhang | .122 | L.F. | .018 | .29 | .63 | .92 |
| Wood board fascia, plywood soffit, 1' overhang | .122 | L.F. | .013 | .20 | .44 | .64 |
| 2' overhang | .122 | L.F. | .019 | .21 | .66 | .87 |
| Gutter, 5" box, aluminum, seamless, painted | .122 | L.F. | .008 | .15 | .32 | .47 |
| Vinyl | .122 | L.F. | .009 | .13 | .32 | .45 |
| Downspout 2" x 3", aluminum, one story house | .042 | L.F. | .002 | .04 | .07 | .11 |
| Two story house | .070 | L.F. | .003 | .07 | .11 | .18 |
| Vinyl, one story house | .042 | L.F. | .002 | .04 | .07 | .11 |
| Two story house | .070 | L.F. | .003 | .07 | .11 | .18 |

| System Description | QUAN. | UNIT | LABOR HOURS | COST PER S.F. | | |
|---|---|---|---|---|---|---|
| | | | | MAT. | INST. | TOTAL |
| **ASPHALT, ROOF SHINGLES, CLASS A** | | | | | | |
| Shingles, inorganic class A 210-235 lb./sq. 4/12 pitch | 1.230 | S.F. | .019 | .40 | .64 | 1.04 |
| Drip edge, metal, 5" wide | .100 | L.F. | .002 | .02 | .07 | .09 |
| Building paper, #15 asphalt felt | 1.300 | S.F. | .002 | .03 | .06 | .09 |
| Soffit & fascia, white painted aluminum, 1' overhang | .080 | L.F. | .012 | .19 | .42 | .61 |
| Rake trim, 1" x 6" | .043 | L.F. | .002 | .03 | .06 | .09 |
| Rake trim, prime and paint | .043 | L.F. | .002 | .01 | .06 | .07 |
| Gutter, seamless, aluminum, painted | .040 | L.F. | .003 | .05 | .10 | .15 |
| Downspouts, painted aluminum | .020 | L.F. | .001 | .02 | .03 | .05 |
| TOTAL | | | .043 | .75 | 1.44 | 2.19 |
| **WOOD, CEDAR SHINGLES, NO. 1 PERFECTIONS, 18" LONG** | | | | | | |
| Shingles, red cedar, No. 1 perfections, 5" exp., 4/12 pitch | 1.230 | S.F. | .035 | 2.11 | 1.25 | 3.36 |
| Drip edge, metal, 5" wide | .100 | L.F. | .002 | .02 | .07 | .09 |
| Building paper, #15 asphalt felt | 1.300 | S.F. | .002 | .03 | .06 | .09 |
| Soffit & fascia, white painted aluminum, 1' overhang | .080 | L.F. | .012 | .19 | .42 | .61 |
| Rake trim, 1" x 6" | .043 | L.F. | .002 | .03 | .06 | .09 |
| Rake trim, prime and paint | .043 | L.F. | .001 | .01 | .03 | .04 |
| Gutter, seamless, aluminum, painted | .040 | L.F. | .003 | .05 | .10 | .15 |
| Downspouts, painted aluminum | .020 | L.F. | .001 | .02 | .03 | .05 |
| TOTAL | | | .058 | 2.46 | 2.02 | 4.48 |

The prices in these systems are based on a square foot of plan area.
All quantities have been adjusted accordingly.

| Description | QUAN. | UNIT | LABOR HOURS | COST PER S.F. | | |
|---|---|---|---|---|---|---|
| | | | | MAT. | INST. | TOTAL |
| | | | | | | |
| | | | | | | |
| | | | | | | |
| | | | | | | |

ROOFING 5

| Shed Roofing Price Sheet | QUAN. | UNIT | LABOR HOURS | COST PER S.F. | | |
|---|---|---|---|---|---|---|
| | | | | MAT. | INST. | TOTAL |
| Shingles, asphalt, inorganic, class A, 210-235 lb./sq., 4/12 pitch | 1.230 | S.F. | .017 | .37 | .59 | .96 |
| 8/12 pitch | 1.330 | S.F. | .019 | .40 | .64 | 1.04 |
| Laminated, multi-layered, 240-260 lb./sq. 4/12 pitch | 1.230 | S.F. | .021 | .52 | .72 | 1.24 |
| 8/12 pitch | 1.330 | S.F. | .023 | .57 | .78 | 1.35 |
| Premium laminated, multi-layered, 260-300 lb./sq. 4/12 pitch | 1.230 | S.F. | .027 | .65 | .93 | 1.58 |
| 8/12 pitch | 1.330 | S.F. | .030 | .71 | 1.01 | 1.72 |
| Clay tile, Spanish tile, red, 4/12 pitch | 1.230 | S.F. | .053 | 3.96 | 1.81 | 5.77 |
| 8/12 pitch | 1.330 | S.F. | .058 | 4.29 | 1.96 | 6.25 |
| Mission tile, red, 4/12 pitch | 1.230 | S.F. | .083 | 8.35 | 2.83 | 11.18 |
| 8/12 pitch | 1.330 | S.F. | .090 | 9.05 | 3.07 | 12.12 |
| French tile, red, 4/12 pitch | 1.230 | S.F. | .071 | 7.55 | 2.41 | 9.96 |
| 8/12 pitch | 1.330 | S.F. | .077 | 8.20 | 2.61 | 10.81 |
| Slate, Buckingham, Virginia, black, 4/12 pitch | 1.230 | S.F. | .055 | 7.15 | 1.86 | 9.01 |
| 8/12 pitch | 1.330 | S.F. | .059 | 7.75 | 2.02 | 9.77 |
| Vermont, black or grey, 4/12 pitch | 1.230 | S.F. | .055 | 4.56 | 1.86 | 6.42 |
| 8/12 pitch | 1.330 | S.F. | .059 | 4.94 | 2.02 | 6.96 |
| Wood, red cedar, No.1 5X, 16" long, 5" exposure, 4/12 pitch | 1.230 | S.F. | .038 | 1.99 | 1.37 | 3.36 |
| 8/12 pitch | 1.330 | S.F. | .042 | 2.16 | 1.48 | 3.64 |
| Fire retardant, 4/12 pitch | 1.230 | S.F. | .038 | 2.39 | 1.37 | 3.76 |
| 8/12 pitch | 1.330 | S.F. | .042 | 2.59 | 1.48 | 4.07 |
| 18" long, 6" exposure, 4/12 pitch | 1.230 | S.F. | .035 | 2.11 | 1.25 | 3.36 |
| 8/12 pitch | 1.330 | S.F. | .038 | 2.29 | 1.35 | 3.64 |
| Fire retardant, 4/12 pitch | 1.230 | S.F. | .035 | 2.49 | 1.25 | 3.74 |
| 8/12 pitch | 1.330 | S.F. | .038 | 2.70 | 1.35 | 4.05 |
| Resquared & rebutted, 18" long, 6" exposure, 4/12 pitch | 1.230 | S.F. | .032 | 2.63 | 1.14 | 3.77 |
| 8/12 pitch | 1.330 | S.F. | .035 | 2.85 | 1.24 | 4.09 |
| Fire retardant, 4/12 pitch | 1.230 | S.F. | .032 | 3.01 | 1.14 | 4.15 |
| 8/12 pitch | 1.330 | S.F. | .035 | 3.26 | 1.24 | 4.50 |
| Wood shakes, hand split, 24" long, 10" exposure, 4/12 pitch | 1.230 | S.F. | .038 | 1.82 | 1.37 | 3.19 |
| 8/12 pitch | 1.330 | S.F. | .042 | 1.98 | 1.48 | 3.46 |
| Fire retardant, 4/12 pitch | 1.230 | S.F. | .038 | 2.22 | 1.37 | 3.59 |
| 8/12 pitch | 1.330 | S.F. | .042 | 2.41 | 1.48 | 3.89 |
| 18" long, 8" exposure, 4/12 pitch | 1.230 | S.F. | .048 | 1.28 | 1.70 | 2.98 |
| 8/12 pitch | 1.330 | S.F. | .052 | 1.39 | 1.85 | 3.24 |
| Fire retardant, 4/12 pitch | 1.230 | S.F. | .048 | 1.68 | 1.70 | 3.38 |
| 8/12 pitch | 1.330 | S.F. | .052 | 1.82 | 1.85 | 3.67 |
| Drip edge, metal, 5" wide | .100 | L.F. | .002 | .02 | .07 | .09 |
| 8" wide | .100 | L.F. | .002 | .04 | .07 | .11 |
| Building paper, #15 asphalt felt | 1.300 | S.F. | .002 | .03 | .06 | .09 |
| Soffit & fascia, aluminum vented, 1' overhang | .080 | L.F. | .012 | .19 | .42 | .61 |
| 2' overhang | .080 | L.F. | .013 | .28 | .46 | .74 |
| Vinyl vented, 1' overhang | .080 | L.F. | .011 | .13 | .38 | .51 |
| 2' overhang | .080 | L.F. | .012 | .19 | .42 | .61 |
| Wood board fascia, plywood soffit, 1' overhang | .080 | L.F. | .010 | .14 | .32 | .46 |
| 2' overhang | .080 | L.F. | .014 | .15 | .49 | .64 |
| Rake, trim, painted, 1" x 6" | .043 | L.F. | .004 | .04 | .12 | .16 |
| 1" x 8" | .043 | L.F. | .004 | .04 | .12 | .16 |
| Gutter, 5" box, aluminum, seamless, painted | .040 | L.F. | .003 | .05 | .10 | .15 |
| Vinyl | .040 | L.F. | .003 | .04 | .10 | .14 |
| Downspout 2" x 3", aluminum, one story house | .020 | L.F. | .001 | .02 | .03 | .05 |
| Two story house | .020 | L.F. | .001 | .03 | .05 | .08 |
| Vinyl, one story house | .020 | L.F. | .001 | .02 | .03 | .05 |
| Two story house | .020 | L.F. | .001 | .03 | .05 | .08 |
| | | | | | | |
| | | | | | | |
| | | | | | | |

**5 ROOFING**

185

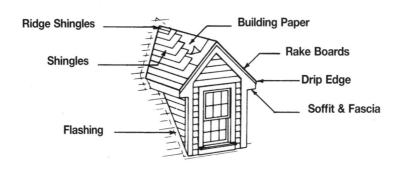

Ridge Shingles — Building Paper
Shingles — Rake Boards
Drip Edge
Soffit & Fascia
Flashing

| System Description | QUAN. | UNIT | LABOR HOURS | COST PER S.F. | | |
|---|---|---|---|---|---|---|
| | | | | MAT. | INST. | TOTAL |
| **ASPHALT, ROOF SHINGLES, CLASS A** | | | | | | |
| Shingles, standard inorganic class A 210-235 lb./sq | 1.400 | S.F. | .020 | .43 | .69 | 1.12 |
| Drip edge, metal, 5″ wide | .220 | L.F. | .004 | .05 | .16 | .21 |
| Building paper, #15 asphalt felt | 1.500 | S.F. | .002 | .04 | .07 | .11 |
| Ridge shingles, asphalt | .280 | L.F. | .007 | .22 | .23 | .45 |
| Soffit & fascia, aluminum, vented | .220 | L.F. | .032 | .52 | 1.14 | 1.66 |
| Flashing, aluminum, mill finish, .013″ thick | 1.500 | S.F. | .083 | .59 | 2.79 | 3.38 |
| TOTAL | | | .148 | 1.85 | 5.08 | 6.93 |
| **WOOD, CEDAR, NO. 1 PERFECTIONS** | | | | | | |
| Shingles, red cedar, No.1 perfections, 18″ long, 5″ exp. | 1.400 | S.F. | .041 | 2.46 | 1.46 | 3.92 |
| Drip edge, metal, 5″ wide | .220 | L.F. | .004 | .05 | .16 | .21 |
| Building paper, #15 asphalt felt | 1.500 | S.F. | .002 | .04 | .07 | .11 |
| Ridge shingles, wood | .280 | L.F. | .008 | .74 | .29 | 1.03 |
| Soffit & fascia, aluminum, vented | .220 | L.F. | .032 | .52 | 1.14 | 1.66 |
| Flashing, aluminum, mill finish, .013″ thick | 1.500 | S.F. | .083 | .59 | 2.79 | 3.38 |
| TOTAL | | | .170 | 4.40 | 5.91 | 10.31 |
| **SLATE, BUCKINGHAM, BLACK** | | | | | | |
| Shingles, Buckingham, Virginia, black | 1.400 | S.F. | .064 | 8.33 | 2.17 | 10.50 |
| Drip edge, metal, 5″ wide | .220 | L.F. | .004 | .05 | .16 | .21 |
| Building paper, #15 asphalt felt | 1.500 | S.F. | .002 | .04 | .07 | .11 |
| Ridge shingles, slate | .280 | L.F. | .011 | 2.62 | .38 | 3.00 |
| Soffit & fascia, aluminum, vented | .220 | L.F. | .032 | .52 | 1.14 | 1.66 |
| Flashing, copper, 16 oz. | 1.500 | S.F. | .104 | 3.80 | 3.53 | 7.33 |
| TOTAL | | | .217 | 15.36 | 7.45 | 22.81 |

The prices in these systems are based on a square foot of plan area under the dormer roof.

| Description | QUAN. | UNIT | LABOR HOURS | COST PER S.F. | | |
|---|---|---|---|---|---|---|
| | | | | MAT. | INST. | TOTAL |
| | | | | | | |
| | | | | | | |
| | | | | | | |
| | | | | | | |
| | | | | | | |

**Important: See the Reference Section for critical supporting data - Reference Nos., Crews & Location Factors**

## Gable Dormer Roofing Price Sheet

| | QUAN. | UNIT | LABOR HOURS | COST PER S.F. MAT. | COST PER S.F. INST. | COST PER S.F. TOTAL |
|---|---|---|---|---|---|---|
| Shingles, asphalt, standard, inorganic, class A, 210-235 lb./sq. | 1.400 | S.F. | .020 | .43 | .69 | 1.12 |
| Laminated, multi-layered, 240-260 lb./sq. | 1.400 | S.F. | .025 | .61 | .84 | 1.45 |
| Premium laminated, multi-layered, 260-300 lb./sq. | 1.400 | S.F. | .032 | .76 | 1.09 | 1.85 |
| Clay tile, Spanish tile, red | 1.400 | S.F. | .062 | 4.62 | 2.11 | 6.73 |
| Mission tile, red | 1.400 | S.F. | .097 | 9.75 | 3.30 | 13.05 |
| French tile, red | 1.400 | S.F. | .083 | 8.80 | 2.81 | 11.61 |
| Slate Buckingham, Virginia, black | 1.400 | S.F. | .064 | 8.35 | 2.17 | 10.52 |
| Vermont, black or grey | 1.400 | S.F. | .064 | 5.30 | 2.17 | 7.47 |
| Wood, red cedar, No.1 5X, 16" long, 5" exposure | 1.400 | S.F. | .045 | 2.32 | 1.60 | 3.92 |
| Fire retardant | 1.400 | S.F. | .045 | 2.78 | 1.60 | 4.38 |
| 18" long, No.1 perfections, 5" exposure | 1.400 | S.F. | .041 | 2.46 | 1.46 | 3.92 |
| Fire retardant | 1.400 | S.F. | .041 | 2.90 | 1.46 | 4.36 |
| Resquared & rebutted, 18" long, 5" exposure | 1.400 | S.F. | .037 | 3.07 | 1.33 | 4.40 |
| Fire retardant | 1.400 | S.F. | .037 | 3.51 | 1.33 | 4.84 |
| Shakes hand split, 24" long, 10" exposure | 1.400 | S.F. | .045 | 2.13 | 1.60 | 3.73 |
| Fire retardant | 1.400 | S.F. | .045 | 2.59 | 1.60 | 4.19 |
| 18" long, 8" exposure | 1.400 | S.F. | .056 | 1.50 | 1.99 | 3.49 |
| Fire retardant | 1.400 | S.F. | .056 | 1.96 | 1.99 | 3.95 |
| Drip edge, metal, 5" wide | .220 | L.F. | .004 | .05 | .16 | .21 |
| 8" wide | .220 | L.F. | .004 | .08 | .16 | .24 |
| Building paper, #15 asphalt felt | 1.500 | S.F. | .002 | .04 | .07 | .11 |
| Clay | .280 | L.F. | .011 | 2.66 | .38 | 3.04 |
| Slate | .280 | L.F. | .011 | 2.62 | .38 | 3.00 |
| Wood | .280 | L.F. | .008 | .74 | .29 | 1.03 |
| Soffit & fascia, aluminum, vented | .220 | L.F. | .032 | .52 | 1.14 | 1.66 |
| Vinyl, vented | .220 | L.F. | .029 | .36 | 1.05 | 1.41 |
| Wood, board fascia, plywood soffit | .220 | L.F. | .026 | .38 | .89 | 1.27 |
| Flashing, aluminum, .013" thick | 1.500 | S.F. | .083 | .59 | 2.79 | 3.38 |
| .032" thick | 1.500 | S.F. | .083 | 1.71 | 2.79 | 4.50 |
| .040" thick | 1.500 | S.F. | .083 | 2.40 | 2.79 | 5.19 |
| .050" thick | 1.500 | S.F. | .083 | 3.03 | 2.79 | 5.82 |
| Copper, 16 oz. | 1.500 | S.F. | .104 | 3.80 | 3.53 | 7.33 |
| 20 oz. | 1.500 | S.F. | .109 | 6.45 | 3.69 | 10.14 |
| 24 oz. | 1.500 | S.F. | .114 | 7.75 | 3.87 | 11.62 |
| 32 oz. | 1.500 | S.F. | .120 | 10.30 | 4.05 | 14.35 |

**5 ROOFING**

| System Description | QUAN. | UNIT | LABOR HOURS | COST PER S.F. | | |
|---|---|---|---|---|---|---|
| | | | | MAT. | INST. | TOTAL |
| **ASPHALT, ROOF SHINGLES, CLASS A** | | | | | | |
| Shingles, standard inorganic class A 210-235 lb./sq. | 1.100 | S.F. | .016 | .34 | .54 | .88 |
| Drip edge, aluminum, 5" wide | .250 | L.F. | .005 | .06 | .18 | .24 |
| Building paper, #15 asphalt felt | 1.200 | S.F. | .002 | .03 | .06 | .09 |
| Soffit & fascia, aluminum, vented, 1' overhang | .250 | L.F. | .036 | .60 | 1.30 | 1.90 |
| Flashing, aluminum, mill finish, 0.013" thick | .800 | L.F. | .044 | .31 | 1.49 | 1.80 |
| TOTAL | | | .103 | 1.34 | 3.57 | 4.91 |
| **WOOD, CEDAR, NO. 1 PERFECTIONS, 18" LONG** | | | | | | |
| Shingles, wood, red cedar, #1 perfections, 5" exposure | 1.100 | S.F. | .032 | 1.94 | 1.14 | 3.08 |
| Drip edge, aluminum, 5" wide | .250 | L.F. | .005 | .06 | .18 | .24 |
| Building paper, #15 asphalt felt | 1.200 | S.F. | .002 | .03 | .06 | .09 |
| Soffit & fascia, aluminum, vented, 1' overhang | .250 | L.F. | .036 | .60 | 1.30 | 1.90 |
| Flashing, aluminum, mill finish, 0.013" thick | .800 | L.F. | .044 | .31 | 1.49 | 1.80 |
| TOTAL | | | .119 | 2.94 | 4.17 | 7.11 |
| **SLATE, BUCKINGHAM, BLACK** | | | | | | |
| Shingles, slate, Buckingham, black | 1.100 | S.F. | .050 | 6.55 | 1.71 | 8.26 |
| Drip edge, aluminum, 5" wide | .250 | L.F. | .005 | .06 | .18 | .24 |
| Building paper, #15 asphalt felt | 1.200 | S.F. | .002 | .03 | .06 | .09 |
| Soffit & fascia, aluminum, vented, 1' overhang | .250 | L.F. | .036 | .60 | 1.30 | 1.90 |
| Flashing, copper, 16 oz. | .800 | L.F. | .056 | 2.02 | 1.88 | 3.90 |
| TOTAL | | | .149 | 9.26 | 5.13 | 14.39 |

The prices in this system are based on a square foot of plan area under the dormer roof.

| Description | QUAN. | UNIT | LABOR HOURS | COST PER S.F. | | |
|---|---|---|---|---|---|---|
| | | | | MAT. | INST. | TOTAL |
| | | | | | | |
| | | | | | | |
| | | | | | | |
| | | | | | | |
| | | | | | | |

**Important: See the Reference Section for critical supporting data - Reference Nos., Crews & Location Factors**

ROOFING 5

| Shed Dormer Roofing Price Sheet | QUAN. | UNIT | LABOR HOURS | COST PER S.F. MAT. | COST PER S.F. INST. | COST PER S.F. TOTAL |
|---|---|---|---|---|---|---|
| Shingles, asphalt, standard, inorganic, class A, 210-235 lb./sq. | 1.100 | S.F. | .016 | .34 | .54 | .88 |
| Laminated, multi-layered, 240-260 lb./sq. | 1.100 | S.F. | .020 | .48 | .66 | 1.14 |
| Premium laminated, multi-layered, 260-300 lb./sq. | 1.100 | S.F. | .025 | .60 | .85 | 1.45 |
| | | | | | | |
| Clay tile, Spanish tile, red | 1.100 | S.F. | .049 | 3.63 | 1.66 | 5.29 |
| Mission tile, red | 1.100 | S.F. | .077 | 7.65 | 2.60 | 10.25 |
| French tile, red | 1.100 | S.F. | .065 | 6.95 | 2.21 | 9.16 |
| | | | | | | |
| Slate Buckingham, Virginia, black | 1.100 | S.F. | .050 | 6.55 | 1.71 | 8.26 |
| Vermont, black or grey | 1.100 | S.F. | .050 | 4.18 | 1.71 | 5.89 |
| Wood, red cedar, No. 1 5X, 16" long, 5" exposure | 1.100 | S.F. | .035 | 1.83 | 1.25 | 3.08 |
| Fire retardant | 1.100 | S.F. | .035 | 2.19 | 1.25 | 3.44 |
| 18" long, No.1 perfections, 5" exposure | 1.100 | S.F. | .032 | 1.94 | 1.14 | 3.08 |
| Fire retardant | 1.100 | S.F. | .032 | 2.29 | 1.14 | 3.43 |
| Resquared & rebutted, 18" long, 5" exposure | 1.100 | S.F. | .029 | 2.41 | 1.05 | 3.46 |
| Fire retardant | 1.100 | S.F. | .029 | 2.76 | 1.05 | 3.81 |
| Shakes hand split, 24" long, 10" exposure | 1.100 | S.F. | .035 | 1.67 | 1.25 | 2.92 |
| Fire retardant | 1.100 | S.F. | .035 | 2.03 | 1.25 | 3.28 |
| 18" long, 8" exposure | 1.100 | S.F. | .044 | 1.18 | 1.56 | 2.74 |
| Fire retardant | 1.100 | S.F. | .044 | 1.54 | 1.56 | 3.10 |
| Drip edge, metal, 5" wide | .250 | L.F. | .005 | .06 | .18 | .24 |
| 8" wide | .250 | L.F. | .005 | .09 | .18 | .27 |
| Building paper, #15 asphalt felt | 1.200 | S.F. | .002 | .03 | .06 | .09 |
| | | | | | | |
| Soffit & fascia, aluminum, vented | .250 | L.F. | .036 | .60 | 1.30 | 1.90 |
| Vinyl, vented | .250 | L.F. | .033 | .41 | 1.19 | 1.60 |
| Wood, board fascia, plywood soffit | .250 | L.F. | .030 | .44 | 1.02 | 1.46 |
| | | | | | | |
| Flashing, aluminum, .013" thick | .800 | L.F. | .044 | .31 | 1.49 | 1.80 |
| .032" thick | .800 | L.F. | .044 | .91 | 1.49 | 2.40 |
| .040" thick | .800 | L.F. | .044 | 1.28 | 1.49 | 2.77 |
| .050" thick | .800 | L.F. | .044 | 1.62 | 1.49 | 3.11 |
| Copper, 16 oz. | .800 | L.F. | .056 | 2.02 | 1.88 | 3.90 |
| 20 oz. | .800 | L.F. | .058 | 3.43 | 1.97 | 5.40 |
| 24 oz. | .800 | L.F. | .061 | 4.12 | 2.06 | 6.18 |
| 32 oz. | .800 | L.F. | .064 | 5.50 | 2.16 | 7.66 |

**5 ROOFING**

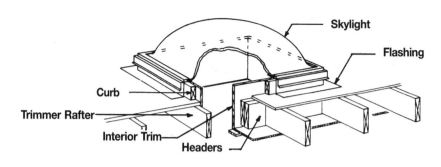

| System Description | QUAN. | UNIT | LABOR HOURS | COST EACH | | |
|---|---|---|---|---|---|---|
| | | | | MAT. | INST. | TOTAL |
| **SKYLIGHT, FIXED, 32″ X 32″** | | | | | | |
| Skylight, fixed bubble, insulating, 32″ x 32″ | 1.000 | Ea. | 1.422 | 103.47 | 46.22 | 149.69 |
| Trimmer rafters, 2″ x 6″ | 28.000 | L.F. | .448 | 17.36 | 15.96 | 33.32 |
| Headers, 2″ x 6″ | 6.000 | L.F. | .267 | 3.72 | 9.48 | 13.20 |
| Curb, 2″ x 4″ | 12.000 | L.F. | .154 | 4.68 | 5.52 | 10.20 |
| Flashing, aluminum, .013″ thick | 13.500 | S.F. | .745 | 5.27 | 25.11 | 30.38 |
| Trim, stock pine, 11/16″ x 2-1/2″ | 12.000 | L.F. | .400 | 10.32 | 14.28 | 24.60 |
| Trim primer coat, oil base, brushwork | 12.000 | L.F. | .148 | .36 | 4.68 | 5.04 |
| Trim paint, 1 coat, brushwork | 12.000 | L.F. | .148 | .36 | 4.68 | 5.04 |
| TOTAL | | | 3.732 | 145.54 | 125.93 | 271.47 |
| **SKYLIGHT, FIXED, 48″ X 48″** | | | | | | |
| Skylight, fixed bubble, insulating, 48″ x 48″ | 1.000 | Ea. | 1.296 | 139.20 | 41.92 | 181.12 |
| Trimmer rafters, 2″ x 6″ | 28.000 | L.F. | .448 | 17.36 | 15.96 | 33.32 |
| Headers, 2″ x 6″ | 8.000 | L.F. | .356 | 4.96 | 12.64 | 17.60 |
| Curb, 2″ x 4″ | 16.000 | L.F. | .205 | 6.24 | 7.36 | 13.60 |
| Flashing, aluminum, .013″ thick | 16.000 | S.F. | .883 | 6.24 | 29.76 | 36.00 |
| Trim, stock pine, 11/16″ x 2-1/2″ | 16.000 | L.F. | .533 | 13.76 | 19.04 | 32.80 |
| Trim primer coat, oil base, brushwork | 16.000 | L.F. | .197 | .48 | 6.24 | 6.72 |
| Trim paint, 1 coat, brushwork | 16.000 | L.F. | .197 | .48 | 6.24 | 6.72 |
| TOTAL | | | 4.115 | 188.72 | 139.16 | 327.88 |
| **SKYWINDOW, OPERATING, 24″ X 48″** | | | | | | |
| Skywindow, operating, thermopane glass, 24″ x 48″ | 1.000 | Ea. | 3.200 | 565.00 | 104.00 | 669.00 |
| Trimmer rafters, 2″ x 6″ | 28.000 | L.F. | .448 | 17.36 | 15.96 | 33.32 |
| Headers, 2″ x 6″ | 8.000 | L.F. | .267 | 3.72 | 9.48 | 13.20 |
| Curb, 2″ x 4″ | 14.000 | L.F. | .179 | 5.46 | 6.44 | 11.90 |
| Flashing, aluminum, .013″ thick | 14.000 | S.F. | .772 | 5.46 | 26.04 | 31.50 |
| Trim, stock pine, 11/16″ x 2-1/2″ | 14.000 | L.F. | .467 | 12.04 | 16.66 | 28.70 |
| Trim primer coat, oil base, brushwork | 14.000 | L.F. | .172 | .42 | 5.46 | 5.88 |
| Trim paint, 1 coat, brushwork | 14.000 | L.F. | .172 | .42 | 5.46 | 5.88 |
| TOTAL | | | 5.677 | 609.88 | 189.50 | 799.38 |

The prices in these systems are on a cost each basis.

| Description | QUAN. | UNIT | LABOR HOURS | COST EACH | | |
|---|---|---|---|---|---|---|
| | | | | MAT. | INST. | TOTAL |
| | | | | | | |
| | | | | | | |

**Important: See the Reference Section for critical supporting data - Reference Nos., Crews & Location Factors**

ROOFING 5

| Skylight/Skywindow Price Sheet | QUAN. | UNIT | LABOR HOURS | COST EACH | | |
|---|---|---|---|---|---|---|
| | | | | MAT. | INST. | TOTAL |
| Skylight, fixed bubble insulating, 24" x 24" | 1.000 | Ea. | .800 | 58.00 | 26.00 | 84.00 |
| 32" x 32" | 1.000 | Ea. | 1.422 | 103.00 | 46.00 | 149.00 |
| 32" x 48" | 1.000 | Ea. | .864 | 93.00 | 28.00 | 121.00 |
| 48" x 48" | 1.000 | Ea. | 1.296 | 139.00 | 42.00 | 181.00 |
| Ventilating bubble insulating, 36" x 36" | 1.000 | Ea. | 2.667 | 395.00 | 86.50 | 481.50 |
| 52" x 52" | 1.000 | Ea. | 2.667 | 590.00 | 86.50 | 676.50 |
| 28" x 52" | 1.000 | Ea. | 3.200 | 460.00 | 104.00 | 564.00 |
| 36" x 52" | 1.000 | Ea. | 3.200 | 500.00 | 104.00 | 604.00 |
| Skywindow, operating, thermopane glass, 24" x 48" | 1.000 | Ea. | 3.200 | 565.00 | 104.00 | 669.00 |
| 32" x 48" | 1.000 | Ea. | 3.556 | 595.00 | 115.00 | 710.00 |
| Trimmer rafters, 2" x 6" | 28.000 | L.F. | .448 | 17.35 | 15.95 | 33.30 |
| 2" x 8" | 28.000 | L.F. | .472 | 26.50 | 16.80 | 43.30 |
| 2" x 10" | 28.000 | L.F. | .711 | 38.50 | 25.00 | 63.50 |
| Headers, 24" window, 2" x 6" | 4.000 | L.F. | .178 | 2.48 | 6.30 | 8.78 |
| 2" x 8" | 4.000 | L.F. | .188 | 3.80 | 6.70 | 10.50 |
| 2" x 10" | 4.000 | L.F. | .200 | 5.50 | 7.10 | 12.60 |
| 32" window, 2" x 6" | 6.000 | L.F. | .267 | 3.72 | 9.50 | 13.22 |
| 2" x 8" | 6.000 | L.F. | .282 | 5.70 | 10.10 | 15.80 |
| 2" x 10" | 6.000 | L.F. | .300 | 8.20 | 10.70 | 18.90 |
| 48" window, 2" x 6" | 8.000 | L.F. | .356 | 4.96 | 12.65 | 17.61 |
| 2" x 8" | 8.000 | L.F. | .376 | 7.60 | 13.45 | 21.05 |
| 2" x 10" | 8.000 | L.F. | .400 | 10.95 | 14.25 | 25.20 |
| Curb, 2" x 4", skylight, 24" x 24" | 8.000 | L.F. | .102 | 3.12 | 3.68 | 6.80 |
| 32" x 32" | 12.000 | L.F. | .154 | 4.68 | 5.50 | 10.18 |
| 32" x 48" | 14.000 | L.F. | .179 | 5.45 | 6.45 | 11.90 |
| 48" x 48" | 16.000 | L.F. | .205 | 6.25 | 7.35 | 13.60 |
| Flashing, aluminum .013" thick, skylight, 24" x 24" | 9.000 | S.F. | .497 | 3.51 | 16.75 | 20.26 |
| 32" x 32" | 13.500 | S.F. | .745 | 5.25 | 25.00 | 30.25 |
| 32" x 48" | 14.000 | S.F. | .772 | 5.45 | 26.00 | 31.45 |
| 48" x 48" | 16.000 | S.F. | .883 | 6.25 | 30.00 | 36.25 |
| Copper 16 oz., skylight, 24" x 24" | 9.000 | S.F. | .626 | 23.00 | 21.00 | 44.00 |
| 32" x 32" | 13.500 | S.F. | .939 | 34.00 | 31.50 | 65.50 |
| 32" x 48" | 14.000 | S.F. | .974 | 35.50 | 33.00 | 68.50 |
| 48" x 48" | 16.000 | S.F. | 1.113 | 40.50 | 37.50 | 78.00 |
| Trim, interior casing painted, 24" x 24" | 8.000 | L.F. | .347 | 6.80 | 12.10 | 18.90 |
| 32" x 32" | 12.000 | L.F. | .520 | 10.20 | 18.10 | 28.30 |
| 32" x 48" | 14.000 | L.F. | .607 | 11.90 | 21.00 | 32.90 |
| 48" x 48" | 16.000 | L.F. | .693 | 13.60 | 24.00 | 37.60 |
| | | | | | | |
| | | | | | | |
| | | | | | | |
| | | | | | | |
| | | | | | | |
| | | | | | | |
| | | | | | | |
| | | | | | | |
| | | | | | | |
| | | | | | | |
| | | | | | | |

**5 ROOFING**

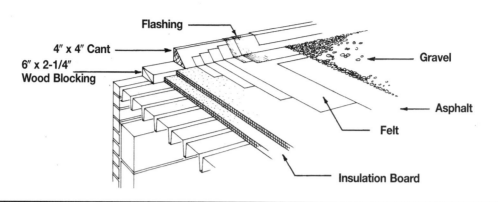

Flashing · 4" x 4" Cant · 6" x 2-1/4" Wood Blocking · Gravel · Asphalt · Felt · Insulation Board

| System Description | QUAN. | UNIT | LABOR HOURS | MAT. | INST. | TOTAL |
|---|---|---|---|---|---|---|
| **ASPHALT, ORGANIC, 4-PLY, INSULATED DECK** | | | | | | |
| Membrane, asphalt, 4-plies #15 felt, gravel surfacing | 1.000 | S.F. | .025 | .60 | .96 | 1.56 |
| Insulation board, 2-layers of 1-1/16" glass fiber | 2.000 | S.F. | .016 | 1.68 | .54 | 2.22 |
| Wood blocking, 2" x 6" | .040 | L.F. | .004 | .07 | .15 | .22 |
| Treated 4" x 4" cant strip | .040 | L.F. | .001 | .06 | .04 | .10 |
| Flashing, aluminum, 0.040" thick | .050 | S.F. | .003 | .08 | .09 | .17 |
| TOTAL | | | .049 | 2.49 | 1.78 | 4.27 |
| **ASPHALT, INORGANIC, 3-PLY, INSULATED DECK** | | | | | | |
| Membrane, asphalt, 3-plies type IV glass felt, gravel surfacing | 1.000 | S.F. | .028 | .59 | 1.06 | 1.65 |
| Insulation board, 2-layers of 1-1/16" glass fiber | 2.000 | S.F. | .016 | 1.68 | .54 | 2.22 |
| Wood blocking, 2" x 6" | .040 | L.F. | .004 | .07 | .15 | .22 |
| Treated 4" x 4" cant strip | .040 | L.F. | .001 | .06 | .04 | .10 |
| Flashing, aluminum, 0.040" thick | .050 | S.F. | .003 | .08 | .09 | .17 |
| TOTAL | | | .052 | 2.48 | 1.88 | 4.36 |
| **COAL TAR, ORGANIC, 4-PLY, INSULATED DECK** | | | | | | |
| Membrane, coal tar, 4-plies #15 felt, gravel surfacing | 1.000 | S.F. | .027 | 1.12 | 1.01 | 2.13 |
| Insulation board, 2-layers of 1-1/16" glass fiber | 2.000 | S.F. | .016 | 1.68 | .54 | 2.22 |
| Wood blocking, 2" x 6" | .040 | L.F. | .004 | .07 | .15 | .22 |
| Treated 4" x 4" cant strip | .040 | L.F. | .001 | .06 | .04 | .10 |
| Flashing, aluminum, 0.040" thick | .050 | S.F. | .003 | .08 | .09 | .17 |
| TOTAL | | | .051 | 3.01 | 1.83 | 4.84 |
| **COAL TAR, INORGANIC, 3-PLY, INSULATED DECK** | | | | | | |
| Membrane, coal tar, 3-plies type IV glass felt, gravel surfacing | 1.000 | S.F. | .029 | .93 | 1.12 | 2.05 |
| Insulation board, 2-layers of 1-1/16" glass fiber | 2.000 | S.F. | .016 | 1.68 | .54 | 2.22 |
| Wood blocking, 2" x 6" | .040 | L.F. | .004 | .07 | .15 | .22 |
| Treated 4" x 4" cant strip | .040 | L.F. | .001 | .06 | .04 | .10 |
| Flashing, aluminum, 0.040" thick | .050 | S.F. | .003 | .08 | .09 | .17 |
| TOTAL | | | .053 | 2.82 | 1.94 | 4.76 |

| Description | QUAN. | UNIT | LABOR HOURS | MAT. | INST. | TOTAL |
|---|---|---|---|---|---|---|
| | | | | | | |
| | | | | | | |

**Important: See the Reference Section for critical supporting data - Reference Nos., Crews & Location Factors**

| Built-Up Roofing Price Sheet | QUAN. | UNIT | LABOR HOURS | COST PER S.F. MAT. | INST. | TOTAL |
|---|---|---|---|---|---|---|
| embrane, asphalt, 4-plies #15 organic felt, gravel surfacing | 1.000 | S.F. | .025 | .60 | .96 | 1.56 |
| Asphalt base sheet & 3-plies #15 asphalt felt | 1.000 | S.F. | .025 | .45 | .96 | 1.41 |
| 3-plies type IV glass fiber felt | 1.000 | S.F. | .028 | .59 | 1.06 | 1.65 |
| 4-plies type IV glass fiber felt | 1.000 | S.F. | .028 | .71 | 1.06 | 1.77 |
| Coal tar, 4-plies #15 organic felt, gravel surfacing | 1.000 | S.F. | .027 | | | |
| 4-plies tarred felt | 1.000 | S.F. | .027 | 1.12 | 1.01 | 2.13 |
| 3-plies type IV glass fiber felt | 1.000 | S.F. | .029 | .93 | 1.12 | 2.05 |
| 4-plies type IV glass fiber felt | 1.000 | S.F. | .027 | 1.27 | 1.01 | 2.28 |
| Roll, asphalt, 1-ply #15 organic felt, 2-plies mineral surfaced | 1.000 | S.F. | .021 | .39 | .78 | 1.17 |
| 3-plies type IV glass fiber, 1-ply mineral surfaced | 1.000 | S.F. | .022 | .59 | .85 | 1.44 |
| sulation boards, glass fiber, 1-1/16" thick | 1.000 | S.F. | .008 | .84 | .27 | 1.11 |
| 2-1/16" thick | 1.000 | S.F. | .010 | 1.23 | .34 | 1.57 |
| 2-7/16" thick | 1.000 | S.F. | .010 | 1.41 | .34 | 1.75 |
| Expanded perlite, 1" thick | 1.000 | S.F. | .010 | .32 | .34 | .66 |
| 1-1/2" thick | 1.000 | S.F. | .010 | .43 | .34 | .77 |
| 2" thick | 1.000 | S.F. | .011 | .63 | .39 | 1.02 |
| Fiberboard, 1" thick | 1.000 | S.F. | .010 | .37 | .34 | .71 |
| 1-1/2" thick | 1.000 | S.F. | .010 | .55 | .34 | .89 |
| 2" thick | 1.000 | S.F. | .010 | .75 | .34 | 1.09 |
| Extruded polystyrene, 15 PSI compressive strength, 2" thick R10 | 1.000 | S.F. | .006 | .40 | .22 | .62 |
| 3" thick R15 | 1.000 | S.F. | .008 | .85 | .27 | 1.12 |
| 4" thick R20 | 1.000 | S.F. | .008 | 1.18 | .27 | 1.45 |
| Tapered for drainage | 1.000 | S.F. | .005 | .39 | .18 | .57 |
| 40 PSI compressive strength, 1" thick R5 | 1.000 | S.F. | .005 | .39 | .18 | .57 |
| 2" thick R10 | 1.000 | S.F. | .006 | .76 | .22 | .98 |
| 3" thick R15 | 1.000 | S.F. | .008 | 1.11 | .27 | 1.38 |
| 4" thick R20 | 1.000 | S.F. | .008 | 1.49 | .27 | 1.76 |
| Fiberboard high density, 1/2" thick R1.3 | 1.000 | S.F. | .008 | .22 | .27 | .49 |
| 1" thick R2.5 | 1.000 | S.F. | .010 | .40 | .34 | .74 |
| 1 1/2" thick R3.8 | 1.000 | S.F. | .010 | .65 | .34 | .99 |
| Polyisocyanurate, 1 1/2" thick R10.87 | 1.000 | S.F. | .006 | .39 | .22 | .61 |
| 2" thick R14.29 | 1.000 | S.F. | .007 | .50 | .25 | .75 |
| 3 1/2" thick R25 | 1.000 | S.F. | .008 | .81 | .27 | 1.08 |
| Tapered for drainage | 1.000 | S.F. | .006 | .42 | .19 | .61 |
| Expanded polystyrene, 1" thick | 1.000 | S.F. | .005 | .21 | .18 | .39 |
| 2" thick R10 | 1.000 | S.F. | .006 | .41 | .22 | .63 |
| 3" thick | 1.000 | S.F. | .006 | .65 | .22 | .87 |
| Wood blocking, treated, 6" x 2" & 4" x 4" cant | .040 | L.F. | .002 | .09 | .09 | .18 |
| 6" x 4-1/2" & 4" x 4" cant | .040 | L.F. | .005 | .15 | .19 | .34 |
| 6" x 5" & 4" x 4" cant | .040 | L.F. | .007 | .19 | .24 | .43 |
| Flashing, aluminum, 0.019" thick | .050 | S.F. | .003 | .04 | .09 | .13 |
| 0.032" thick | .050 | S.F. | .003 | .06 | .09 | .15 |
| 0.040" thick | .050 | S.F. | .003 | .08 | .09 | .17 |
| Copper sheets, 16 oz., under 500 lbs. | .050 | S.F. | .003 | .13 | .12 | .25 |
| Over 500 lbs. | .050 | S.F. | .003 | .16 | .09 | .25 |
| 20 oz., under 500 lbs. | .050 | S.F. | .004 | .21 | .12 | .33 |
| Over 500 lbs. | .050 | S.F. | .003 | .20 | .09 | .29 |
| Stainless steel, 32 gauge | .050 | S.F. | .003 | .12 | .09 | .21 |
| 28 gauge | .050 | S.F. | .003 | .14 | .09 | .23 |
| 26 gauge | .050 | S.F. | .003 | .18 | .09 | .27 |
| 24 gauge | .050 | S.F. | .003 | .23 | .09 | .32 |

**5 ROOFING**

or information about Means Estimating Seminars, see yellow pages 11 and 12 in back of book

# Division 6
# Interiors

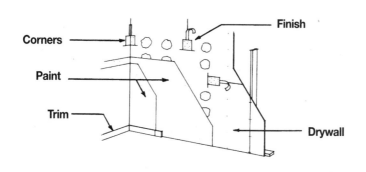

| System Description | QUAN. | UNIT | LABOR HOURS | COST PER S.F. | | |
|---|---|---|---|---|---|---|
| | | | | MAT. | INST. | TOTAL |
| **1/2" SHEETROCK, TAPED & FINISHED** | | | | | | |
| Drywall, 1/2" thick, standard | 1.000 | S.F. | .008 | .23 | .28 | .51 |
| Finish, taped & finished joints | 1.000 | S.F. | .008 | .04 | .28 | .32 |
| Corners, taped & finished, 32 L.F. per 12' x 12' room | .083 | L.F. | .001 | .01 | .05 | .06 |
| Painting, primer & 2 coats | 1.000 | S.F. | .011 | .15 | .34 | .49 |
| Paint trim, to 6" wide, primer + 1 coat enamel | .125 | L.F. | .001 | .01 | .04 | .05 |
| Trim, baseboard | .125 | L.F. | .005 | .18 | .18 | .36 |
| TOTAL | | | .034 | .62 | 1.17 | 1.79 |
| **THINCOAT, SKIM-COAT, ON 1/2" BACKER DRYWALL** | | | | | | |
| Drywall, 1/2" thick, thincoat backer | 1.000 | S.F. | .008 | .23 | .28 | .51 |
| Thincoat plaster | 1.000 | S.F. | .011 | .08 | .36 | .44 |
| Corners, taped & finished, 32 L.F. per 12' x 12' room | .083 | L.F. | .001 | .01 | .05 | .06 |
| Painting, primer & 2 coats | 1.000 | S.F. | .011 | .15 | .34 | .49 |
| Paint trim, to 6" wide, primer + 1 coat enamel | .125 | L.F. | .001 | .01 | .04 | .05 |
| Trim, baseboard | .125 | L.F. | .005 | .18 | .18 | .36 |
| TOTAL | | | .037 | .66 | 1.25 | 1.91 |
| **5/8" SHEETROCK, TAPED & FINISHED** | | | | | | |
| Drywall, 5/8" thick, standard | 1.000 | S.F. | .008 | .24 | .28 | .52 |
| Finish, taped & finished joints | 1.000 | S.F. | .008 | .04 | .28 | .32 |
| Corners, taped & finished, 32 L.F. per 12' x 12' room | .083 | L.F. | .001 | .01 | .05 | .06 |
| Painting, primer & 2 coats | 1.000 | S.F. | .011 | .15 | .34 | .49 |
| Trim, baseboard | .125 | L.F. | .005 | .18 | .18 | .36 |
| Paint trim, to 6" wide, primer + 1 coat enamel | .125 | L.F. | .001 | .01 | .04 | .05 |
| TOTAL | | | .034 | .63 | 1.17 | 1.80 |

The costs in this system are based on a square foot of wall.
Do not deduct for openings.

| Description | QUAN. | UNIT | LABOR HOURS | COST PER S.F. | | |
|---|---|---|---|---|---|---|
| | | | | MAT. | INST. | TOTAL |
| | | | | | | |
| | | | | | | |
| | | | | | | |

**Important: See the Reference Section for critical supporting data - Reference Nos., Crews & Location Factors**

| Drywall & Thincoat Wall Price Sheet | QUAN. | UNIT | LABOR HOURS | COST PER S.F. | | |
|---|---|---|---|---|---|---|
| | | | | MAT. | INST. | TOTAL |
| Drywall-sheetrock, 1/2" thick, standard | 1.000 | S.F. | .008 | .23 | .28 | .51 |
| Fire resistant | 1.000 | S.F. | .008 | .24 | .28 | .52 |
| Water resistant | 1.000 | S.F. | .008 | .24 | .28 | .52 |
| 5/8" thick, standard | 1.000 | S.F. | .008 | .24 | .28 | .52 |
| Fire resistant | 1.000 | S.F. | .008 | .24 | .28 | .52 |
| Water resistant | 1.000 | S.F. | .008 | .30 | .28 | .58 |
| Drywall backer for thincoat system, 1/2" thick | 1.000 | S.F. | .008 | .23 | .28 | .51 |
| 5/8" thick | 1.000 | S.F. | .008 | .24 | .28 | .52 |
| Finish drywall, taped & finished | 1.000 | S.F. | .008 | .04 | .28 | .32 |
| Texture spray | 1.000 | S.F. | .010 | .06 | .33 | .39 |
| Thincoat plaster, including tape | 1.000 | S.F. | .011 | .08 | .36 | .44 |
| Corners drywall, taped & finished, 32 L.F. per 4' x 4' room | .250 | L.F. | .004 | .02 | .13 | .15 |
| 6' x 6' room | .110 | L.F. | .002 | .01 | .06 | .07 |
| 10' x 10' room | .100 | L.F. | .001 | .01 | .05 | .06 |
| 12' x 12' room | .083 | L.F. | .001 | .01 | .04 | .05 |
| 16' x 16' room | .063 | L.F. | .001 | | .03 | .03 |
| Thincoat system, 32 L.F. per 4' x 4' room | .250 | L.F. | .003 | .02 | .10 | .12 |
| 6' x 6' room | .110 | L.F. | .001 | .01 | .04 | .05 |
| 10' x 10' room | .100 | L.F. | .001 | .01 | .03 | .04 |
| 12' x 12' room | .083 | L.F. | .001 | .01 | .03 | .04 |
| 16' x 16' room | .063 | L.F. | .001 | .01 | .02 | .03 |
| Painting, primer, & 1 coat | 1.000 | S.F. | .008 | .10 | .26 | .36 |
| & 2 coats | 1.000 | S.F. | .011 | .15 | .34 | .49 |
| Wallpaper, $7/double roll | 1.000 | S.F. | .013 | .31 | .40 | .71 |
| $17/double roll | 1.000 | S.F. | .015 | .67 | .48 | 1.15 |
| $40/double roll | 1.000 | S.F. | .018 | 1.61 | .58 | 2.19 |
| Tile, ceramic adhesive thin set, 4 1/4" x 4 1/4" tiles | 1.000 | S.F. | .084 | 2.30 | 2.50 | 4.80 |
| 6" x 6" tiles | 1.000 | S.F. | .080 | 2.83 | 2.37 | 5.20 |
| Pregrouted sheets | 1.000 | S.F. | .067 | 4.57 | 1.98 | 6.55 |
| Trim, painted or stained, baseboard | .125 | L.F. | .006 | .19 | .22 | .41 |
| Base shoe | .125 | L.F. | .005 | .11 | .19 | .30 |
| Chair rail | .125 | L.F. | .005 | .13 | .17 | .30 |
| Cornice molding | .125 | L.F. | .004 | .10 | .15 | .25 |
| Cove base, vinyl | .125 | L.F. | .003 | .06 | .11 | .17 |
| Paneling, not including furring or trim | | | | | | |
| Plywood, prefinished, 1/4" thick, 4' x 8' sheets, vert. grooves | | | | | | |
| Birch faced, minimum | 1.000 | S.F. | .032 | .81 | 1.14 | 1.95 |
| Average | 1.000 | S.F. | .038 | 1.23 | 1.36 | 2.59 |
| Maximum | 1.000 | S.F. | .046 | 1.80 | 1.63 | 3.43 |
| Mahogany, African | 1.000 | S.F. | .040 | 2.31 | 1.42 | 3.73 |
| Philippine (lauan) | 1.000 | S.F. | .032 | .99 | 1.14 | 2.13 |
| Oak or cherry, minimum | 1.000 | S.F. | .032 | 1.94 | 1.14 | 3.08 |
| Maximum | 1.000 | S.F. | .040 | 2.97 | 1.42 | 4.39 |
| Rosewood | 1.000 | S.F. | .050 | 4.21 | 1.78 | 5.99 |
| Teak | 1.000 | S.F. | .040 | 2.97 | 1.42 | 4.39 |
| Chestnut | 1.000 | S.F. | .043 | 4.39 | 1.52 | 5.91 |
| Pecan | 1.000 | S.F. | .040 | 1.89 | 1.42 | 3.31 |
| Walnut, minimum | 1.000 | S.F. | .032 | 2.53 | 1.14 | 3.67 |
| Maximum | 1.000 | S.F. | .040 | 4.80 | 1.42 | 6.22 |

**INTERIORS**

**6**

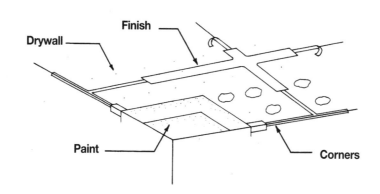

Finish

Drywall

Paint

Corners

| System Description | QUAN. | UNIT | LABOR HOURS | COST PER S.F. | | |
|---|---|---|---|---|---|---|
| | | | | MAT. | INST. | TOTAL |
| **1/2" SHEETROCK, TAPED & FINISHED** | | | | | | |
| Drywall, 1/2" thick, standard | 1.000 | S.F. | .008 | .23 | .28 | .51 |
| Finish, taped & finished | 1.000 | S.F. | .008 | .04 | .28 | .32 |
| Corners, taped & finished, 12' x 12' room | .333 | L.F. | .005 | .02 | .17 | .19 |
| Paint, primer & 2 coats | 1.000 | S.F. | .011 | .15 | .34 | .49 |
| TOTAL | | | .032 | .44 | 1.07 | 1.51 |
| **THINCOAT, SKIM COAT ON 1/2" BACKER DRYWALL** | | | | | | |
| Drywall, 1/2" thick, thincoat backer | 1.000 | S.F. | .008 | .23 | .28 | .51 |
| Thincoat plaster | 1.000 | S.F. | .011 | .08 | .36 | .44 |
| Corners, taped & finished, 12' x 12' room | .333 | L.F. | .005 | .02 | .17 | .19 |
| Paint, primer & 2 coats | 1.000 | S.F. | .011 | .15 | .34 | .49 |
| TOTAL | | | .035 | .48 | 1.15 | 1.63 |
| **WATER-RESISTANT SHEETROCK, 1/2" THICK, TAPED & FINISHED** | | | | | | |
| Drywall, 1/2" thick, water-resistant | 1.000 | S.F. | .008 | .24 | .28 | .52 |
| Finish, taped & finished | 1.000 | S.F. | .008 | .04 | .28 | .32 |
| Corners, taped & finished, 12' x 12' room | .333 | L.F. | .005 | .02 | .17 | .19 |
| Paint, primer & 2 coats | 1.000 | S.F. | .011 | .15 | .34 | .49 |
| TOTAL | | | .032 | .45 | 1.07 | 1.52 |
| **5/8" SHEETROCK, TAPED & FINISHED** | | | | | | |
| Drywall, 5/8" thick, standard | 1.000 | S.F. | .008 | .24 | .28 | .52 |
| Finish, taped & finished | 1.000 | S.F. | .008 | .04 | .28 | .32 |
| Corners, taped & finished, 12' x 12' room | .333 | L.F. | .005 | .02 | .17 | .19 |
| Paint, primer & 2 coats | 1.000 | S.F. | .011 | .15 | .34 | .49 |
| TOTAL | | | .032 | .45 | 1.07 | 1.52 |

The costs in this system are based on a square foot of ceiling.

| Description | QUAN. | UNIT | LABOR HOURS | COST PER S.F. | | |
|---|---|---|---|---|---|---|
| | | | | MAT. | INST. | TOTAL |
| | | | | | | |
| | | | | | | |
| | | | | | | |

**Important: See the Reference Section for critical supporting data - Reference Nos., Crews & Location Factors**

INTERIORS

6

| Drywall & Thincoat Ceilings | QUAN. | UNIT | LABOR HOURS | COST PER S.F. | | |
|---|---|---|---|---|---|---|
| | | | | MAT. | INST. | TOTAL |
| Drywall-sheetrock, 1/2" thick, standard | 1.000 | S.F. | .008 | .23 | .28 | .51 |
| Fire resistant | 1.000 | S.F. | .008 | .24 | .28 | .52 |
| Water resistant | 1.000 | S.F. | .008 | .24 | .28 | .52 |
| 5/8" thick, standard | 1.000 | S.F. | .008 | .24 | .28 | .52 |
| Fire resistant | 1.000 | S.F. | .008 | .24 | .28 | .52 |
| Water resistant | 1.000 | S.F. | .008 | .30 | .28 | .58 |
| Drywall backer for thincoat system, 1/2" thick | 1.000 | S.F. | .016 | .47 | .56 | 1.03 |
| 5/8" thick | 1.000 | S.F. | .016 | .48 | .56 | 1.04 |
| Finish drywall, taped & finished | 1.000 | S.F. | .008 | .04 | .28 | .32 |
| Texture spray | 1.000 | S.F. | .010 | .06 | .33 | .39 |
| Thincoat plaster | 1.000 | S.F. | .011 | .08 | .36 | .44 |
| Corners taped & finished, 4' x 4' room | 1.000 | L.F. | .015 | .07 | .52 | .59 |
| 6' x 6' room | .667 | L.F. | .010 | .05 | .35 | .40 |
| 10' x 10' room | .400 | L.F. | .006 | .03 | .21 | .24 |
| 12' x 12' room | .333 | L.F. | .005 | .02 | .17 | .19 |
| 16' x 16' room | .250 | L.F. | .003 | .01 | .10 | .11 |
| Thincoat system, 4' x 4' room | 1.000 | L.F. | .011 | .08 | .36 | .44 |
| 6' x 6' room | .667 | L.F. | .007 | .05 | .24 | .29 |
| 10' x 10' room | .400 | L.F. | .004 | .03 | .15 | .18 |
| 12' x 12' room | .333 | L.F. | .004 | .03 | .12 | .15 |
| 16' x 16' room | .250 | L.F. | .002 | .01 | .06 | .07 |
| Painting, primer & 1 coat | 1.000 | S.F. | .008 | .10 | .26 | .36 |
| & 2 coats | 1.000 | S.F. | .011 | .15 | .34 | .49 |
| Wallpaper, double roll, solid pattern, avg. workmanship | 1.000 | S.F. | .013 | .31 | .40 | .71 |
| Basic pattern, avg. workmanship | 1.000 | S.F. | .015 | .67 | .48 | 1.15 |
| Basic pattern, quality workmanship | 1.000 | S.F. | .018 | 1.61 | .58 | 2.19 |
| Tile, ceramic adhesive thin set, 4 1/4" x 4 1/4" tiles | 1.000 | S.F. | .084 | 2.30 | 2.50 | 4.80 |
| 6" x 6" tiles | 1.000 | S.F. | .080 | 2.83 | 2.37 | 5.20 |
| Pregrouted sheets | 1.000 | S.F. | .067 | 4.57 | 1.98 | 6.55 |
| | | | | | | |
| | | | | | | |
| | | | | | | |
| | | | | | | |
| | | | | | | |
| | | | | | | |
| | | | | | | |
| | | | | | | |
| | | | | | | |
| | | | | | | |
| | | | | | | |
| | | | | | | |
| | | | | | | |

6 INTERIORS

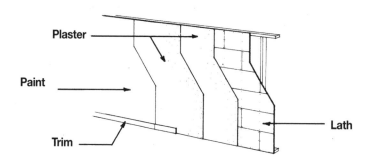

Plaster

Paint

Trim

Lath

| System Description | QUAN. | UNIT | LABOR HOURS | COST PER S.F. | | |
|---|---|---|---|---|---|---|
| | | | | MAT. | INST. | TOTAL |
| **PLASTER ON GYPSUM LATH** | | | | | | |
| Plaster, gypsum or perlite, 2 coats | 1.000 | S.F. | .053 | .38 | 1.74 | 2.12 |
| Lath, 3/8" gypsum | 1.000 | S.F. | .010 | .43 | .34 | .77 |
| Corners, expanded metal, 32 L.F. per 12' x 12' room | .083 | L.F. | .002 | .01 | .06 | .07 |
| Painting, primer & 2 coats | 1.000 | S.F. | .011 | .15 | .34 | .49 |
| Paint trim, to 6" wide, primer + 1 coat enamel | .125 | L.F. | .001 | .01 | .04 | .05 |
| Trim, baseboard | .125 | L.F. | .005 | .18 | .18 | .36 |
| | | | | | | |
| TOTAL | | | .082 | 1.16 | 2.70 | 3.86 |
| **PLASTER ON METAL LATH** | | | | | | |
| Plaster, gypsum or perlite, 2 coats | 1.000 | S.F. | .053 | .38 | 1.74 | 2.12 |
| Lath, 2.5 Lb. diamond, metal | 1.000 | S.F. | .010 | .17 | .34 | .51 |
| Corners, expanded metal, 32 L.F. per 12' x 12' room | .083 | L.F. | .002 | .01 | .06 | .07 |
| Painting, primer & 2 coats | 1.000 | S.F. | .011 | .15 | .34 | .49 |
| Paint trim, to 6" wide, primer + 1 coat enamel | .125 | L.F. | .001 | .01 | .04 | .05 |
| Trim, baseboard | .125 | L.F. | .005 | .18 | .18 | .36 |
| | | | | | | |
| TOTAL | | | .082 | .90 | 2.70 | 3.60 |
| **STUCCO ON METAL LATH** | | | | | | |
| Stucco, 2 coats | 1.000 | S.F. | .041 | .23 | 1.34 | 1.57 |
| Lath, 2.5 Lb. diamond, metal | 1.000 | S.F. | .010 | .17 | .34 | .51 |
| Corners, expanded metal, 32 L.F. per 12' x 12' room | .083 | L.F. | .002 | .01 | .06 | .07 |
| Painting, primer & 2 coats | 1.000 | S.F. | .011 | .15 | .34 | .49 |
| Paint trim, to 6" wide, primer + 1 coat enamel | .125 | L.F. | .001 | .01 | .04 | .05 |
| Trim, baseboard | .125 | L.F. | .005 | .18 | .18 | .36 |
| | | | | | | |
| TOTAL | | | .070 | .75 | 2.30 | 3.05 |

The costs in these systems are based on a per square foot of wall area.
Do not deduct for openings.

| Description | QUAN. | UNIT | LABOR HOURS | COST PER S.F. | | |
|---|---|---|---|---|---|---|
| | | | | MAT. | INST. | TOTAL |
| | | | | | | |
| | | | | | | |
| | | | | | | |

INTERIORS 6

| Plaster & Stucco Wall Price Sheet | QUAN. | UNIT | LABOR HOURS | COST PER S.F. | | |
|---|---|---|---|---|---|---|
| | | | | MAT. | INST. | TOTAL |
| Plaster, gypsum or perlite, 2 coats | 1.000 | S.F. | .053 | .38 | 1.74 | 2.12 |
| 3 coats | 1.000 | S.F. | .065 | .54 | 2.12 | 2.66 |
| Lath, gypsum, standard, 3/8" thick | 1.000 | S.F. | .010 | .43 | .34 | .77 |
| 1/2" thick | 1.000 | S.F. | .013 | .44 | .42 | .86 |
| Fire resistant, 3/8" thick | 1.000 | S.F. | .013 | .43 | .42 | .85 |
| 1/2" thick | 1.000 | S.F. | .014 | .48 | .45 | .93 |
| Metal, diamond, 2.5 Lb. | 1.000 | S.F. | .010 | .17 | .34 | .51 |
| 3.4 Lb. | 1.000 | S.F. | .012 | .25 | .39 | .64 |
| Rib, 2.75 Lb. | 1.000 | S.F. | .012 | .25 | .39 | .64 |
| 3.4 Lb. | 1.000 | S.F. | .013 | .36 | .42 | .78 |
| Corners, expanded metal, 32 L.F. per 4' x 4' room | .250 | L.F. | .005 | .03 | .18 | .21 |
| 6' x 6' room | .110 | L.F. | .002 | .01 | .08 | .09 |
| 10' x 10' room | .100 | L.F. | .002 | .01 | .07 | .08 |
| 12' x 12' room | .083 | L.F. | .002 | .01 | .06 | .07 |
| 16' x 16' room | .063 | L.F. | .001 | .01 | .04 | .05 |
| | | | | | | |
| Painting, primer & 1 coats | 1.000 | S.F. | .008 | .10 | .26 | .36 |
| Primer & 2 coats | 1.000 | S.F. | .011 | .15 | .34 | .49 |
| Wallpaper, low price double roll | 1.000 | S.F. | .013 | .31 | .40 | .71 |
| Medium price double roll | 1.000 | S.F. | .015 | .67 | .48 | 1.15 |
| High price double roll | 1.000 | S.F. | .018 | 1.61 | .58 | 2.19 |
| | | | | | | |
| Tile, ceramic thin set, 4-1/4" x 4-1/4" tiles | 1.000 | S.F. | .084 | 2.30 | 2.50 | 4.80 |
| 6" x 6" tiles | 1.000 | S.F. | .080 | 2.83 | 2.37 | 5.20 |
| Pregrouted sheets | 1.000 | S.F. | .067 | 4.57 | 1.98 | 6.55 |
| | | | | | | |
| Trim, painted or stained, baseboard | .125 | L.F. | .006 | .19 | .22 | .41 |
| Base shoe | .125 | L.F. | .005 | .11 | .19 | .30 |
| Chair rail | .125 | L.F. | .005 | .13 | .17 | .30 |
| Cornice molding | .125 | L.F. | .004 | .10 | .15 | .25 |
| Cove base, vinyl | .125 | L.F. | .003 | .06 | .11 | .17 |
| | | | | | | |
| Paneling not including furring or trim | | | | | | |
| Plywood, prefinished, 1/4" thick, 4' x 8' sheets, vert. grooves | | | | | | |
| Birch faced, minimum | 1.000 | S.F. | .032 | .81 | 1.14 | 1.95 |
| Average | 1.000 | S.F. | .038 | 1.23 | 1.36 | 2.59 |
| Maximum | 1.000 | S.F. | .046 | 1.80 | 1.63 | 3.43 |
| Mahogany, African | 1.000 | S.F. | .040 | 2.31 | 1.42 | 3.73 |
| Philippine (lauan) | 1.000 | S.F. | .032 | .99 | 1.14 | 2.13 |
| Oak or cherry, minimum | 1.000 | S.F. | .032 | 1.94 | 1.14 | 3.08 |
| Maximum | 1.000 | S.F. | .040 | 2.97 | 1.42 | 4.39 |
| Rosewood | 1.000 | S.F. | .050 | 4.21 | 1.78 | 5.99 |
| Teak | 1.000 | S.F. | .040 | 2.97 | 1.42 | 4.39 |
| Chestnut | 1.000 | S.F. | .043 | 4.39 | 1.52 | 5.91 |
| Pecan | 1.000 | S.F. | .040 | 1.89 | 1.42 | 3.31 |
| Walnut, minimum | 1.000 | S.F. | .032 | 2.53 | 1.14 | 3.67 |
| Maximum | 1.000 | S.F. | .040 | 4.80 | 1.42 | 6.22 |
| | | | | | | |
| | | | | | | |
| | | | | | | |
| | | | | | | |
| | | | | | | |
| | | | | | | |
| | | | | | | |

**6 INTERIORS**

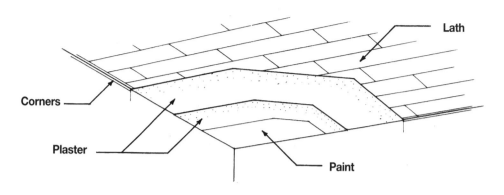

| System Description | QUAN. | UNIT | LABOR HOURS | COST PER S.F. | | |
|---|---|---|---|---|---|---|
| | | | | MAT. | INST. | TOTAL |
| **PLASTER ON GYPSUM LATH** | | | | | | |
| Plaster, gypsum or perlite, 2 coats | 1.000 | S.F. | .061 | .38 | 1.99 | 2.37 |
| Gypsum lath, plain or perforated, nailed, 3/8" thick | 1.000 | S.F. | .010 | .43 | .34 | .77 |
| Gypsum lath, ceiling installation adder | 1.000 | S.F. | .004 | | .13 | .13 |
| Corners, expanded metal, 12' x 12' room | .330 | L.F. | .007 | .04 | .23 | .27 |
| Painting, primer & 2 coats | 1.000 | S.F. | .011 | .15 | .34 | .49 |
| TOTAL | | | .093 | 1.00 | 3.03 | 4.03 |
| **PLASTER ON METAL LATH** | | | | | | |
| Plaster, gypsum or perlite, 2 coats | 1.000 | S.F. | .061 | .38 | 1.99 | 2.37 |
| Lath, 2.5 Lb. diamond, metal | 1.000 | S.F. | .012 | .17 | .39 | .56 |
| Corners, expanded metal, 12' x 12' room | .330 | L.F. | .007 | .04 | .23 | .27 |
| Painting, primer & 2 coats | 1.000 | S.F. | .011 | .15 | .34 | .49 |
| TOTAL | | | .091 | .74 | 2.95 | 3.69 |
| **STUCCO ON GYPSUM LATH** | | | | | | |
| Stucco, 2 coats | 1.000 | S.F. | .041 | .23 | 1.34 | 1.57 |
| Gypsum lath, plain or perforated, nailed, 3/8" thick | 1.000 | S.F. | .010 | .43 | .34 | .77 |
| Gypsum lath, ceiling installation adder | 1.000 | S.F. | .004 | | .13 | .13 |
| Corners, expanded metal, 12' x 12' room | .330 | L.F. | .007 | .04 | .23 | .27 |
| Painting, primer & 2 coats | 1.000 | S.F. | .011 | .15 | .34 | .49 |
| TOTAL | | | .073 | .85 | 2.38 | 3.23 |
| **STUCCO ON METAL LATH** | | | | | | |
| Stucco, 2 coats | 1.000 | S.F. | .041 | .23 | 1.34 | 1.57 |
| Lath, 2.5 Lb. diamond, metal | 1.000 | S.F. | .012 | .17 | .39 | .56 |
| Corners, expanded metal, 12' x 12' room | .330 | L.F. | .007 | .04 | .23 | .27 |
| Painting, primer & 2 coats | 1.000 | S.F. | .011 | .15 | .34 | .49 |
| TOTAL | | | .071 | .59 | 2.30 | 2.89 |

The costs in these systems are based on a square foot of ceiling area.

| Description | QUAN. | UNIT | LABOR HOURS | COST PER S.F. | | |
|---|---|---|---|---|---|---|
| | | | | MAT. | INST. | TOTAL |
| | | | | | | |
| | | | | | | |

**Important: See the Reference Section for critical supporting data - Reference Nos., Crews & Location Factors**

| Plaster & Stucco Ceiling Price Sheet | QUAN. | UNIT | LABOR HOURS | COST PER S.F. | | |
|---|---|---|---|---|---|---|
| | | | | MAT. | INST. | TOTAL |
| Plaster, gypsum or perlite, 2 coats | 1.000 | S.F. | .061 | .38 | 1.99 | 2.37 |
| 3 coats | 1.000 | S.F. | .065 | .54 | 2.12 | 2.66 |
| Lath, gypsum, standard, 3/8" thick | 1.000 | S.F. | .014 | .43 | .47 | .90 |
| 1/2" thick | 1.000 | S.F. | .015 | .43 | .49 | .92 |
| Fire resistant, 3/8" thick | 1.000 | S.F. | .017 | .43 | .55 | .98 |
| 1/2" thick | 1.000 | S.F. | .018 | .48 | .58 | 1.06 |
| Metal, diamond, 2.5 Lb. | 1.000 | S.F. | .012 | .17 | .39 | .56 |
| 3.4 Lb. | 1.000 | S.F. | .015 | .25 | .49 | .74 |
| Rib, 2.75 Lb. | 1.000 | S.F. | .012 | .25 | .39 | .64 |
| 3.4 Lb. | 1.000 | S.F. | .013 | .36 | .42 | .78 |
| Corners expanded metal, 4' x 4' room | 1.000 | L.F. | .020 | .11 | .71 | .82 |
| 6' x 6' room | .667 | L.F. | .013 | .07 | .48 | .55 |
| 10' x 10' room | .400 | L.F. | .008 | .04 | .28 | .32 |
| 12' x 12' room | .333 | L.F. | .007 | .04 | .23 | .27 |
| 16' x 16' room | .250 | L.F. | .004 | .02 | .13 | .15 |
| Painting, primer & 1 coat | 1.000 | S.F. | .008 | .10 | .26 | .36 |
| Primer & 2 coats | 1.000 | S.F. | .011 | .15 | .34 | .49 |

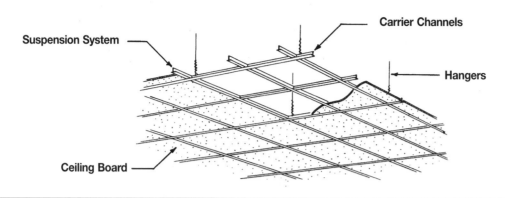

| System Description | QUAN. | UNIT | LABOR HOURS | COST PER S.F. | | |
|---|---|---|---|---|---|---|
| | | | | MAT. | INST. | TOTAL |
| **2' X 2' GRID, FILM FACED FIBERGLASS, 5/8" THICK** | | | | | | |
| Suspension system, 2' x 2' grid, T bar | 1.000 | S.F. | .012 | .82 | .44 | 1.26 |
| Ceiling board, film faced fiberglass, 5/8" thick | 1.000 | S.F. | .013 | .57 | .46 | 1.03 |
| Carrier channels, 1-1/2" x 3/4" | 1.000 | S.F. | .017 | .24 | .61 | .85 |
| Hangers, #12 wire | 1.000 | S.F. | .002 | .03 | .06 | .09 |
| TOTAL | | | .044 | 1.66 | 1.57 | 3.23 |
| **2' X 4' GRID, FILM FACED FIBERGLASS, 5/8" THICK** | | | | | | |
| Suspension system, 2' x 4' grid, T bar | 1.000 | S.F. | .010 | .71 | .36 | 1.07 |
| Ceiling board, film faced fiberglass, 5/8" thick | 1.000 | S.F. | .013 | .57 | .46 | 1.03 |
| Carrier channels, 1-1/2" x 3/4" | 1.000 | S.F. | .017 | .24 | .61 | .85 |
| Hangers, #12 wire | 1.000 | S.F. | .002 | .03 | .06 | .09 |
| TOTAL | | | .042 | 1.55 | 1.49 | 3.04 |
| **2' X 2' GRID, MINERAL FIBER, REVEAL EDGE, 1" THICK** | | | | | | |
| Suspension system, 2' x 2' grid, T bar | 1.000 | S.F. | .012 | .82 | .44 | 1.26 |
| Ceiling board, mineral fiber, reveal edge, 1" thick | 1.000 | S.F. | .013 | 1.31 | .47 | 1.78 |
| Carrier channels, 1-1/2" x 3/4" | 1.000 | S.F. | .017 | .24 | .61 | .85 |
| Hangers, #12 wire | 1.000 | S.F. | .002 | .03 | .06 | .09 |
| TOTAL | | | .044 | 2.40 | 1.58 | 3.98 |
| **2' X 4' GRID, MINERAL FIBER, REVEAL EDGE, 1" THICK** | | | | | | |
| Suspension system, 2' x 4' grid, T bar | 1.000 | S.F. | .010 | .71 | .36 | 1.07 |
| Ceiling board, mineral fiber, reveal edge, 1" thick | 1.000 | S.F. | .013 | 1.31 | .47 | 1.78 |
| Carrier channels, 1-1/2" x 3/4" | 1.000 | S.F. | .017 | .24 | .61 | .85 |
| Hangers, #12 wire | 1.000 | S.F. | .002 | .03 | .06 | .09 |
| TOTAL | | | .042 | 2.29 | 1.50 | 3.79 |

| Description | QUAN. | UNIT | LABOR HOURS | COST PER S.F. | | |
|---|---|---|---|---|---|---|
| | | | | MAT. | INST. | TOTAL |
| | | | | | | |
| | | | | | | |
| | | | | | | |

INTERIORS 6

**Important: See the Reference Section for critical supporting data - Reference Nos., Crews & Location Factors**

# Suspended Ceiling Price Sheet

| Suspended Ceiling Price Sheet | QUAN. | UNIT | LABOR HOURS | COST PER S.F. | | |
|---|---|---|---|---|---|---|
| | | | | MAT. | INST. | TOTAL |
| Suspension systems, T bar, 2' x 2' grid | 1.000 | S.F. | .012 | .82 | .44 | 1.26 |
| 2' x 4' grid | 1.000 | S.F. | .010 | .71 | .36 | 1.07 |
| Concealed Z bar, 12" module | 1.000 | S.F. | .015 | .40 | .55 | .95 |
| | | | | | | |
| Ceiling boards, fiberglass, film faced, 2' x 2' or 2' x 4', 5/8" thick | 1.000 | S.F. | .013 | .57 | .46 | 1.03 |
| 3/4" thick | 1.000 | S.F. | .013 | 1.24 | .47 | 1.71 |
| 3" thick thermal R11 | 1.000 | S.F. | .018 | 1.38 | .63 | 2.01 |
| Glass cloth faced, 3/4" thick | 1.000 | S.F. | .016 | 1.85 | .57 | 2.42 |
| 1" thick | 1.000 | S.F. | .016 | 1.90 | .59 | 2.49 |
| 1-1/2" thick, nubby face | 1.000 | S.F. | .017 | 2.43 | .60 | 3.03 |
| Mineral fiber boards, 5/8" thick, aluminum face 2' x 2' | 1.000 | S.F. | .013 | 1.53 | .47 | 2.00 |
| 2' x 4' | 1.000 | S.F. | .012 | 1.00 | .44 | 1.44 |
| Standard faced, 2' x 2' or 2' x 4' | 1.000 | S.F. | .012 | .67 | .42 | 1.09 |
| Plastic coated face, 2' x 2' or 2' x 4' | 1.000 | S.F. | .020 | 1.05 | .71 | 1.76 |
| Fire rated, 2 hour rating, 5/8" thick | 1.000 | S.F. | .012 | .91 | .42 | 1.33 |
| | | | | | | |
| Tegular edge, 2' x 2' or 2' x 4', 5/8" thick, fine textured | 1.000 | S.F. | .013 | 1.13 | .61 | 1.74 |
| Rough textured | 1.000 | S.F. | .015 | 1.47 | .61 | 2.08 |
| 3/4" thick, fine textured | 1.000 | S.F. | .016 | 1.61 | .63 | 2.24 |
| Rough textured | 1.000 | S.F. | .018 | 1.80 | .63 | 2.43 |
| Luminous panels, prismatic, acrylic | 1.000 | S.F. | .020 | 1.88 | .71 | 2.59 |
| Polystyrene | 1.000 | S.F. | .020 | .96 | .71 | 1.67 |
| Flat or ribbed, acrylic | 1.000 | S.F. | .020 | 3.28 | .71 | 3.99 |
| Polystyrene | 1.000 | S.F. | .020 | 2.24 | .71 | 2.95 |
| Drop pan, white, acrylic | 1.000 | S.F. | .020 | 4.81 | .71 | 5.52 |
| Polystyrene | 1.000 | S.F. | .020 | 4.02 | .71 | 4.73 |
| Carrier channels, 4'-0" on center, 3/4" x 1-1/2" | 1.000 | S.F. | .017 | .24 | .61 | .85 |
| 1-1/2" x 3-1/2" | 1.000 | S.F. | .017 | .44 | .61 | 1.05 |
| Hangers, #12 wire | 1.000 | S.F. | .002 | .03 | .06 | .09 |
| | | | | | | |
| | | | | | | |
| | | | | | | |
| | | | | | | |
| | | | | | | |
| | | | | | | |
| | | | | | | |
| | | | | | | |
| | | | | | | |
| | | | | | | |
| | | | | | | |
| | | | | | | |
| | | | | | | |
| | | | | | | |
| | | | | | | |
| | | | | | | |
| | | | | | | |
| | | | | | | |
| | | | | | | |

6 INTERIORS

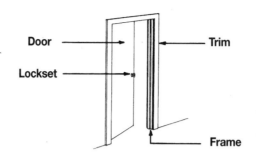

| System Description | QUAN. | UNIT | LABOR HOURS | COST EACH | | |
|---|---|---|---|---|---|---|
| | | | | MAT. | INST. | TOTAL |
| **LAUAN, FLUSH DOOR, HOLLOW CORE** | | | | | | |
| Door, flush, lauan, hollow core, 2'-8" wide x 6'-8" high | 1.000 | Ea. | .889 | 32.50 | 31.50 | 64.00 |
| Frame, pine, 4-5/8" jamb | 17.000 | L.F. | .725 | 96.90 | 25.84 | 122.74 |
| Trim, stock pine, 11/16" x 2-1/2" | 34.000 | L.F. | 1.133 | 25.50 | 40.46 | 65.96 |
| Paint trim, to 6" wide, primer + 1 coat enamel | 34.000 | L.F. | .340 | 3.40 | 10.88 | 14.28 |
| Butt hinges, chrome, 3-1/2" x 3-1/2" | 1.500 | Pr. | | 24.60 | | 24.60 |
| Lockset, passage | 1.000 | Ea. | .500 | 13.85 | 17.80 | 31.65 |
| Prime door & frame, oil, brushwork | 2.000 | Face | 1.600 | 4.52 | 51.00 | 55.52 |
| Paint door and frame, oil, 2 coats | 2.000 | Face | 2.667 | 7.28 | 85.00 | 92.28 |
| | | | | | | |
| TOTAL | | | 7.854 | 208.55 | 262.48 | 471.03 |
| **BIRCH, FLUSH DOOR, HOLLOW CORE** | | | | | | |
| Door, flush, birch, hollow core, 2'-8" wide x 6'-8" high | 1.000 | Ea. | .889 | 46.50 | 31.50 | 78.00 |
| Frame, pine, 4-5/8" jamb | 17.000 | L.F. | .725 | 96.90 | 25.84 | 122.74 |
| Trim, stock pine, 11/16" x 2-1/2" | 34.000 | L.F. | 1.133 | 25.50 | 40.46 | 65.96 |
| Butt hinges, chrome, 3-1/2" x 3-1/2" | 1.500 | Pr. | | 24.60 | | 24.60 |
| Lockset, passage | 1.000 | Ea. | .500 | 13.85 | 17.80 | 31.65 |
| Prime door & frame, oil, brushwork | 2.000 | Face | 1.600 | 4.52 | 51.00 | 55.52 |
| Paint door and frame, oil, 2 coats | 2.000 | Face | 2.667 | 7.28 | 85.00 | 92.28 |
| | | | | | | |
| TOTAL | | | 7.514 | 219.15 | 251.60 | 470.75 |
| **RAISED PANEL, SOLID, PINE DOOR** | | | | | | |
| Door, pine, raised panel, 2'-8" wide x 6'-8" high | 1.000 | Ea. | .889 | 157.00 | 31.50 | 188.50 |
| Frame, pine, 4-5/8" jamb | 17.000 | L.F. | .725 | 96.90 | 25.84 | 122.74 |
| Trim, stock pine, 11/16" x 2-1/2" | 34.000 | L.F. | 1.133 | 25.50 | 40.46 | 65.96 |
| Butt hinges, bronze, 3-1/2" x 3-1/2" | 1.500 | Pr. | | 28.05 | | 28.05 |
| Lockset, passage | 1.000 | Ea. | .500 | 13.85 | 17.80 | 31.65 |
| Prime door & frame, oil, brushwork | 2.000 | | 1.600 | 4.52 | 51.00 | 55.52 |
| Paint door and frame, oil, 2 coats | 2.000 | | 2.667 | 7.28 | 85.00 | 92.28 |
| | | | | | | |
| TOTAL | | | 7.514 | 333.10 | 251.60 | 584.70 |

The costs in these systems are based on a cost per each door.

| Description | QUAN. | UNIT | LABOR HOURS | COST EACH | | |
|---|---|---|---|---|---|---|
| | | | | MAT. | INST. | TOTAL |
| | | | | | | |
| | | | | | | |
| | | | | | | |

**Important: See the Reference Section for critical supporting data - Reference Nos., Crews & Location Factors**

| Interior Door Price Sheet | QUAN. | UNIT | LABOR HOURS | COST EACH | | |
|---|---|---|---|---|---|---|
| | | | | MAT. | INST. | TOTAL |
| Door, hollow core, lauan 1-3/8" thick, 6'-8" high x 1'-6" wide | 1.000 | Ea. | .889 | 25.50 | 31.50 | 57.00 |
| 2'-0" wide | 1.000 | Ea. | .889 | 28.00 | 31.50 | 59.50 |
| 2'-6" wide | 1.000 | Ea. | .889 | 31.50 | 31.50 | 63.00 |
| 2'-8" wide | 1.000 | Ea. | .889 | 32.50 | 31.50 | 64.00 |
| 3'-0" wide | 1.000 | Ea. | .941 | 35.00 | 33.50 | 68.50 |
| Birch 1-3/8" thick, 6'-8" high x 1'-6" wide | 1.000 | Ea. | .889 | 33.50 | 31.50 | 65.00 |
| 2'-0" wide | 1.000 | Ea. | .889 | 40.00 | 31.50 | 71.50 |
| 2'-6" wide | 1.000 | Ea. | .889 | 46.50 | 31.50 | 78.00 |
| 2'-8" wide | 1.000 | Ea. | .889 | 46.50 | 31.50 | 78.00 |
| 3'-0" wide | 1.000 | Ea. | .941 | 52.50 | 33.50 | 86.00 |
| Louvered pine 1-3/8" thick, 6'-8" high x 1'-6" wide | 1.000 | Ea. | .842 | 101.00 | 30.00 | 131.00 |
| 2'-0" wide | 1.000 | Ea. | .889 | 127.00 | 31.50 | 158.50 |
| 2'-6" wide | 1.000 | Ea. | .889 | 138.00 | 31.50 | 169.50 |
| 2'-8" wide | 1.000 | Ea. | .889 | 146.00 | 31.50 | 177.50 |
| 3'-0" wide | 1.000 | Ea. | .941 | 156.00 | 33.50 | 189.50 |
| Paneled pine 1-3/8" thick, 6'-8" high x 1'-6" wide | 1.000 | Ea. | .842 | 112.00 | 30.00 | 142.00 |
| 2'-0" wide | 1.000 | Ea. | .889 | 129.00 | 31.50 | 160.50 |
| 2'-6" wide | 1.000 | Ea. | .889 | 145.00 | 31.50 | 176.50 |
| 2'-8" wide | 1.000 | Ea. | .889 | 157.00 | 31.50 | 188.50 |
| 3'-0" wide | 1.000 | Ea. | .941 | 167.00 | 33.50 | 200.50 |
| Frame, pine, 1'-6" thru 2'-0" wide door, 3-5/8" deep | 16.000 | L.F. | .683 | 68.00 | 24.50 | 92.50 |
| 4-5/8" deep | 16.000 | L.F. | .683 | 91.00 | 24.50 | 115.50 |
| 5-5/8" deep | 16.000 | L.F. | .683 | 85.00 | 24.50 | 109.50 |
| 2'-6" thru 3'0" wide door, 3-5/8" deep | 17.000 | L.F. | .725 | 72.00 | 26.00 | 98.00 |
| 4-5/8" deep | 17.000 | L.F. | .725 | 97.00 | 26.00 | 123.00 |
| 5-5/8" deep | 17.000 | L.F. | .725 | 90.00 | 26.00 | 116.00 |
| Trim, casing, painted, both sides, 1'-6" thru 2'-6" wide door | 32.000 | L.F. | 1.855 | 26.00 | 63.00 | 89.00 |
| 2'-6" thru 3'-0" wide door | 34.000 | L.F. | 1.971 | 27.50 | 67.00 | 94.50 |
| Butt hinges 3-1/2" x 3-1/2", steel plated, chrome | 1.500 | Pr. | | 24.50 | | 24.50 |
| Bronze | 1.500 | Pr. | | 28.00 | | 28.00 |
| Locksets, passage, minimum | 1.000 | Ea. | .500 | 13.85 | 17.80 | 31.65 |
| Maximum | 1.000 | Ea. | .575 | 15.95 | 20.50 | 36.45 |
| Privacy, miniumum | 1.000 | Ea. | .625 | 17.30 | 22.50 | 39.80 |
| Maximum | 1.000 | Ea. | .675 | 18.70 | 24.00 | 42.70 |
| Paint 2 sides, primer & 2 cts., flush door, 1'-6" to 2'-0" wide | 2.000 | Face | 5.547 | 13.60 | 177.00 | 190.60 |
| 2'-6" thru 3'-0" wide | 2.000 | Face | 6.933 | 17.00 | 221.00 | 238.00 |
| Louvered door, 1'-6" thru 2'-0" wide | 2.000 | Face | 6.400 | 13.10 | 204.00 | 217.10 |
| 2'-6" thru 3'-0" wide | 2.000 | Face | 8.000 | 16.35 | 255.00 | 271.35 |
| Paneled door, 1'-6" thru 2'-0" wide | 2.000 | Face | 6.400 | 13.10 | 204.00 | 217.10 |
| 2'-6" thru 3'-0" wide | 2.000 | Face | 8.000 | 16.35 | 255.00 | 271.35 |
| | | | | | | |
| | | | | | | |
| | | | | | | |
| | | | | | | |
| | | | | | | |
| | | | | | | |
| | | | | | | |
| | | | | | | |
| | | | | | | |
| | | | | | | |
| | | | | | | |

**6 INTERIORS**

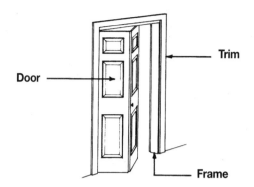

Door

Trim

Frame

| System Description | QUAN. | UNIT | LABOR HOURS | COST EACH | | |
|---|---|---|---|---|---|---|
| | | | | MAT. | INST. | TOTAL |
| **BI-PASSING, FLUSH, LAUAN, HOLLOW CORE, 4'-0" X 6'-8"** | | | | | | |
| Door, flush, lauan, hollow core, 4'-0" x 6'-8" opening | 1.000 | Ea. | 1.333 | 174.00 | 47.50 | 221.50 |
| Frame, pine, 4-5/8" jamb | 18.000 | L.F. | .768 | 102.60 | 27.36 | 129.96 |
| Trim, stock pine, 11/16" x 2-1/2" | 36.000 | L.F. | 1.200 | 27.00 | 42.84 | 69.84 |
| Prime door & frame, oil, brushwork | 2.000 | Face | 1.600 | 4.52 | 51.00 | 55.52 |
| Paint door and frame, oil, 2 coats | 2.000 | Face | 2.667 | 7.28 | 85.00 | 92.28 |
| TOTAL | | | 7.568 | 315.40 | 253.70 | 569.10 |
| **BI-PASSING, FLUSH, BIRCH, HOLLOW CORE, 6'-0" X 6'-8"** | | | | | | |
| Door, flush, birch, hollow core, 6'-0" x 6'-8" opening | 1.000 | Ea. | 1.600 | 235.00 | 57.00 | 292.00 |
| Frame, pine, 4-5/8" jamb | 19.000 | L.F. | .811 | 108.30 | 28.88 | 137.18 |
| Trim, stock pine, 11/16" x 2-1/2" | 38.000 | L.F. | 1.267 | 28.50 | 45.22 | 73.72 |
| Prime door & frame, oil, brushwork | 2.000 | Face | 2.000 | 5.65 | 63.75 | 69.40 |
| Paint door and frame, oil, 2 coats | 2.000 | Face | 3.333 | 9.10 | 106.25 | 115.35 |
| TOTAL | | | 9.011 | 386.55 | 301.10 | 687.65 |
| **BI-FOLD, PINE, PANELED, 3'-0" X 6'-8"** | | | | | | |
| Door, pine, paneled, 3'-0" x 6'-8" opening | 1.000 | Ea. | 1.231 | 126.00 | 44.00 | 170.00 |
| Frame, pine, 4-5/8" jamb | 17.000 | L.F. | .725 | 96.90 | 25.84 | 122.74 |
| Trim, stock pine, 11/16" x 2-1/2" | 34.000 | L.F. | 1.133 | 25.50 | 40.46 | 65.96 |
| Prime door & frame, oil, brushwork | 2.000 | Face | 1.600 | 4.52 | 51.00 | 55.52 |
| Paint door and frame, oil, 2 coats | 2.000 | Face | 2.667 | 7.28 | 85.00 | 92.28 |
| TOTAL | | | 7.356 | 260.20 | 246.30 | 506.50 |
| **BI-FOLD, PINE, LOUVERED, 6'-0" X 6'-8"** | | | | | | |
| Door, pine, louvered, 6'-0" x 6'-8" opening | 1.000 | Ea. | 1.600 | 247.00 | 57.00 | 304.00 |
| Frame, pine, 4-5/8" jamb | 19.000 | L.F. | .811 | 108.30 | 28.88 | 137.18 |
| Trim, stock pine, 11/16" x 2-1/2" | 38.000 | L.F. | 1.267 | 28.50 | 45.22 | 73.72 |
| Prime door & frame, oil, brushwork | 2.500 | Face | 2.000 | 5.65 | 63.75 | 69.40 |
| Paint door and frame, oil, 2 coats | 2.500 | Face | 3.333 | 9.10 | 106.25 | 115.35 |
| TOTAL | | | 9.011 | 398.55 | 301.10 | 699.65 |

The costs in this system are based on a cost per each door.

| Description | QUAN. | UNIT | LABOR HOURS | COST EACH | | |
|---|---|---|---|---|---|---|
| | | | | MAT. | INST. | TOTAL |
| | | | | | | |
| | | | | | | |

INTERIORS 6

**Important: See the Reference Section for critical supporting data - Reference Nos., Crews & Location Factors**

| Closet Door Price Sheet | QUAN. | UNIT | LABOR HOURS | COST EACH | | |
|---|---|---|---|---|---|---|
| | | | | MAT. | INST. | TOTAL |
| Doors, bi-passing, pine, louvered, 4'-0" x 6'-8" opening | 1.000 | Ea. | 1.333 | 390.00 | 47.50 | 437.50 |
| 6'-0" x 6'-8" opening | 1.000 | Ea. | 1.600 | 480.00 | 57.00 | 537.00 |
| Paneled, 4'-0" x 6'-8" opening | 1.000 | Ea. | 1.333 | 375.00 | 47.50 | 422.50 |
| 6'-0" x 6'-8" opening | 1.000 | Ea. | 1.600 | 465.00 | 57.00 | 522.00 |
| Flush, birch, hollow core, 4'-0" x 6'-8" opening | 1.000 | Ea. | 1.333 | 196.00 | 47.50 | 243.50 |
| 6'-0" x 6'-8" opening | 1.000 | Ea. | 1.600 | 235.00 | 57.00 | 292.00 |
| Flush, lauan, hollow core, 4'-0" x 6'-8" opening | 1.000 | Ea. | 1.333 | 174.00 | 47.50 | 221.50 |
| 6'-0" x 6'-8" opening | 1.000 | Ea. | 1.600 | 201.00 | 57.00 | 258.00 |
| Bi-fold, pine, louvered, 3'-0" x 6'-8" opening | 1.000 | Ea. | 1.231 | 126.00 | 44.00 | 170.00 |
| 6'-0" x 6'-8" opening | 1.000 | Ea. | 1.600 | 247.00 | 57.00 | 304.00 |
| Paneled, 3'-0" x 6'-8" opening | 1.000 | Ea. | 1.231 | 126.00 | 44.00 | 170.00 |
| 6'-0" x 6'-8" opening | 1.000 | Ea. | 1.600 | 247.00 | 57.00 | 304.00 |
| Flush, birch, hollow core, 3'-0" x 6'-8" opening | 1.000 | Ea. | 1.231 | 49.50 | 44.00 | 93.50 |
| 6'-0" x 6'-8" opening | 1.000 | Ea. | 1.600 | 108.00 | 57.00 | 165.00 |
| Flush, lauan, hollow core, 3'-0" x 6'8" opening | 1.000 | Ea. | 1.231 | 158.00 | 44.00 | 202.00 |
| 6'-0" x 6'-8" opening | 1.000 | Ea. | 1.600 | 310.00 | 57.00 | 367.00 |
| Frame pine, 3'-0" door, 3-5/8" deep | 17.000 | L.F. | .725 | 72.00 | 26.00 | 98.00 |
| 4-5/8" deep | 17.000 | L.F. | .725 | 97.00 | 26.00 | 123.00 |
| 5-5/8" deep | 17.000 | L.F. | .725 | 90.00 | 26.00 | 116.00 |
| 4'-0" door, 3-5/8" deep | 18.000 | L.F. | .768 | 76.50 | 27.50 | 104.00 |
| 4-5/8" deep | 18.000 | L.F. | .768 | 103.00 | 27.50 | 130.50 |
| 5-5/8" deep | 18.000 | L.F. | .768 | 95.50 | 27.50 | 123.00 |
| 6'-0" door, 3-5/8" deep | 19.000 | L.F. | .811 | 80.50 | 29.00 | 109.50 |
| 4-5/8" deep | 19.000 | L.F. | .811 | 108.00 | 29.00 | 137.00 |
| 5-5/8" deep | 19.000 | L.F. | .811 | 101.00 | 29.00 | 130.00 |
| Trim both sides, painted 3'-0" x 6'-8" door | 34.000 | L.F. | 1.971 | 27.50 | 67.00 | 94.50 |
| 4'-0" x 6'-8" door | 36.000 | L.F. | 2.086 | 29.00 | 71.00 | 100.00 |
| 6'-0" x 6'-8" door | 38.000 | L.F. | 2.203 | 31.00 | 75.00 | 106.00 |
| Paint 2 sides, primer & 2 cts., flush door & frame, 3' x 6'-8" opng | 2.000 | Face | 2.914 | 8.85 | 102.00 | 110.85 |
| 4'-0" x 6'-8" opening | 2.000 | Face | 3.886 | 11.80 | 136.00 | 147.80 |
| 6'-0" x 6'-8" opening | 2.000 | Face | 4.857 | 14.75 | 170.00 | 184.75 |
| Paneled door & frame, 3'-0" x 6'-8" opening | 2.000 | Face | 6.000 | 12.30 | 191.00 | 203.30 |
| 4'-0" x 6'-8" opening | 2.000 | Face | 8.000 | 16.35 | 255.00 | 271.35 |
| 6'-0" x 6'-8" opening | 2.000 | Face | 10.000 | 20.50 | 320.00 | 340.50 |
| Louvered door & frame, 3'-0" x 6'-8" opening | 2.000 | Face | 6.000 | 12.30 | 191.00 | 203.30 |
| 4'-0" x 6'-8" opening | 2.000 | Face | 8.000 | 16.35 | 255.00 | 271.35 |
| 6'-0" x 6'-8" opening | 2.000 | Face | 10.000 | 20.50 | 320.00 | 340.50 |
| | | | | | | |
| | | | | | | |
| | | | | | | |
| | | | | | | |
| | | | | | | |
| | | | | | | |
| | | | | | | |
| | | | | | | |

**6**

| System Description | QUAN. | UNIT | LABOR HOURS | COST PER S.F. | | |
|---|---|---|---|---|---|---|
| | | | | MAT. | INST. | TOTAL |
| Carpet, direct glue-down, nylon, level loop, 26 oz. | 1.000 | S.F. | .018 | 1.71 | .59 | 2.30 |
| 32 oz. | 1.000 | S.F. | .018 | 2.41 | .59 | 3.00 |
| 40 oz. | 1.000 | S.F. | .018 | 3.59 | .59 | 4.18 |
| Nylon, plush, 20 oz. | 1.000 | S.F. | .018 | 1.13 | .59 | 1.72 |
| 24 oz. | 1.000 | S.F. | .018 | 1.20 | .59 | 1.79 |
| 30 oz. | 1.000 | S.F. | .018 | 1.78 | .59 | 2.37 |
| 36 oz. | 1.000 | S.F. | .018 | 2.25 | .59 | 2.84 |
| 42 oz. | 1.000 | S.F. | .022 | 2.35 | .71 | 3.06 |
| 48 oz. | 1.000 | S.F. | .022 | 3.10 | .71 | 3.81 |
| 54 oz. | 1.000 | S.F. | .022 | 3.50 | .71 | 4.21 |
| Olefin, 15 oz. | 1.000 | S.F. | .018 | .61 | .59 | 1.20 |
| 22 oz. | 1.000 | S.F. | .018 | .72 | .59 | 1.31 |
| Tile, foam backed, needle punch | 1.000 | S.F. | .014 | 2.65 | .46 | 3.11 |
| Tufted loop or shag | 1.000 | S.F. | .014 | 1.12 | .46 | 1.58 |
| Wool, 36 oz., level loop | 1.000 | S.F. | .018 | 7.90 | .59 | 8.49 |
| 32 oz., patterned | 1.000 | S.F. | .020 | 7.80 | .66 | 8.46 |
| 48 oz., patterned | 1.000 | S.F. | .020 | 7.95 | .66 | 8.61 |
| Padding, sponge rubber cushion, minimum | 1.000 | S.F. | .006 | .35 | .20 | .55 |
| Maximum | 1.000 | S.F. | .006 | .95 | .20 | 1.15 |
| Felt, 32 oz. to 56 oz., minimum | 1.000 | S.F. | .006 | .40 | .20 | .60 |
| Maximum | 1.000 | S.F. | .006 | .74 | .20 | .94 |
| Bonded urethane, 3/8" thick, minimum | 1.000 | S.F. | .006 | .43 | .20 | .63 |
| Maximum | 1.000 | S.F. | .006 | .74 | .20 | .94 |
| Prime urethane, 1/4" thick, minimum | 1.000 | S.F. | .006 | .25 | .20 | .45 |
| Maximum | 1.000 | S.F. | .006 | .46 | .20 | .66 |
| Stairs, for stairs, add to above carpet prices | 1.000 | Riser | .267 | | 8.75 | 8.75 |
| Underlayment plywood, 3/8" thick | 1.000 | S.F. | .011 | .58 | .38 | .96 |
| 1/2" thick | 1.000 | S.F. | .011 | .70 | .39 | 1.09 |
| 5/8" thick | 1.000 | S.F. | .011 | .77 | .41 | 1.18 |
| 3/4" thick | 1.000 | S.F. | .012 | .94 | .44 | 1.38 |
| Particle board, 3/8" thick | 1.000 | S.F. | .011 | .42 | .38 | .80 |
| 1/2" thick | 1.000 | S.F. | .011 | .44 | .39 | .83 |
| 5/8" thick | 1.000 | S.F. | .011 | .59 | .41 | 1.00 |
| 3/4" thick | 1.000 | S.F. | .012 | .64 | .44 | 1.08 |
| Hardboard, 4' x 4', 0.215" thick | 1.000 | S.F. | .011 | .45 | .38 | .83 |

**Important: See the Reference Section for critical supporting data - Reference Nos., Crews & Location Factors**

| System Description | QUAN. | UNIT | LABOR HOURS | COST PER S.F. | | |
|---|---|---|---|---|---|---|
| | | | | MAT. | INST. | TOTAL |
| Resilient flooring, asphalt tile on concrete, 1/8" thick | | | | | | |
|     Color group B | 1.000 | S.F. | .020 | 1.06 | .66 | 1.72 |
|     Color group C & D | 1.000 | S.F. | .020 | 1.17 | .66 | 1.83 |
| | | | | | | |
|    Asphalt tile on wood subfloor, 1/8" thick | | | | | | |
|     Color group B | 1.000 | S.F. | .020 | 1.26 | .66 | 1.92 |
|     Color group C & D | 1.000 | S.F. | .020 | 1.37 | .66 | 2.03 |
| | | | | | | |
| Vinyl composition tile, 12" x 12", 1/16" thick | 1.000 | S.F. | .016 | .80 | .53 | 1.33 |
|     Embossed | 1.000 | S.F. | .016 | .98 | .53 | 1.51 |
|     Marbleized | 1.000 | S.F. | .016 | .98 | .53 | 1.51 |
|     Plain | 1.000 | S.F. | .016 | 1.10 | .53 | 1.63 |
|    .080" thick, embossed | 1.000 | S.F. | .016 | 1.01 | .53 | 1.54 |
|     Marbleized | 1.000 | S.F. | .016 | 1.11 | .53 | 1.64 |
|     Plain | 1.000 | S.F. | .016 | 1.62 | .53 | 2.15 |
| | | | | | | |
|    1/8" thick, marbleized | 1.000 | S.F. | .016 | 1.08 | .53 | 1.61 |
|     Plain | 1.000 | S.F. | .016 | 2.05 | .53 | 2.58 |
| Vinyl tile, 12" x 12", .050" thick, minimum | 1.000 | S.F. | .016 | 1.75 | .53 | 2.28 |
|     Maximum | 1.000 | S.F. | .016 | 3.41 | .53 | 3.94 |
|    1/8" thick, minimum | 1.000 | S.F. | .016 | 2.20 | .53 | 2.73 |
|     Maximum | 1.000 | S.F. | .016 | 4.93 | .53 | 5.46 |
|    1/8" thick, solid colors | 1.000 | S.F. | .016 | 3.52 | .53 | 4.05 |
|     Florentine pattern | 1.000 | S.F. | .016 | 4.05 | .53 | 4.58 |
|     Marbleized or travertine pattern | 1.000 | S.F. | .016 | 8.30 | .53 | 8.83 |
| | | | | | | |
| Vinyl sheet goods, backed, .070" thick, minimum | 1.000 | S.F. | .032 | 2.04 | 1.05 | 3.09 |
|     Maximum | 1.000 | S.F. | .040 | 2.76 | 1.32 | 4.08 |
|    .093" thick, minimum | 1.000 | S.F. | .035 | 2.20 | 1.14 | 3.34 |
|     Maximum | 1.000 | S.F. | .040 | 3.14 | 1.32 | 4.46 |
|    .125" thick, minimum | 1.000 | S.F. | .035 | 2.48 | 1.14 | 3.62 |
|     Maximum | 1.000 | S.F. | .040 | 3.93 | 1.32 | 5.25 |
| Wood, oak, finished in place, 25/32" x 2-1/2" clear | 1.000 | S.F. | .074 | 3.69 | 2.38 | 6.07 |
|     Select | 1.000 | S.F. | .074 | 3.77 | 2.38 | 6.15 |
|     No. 1 common | 1.000 | S.F. | .074 | 4.57 | 2.38 | 6.95 |
|    Prefinished, oak, 2-1/2" wide | 1.000 | S.F. | .047 | 6.45 | 1.68 | 8.13 |
|     3-1/4" wide | 1.000 | S.F. | .043 | 8.25 | 1.54 | 9.79 |
|    Ranch plank, oak, random width | 1.000 | S.F. | .055 | 8.00 | 1.96 | 9.96 |
|    Parquet, 5/16" thick, finished in place, oak, minimum | 1.000 | S.F. | .077 | 3.68 | 2.48 | 6.16 |
|     Maximum | 1.000 | S.F. | .107 | 6.35 | 3.55 | 9.90 |
|     Teak, minimum | 1.000 | S.F. | .077 | 5.30 | 2.48 | 7.78 |
|     Maximum | 1.000 | S.F. | .107 | 8.70 | 3.55 | 12.25 |
| Sleepers, treated, 16" O.C., 1" x 2" | 1.000 | S.F. | .007 | .11 | .24 | .35 |
|     1" x 3" | 1.000 | S.F. | .008 | .20 | .28 | .48 |
|     2" x 4" | 1.000 | S.F. | .011 | .55 | .38 | .93 |
|     2" x 6" | 1.000 | S.F. | .012 | 1.23 | .44 | 1.67 |
| Subfloor, plywood, 1/2" thick | 1.000 | S.F. | .011 | .54 | .38 | .92 |
|     5/8" thick | 1.000 | S.F. | .012 | .62 | .42 | 1.04 |
|     3/4" thick | 1.000 | S.F. | .013 | .73 | .46 | 1.19 |
| Ceramic tile, color group 2, 1" x 1" | 1.000 | S.F. | .087 | 4.61 | 2.59 | 7.20 |
|     2" x 2" or 2" x 1" | 1.000 | S.F. | .084 | 4.37 | 2.50 | 6.87 |
|    Color group 1, 8" x 8" | | S.F. | .064 | 3.39 | 1.90 | 5.29 |
|     12" x 12" | | S.F. | .049 | 4.25 | 1.46 | 5.71 |
|     16" x 16" | | S.F. | .029 | 5.45 | .86 | 6.31 |

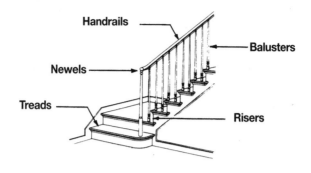

Handrails

Balusters

Newels

Treads

Risers

| System Description | QUAN. | UNIT | LABOR HOURS | COST EACH | | |
|---|---|---|---|---|---|---|
| | | | | MAT. | INST. | TOTAL |
| **7 RISERS, OAK TREADS, BOX STAIRS** | | | | | | |
| Treads, oak, 9-1/2" x 1-1/16" thick | 6.000 | Ea. | 2.667 | 153.00 | 94.80 | 247.80 |
| Risers, 3/4" thick, beech | 7.000 | Ea. | 2.625 | 133.35 | 93.45 | 226.80 |
| Balusters, birch, 30" high | 12.000 | Ea. | 3.429 | 83.40 | 121.80 | 205.20 |
| Newels, 3-1/4" wide | 2.000 | Ea. | 2.286 | 74.00 | 81.00 | 155.00 |
| Handrails, oak laminated | 7.000 | L.F. | .933 | 50.05 | 33.25 | 83.30 |
| Stringers, 2" x 10", 3 each | 21.000 | L.F. | .306 | 8.19 | 10.92 | 19.11 |
| TOTAL | | | 12.246 | 501.99 | 435.22 | 937.21 |
| **14 RISERS, OAK TREADS, BOX STAIRS** | | | | | | |
| Treads, oak, 9-1/2" x 1-1/16" thick | 13.000 | Ea. | 5.778 | 331.50 | 205.40 | 536.90 |
| Risers, 3/4" thick, beech | 14.000 | Ea. | 5.250 | 266.70 | 186.90 | 453.60 |
| Balusters, birch, 30" high | 26.000 | Ea. | 7.428 | 180.70 | 263.90 | 444.60 |
| Newels, 3-1/4" wide | 2.000 | Ea. | 2.286 | 74.00 | 81.00 | 155.00 |
| Handrails, oak, laminated | 14.000 | L.F. | 1.867 | 100.10 | 66.50 | 166.60 |
| Stringers, 2" x 10", 3 each | 42.000 | L.F. | 5.169 | 57.54 | 183.96 | 241.50 |
| TOTAL | | | 27.778 | 1010.54 | 987.66 | 1998.20 |
| **14 RISERS, PINE TREADS, BOX STAIRS** | | | | | | |
| Treads, pine, 9-1/2" x 3/4" thick | 13.000 | Ea. | 5.778 | 197.60 | 205.40 | 403.00 |
| Risers, 3/4" thick, pine | 14.000 | Ea. | 5.091 | 129.36 | 181.44 | 310.80 |
| Balusters, pine, 30" high | 26.000 | Ea. | 7.428 | 117.52 | 263.90 | 381.42 |
| Newels, 3-1/4" wide | 2.000 | Ea. | 2.286 | 74.00 | 81.00 | 155.00 |
| Handrails, oak, laminated | 14.000 | L.F. | 1.867 | 100.10 | 66.50 | 166.60 |
| Stringers, 2" x 10", 3 each | 42.000 | L.F. | 5.169 | 57.54 | 183.96 | 241.50 |
| TOTAL | | | 27.619 | 676.12 | 982.20 | 1658.32 |

| Description | QUAN. | UNIT | LABOR HOURS | COST EACH | | |
|---|---|---|---|---|---|---|
| | | | | MAT. | INST. | TOTAL |
| | | | | | | |
| | | | | | | |
| | | | | | | |
| | | | | | | |
| | | | | | | |

**Important: See the Reference Section for critical supporting data - Reference Nos., Crews & Location Factors**

INTERIORS 6

## Stairway Price Sheet

| Stairway Price Sheet | QUAN. | UNIT | LABOR HOURS | COST EACH MAT. | COST EACH INST. | COST EACH TOTAL |
|---|---|---|---|---|---|---|
| reads, oak, 1-1/16" x 9-1/2", 3' long, 7 riser stair | 6.000 | Ea. | 2.667 | 153.00 | 95.00 | 248.00 |
| 14 riser stair | 13.000 | Ea. | 5.778 | 330.00 | 205.00 | 535.00 |
| 1-1/16" x 11-1/2", 3' long, 7 riser stair | 6.000 | Ea. | 2.667 | 156.00 | 95.00 | 251.00 |
| 14 riser stair | 13.000 | Ea. | 5.778 | 340.00 | 205.00 | 545.00 |
| Pine, 3/4" x 9-1/2", 3' long, 7 riser stair | 6.000 | Ea. | 2.667 | 91.00 | 95.00 | 186.00 |
| 14 riser stair | 13.000 | Ea. | 5.778 | 198.00 | 205.00 | 403.00 |
| 3/4" x 11-1/4", 3' long, 7 riser stair | 6.000 | Ea. | 2.667 | 109.00 | 95.00 | 204.00 |
| 14 riser stair | 13.000 | Ea. | 5.778 | 236.00 | 205.00 | 441.00 |
| isers, oak, 3/4" x 7-1/2" high, 7 riser stair | 7.000 | Ea. | 2.625 | 116.00 | 93.50 | 209.50 |
| 14 riser stair | 14.000 | Ea. | 5.250 | 231.00 | 187.00 | 418.00 |
| Beech, 3/4" x 7-1/2" high, 7 riser stair | 7.000 | Ea. | 2.625 | 133.00 | 93.50 | 226.50 |
| 14 riser stair | 14.000 | Ea. | 5.250 | 267.00 | 187.00 | 454.00 |
| aluster, turned, 30" high, pine, 7 riser stair | 12.000 | Ea. | 3.429 | 54.00 | 122.00 | 176.00 |
| 14 riser stair | 26.000 | Ea. | 7.428 | 118.00 | 264.00 | 382.00 |
| 30" birch, 7 riser stair | 12.000 | Ea. | 3.429 | 83.50 | 122.00 | 205.50 |
| 14 riser stair | 26.000 | Ea. | 7.428 | 181.00 | 264.00 | 445.00 |
| 42" pine, 7 riser stair | 12.000 | Ea. | 3.556 | 79.00 | 127.00 | 206.00 |
| 14 riser stair | 26.000 | Ea. | 7.704 | 172.00 | 274.00 | 446.00 |
| 42" birch, 7 riser stair | 12.000 | Ea. | 3.556 | 101.00 | 127.00 | 228.00 |
| 14 riser stair | 26.000 | Ea. | 7.704 | 218.00 | 274.00 | 492.00 |
| ewels, 3-1/4" wide, starting, 7 riser stair | 2.000 | Ea. | 2.286 | 74.00 | 81.00 | 155.00 |
| 14 riser stair | 2.000 | Ea. | 2.286 | 74.00 | 81.00 | 155.00 |
| Landing, 7 riser stair | 2.000 | Ea. | 3.200 | 162.00 | 114.00 | 276.00 |
| 14 riser stair | 2.000 | Ea. | 3.200 | 162.00 | 114.00 | 276.00 |
| andrails, oak, laminated, 7 riser stair | 7.000 | L.F. | .933 | 50.00 | 33.50 | 83.50 |
| 14 riser stair | 14.000 | L.F. | 1.867 | 100.00 | 66.50 | 166.50 |
| tringers, fir, 2" x 10" 7 riser stair | 21.000 | L.F. | 2.585 | 29.00 | 92.00 | 121.00 |
| 14 riser stair | 42.000 | L.F. | 5.169 | 57.50 | 184.00 | 241.50 |
| 2" x 12", 7 riser stair | 21.000 | L.F. | 2.585 | 39.50 | 92.00 | 131.50 |
| 14 riser stair | 42.000 | L.F. | 5.169 | 78.50 | 184.00 | 262.50 |

## Special Stairways

| Special Stairways | QUAN. | UNIT | LABOR HOURS | COST EACH MAT. | COST EACH INST. | COST EACH TOTAL |
|---|---|---|---|---|---|---|
| asement stairs, prefabricated, open risers | 1.000 | Flight | 4.000 | 630.00 | 142.00 | 772.00 |
| urved stairways, 3'-3" wide, prefabricated oak, 9' high | 1.000 | Flight | 22.857 | 7025.00 | 815.00 | 7840.00 |
| 10' high | 1.000 | Flight | 22.857 | 7925.00 | 815.00 | 8740.00 |
| Open two sides, 9' high | 1.000 | Flight | 32.000 | 11000.00 | 1150.00 | 12150.00 |
| 10' high | 1.000 | Flight | 32.000 | 11900.00 | 1150.00 | 13050.00 |
| piral stairs, oak, 4'-6" diameter, prefabricated, 9' high | 1.000 | Flight | 10.667 | 4400.00 | 380.00 | 4780.00 |
| Aluminum, 5'-0" diameter stock unit | 1.000 | Flight | 9.956 | 3000.00 | 485.00 | 3485.00 |
| Custom unit | 1.000 | Flight | 9.956 | 5675.00 | 485.00 | 6160.00 |
| Cast iron, 4'-0" diameter, minimum | 1.000 | Flight | 9.956 | 2675.00 | 485.00 | 3160.00 |
| Maximum | 1.000 | Flight | 17.920 | 3650.00 | 875.00 | 4525.00 |
| teel, industrial, pre-erected, 3'-6" wide, bar rail | 1.000 | Flight | 7.724 | 2675.00 | 595.00 | 3270.00 |
| Picket rail | 1.000 | Flight | 7.724 | 3000.00 | 595.00 | 3595.00 |

**or information about Means Estimating Seminars, see yellow pages 11 and 12 in back of book**

# Division 7
# Specialties

Soffit Drywall — Soffit Framing — Top Cabinets — Counter Top — Bottom Cabinets

| System Description | QUAN. | UNIT | LABOR HOURS | COST PER L.F. | | |
|---|---|---|---|---|---|---|
| | | | | MAT. | INST. | TOTAL |
| **KITCHEN, ECONOMY GRADE** | | | | | | |
| Top cabinets, economy grade | 1.000 | L.F. | .171 | 30.08 | 6.08 | 36.16 |
| Bottom cabinets, economy grade | 1.000 | L.F. | .256 | 45.12 | 9.12 | 54.24 |
| Counter top, laminated plastic, post formed | 1.000 | L.F. | .267 | 9.40 | 9.50 | 18.90 |
| Blocking, wood, 2" x 4" | 1.000 | L.F. | .032 | .39 | 1.14 | 1.53 |
| Soffit, framing, wood, 2" x 4" | 4.000 | L.F. | .071 | 1.56 | 2.52 | 4.08 |
| Soffit drywall | 2.000 | S.F. | .047 | .56 | 1.68 | 2.24 |
| Drywall painting | 2.000 | S.F. | .013 | .10 | .48 | .58 |
| TOTAL | | | .857 | 87.21 | 30.52 | 117.73 |
| **AVERAGE GRADE** | | | | | | |
| Top cabinets, average grade | 1.000 | L.F. | .213 | 37.60 | 7.60 | 45.20 |
| Bottom cabinets, average grade | 1.000 | L.F. | .320 | 56.40 | 11.40 | 67.80 |
| Counter top, laminated plastic, square edge, incl. backsplash | 1.000 | L.F. | .267 | 28.50 | 9.50 | 38.00 |
| Blocking, wood, 2" x 4" | 1.000 | L.F. | .032 | .39 | 1.14 | 1.53 |
| Soffit framing, wood, 2" x 4" | 4.000 | L.F. | .071 | 1.56 | 2.52 | 4.08 |
| Soffit drywall | 2.000 | S.F. | .047 | .56 | 1.68 | 2.24 |
| Drywall painting | 2.000 | S.F. | .013 | .10 | .48 | .58 |
| TOTAL | | | .963 | 125.11 | 34.32 | 159.43 |
| **CUSTOM GRADE** | | | | | | |
| Top cabinets, custom grade | 1.000 | L.F. | .256 | 101.20 | 9.20 | 110.40 |
| Bottom cabinets, custom grade | 1.000 | L.F. | .384 | 151.80 | 13.80 | 165.60 |
| Counter top, laminated plastic, square edge, incl. backsplash | 1.000 | L.F. | .267 | 28.50 | 9.50 | 38.00 |
| Blocking, wood, 2" x 4" | 1.000 | L.F. | .032 | .39 | 1.14 | 1.53 |
| Soffit framing, wood, 2" x 4" | 4.000 | L.F. | .071 | 1.56 | 2.52 | 4.08 |
| Soffit drywall | 2.000 | S.F. | .047 | .56 | 1.68 | 2.24 |
| Drywall painting | 2.000 | S.F. | .013 | .10 | .48 | .58 |
| TOTAL | | | 1.070 | 284.11 | 38.32 | 322.43 |

| Description | QUAN. | UNIT | LABOR HOURS | COST PER L.F. | | |
|---|---|---|---|---|---|---|
| | | | | MAT. | INST. | TOTAL |
| | | | | | | |
| | | | | | | |
| | | | | | | |

| Kitchen Price Sheet | QUAN. | UNIT | LABOR HOURS | COST PER L.F. | | |
|---|---|---|---|---|---|---|
| | | | | MAT. | INST. | TOTAL |
| Top cabinets, economy grade | 1.000 | L.F. | .171 | 30.00 | 6.10 | 36.10 |
| Average grade | 1.000 | L.F. | .213 | 37.50 | 7.60 | 45.10 |
| Custom grade | 1.000 | L.F. | .256 | 101.00 | 9.20 | 110.20 |
| | | | | | | |
| Bottom cabinets, economy grade | 1.000 | L.F. | .256 | 45.00 | 9.10 | 54.10 |
| Average grade | 1.000 | L.F. | .320 | 56.50 | 11.40 | 67.90 |
| Custom grade | 1.000 | L.F. | .384 | 152.00 | 13.80 | 165.80 |
| | | | | | | |
| Counter top, laminated plastic, 7/8" thick, no splash | 1.000 | L.F. | .267 | 17.30 | 9.50 | 26.80 |
| With backsplash | 1.000 | L.F. | .267 | 22.50 | 9.50 | 32.00 |
| 1-1/4" thick, no splash | 1.000 | L.F. | .286 | 20.50 | 10.15 | 30.65 |
| With backsplash | 1.000 | L.F. | .286 | 25.50 | 10.15 | 35.65 |
| Post formed, laminated plastic | 1.000 | L.F. | .267 | 9.40 | 9.50 | 18.90 |
| | | | | | | |
| Marble, with backsplash, minimum | 1.000 | L.F. | .471 | 33.00 | 16.80 | 49.80 |
| Maximum | 1.000 | L.F. | .615 | 84.00 | 22.00 | 106.00 |
| Maple, solid laminated, no backsplash | 1.000 | L.F. | .286 | 34.00 | 10.15 | 44.15 |
| With backsplash | 1.000 | L.F. | .286 | 38.50 | 10.15 | 48.65 |
| Blocking, wood, 2" x 4" | 1.000 | L.F. | .032 | .39 | 1.14 | 1.53 |
| 2" x 6" | 1.000 | L.F. | .036 | .62 | 1.28 | 1.90 |
| 2" x 8" | 1.000 | L.F. | .040 | .95 | 1.42 | 2.37 |
| | | | | | | |
| Soffit framing, wood, 2" x 3" | 4.000 | L.F. | .064 | 1.08 | 2.28 | 3.36 |
| 2" x 4" | 4.000 | L.F. | .071 | 1.56 | 2.52 | 4.08 |
| Soffit, drywall, painted | 2.000 | S.F. | .060 | .66 | 2.16 | 2.82 |
| | | | | | | |
| Paneling, standard | 2.000 | S.F. | .064 | 1.62 | 2.28 | 3.90 |
| Deluxe | 2.000 | S.F. | .091 | 3.60 | 3.26 | 6.86 |
| Sinks, porcelain on cast iron, single bowl, 21" x 24" | 1.000 | Ea. | 10.334 | 310.00 | 365.00 | 675.00 |
| 21" x 30" | 1.000 | Ea. | 10.334 | 365.00 | 365.00 | 730.00 |
| Double bowl, 20" x 32" | 1.000 | Ea. | 10.810 | 410.00 | 380.00 | 790.00 |
| | | | | | | |
| Stainless steel, single bowl, 16" x 20" | 1.000 | Ea. | 10.334 | 405.00 | 365.00 | 770.00 |
| 22" x 25" | 1.000 | Ea. | 10.334 | 435.00 | 365.00 | 800.00 |
| Double bowl, 20" x 32" | 1.000 | Ea. | 10.810 | 250.00 | 380.00 | 630.00 |

**7 SPECIALTIES**

| System Description | QUAN. | UNIT | LABOR HOURS | COST EACH MAT. | COST EACH INST. | COST EACH TOTAL |
|---|---|---|---|---|---|---|
| All appliances include plumbing and electrical rough-in & hook-ups | | | | | | |
| Range, free standing, minimum | 1.000 | Ea. | 3.600 | 345.00 | 119.00 | 464.00 |
| Maximum | 1.000 | Ea. | 6.000 | 1400.00 | 181.00 | 1581.00 |
| Built-in, minimum | 1.000 | Ea. | 3.333 | 415.00 | 129.00 | 544.00 |
| Maximum | 1.000 | Ea. | 10.000 | 1250.00 | 365.00 | 1615.00 |
| Counter top range, 4-burner, minimum | 1.000 | Ea. | 3.333 | 276.00 | 129.00 | 405.00 |
| Maximum | 1.000 | Ea. | 4.667 | 550.00 | 181.00 | 731.00 |
| Compactor, built-in, minimum | 1.000 | Ea. | 2.215 | 405.00 | 81.00 | 486.00 |
| Maximum | 1.000 | Ea. | 3.282 | 440.00 | 119.00 | 559.00 |
| Dishwasher, built-in, minimum | 1.000 | Ea. | 6.735 | 350.00 | 264.00 | 614.00 |
| Maximum | 1.000 | Ea. | 9.235 | 380.00 | 360.00 | 740.00 |
| Garbage disposer, minimum | 1.000 | Ea. | 2.810 | 71.00 | 110.00 | 181.00 |
| Maximum | 1.000 | Ea. | 2.810 | 229.00 | 110.00 | 339.00 |
| Microwave oven, minimum | 1.000 | Ea. | 2.615 | 110.00 | 102.00 | 212.00 |
| Maximum | 1.000 | Ea. | 4.615 | 455.00 | 179.00 | 634.00 |
| Range hood, ducted, minimum | 1.000 | Ea. | 4.658 | 69.00 | 171.00 | 240.00 |
| Maximum | 1.000 | Ea. | 5.991 | 525.00 | 219.00 | 744.00 |
| Ductless, minimum | 1.000 | Ea. | 2.615 | 68.00 | 96.50 | 164.50 |
| Maximum | 1.000 | Ea. | 3.948 | 525.00 | 145.00 | 670.00 |
| Refrigerator, 16 cu.ft., minimum | 1.000 | Ea. | 2.000 | 620.00 | 52.00 | 672.00 |
| Maximum | 1.000 | Ea. | 3.200 | 990.00 | 83.00 | 1073.00 |
| 16 cu.ft. with icemaker, minimum | 1.000 | Ea. | 4.210 | 770.00 | 133.00 | 903.00 |
| Maximum | 1.000 | Ea. | 5.410 | 1150.00 | 164.00 | 1314.00 |
| 19 cu.ft., minimum | 1.000 | Ea. | 2.667 | 760.00 | 69.00 | 829.00 |
| Maximum | 1.000 | Ea. | 4.667 | 1325.00 | 121.00 | 1446.00 |
| 19 cu.ft. with icemaker, minimum | 1.000 | Ea. | 5.143 | 980.00 | 157.00 | 1137.00 |
| Maximum | 1.000 | Ea. | 7.143 | 1550.00 | 209.00 | 1759.00 |
| Sinks, porcelain on cast iron single bowl, 21" x 24" | 1.000 | Ea. | 10.334 | 310.00 | 365.00 | 675.00 |
| 21" x 30" | 1.000 | Ea. | 10.334 | 365.00 | 365.00 | 730.00 |
| Double bowl, 20" x 32" | 1.000 | Ea. | 10.810 | 410.00 | 380.00 | 790.00 |
| Stainless steel, single bowl 16" x 20" | 1.000 | Ea. | 10.334 | 405.00 | 365.00 | 770.00 |
| 22" x 25" | 1.000 | Ea. | 10.334 | 435.00 | 365.00 | 800.00 |
| Double bowl, 20" x 32" | 1.000 | Ea. | 10.810 | 250.00 | 380.00 | 630.00 |
| Water heater, electric, 30 gallon | 1.000 | Ea. | 3.636 | 277.00 | 142.00 | 419.00 |
| 40 gallon | 1.000 | Ea. | 4.000 | 305.00 | 157.00 | 462.00 |
| Gas, 30 gallon | 1.000 | Ea. | 4.000 | 365.00 | 157.00 | 522.00 |
| 75 gallon | 1.000 | Ea. | 5.333 | 690.00 | 209.00 | 899.00 |
| Wall, packaged terminal heater/air conditioner cabinet, wall sleeve, louver, electric heat, thermostat, manual changeover, 208V | | | | | | |
| 6000 BTUH cooling, 8800 BTU heating | 1.000 | Ea. | 2.667 | 1150.00 | 94.50 | 1244.50 |
| 9000 BTUH cooling, 13,900 BTU heating | 1.000 | Ea. | 3.200 | 1150.00 | 113.00 | 1263.00 |
| 12,000 BTUH cooling, 13,900 BTU heating | 1.000 | Ea. | 4.000 | 1200.00 | 142.00 | 1342.00 |
| 15,000 BTUH cooling, 13,900 BTU heating | 1.000 | Ea. | 5.333 | 1275.00 | 189.00 | 1464.00 |

| System Description | QUAN. | UNIT | LABOR HOURS | COST EACH | | |
|---|---|---|---|---|---|---|
| | | | | MAT. | INST. | TOTAL |
| Curtain rods, stainless, 1" diameter, 3' long | 1.000 | Ea. | .615 | 29.50 | 22.00 | 51.50 |
| 5' long | 1.000 | Ea. | .615 | 29.50 | 22.00 | 51.50 |
| Grab bar, 1" diameter, 12" long | 1.000 | Ea. | .283 | 20.00 | 10.05 | 30.05 |
| 36" long | 1.000 | Ea. | .340 | 24.50 | 12.10 | 36.60 |
| 1-1/4" diameter, 12" long | 1.000 | Ea. | .333 | 23.50 | 11.85 | 35.35 |
| 36" long | 1.000 | Ea. | .400 | 29.00 | 14.25 | 43.25 |
| 1-1/2" diameter, 12" long | 1.000 | Ea. | .383 | 27.00 | 13.65 | 40.65 |
| 36" long | 1.000 | Ea. | .460 | 33.50 | 16.40 | 49.90 |
| Mirror, 18" x 24" | 1.000 | Ea. | .400 | 70.00 | 14.25 | 84.25 |
| 72" x 24" | 1.000 | Ea. | 1.333 | 264.00 | 47.50 | 311.50 |
| Medicine chest with mirror, 18" x 24" | 1.000 | Ea. | .400 | 109.00 | 14.25 | 123.25 |
| 36" x 24" | 1.000 | Ea. | .600 | 164.00 | 21.50 | 185.50 |
| Toilet tissue dispenser, surface mounted, minimum | 1.000 | Ea. | .267 | 12.95 | 9.50 | 22.45 |
| Maximum | 1.000 | Ea. | .400 | 19.45 | 14.25 | 33.70 |
| Flush mounted, minimum | 1.000 | Ea. | .293 | 14.25 | 10.45 | 24.70 |
| Maximum | 1.000 | Ea. | .427 | 20.50 | 15.20 | 35.70 |
| Towel bar, 18" long, minimum | 1.000 | Ea. | .278 | 28.00 | 9.90 | 37.90 |
| Maximum | 1.000 | Ea. | .348 | 35.00 | 12.40 | 47.40 |
| 24" long, minimum | 1.000 | Ea. | .313 | 31.50 | 11.15 | 42.65 |
| Maximum | 1.000 | Ea. | .383 | 38.50 | 13.65 | 52.15 |
| 36" long, minimum | 1.000 | Ea. | .381 | 58.00 | 13.55 | 71.55 |
| Maximum | 1.000 | Ea. | .419 | 64.00 | 14.90 | 78.90 |

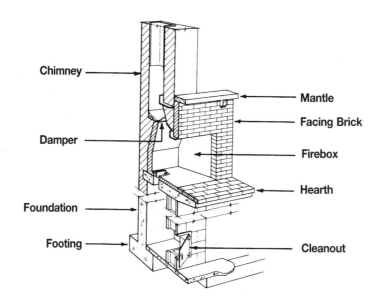

Chimney — Mantle — Facing Brick — Damper — Firebox — Foundation — Hearth — Footing — Cleanout

| System Description | QUAN. | UNIT | LABOR HOURS | COST EACH | | |
|---|---|---|---|---|---|---|
| | | | | MAT. | INST. | TOTAL |
| **MASONRY FIREPLACE** | | | | | | |
| Footing, 8" thick, concrete, 4' x 7' | .700 | C.Y. | 2.110 | 76.30 | 65.79 | 142.09 |
| Foundation, concrete block, 32" x 60" x 4' deep | 1.000 | Ea. | 5.275 | 103.20 | 171.00 | 274.20 |
| Fireplace, brick firebox, 30" x 29" opening | 1.000 | Ea. | 40.000 | 420.00 | 1275.00 | 1695.00 |
| Damper, cast iron, 30" opening | 1.000 | Ea. | 1.333 | 78.00 | 47.50 | 125.50 |
| Facing brick, standard size brick, 6' x 5' | 30.000 | S.F. | 5.217 | 88.80 | 169.50 | 258.30 |
| Hearth, standard size brick, 3' x 6' | 1.000 | Ea. | 8.000 | 154.00 | 253.00 | 407.00 |
| Chimney, standard size brick, 8" x 12" flue, one story house | 12.000 | V.L.F. | 12.000 | 294.00 | 378.00 | 672.00 |
| Mantle, 4" x 8", wood | 6.000 | L.F. | 1.333 | 27.96 | 47.40 | 75.36 |
| Cleanout, cast iron, 8" x 8" | 1.000 | Ea. | .667 | 26.00 | 24.00 | 50.00 |
| | | | | | | |
| TOTAL | | | 75.935 | 1268.26 | 2431.19 | 3699.45 |

The costs in this system are on a cost each basis.

| Description | QUAN. | UNIT | LABOR HOURS | COST EACH | | |
|---|---|---|---|---|---|---|
| | | | | MAT. | INST. | TOTAL |
| | | | | | | |
| | | | | | | |
| | | | | | | |
| | | | | | | |
| | | | | | | |
| | | | | | | |
| | | | | | | |
| | | | | | | |
| | | | | | | |

**Important: See the Reference Section for critical supporting data - Reference Nos., Crews & Location Factors**

| Masonry Fireplace Price Sheet | QUAN. | UNIT | LABOR HOURS | COST EACH | | |
|---|---|---|---|---|---|---|
| | | | | MAT. | INST. | TOTAL |
| Footing 8″ thick, 3′ x 6′ | .440 | C.Y. | 1.326 | 48.00 | 41.00 | 89.00 |
| 4′ x 7′ | .700 | C.Y. | 2.110 | 76.50 | 66.00 | 142.50 |
| 5′ x 8′ | 1.000 | C.Y. | 3.014 | 109.00 | 94.00 | 203.00 |
| 1′ thick, 3′ x 6′ | .670 | C.Y. | 2.020 | 73.00 | 63.00 | 136.00 |
| 4′ x 7′ | 1.030 | C.Y. | 3.105 | 112.00 | 97.00 | 209.00 |
| 5′ x 8′ | 1.480 | C.Y. | 4.461 | 161.00 | 139.00 | 300.00 |
| Foundation-concrete block, 24″ x 48″, 4′ deep | 1.000 | Ea. | 4.267 | 82.50 | 137.00 | 219.50 |
| 8′ deep | 1.000 | Ea. | 8.533 | 165.00 | 274.00 | 439.00 |
| 24″ x 60″, 4′ deep | 1.000 | Ea. | 4.978 | 96.50 | 160.00 | 256.50 |
| 8′ deep | 1.000 | Ea. | 9.956 | 193.00 | 320.00 | 513.00 |
| 32″ x 48″, 4′ deep | 1.000 | Ea. | 4.711 | 91.00 | 151.00 | 242.00 |
| 8′ deep | 1.000 | Ea. | 9.422 | 182.00 | 300.00 | 482.00 |
| 32″ x 60″, 4′ deep | 1.000 | Ea. | 5.333 | 103.00 | 171.00 | 274.00 |
| 8′ deep | 1.000 | Ea. | 10.845 | 210.00 | 350.00 | 560.00 |
| 32″ x 72″, 4′ deep | 1.000 | Ea. | 6.133 | 119.00 | 197.00 | 316.00 |
| 8′ deep | 1.000 | Ea. | 12.267 | 237.00 | 395.00 | 632.00 |
| Fireplace, brick firebox 30″ x 29″ opening | 1.000 | Ea. | 40.000 | 420.00 | 1275.00 | 1695.00 |
| 48″ x 30″ opening | 1.000 | Ea. | 60.000 | 630.00 | 1925.00 | 2555.00 |
| Steel fire box with registers, 25″ opening | 1.000 | Ea. | 26.667 | 795.00 | 860.00 | 1655.00 |
| 48″ opening | 1.000 | Ea. | 44.000 | 1200.00 | 1425.00 | 2625.00 |
| Damper, cast iron, 30″ opening | 1.000 | Ea. | 1.333 | 78.00 | 47.50 | 125.50 |
| 36″ opening | 1.000 | Ea. | 1.556 | 91.00 | 55.50 | 146.50 |
| Steel, 30″ opening | 1.000 | Ea. | 1.333 | 70.00 | 47.50 | 117.50 |
| 36″ opening | 1.000 | Ea. | 1.556 | 81.50 | 55.50 | 137.00 |
| Facing for fireplace, standard size brick, 6′ x 5′ | 30.000 | S.F. | 5.217 | 89.00 | 170.00 | 259.00 |
| 7′ x 5′ | 35.000 | S.F. | 6.087 | 104.00 | 198.00 | 302.00 |
| 8′ x 6′ | 48.000 | S.F. | 8.348 | 142.00 | 271.00 | 413.00 |
| Fieldstone, 6′ x 5′ | 30.000 | S.F. | 5.217 | 420.00 | 170.00 | 590.00 |
| 7′ x 5′ | 35.000 | S.F. | 6.087 | 490.00 | 198.00 | 688.00 |
| 8′ x 6′ | 48.000 | S.F. | 8.348 | 670.00 | 271.00 | 941.00 |
| Sheetrock on metal, studs, 6′ x 5′ | 30.000 | S.F. | .980 | 14.70 | 34.50 | 49.20 |
| 7′ x 5′ | 35.000 | S.F. | 1.143 | 17.15 | 40.50 | 57.65 |
| 8′ x 6′ | 48.000 | S.F. | 1.568 | 23.50 | 55.00 | 78.50 |
| Hearth, standard size brick, 3′ x 6′ | 1.000 | Ea. | 8.000 | 154.00 | 253.00 | 407.00 |
| 3′ x 7′ | 1.000 | Ea. | 9.280 | 179.00 | 293.00 | 472.00 |
| 3′ x 8′ | 1.000 | Ea. | 10.640 | 205.00 | 335.00 | 540.00 |
| Stone, 3′ x 6′ | 1.000 | Ea. | 8.000 | 166.00 | 253.00 | 419.00 |
| 3′ x 7′ | 1.000 | Ea. | 9.280 | 193.00 | 293.00 | 486.00 |
| 3′ x 8′ | 1.000 | Ea. | 10.640 | 221.00 | 335.00 | 556.00 |
| Chimney, standard size brick , 8″ x 12″ flue, one story house | 12.000 | V.L.F. | 12.000 | 294.00 | 380.00 | 674.00 |
| Two story house | 20.000 | V.L.F. | 20.000 | 490.00 | 630.00 | 1120.00 |
| Mantle wood, beams, 4″ x 8″ | 6.000 | L.F. | 1.333 | 28.00 | 47.50 | 75.50 |
| 4″ x 10″ | 6.000 | L.F. | 1.371 | 35.00 | 49.00 | 84.00 |
| Ornate, prefabricated, 6′ x 3′-6″ opening, minimum | 1.000 | Ea. | 1.600 | 142.00 | 57.00 | 199.00 |
| Maximum | 1.000 | Ea. | 1.600 | 173.00 | 57.00 | 230.00 |
| Cleanout, door and frame, cast iron, 8″ x 8″ | 1.000 | Ea. | .667 | 26.00 | 24.00 | 50.00 |
| 12″ x 12″ | 1.000 | Ea. | .800 | 36.50 | 28.50 | 65.00 |
| | | | | | | |
| | | | | | | |
| | | | | | | |
| | | | | | | |
| | | | | | | |
| | | | | | | |

**7 SPECIALTIES**

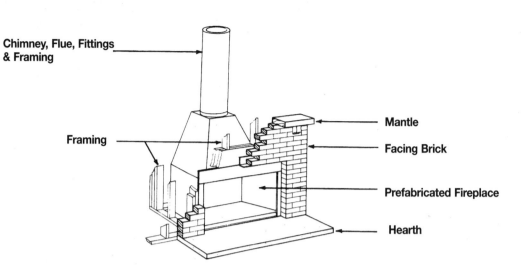

Chimney, Flue, Fittings & Framing

Framing

Mantle

Facing Brick

Prefabricated Fireplace

Hearth

| System Description | QUAN. | UNIT | LABOR HOURS | COST EACH | | |
|---|---|---|---|---|---|---|
| | | | | MAT. | INST. | TOTAL |
| **PREFABRICATED FIREPLACE** | | | | | | |
| Prefabricated fireplace, metal, minimum | 1.000 | Ea. | 6.154 | 1075.00 | 219.00 | 1294.00 |
| Framing, 2" x 4" studs, 6' x 5' | 35.000 | L.F. | .509 | 13.65 | 18.20 | 31.85 |
| Fire resistant gypsum drywall, unfinished | 40.000 | S.F. | .320 | 9.60 | 11.20 | 20.80 |
| Drywall finishing adder | 40.000 | S.F. | .320 | 1.60 | 11.20 | 12.80 |
| Facing, brick, standard size brick, 6' x 5' | 30.000 | S.F. | 5.217 | 88.80 | 169.50 | 258.30 |
| Hearth, standard size brick, 3' x 6' | 1.000 | Ea. | 8.000 | 154.00 | 253.00 | 407.00 |
| Chimney, one story house, framing, 2" x 4" studs | 80.000 | L.F. | 1.164 | 31.20 | 41.60 | 72.80 |
| Sheathing, plywood, 5/8" thick | 32.000 | S.F. | .758 | 75.20 | 26.88 | 102.08 |
| Flue, 10" metal, insulated pipe | 12.000 | V.L.F. | 4.000 | 270.00 | 140.40 | 410.40 |
| Fittings, ceiling support | 1.000 | Ea. | .667 | 131.00 | 23.50 | 154.50 |
| Fittings, joist shield | 1.000 | Ea. | .667 | 73.50 | 23.50 | 97.00 |
| Fittings, roof flashing | 1.000 | Ea. | .667 | 149.00 | 23.50 | 172.50 |
| Mantle beam, wood, 4" x 8" | 6.000 | L.F. | 1.333 | 27.96 | 47.40 | 75.36 |
| | | | | | | |
| TOTAL | | | 29.776 | 2100.51 | 1008.88 | 3109.39 |

The costs in this system are on a cost each basis.

| Description | QUAN. | UNIT | LABOR HOURS | COST EACH | | |
|---|---|---|---|---|---|---|
| | | | | MAT. | INST. | TOTAL |
| | | | | | | |
| | | | | | | |
| | | | | | | |
| | | | | | | |
| | | | | | | |
| | | | | | | |
| | | | | | | |
| | | | | | | |
| | | | | | | |

**Important: See the Reference Section for critical supporting data - Reference Nos., Crews & Location Factors**

| Prefabricated Fireplace Price Sheet | QUAN. | UNIT | LABOR HOURS | COST EACH | | |
|---|---|---|---|---|---|---|
| | | | | MAT. | INST. | TOTAL |
| Prefabricated fireplace, minimum | 1.000 | Ea. | 6.154 | 1075.00 | 219.00 | 1294.00 |
| Average | 1.000 | Ea. | 8.000 | 1300.00 | 285.00 | 1585.00 |
| Maximum | 1.000 | Ea. | 8.889 | 3125.00 | 315.00 | 3440.00 |
| Framing, 2" x 4" studs, fireplace, 6' x 5' | 35.000 | L.F. | .509 | 13.65 | 18.20 | 31.85 |
| 7' x 5' | 40.000 | L.F. | .582 | 15.60 | 21.00 | 36.60 |
| 8' x 6' | 45.000 | L.F. | .655 | 17.55 | 23.50 | 41.05 |
| Sheetrock, 1/2" thick, fireplace, 6' x 5' | 40.000 | S.F. | .640 | 11.20 | 22.50 | 33.70 |
| 7' x 5' | 45.000 | S.F. | .720 | 12.60 | 25.00 | 37.60 |
| 8' x 6' | 50.000 | S.F. | .800 | 14.00 | 28.00 | 42.00 |
| Facing for fireplace, brick, 6' x 5' | 30.000 | S.F. | 5.217 | 89.00 | 170.00 | 259.00 |
| 7' x 5' | 35.000 | S.F. | 6.087 | 104.00 | 198.00 | 302.00 |
| 8' x 6' | 48.000 | S.F. | 8.348 | 142.00 | 271.00 | 413.00 |
| Fieldstone, 6' x 5' | 30.000 | S.F. | 5.217 | 450.00 | 170.00 | 620.00 |
| 7' x 5' | 35.000 | S.F. | 6.087 | 525.00 | 198.00 | 723.00 |
| 8' x 6' | 48.000 | S.F. | 8.348 | 725.00 | 271.00 | 996.00 |
| Hearth, standard size brick, 3' x 6' | 1.000 | Ea. | 8.000 | 154.00 | 253.00 | 407.00 |
| 3' x 7' | 1.000 | Ea. | 9.280 | 179.00 | 293.00 | 472.00 |
| 3' x 8' | 1.000 | Ea. | 10.640 | 205.00 | 335.00 | 540.00 |
| Stone, 3' x 6' | 1.000 | Ea. | 8.000 | 166.00 | 253.00 | 419.00 |
| 3' x 7' | 1.000 | Ea. | 9.280 | 193.00 | 293.00 | 486.00 |
| 3' x 8' | 1.000 | Ea. | 10.640 | 221.00 | 335.00 | 556.00 |
| Chimney, framing, 2" x 4", one story house | 80.000 | L.F. | 1.164 | 31.00 | 41.50 | 72.50 |
| Two story house | 120.000 | L.F. | 1.746 | 47.00 | 62.50 | 109.50 |
| Sheathing, plywood, 5/8" thick | 32.000 | S.F. | .758 | 75.00 | 27.00 | 102.00 |
| Stucco on plywood | 32.000 | S.F. | 1.125 | 43.50 | 38.50 | 82.00 |
| Flue, 10" metal pipe, insulated, one story house | 12.000 | V.L.F. | 4.000 | 270.00 | 140.00 | 410.00 |
| Two story house | 20.000 | V.L.F. | 6.667 | 450.00 | 234.00 | 684.00 |
| Fittings, ceiling support | 1.000 | Ea. | .667 | 131.00 | 23.50 | 154.50 |
| Fittings joist sheild, one story house | 1.000 | Ea. | .667 | 73.50 | 23.50 | 97.00 |
| Two story house | 2.000 | Ea. | 1.333 | 147.00 | 47.00 | 194.00 |
| Fittings roof flashing | 1.000 | Ea. | .667 | 149.00 | 23.50 | 172.50 |
| Mantle, wood beam, 4" x 8" | 6.000 | L.F. | 1.333 | 28.00 | 47.50 | 75.50 |
| 4" x 10" | 6.000 | L.F. | 1.371 | 35.00 | 49.00 | 84.00 |
| Ornate prefabricated, 6' x 3'-6" opening, minimum | 1.000 | Ea. | 1.600 | 142.00 | 57.00 | 199.00 |
| Maximum | 1.000 | Ea. | 1.600 | 173.00 | 57.00 | 230.00 |
| | | | | | | |
| | | | | | | |
| | | | | | | |
| | | | | | | |
| | | | | | | |
| | | | | | | |
| | | | | | | |
| | | | | | | |
| | | | | | | |
| | | | | | | |

| System Description | QUAN. | UNIT | LABOR HOURS | COST EACH | | |
|---|---|---|---|---|---|---|
| | | | | MAT. | INST. | TOTAL |
| Economy, lean to, shell only, not including 2' stub wall, fndtn, flrs, heat | | | | | | |
| 4' x 16' | 1.000 | Ea. | 26.212 | 2000.00 | 935.00 | 2935.00 |
| 4' x 24' | 1.000 | Ea. | 30.259 | 2325.00 | 1075.00 | 3400.00 |
| 6' x 10' | 1.000 | Ea. | 16.552 | 1675.00 | 590.00 | 2265.00 |
| 6' x 16' | 1.000 | Ea. | 23.034 | 2350.00 | 820.00 | 3170.00 |
| 6' x 24' | 1.000 | Ea. | 29.793 | 3025.00 | 1050.00 | 4075.00 |
| 8' x 10' | 1.000 | Ea. | 22.069 | 2250.00 | 785.00 | 3035.00 |
| 8' x 16' | 1.000 | Ea. | 38.400 | 3900.00 | 1375.00 | 5275.00 |
| 8' x 24' | 1.000 | Ea. | 49.655 | 5050.00 | 1775.00 | 6825.00 |
| Free standing, 8' x 8' | 1.000 | Ea. | 17.356 | 2625.00 | 620.00 | 3245.00 |
| 8' x 16' | 1.000 | Ea. | 30.211 | 4575.00 | 1075.00 | 5650.00 |
| 8' x 24' | 1.000 | Ea. | 39.051 | 5900.00 | 1400.00 | 7300.00 |
| 10' x 10' | 1.000 | Ea. | 18.824 | 3150.00 | 670.00 | 3820.00 |
| 10' x 16' | 1.000 | Ea. | 24.095 | 4025.00 | 860.00 | 4885.00 |
| 10' x 24' | 1.000 | Ea. | 31.624 | 5300.00 | 1125.00 | 6425.00 |
| 14' x 10' | 1.000 | Ea. | 20.741 | 3925.00 | 735.00 | 4660.00 |
| 14' x 16' | 1.000 | Ea. | 24.889 | 4700.00 | 880.00 | 5580.00 |
| 14' x 24' | 1.000 | Ea. | 33.349 | 6300.00 | 1175.00 | 7475.00 |
| Standard,lean to,shell only,not incl.2'stub wall, fndtn,flrs, heat 4'x10' | 1.000 | Ea. | 28.235 | 2150.00 | 1000.00 | 3150.00 |
| 4' x 16' | 1.000 | Ea. | 39.341 | 3000.00 | 1400.00 | 4400.00 |
| 4' x 24' | 1.000 | Ea. | 45.412 | 3475.00 | 1625.00 | 5100.00 |
| 6' x 10' | 1.000 | Ea. | 24.827 | 2525.00 | 880.00 | 3405.00 |
| 6' x 16' | 1.000 | Ea. | 34.538 | 3500.00 | 1225.00 | 4725.00 |
| 6' x 24' | 1.000 | Ea. | 44.689 | 4525.00 | 1600.00 | 6125.00 |
| 8' x 10' | 1.000 | Ea. | 33.103 | 3350.00 | 1175.00 | 4525.00 |
| 8' x 16' | 1.000 | Ea. | 57.600 | 5850.00 | 2050.00 | 7900.00 |
| 8' x 24' | 1.000 | Ea. | 74.482 | 7550.00 | 2650.00 | 10200.00 |
| Free standing, 8' x 8' | 1.000 | Ea. | 26.034 | 3925.00 | 925.00 | 4850.00 |
| 8' x 16' | 1.000 | Ea. | 45.316 | 6850.00 | 1625.00 | 8475.00 |
| 8' x 24' | 1.000 | Ea. | 58.577 | 8850.00 | 2075.00 | 10925.00 |
| 10' x 10' | 1.000 | Ea. | 28.236 | 4725.00 | 1000.00 | 5725.00 |
| 10' x 16' | 1.000 | Ea. | 36.142 | 6050.00 | 1275.00 | 7325.00 |
| 10' x 24' | 1.000 | Ea. | 47.436 | 7950.00 | 1700.00 | 9650.00 |
| 14' x 10' | 1.000 | Ea. | 31.112 | 5875.00 | 1100.00 | 6975.00 |
| 14' x 16' | 1.000 | Ea. | 37.334 | 7050.00 | 1325.00 | 8375.00 |
| 14' x 24' | 1.000 | Ea. | 50.030 | 9450.00 | 1775.00 | 11225.00 |
| Deluxe,lean to,shell only,not incl.2'stub wall, fndtn, flrs or heat, 4'x10' | 1.000 | Ea. | 20.645 | 3750.00 | 735.00 | 4485.00 |
| 4' x 16' | 1.000 | Ea. | 33.032 | 5975.00 | 1175.00 | 7150.00 |
| 4' x 24' | 1.000 | Ea. | 49.548 | 8975.00 | 1750.00 | 10725.00 |
| 6' x 10' | 1.000 | Ea. | 30.968 | 5600.00 | 1100.00 | 6700.00 |
| 6' x 16' | 1.000 | Ea. | 49.548 | 8975.00 | 1750.00 | 10725.00 |
| 6' x 24' | 1.000 | Ea. | 74.323 | 13500.00 | 2650.00 | 16150.00 |
| 8' x 10' | 1.000 | Ea. | 41.290 | 7475.00 | 1475.00 | 8950.00 |
| 8' x 16' | 1.000 | Ea. | 66.065 | 12000.00 | 2350.00 | 14350.00 |
| 8' x 24' | 1.000 | Ea. | 99.097 | 18000.00 | 3525.00 | 21525.00 |
| Freestanding, 8' x 8' | 1.000 | Ea. | 18.618 | 5150.00 | 660.00 | 5810.00 |
| 8' x 16' | 1.000 | Ea. | 37.236 | 10300.00 | 1325.00 | 11625.00 |
| 8' x 24' | 1.000 | Ea. | 55.855 | 15500.00 | 1975.00 | 17475.00 |
| 10' x 10' | 1.000 | Ea. | 29.091 | 8050.00 | 1025.00 | 9075.00 |
| 10' x 16' | 1.000 | Ea. | 46.546 | 12900.00 | 1650.00 | 14550.00 |
| 10' x 24' | 1.000 | Ea. | 69.818 | 19300.00 | 2475.00 | 21775.00 |
| 14' x 10' | 1.000 | Ea. | 40.727 | 11300.00 | 1450.00 | 12750.00 |
| 14' x 16' | 1.000 | Ea. | 65.164 | 18000.00 | 2325.00 | 20325.00 |
| 14' x 24' | 1.000 | Ea. | 97.746 | 27000.00 | 3475.00 | 30475.00 |

SPECIALTIES 7

| System Description | QUAN. | UNIT | LABOR HOURS | COST EACH | | |
|---|---|---|---|---|---|---|
| | | | | MAT. | INST. | TOTAL |
| Swimming pools, vinyl lined, metal sides, sand bottom, 12' x 28' | 1.000 | Ea. | 50.177 | 3125.00 | 1825.00 | 4950.00 |
| 12' x 32' | 1.000 | Ea. | 55.366 | 3450.00 | 2000.00 | 5450.00 |
| 12' x 36' | 1.000 | Ea. | 60.061 | 3750.00 | 2175.00 | 5925.00 |
| 16' x 32' | 1.000 | Ea. | 66.798 | 4175.00 | 2425.00 | 6600.00 |
| 16' x 36' | 1.000 | Ea. | 71.190 | 4450.00 | 2575.00 | 7025.00 |
| 16' x 40' | 1.000 | Ea. | 74.703 | 4650.00 | 2700.00 | 7350.00 |
| 20' x 36' | 1.000 | Ea. | 77.860 | 4850.00 | 2825.00 | 7675.00 |
| 20' x 40' | 1.000 | Ea. | 82.135 | 5125.00 | 2975.00 | 8100.00 |
| 20' x 44' | 1.000 | Ea. | 90.348 | 5650.00 | 3275.00 | 8925.00 |
| 24' x 40' | 1.000 | Ea. | 98.562 | 6150.00 | 3575.00 | 9725.00 |
| 24' x 44' | 1.000 | Ea. | 108.418 | 6775.00 | 3950.00 | 10725.00 |
| 24' x 48' | 1.000 | Ea. | 118.274 | 7375.00 | 4275.00 | 11650.00 |
| Vinyl lined, concrete sides, 12' x 28' | 1.000 | Ea. | 79.447 | 4950.00 | 2900.00 | 7850.00 |
| 12' x 32' | 1.000 | Ea. | 88.818 | 5550.00 | 3225.00 | 8775.00 |
| 12' x 36' | 1.000 | Ea. | 97.656 | 6100.00 | 3550.00 | 9650.00 |
| 16' x 32' | 1.000 | Ea. | 111.393 | 6950.00 | 4025.00 | 10975.00 |
| 16' x 36' | 1.000 | Ea. | 121.354 | 7575.00 | 4400.00 | 11975.00 |
| 16' x 40' | 1.000 | Ea. | 130.445 | 8150.00 | 4725.00 | 12875.00 |
| 28' x 36' | 1.000 | Ea. | 140.585 | 8775.00 | 5100.00 | 13875.00 |
| 20' x 40' | 1.000 | Ea. | 149.336 | 9325.00 | 5425.00 | 14750.00 |
| 20' x 44' | 1.000 | Ea. | 164.270 | 10300.00 | 5975.00 | 16275.00 |
| 24' x 40' | 1.000 | Ea. | 179.203 | 11200.00 | 6500.00 | 17700.00 |
| 24' x 44' | 1.000 | Ea. | 197.124 | 12300.00 | 7150.00 | 19450.00 |
| 24' x 48' | 1.000 | Ea. | 215.044 | 13400.00 | 7800.00 | 21200.00 |
| Gunite, bottom and sides, 12' x 28' | 1.000 | Ea. | 129.767 | 6500.00 | 4700.00 | 11200.00 |
| 12' x 32' | 1.000 | Ea. | 142.164 | 7125.00 | 5175.00 | 12300.00 |
| 12' x 36' | 1.000 | Ea. | 153.028 | 7675.00 | 5575.00 | 13250.00 |
| 16' x 32' | 1.000 | Ea. | 167.743 | 8400.00 | 6100.00 | 14500.00 |
| 16' x 36' | 1.000 | Ea. | 176.421 | 8850.00 | 6400.00 | 15250.00 |
| 16' x 40' | 1.000 | Ea. | 182.368 | 9125.00 | 6600.00 | 15725.00 |
| 20' x 36' | 1.000 | Ea. | 187.949 | 9425.00 | 6825.00 | 16250.00 |
| 20' x 40' | 1.000 | Ea. | 179.200 | 12400.00 | 6500.00 | 18900.00 |
| 20' x 44' | 1.000 | Ea. | 197.120 | 13700.00 | 7150.00 | 20850.00 |
| 24' x 40' | 1.000 | Ea. | 215.040 | 14900.00 | 7775.00 | 22675.00 |
| 24' x 44' | 1.000 | Ea. | 273.244 | 13700.00 | 9925.00 | 23625.00 |
| 24' x 48' | 1.000 | Ea. | 298.077 | 14900.00 | 10800.00 | 25700.00 |

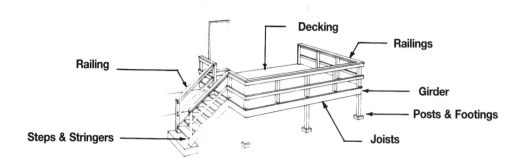

| System Description | QUAN. | UNIT | LABOR HOURS | COST PER S.F. | | |
|---|---|---|---|---|---|---|
| | | | | MAT. | INST. | TOTAL |
| **8' X 12' DECK, PRESSURE TREATED LUMBER, JOISTS 16" O.C.** | | | | | | |
| Decking, 2" x 6" lumber | 2.080 | L.F. | .027 | 1.29 | .96 | 2.25 |
| Lumber preservative | 2.080 | L.F. | | .27 | | .27 |
| Joists, 2" x 8", 16" O.C. | 1.000 | L.F. | .015 | .95 | .52 | 1.47 |
| Lumber preservative | 1.000 | L.F. | | .17 | | .17 |
| Girder, 2" x 10" | .125 | L.F. | .002 | .17 | .08 | .25 |
| Lumber preservative | .125 | L.F. | | .03 | | .03 |
| Hand excavation for footings | .250 | L.F. | .006 | | .15 | .15 |
| Concrete footings | .250 | L.F. | .006 | .22 | .19 | .41 |
| 4" x 4" Posts | .250 | L.F. | .010 | .36 | .37 | .73 |
| Lumber preservative | .250 | L.F. | | .04 | | .04 |
| Stairs, 2" x 10" stringers, 2" x 10" steps | 1.000 | Set | .020 | 3.15 | .71 | 3.86 |
| Railings, 2" x 4" | 1.000 | L.F. | .026 | .39 | .92 | 1.31 |
| Lumber preservative | 1.000 | L.F. | | .09 | | .09 |
| TOTAL | | | .112 | 7.13 | 3.90 | 11.03 |
| **12' X 16' DECK, PRESSURE TREATED LUMBER, JOISTS 24" O.C.** | | | | | | |
| Decking, 2" x 6" | 2.080 | L.F. | .027 | 1.29 | .96 | 2.25 |
| Lumber preservative | 2.080 | L.F. | | .27 | | .27 |
| Joists, 2" x 10", 24" O.C. | .800 | L.F. | .014 | 1.10 | .50 | 1.60 |
| Lumber preservative | .800 | L.F. | | .17 | | .17 |
| Girder, 2" x 10" | .083 | L.F. | .001 | .11 | .05 | .16 |
| Lumber preservative | .083 | L.F. | | .02 | | .02 |
| Hand excavation for footings | .122 | L.F. | .006 | | .15 | .15 |
| Concrete footings | .122 | L.F. | .006 | .22 | .19 | .41 |
| 4" x 4" Posts | .122 | L.F. | .005 | .18 | .18 | .36 |
| Lumber preservative | .122 | L.F. | | .02 | | .02 |
| Stairs, 2" x 10" stringers, 2" x 10" steps | 1.000 | Set | .012 | 1.89 | .43 | 2.32 |
| Railings, 2" x 4" | .670 | L.F. | .017 | .26 | .62 | .88 |
| Lumber preservative | .670 | L.F. | | .06 | | .06 |
| TOTAL | | | .088 | 5.59 | 3.08 | 8.67 |
| **12' X 24' DECK, REDWOOD OR CEDAR, JOISTS 16" O.C.** | | | | | | |
| Decking, 2" x 6" redwood | 2.080 | L.F. | .027 | 4.64 | .96 | 5.60 |
| Joists, 2" x 10", 16" O.C. | 1.000 | L.F. | .018 | 3.85 | .63 | 4.48 |
| Girder, 2" x 10" | .083 | L.F. | .001 | .32 | .05 | .37 |
| Hand excavation for footings | .111 | L.F. | .006 | | .15 | .15 |
| Concrete footings | .111 | L.F. | .006 | .22 | .19 | .41 |
| Lumber preservative | .111 | L.F. | | .02 | | .02 |
| Post, 4" x 4", including concrete footing | .111 | L.F. | .009 | .68 | .32 | 1.00 |
| Stairs, 2" x 10" stringers, 2" x 10" steps | 1.000 | Set | .012 | 1.61 | .43 | 2.04 |
| Railings, 2" x 4" | .540 | L.F. | .005 | .80 | .17 | .97 |
| TOTAL | | | .084 | 12.14 | 2.90 | 15.04 |

The costs in this system are on a square foot basis.

## Wood Deck Price Sheet

| | QUAN. | UNIT | LABOR HOURS | MAT. | INST. | TOTAL |
|---|---|---|---|---|---|---|
| Decking, treated lumber, 1" x 4" | 3.430 | L.F. | .031 | 2.38 | 1.12 | 3.50 |
| 1" x 6" | 2.180 | L.F. | .033 | 2.50 | 1.18 | 3.68 |
| 2" x 4" | 3.200 | L.F. | .041 | 1.54 | 1.47 | 3.01 |
| 2" x 6" | 2.080 | L.F. | .027 | 1.56 | .96 | 2.52 |
| Redwood or cedar,, 1" x 4" | 3.430 | L.F. | .035 | 2.78 | 1.23 | 4.01 |
| 1" x 6" | 2.180 | L.F. | .036 | 2.91 | 1.29 | 4.20 |
| 2" x 4" | 3.200 | L.F. | .028 | 4.91 | 1.01 | 5.92 |
| 2" x 6" | 2.080 | L.F. | .027 | 4.64 | .96 | 5.60 |
| Joists for deck, treated lumber, 2" x 8", 16" O.C. | 1.000 | L.F. | .015 | 1.12 | .52 | 1.64 |
| 24" O.C. | .800 | L.F. | .012 | .90 | .42 | 1.32 |
| 2" x 10", 16" O.C. | 1.000 | L.F. | .018 | 1.59 | .63 | 2.22 |
| 24" O.C. | .800 | L.F. | .014 | 1.27 | .50 | 1.77 |
| Redwood or cedar, 2" x 8", 16" O.C. | 1.000 | L.F. | .015 | 3.14 | .52 | 3.66 |
| 24" O.C. | .800 | L.F. | .012 | 2.51 | .42 | 2.93 |
| 2" x 10", 16" O.C. | 1.000 | L.F. | .018 | 3.85 | .63 | 4.48 |
| 24" O.C. | .800 | L.F. | .014 | 3.08 | .50 | 3.58 |
| Girder for joists, treated lumber, 2" x 10", 8' x 12' deck | .125 | L.F. | .002 | .20 | .08 | .28 |
| 12' x 16' deck | .083 | L.F. | .001 | .13 | .05 | .18 |
| 12' x 24' deck | .083 | L.F. | .001 | .13 | .05 | .18 |
| Redwood or cedar, 2" x 10", 8' x 12' deck | .125 | L.F. | .002 | .48 | .08 | .56 |
| 12' x 16' deck | .083 | L.F. | .001 | .32 | .05 | .37 |
| 12' x 24' deck | .083 | L.F. | .001 | .32 | .05 | .37 |
| Posts, 4" x 4", including concrete footing, 8' x 12' deck | .250 | L.F. | .022 | .62 | .71 | 1.33 |
| 12' x 16' deck | .122 | L.F. | .017 | .42 | .52 | .94 |
| 12' x 24' deck | .111 | L.F. | .017 | .40 | .50 | .90 |
| Stairs 2" x 10" stringers, treated lumber, 8' x 12' deck | 1.000 | Set | .020 | 3.15 | .71 | 3.86 |
| 12' x 16' deck | 1.000 | Set | .012 | 1.89 | .43 | 2.32 |
| 12' x 24' deck | 1.000 | Set | .008 | 1.26 | .28 | 1.54 |
| Redwood or cedar, 8' x 12' deck | 1.000 | Set | .040 | 5.35 | 1.42 | 6.77 |
| 12' x 16' deck | 1.000 | Set | .020 | 2.68 | .71 | 3.39 |
| 12' x 24' deck | 1.000 | Set | .012 | 1.61 | .43 | 2.04 |
| Railings 2" x 4", treated lumber, 8' x 12' deck | 1.000 | L.F. | .026 | .48 | .92 | 1.40 |
| 12' x 16' deck | .670 | L.F. | .017 | .32 | .62 | .94 |
| 12' x 24' deck | .540 | L.F. | .014 | .26 | .50 | .76 |
| Redwood or cedar, 8' x 12' deck | 1.000 | L.F. | .009 | 1.49 | .31 | 1.80 |
| 12' x 16' deck | .670 | L.F. | .006 | .98 | .20 | 1.18 |
| 12' x 24' deck | .540 | L.F. | .005 | .80 | .17 | .97 |

7 SPECIALTIES

# Division 8
# Mechanical

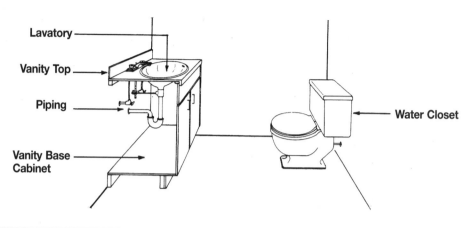

Lavatory

Vanity Top

Piping

Vanity Base Cabinet

Water Closet

| System Description | QUAN. | UNIT | LABOR HOURS | COST EACH | | |
|---|---|---|---|---|---|---|
| | | | | MAT. | INST. | TOTAL |
| **LAVATORY INSTALLED WITH VANITY, PLUMBING IN 2 WALLS** | | | | | | |
| Water closet, floor mounted, 2 piece, close coupled, white | 1.000 | Ea. | 3.019 | 154.00 | 106.00 | 260.00 |
| Rough-in, vent, 2" diameter DWV piping | 1.000 | Ea. | .955 | 25.60 | 33.60 | 59.20 |
| Waste, 4" diameter DWV piping | 1.000 | Ea. | .828 | 31.20 | 29.10 | 60.30 |
| Supply, 1/2" diameter type "L" copper supply piping | 1.000 | Ea. | .593 | 10.14 | 23.22 | 33.36 |
| Lavatory, 20" x 18", P.E. cast iron white | 1.000 | Ea. | 2.500 | 185.00 | 88.00 | 273.00 |
| Rough-in, vent, 1-1/2" diameter DWV piping | 1.000 | Ea. | .901 | 25.00 | 31.80 | 56.80 |
| Waste, 2" diameter DWV piping | 1.000 | Ea. | .955 | 25.60 | 33.60 | 59.20 |
| Supply, 1/2" diameter type "L" copper supply piping | 1.000 | Ea. | .988 | 16.90 | 38.70 | 55.60 |
| Piping, supply, 1/2" diameter type "L" copper supply piping | 10.000 | L.F. | .988 | 16.90 | 38.70 | 55.60 |
| Waste, 4" diameter DWV piping | 7.000 | L.F. | 1.931 | 72.80 | 67.90 | 140.70 |
| Vent, 2" diameter DWV piping | 12.000 | L.F. | 2.866 | 76.80 | 100.80 | 177.60 |
| Vanity base cabinet, 2 door, 30" wide | 1.000 | Ea. | 1.000 | 204.00 | 35.50 | 239.50 |
| Vanity top, plastic & laminated, square edge | 2.670 | L.F. | .712 | 76.10 | 25.37 | 101.47 |
| TOTAL | | | 18.236 | 920.04 | 652.29 | 1572.33 |
| **LAVATORY WITH WALL-HUNG LAVATORY, PLUMBING IN 2 WALLS** | | | | | | |
| Water closet, floor mounted, 2 piece close coupled, white | 1.000 | Ea. | 3.019 | 154.00 | 106.00 | 260.00 |
| Rough-in, vent, 2" diameter DWV piping | 1.000 | Ea. | .955 | 25.60 | 33.60 | 59.20 |
| Waste, 4" diameter DWV piping | 1.000 | Ea. | .828 | 31.20 | 29.10 | 60.30 |
| Supply, 1/2" diameter type "L" copper supply piping | 1.000 | Ea. | .593 | 10.14 | 23.22 | 33.36 |
| Lavatory, 20" x 18", P.E. cast iron, wall hung, white | 1.000 | Ea. | 2.000 | 218.00 | 70.50 | 288.50 |
| Rough-in, vent, 1-1/2" diameter DWV piping | 1.000 | Ea. | .901 | 25.00 | 31.80 | 56.80 |
| Waste, 2" diameter DWV piping | 1.000 | Ea. | .955 | 25.60 | 33.60 | 59.20 |
| Supply, 1/2" diameter type "L" copper supply piping | 1.000 | Ea. | .988 | 16.90 | 38.70 | 55.60 |
| Piping, supply, 1/2" diameter type "L" copper supply piping | 10.000 | L.F. | .988 | 16.90 | 38.70 | 55.60 |
| Waste, 4" diameter DWV piping | 7.000 | L.F. | 1.931 | 72.80 | 67.90 | 140.70 |
| Vent, 2" diameter DWV piping | 12.000 | L.F. | 2.866 | 76.80 | 100.80 | 177.60 |
| Carrier, steel for studs, no arms | 1.000 | Ea. | 1.143 | 29.00 | 44.50 | 73.50 |
| TOTAL | | | 17.167 | 701.94 | 618.42 | 1320.36 |

| Description | QUAN. | UNIT | LABOR HOURS | COST EACH | | |
|---|---|---|---|---|---|---|
| | | | | MAT. | INST. | TOTAL |
| | | | | | | |
| | | | | | | |

**Important: See the Reference Section for critical supporting data - Reference Nos., Crews & Location Factors**

# Two Fixture Lavatory Price Sheet

| | QUAN. | UNIT | LABOR HOURS | COST EACH MAT. | COST EACH INST. | COST EACH TOTAL |
|---|---|---|---|---|---|---|
| Water closet, close coupled standard 2 piece, white | 1.000 | Ea. | 3.019 | 154.00 | 106.00 | 260.00 |
| Color | 1.000 | Ea. | 3.019 | 173.00 | 106.00 | 279.00 |
| One piece elongated bowl, white | 1.000 | Ea. | 3.019 | 545.00 | 106.00 | 651.00 |
| Color | 1.000 | Ea. | 3.019 | 755.00 | 106.00 | 861.00 |
| Low profile, one piece elongated bowl, white | 1.000 | Ea. | 3.019 | 695.00 | 106.00 | 801.00 |
| Color | 1.000 | Ea. | 3.019 | 905.00 | 106.00 | 1011.00 |
| Rough-in for water closet | | | | | | |
| 1/2" copper supply, 4" cast iron waste, 2" cast iron vent | 1.000 | Ea. | 2.376 | 67.00 | 86.00 | 153.00 |
| 4" PVC waste, 2" PVC vent | 1.000 | Ea. | 2.678 | 38.00 | 96.50 | 134.50 |
| 4" copper waste , 2" copper vent | 1.000 | Ea. | 2.520 | 73.50 | 94.00 | 167.50 |
| 3" cast iron waste, 1-1/2" cast iron vent | 1.000 | Ea. | 2.244 | 60.00 | 81.50 | 141.50 |
| 3" PVC waste, 1-1/2" PVC vent | 1.000 | Ea. | 2.388 | 34.00 | 90.00 | 124.00 |
| 3" copper waste, 1-1/2" copper vent | 1.000 | Ea. | 2.524 | 68.00 | 91.00 | 159.00 |
| 1/2" PVC supply, 4" PVC waste, 2" PVC vent | 1.000 | Ea. | 2.974 | 46.50 | 108.00 | 154.50 |
| 3" PVC waste, 1-1/2" PVC vent | 1.000 | Ea. | 2.684 | 43.00 | 102.00 | 145.00 |
| 1/2" steel supply, 4" cast iron waste, 2" cast iron vent | 1.000 | Ea. | 2.545 | 67.00 | 92.50 | 159.50 |
| 4" cast iron waste, 2" steel vent | 1.000 | Ea. | 2.590 | 59.00 | 94.00 | 153.00 |
| 4" PVC waste, 2" PVC vent | 1.000 | Ea. | 2.847 | 37.50 | 103.00 | 140.50 |
| Lavatory, vanity top mounted, P.E. on cast iron 20" x 18" white | 1.000 | Ea. | 2.500 | 185.00 | 88.00 | 273.00 |
| Color | 1.000 | Ea. | 2.500 | 213.00 | 88.00 | 301.00 |
| Steel, enameled 10" x 17" white | 1.000 | Ea. | 2.759 | 99.00 | 97.00 | 196.00 |
| Color | 1.000 | Ea. | 2.500 | 105.00 | 88.00 | 193.00 |
| Vitreous china 20" x 16", white | 1.000 | Ea. | 2.963 | 201.00 | 104.00 | 305.00 |
| Color | 1.000 | Ea. | 2.963 | 201.00 | 104.00 | 305.00 |
| Wall hung, P.E. on cast iron, 20" x 18", white | 1.000 | Ea. | 2.000 | 218.00 | 70.50 | 288.50 |
| Color | 1.000 | Ea. | 2.000 | 255.00 | 70.50 | 325.50 |
| Vitreous china 19" x 17", white | 1.000 | Ea. | 2.286 | 141.00 | 80.50 | 221.50 |
| Color | 1.000 | Ea. | 2.286 | 162.00 | 80.50 | 242.50 |
| Rough-in supply waste and vent for lavatory | | | | | | |
| 1/2" copper supply, 2" cast iron waste, 1-1/2" cast iron vent | 1.000 | Ea. | 2.844 | 67.50 | 104.00 | 171.50 |
| 2" PVC waste, 1-1/2" PVC vent | 1.000 | Ea. | 2.962 | 39.50 | 112.00 | 151.50 |
| 2" copper waste, 1-1/2" copper vent | 1.000 | Ea. | 2.308 | 52.00 | 90.50 | 142.50 |
| 1-1/2" PVC waste, 1-1/4" PVC vent | 1.000 | Ea. | 2.639 | 38.00 | 103.00 | 141.00 |
| 1-1/2" copper waste, 1-1/4" copper vent | 1.000 | Ea. | 2.114 | 45.00 | 82.50 | 127.50 |
| 1/2" PVC supply, 2" PVC waste, 1-1/2" PVC vent | 1.000 | Ea. | 3.456 | 54.00 | 131.00 | 185.00 |
| 1-1/2" PVC waste, 1-1/4" PVC vent | 1.000 | Ea. | 3.133 | 53.00 | 123.00 | 176.00 |
| 1/2" steel supply, 2" cast iron waste, 1-1/2" cast iron vent | 1.000 | Ea. | 3.126 | 67.00 | 115.00 | 182.00 |
| 2" cast iron waste, 2" steel vent | 1.000 | Ea. | 3.225 | 60.50 | 119.00 | 179.50 |
| 2" PVC waste, 1-1/2" PVC vent | 1.000 | Ea. | 3.244 | 39.00 | 123.00 | 162.00 |
| 1-1/2" PVC waste, 1-1/4" PVC vent | 1.000 | Ea. | 2.921 | 38.00 | 114.00 | 152.00 |
| Piping, supply, 1/2" copper, type "L" | 10.000 | L.F. | .988 | 16.90 | 38.50 | 55.40 |
| 1/2" steel | 10.000 | L.F. | 1.270 | 16.60 | 49.50 | 66.10 |
| 1/2" PVC | 10.000 | L.F. | 1.482 | 31.50 | 58.00 | 89.50 |
| Waste, 4" cast iron | 7.000 | L.F. | 1.931 | 73.00 | 68.00 | 141.00 |
| 4" copper | 7.000 | L.F. | 2.800 | 102.00 | 98.50 | 200.50 |
| 4" PVC | 7.000 | L.F. | 2.333 | 37.50 | 82.50 | 120.00 |
| Vent, 2" cast iron | 12.000 | L.F. | 2.866 | 77.00 | 101.00 | 178.00 |
| 2" copper | 12.000 | L.F. | 2.182 | 59.00 | 85.00 | 144.00 |
| 2" PVC | 12.000 | L.F. | 3.254 | 35.00 | 115.00 | 150.00 |
| 2" steel | 12.000 | Ea. | 3.000 | 54.00 | 106.00 | 160.00 |
| Vanity base cabinet, 2 door, 24" x 30" | 1.000 | Ea. | 1.000 | 204.00 | 35.50 | 239.50 |
| 24" x 36" | 1.000 | Ea. | 1.200 | 272.00 | 42.50 | 314.50 |
| Vanity top, laminated plastic, square edge 25" x 32" | 2.670 | L.F. | .712 | 76.00 | 25.50 | 101.50 |
| 25" x 38" | 3.170 | L.F. | .845 | 90.50 | 30.00 | 120.50 |
| Post formed, laminated plastic, 25" x 32" | 2.670 | L.F. | .712 | 25.00 | 25.50 | 50.50 |
| 25" x 38" | 3.170 | L.F. | .845 | 30.00 | 30.00 | 60.00 |
| Cultured marble, 25" x 32" with bowl | 1.000 | Ea. | 2.500 | 153.00 | 88.00 | 241.00 |
| 25" x 38" with bowl | 1.000 | Ea. | 2.500 | 187.00 | 88.00 | 275.00 |
| Carrier for lavatory, steel for studs | 1.000 | Ea. | 1.143 | 29.00 | 44.50 | 73.50 |
| Wood 2" x 8" blocking | 1.330 | L.F. | .053 | 1.26 | 1.89 | 3.15 |

**MECHANICAL**

**8**

231

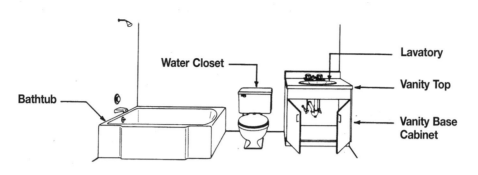

Bathtub    Water Closet    Lavatory    Vanity Top    Vanity Base Cabinet

| System Description | QUAN. | UNIT | LABOR HOURS | COST EACH | | |
|---|---|---|---|---|---|---|
| | | | | MAT. | INST. | TOTAL |
| **BATHROOM INSTALLED WITH VANITY** | | | | | | |
| Water closet, floor mounted, 2 piece, close coupled, white | 1.000 | Ea. | 3.019 | 154.00 | 106.00 | 260.00 |
| Rough-in, waste, 4" diameter DWV piping | 1.000 | Ea. | .828 | 31.20 | 29.10 | 60.30 |
| Vent, 2" diameter DWV piping | 1.000 | Ea. | .955 | 25.60 | 33.60 | 59.20 |
| Supply, 1/2" diameter type "L" copper supply piping | 1.000 | Ea. | .593 | 10.14 | 23.22 | 33.36 |
| Lavatory, 20" x 18", P.E. cast iron with accessories, white | 1.000 | Ea. | 2.500 | 185.00 | 88.00 | 273.00 |
| Rough-in, supply, 1/2" diameter type "L" copper supply piping | 1.000 | Ea. | .988 | 16.90 | 38.70 | 55.60 |
| Waste, 1-1/2" diameter DWV piping | 1.000 | Ea. | 1.803 | 50.00 | 63.60 | 113.60 |
| Bathtub, P.E. cast iron, 5' long with accessories, white | 1.000 | Ea. | 3.636 | 410.00 | 128.00 | 538.00 |
| Rough-in, waste, 4" diameter DWV piping | 1.000 | Ea. | .828 | 31.20 | 29.10 | 60.30 |
| Vent, 1-1/2" diameter DWV piping | 1.000 | Ea. | .593 | 15.28 | 23.20 | 38.48 |
| Supply, 1/2" diameter type "L" copper supply piping | 1.000 | Ea. | .988 | 16.90 | 38.70 | 55.60 |
| Piping, supply, 1/2" diameter type "L" copper supply piping | 20.000 | L.F. | 1.975 | 33.80 | 77.40 | 111.20 |
| Waste, 4" diameter DWV piping | 9.000 | L.F. | 2.483 | 93.60 | 87.30 | 180.90 |
| Vent, 2" diameter DWV piping | 6.000 | L.F. | 1.500 | 27.12 | 52.80 | 79.92 |
| Vanity base cabinet, 2 door, 30" wide | 1.000 | Ea. | 1.000 | 204.00 | 35.50 | 239.50 |
| Vanity top, plastic laminated square edge | 2.670 | L.F. | .712 | 58.74 | 25.37 | 84.11 |
| TOTAL | | | 24.401 | 1363.48 | 879.59 | 2243.07 |
| **BATHROOM WITH WALL HUNG LAVATORY** | | | | | | |
| Water closet, floor mounted, 2 piece, close coupled, white | 1.000 | Ea. | 3.019 | 154.00 | 106.00 | 260.00 |
| Rough-in, vent, 2" diameter DWV piping | 1.000 | Ea. | .955 | 25.60 | 33.60 | 59.20 |
| Waste, 4" diameter DWV piping | 1.000 | Ea. | .828 | 31.20 | 29.10 | 60.30 |
| Supply, 1/2" diameter type "L" copper supply piping | 1.000 | Ea. | .593 | 10.14 | 23.22 | 33.36 |
| Lavatory, 20" x 18" P.E. cast iron, wall hung, white | 1.000 | Ea. | 2.000 | 218.00 | 70.50 | 288.50 |
| Rough-in, waste, 1-1/2" diameter DWV piping | 1.000 | Ea. | 1.803 | 50.00 | 63.60 | 113.60 |
| Supply, 1/2" diameter type "L" copper supply piping | 1.000 | Ea. | .988 | 16.90 | 38.70 | 55.60 |
| Bathtub, P.E. cast iron, 5' long with accessories, white | 1.000 | Ea. | 3.636 | 410.00 | 128.00 | 538.00 |
| Rough-in, waste, 4" diameter DWV piping | 1.000 | Ea. | .828 | 31.20 | 29.10 | 60.30 |
| Supply, 1/2" diameter type "L" copper supply piping | 1.000 | Ea. | .988 | 16.90 | 38.70 | 55.60 |
| Vent, 1-1/2" diameter DWV piping | 1.000 | Ea. | 1.482 | 38.20 | 58.00 | 96.20 |
| Piping, supply, 1/2" diameter type "L" copper supply piping | 20.000 | L.F. | 1.975 | 33.80 | 77.40 | 111.20 |
| Waste, 4" diameter DWV piping | 9.000 | L.F. | 2.483 | 93.60 | 87.30 | 180.90 |
| Vent, 2" diameter DWV piping | 6.000 | L.F. | 1.500 | 27.12 | 52.80 | 79.92 |
| Carrier, steel, for studs, no arms | 1.000 | Ea. | 1.143 | 29.00 | 44.50 | 73.50 |
| TOTAL | | | 24.221 | 1185.66 | 880.52 | 2066.18 |

The costs in this system are a cost each basis, all necessary piping is included

| Description | QUAN. | UNIT | LABOR HOURS | COST EACH | | |
|---|---|---|---|---|---|---|
| | | | | MAT. | INST. | TOTAL |
| | | | | | | |

   **Important: See the Reference Section for critical supporting data - Reference Nos., Crews & Location Factors**

| Three Fixture Bathroom Price Sheet | QUAN. | UNIT | LABOR HOURS | COST EACH | | |
|---|---|---|---|---|---|---|
| | | | | MAT. | INST. | TOTAL |
| Water closet, close coupled standard 2 piece, white | 1.000 | Ea. | 3.019 | 154.00 | 106.00 | 260.00 |
| Color | 1.000 | Ea. | 3.019 | 173.00 | 106.00 | 279.00 |
| One piece, elongated bowl, white | 1.000 | Ea. | 3.019 | 545.00 | 106.00 | 651.00 |
| Color | 1.000 | Ea. | 3.019 | 755.00 | 106.00 | 861.00 |
| Low profile, one piece elongated bowl, white | 1.000 | Ea. | 3.019 | 695.00 | 106.00 | 801.00 |
| Color | 1.000 | Ea. | 3.019 | 905.00 | 106.00 | 1011.00 |
| Rough-in, for water closet | | | | | | |
| 1/2" copper supply, 4" cast iron waste, 2" cast iron vent | 1.000 | Ea. | 2.376 | 67.00 | 86.00 | 153.00 |
| 4" PVC/DWV waste, 2" PVC vent | 1.000 | Ea. | 2.678 | 38.00 | 96.50 | 134.50 |
| 4" copper waste, 2" copper vent | 1.000 | Ea. | 2.520 | 73.50 | 94.00 | 167.50 |
| 3" cast iron waste, 1-1/2" cast iron vent | 1.000 | Ea. | 2.244 | 60.00 | 81.50 | 141.50 |
| 3" PVC waste, 1-1/2" PVC vent | 1.000 | Ea. | 2.388 | 34.00 | 90.00 | 124.00 |
| 3" copper waste, 1-1/2" copper vent | 1.000 | Ea. | 2.014 | 50.50 | 75.50 | 126.00 |
| 1/2" PVC supply, 4" PVC waste, 2" PVC vent | 1.000 | Ea. | 2.974 | 46.50 | 108.00 | 154.50 |
| 3" PVC waste, 1-1/2" PVC supply | 1.000 | Ea. | 2.684 | 43.00 | 102.00 | 145.00 |
| 1/2" steel supply, 4" cast iron waste, 2" cast iron vent | 1.000 | Ea. | 2.545 | 67.00 | 92.50 | 159.50 |
| 4" cast iron waste, 2" steel vent | 1.000 | Ea. | 2.590 | 59.00 | 94.00 | 153.00 |
| 4" PVC waste, 2" PVC vent | 1.000 | Ea. | 2.847 | 37.50 | 103.00 | 140.50 |
| Lavatory, wall hung, P.E. cast iron 20" x 18", white | 1.000 | Ea. | 2.000 | 218.00 | 70.50 | 288.50 |
| Color | 1.000 | Ea. | 2.000 | 255.00 | 70.50 | 325.50 |
| Vitreous china 19" x 17", white | 1.000 | Ea. | 2.286 | 141.00 | 80.50 | 221.50 |
| Color | 1.000 | Ea. | 2.286 | 162.00 | 80.50 | 242.50 |
| Lavatory, for vanity top, P.E. cast iron 20" x 18"", white | 1.000 | Ea. | 2.500 | 185.00 | 88.00 | 273.00 |
| Color | 1.000 | Ea. | 2.500 | 213.00 | 88.00 | 301.00 |
| Steel, enameled 20" x 17", white | 1.000 | Ea. | 2.759 | 99.00 | 97.00 | 196.00 |
| Color | 1.000 | Ea. | 2.500 | 105.00 | 88.00 | 193.00 |
| Vitreous china 20" x 16", white | 1.000 | Ea. | 2.963 | 201.00 | 104.00 | 305.00 |
| Color | 1.000 | Ea. | 2.963 | 201.00 | 104.00 | 305.00 |
| Rough-in, for lavatory | | | | | | |
| 1/2" copper supply, 1-1/2" C.I. waste, 1-1/2" C.I. vent | 1.000 | Ea. | 2.791 | 67.00 | 102.00 | 169.00 |
| 1-1/2" PVC waste, 1-1/4" PVC vent | 1.000 | Ea. | 2.639 | 38.00 | 103.00 | 141.00 |
| 1/2" steel supply, 1-1/4" cast iron waste, 1-1/4" steel vent | 1.000 | Ea. | 2.890 | 54.00 | 107.00 | 161.00 |
| 1-1/4" PVC@ waste, 1-1/4" PVC vent | 1.000 | Ea. | 2.794 | 37.50 | 109.00 | 146.50 |
| 1/2" PVC supply, 1-1/2" PVC waste, 1-1/2" PVC vent | 1.000 | Ea. | 3.260 | 53.50 | 128.00 | 181.50 |
| Bathtub, P.E. cast iron, 5' long corner with fittings, white | 1.000 | Ea. | 3.636 | 410.00 | 128.00 | 538.00 |
| Color | 1.000 | Ea. | 3.636 | 470.00 | 128.00 | 598.00 |
| Rough-in, for bathtub | | | | | | |
| 1/2" copper supply, 4" cast iron waste, 1-1/2" copper vent | 1.000 | Ea. | 2.409 | 63.50 | 91.00 | 154.50 |
| 4" PVC waste, 1-1/2" PVC vent | 1.000 | Ea. | 2.877 | 44.00 | 109.00 | 153.00 |
| 1/2" steel supply, 4" cast iron waste, 1-1/2" steel vent | 1.000 | Ea. | 2.898 | 61.50 | 107.00 | 168.50 |
| 4" PVC waste, 1-1/2" PVC vent | 1.000 | Ea. | 3.159 | 43.50 | 120.00 | 163.50 |
| 1/2" PVC supply, 4" PVC waste, 1-1/2" PVC vent | 1.000 | Ea. | 3.371 | 58.50 | 128.00 | 186.50 |
| Piping, supply 1/2" copper | 20.000 | L.F. | 1.975 | 34.00 | 77.50 | 111.50 |
| 1/2" steel | 20.000 | L.F. | 2.540 | 33.00 | 99.50 | 132.50 |
| 1/2" PVC | 20.000 | L.F. | 2.963 | 63.00 | 116.00 | 179.00 |
| Piping, waste, 4" cast iron no hub | 9.000 | L.F. | 2.483 | 93.50 | 87.50 | 181.00 |
| 4" PVC/DWV | 9.000 | L.F. | 3.000 | 48.00 | 106.00 | 154.00 |
| 4" copper/DWV | 9.000 | L.F. | 3.600 | 131.00 | 127.00 | 258.00 |
| Piping, vent 2" cast iron no hub | 6.000 | L.F. | 1.433 | 38.50 | 50.50 | 89.00 |
| 2" copper/DWV | 6.000 | L.F. | 1.091 | 29.50 | 42.50 | 72.00 |
| 2" PVC/DWV | 6.000 | L.F. | 1.627 | 17.45 | 57.50 | 74.95 |
| 2" steel, galvanized | 6.000 | L.F. | 1.500 | 27.00 | 53.00 | 80.00 |
| Vanity base cabinet, 2 door, 24" x 30" | 1.000 | Ea. | 1.000 | 204.00 | 35.50 | 239.50 |
| 24" x 36" | 1.000 | Ea. | 1.200 | 272.00 | 42.50 | 314.50 |
| Vanity top, laminated plastic square edge 25" x 32" | 2.670 | L.F. | .712 | 58.50 | 25.50 | 84.00 |
| 25" x 38" | 3.160 | L.F. | .843 | 69.50 | 30.00 | 99.50 |
| Cultured marble, 25" x 32", with bowl | 1.000 | Ea. | 2.500 | 153.00 | 88.00 | 241.00 |
| 25" x 38", with bowl | 1.000 | Ea. | 2.500 | 187.00 | 88.00 | 275.00 |
| Carrier, for lavatory, steel for studs, no arms | 1.000 | Ea. | 1.143 | 29.00 | 44.50 | 73.50 |
| Wood, 2" x 8" blocking | 1.300 | L.F. | .052 | 1.24 | 1.85 | 3.09 |

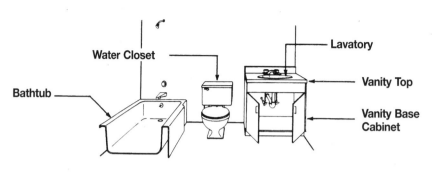

| System Description | QUAN. | UNIT | LABOR HOURS | COST EACH | | |
|---|---|---|---|---|---|---|
| | | | | MAT. | INST. | TOTAL |
| **BATHROOM WITH LAVATORY INSTALLED IN VANITY** | | | | | | |
| Water closet, floor mounted, 2 piece, close coupled, white | 1.000 | Ea. | 3.019 | 154.00 | 106.00 | 260.00 |
| Rough-in, waste, 4" diameter DWV piping | 1.000 | Ea. | .828 | 31.20 | 29.10 | 60.30 |
| Vent, 2" diameter DWV piping | 1.000 | Ea. | .955 | 25.60 | 33.60 | 59.20 |
| Supply, 1/2" diameter type "L" copper supply piping | 1.000 | Ea. | .593 | 10.14 | 23.22 | 33.36 |
| Lavatory, 20" x 18", P.E. cast iron with accessories, white | 1.000 | Ea. | 2.500 | 185.00 | 88.00 | 273.00 |
| Rough-in, waste, 1-1/2" diameter DWV piping | 1.000 | Ea. | 1.803 | 50.00 | 63.60 | 113.60 |
| Supply, 1/2" diameter type "L" copper supply piping | 1.000 | Ea. | .988 | 16.90 | 38.70 | 55.60 |
| Bathtub, P.E. cast iron 5' long with accessories, white | 1.000 | Ea. | 3.636 | 410.00 | 128.00 | 538.00 |
| Rough-in, waste, 4" diameter DWV piping | 1.000 | Ea. | .828 | 31.20 | 29.10 | 60.30 |
| Vent, 1-1/2" diameter DWV piping | 1.000 | Ea. | .593 | 15.28 | 23.20 | 38.48 |
| Supply, 1/2" diameter type "L" copper supply piping | 1.000 | Ea. | .988 | 16.90 | 38.70 | 55.60 |
| Piping, supply, 1/2" diameter type "L" copper supply piping | 10.000 | L.F. | .988 | 16.90 | 38.70 | 55.60 |
| Waste, 4" diameter DWV piping | 6.000 | L.F. | 1.655 | 62.40 | 58.20 | 120.60 |
| Vent, 2" diameter DWV piping | 6.000 | L.F. | 1.500 | 27.12 | 52.80 | 79.92 |
| Vanity base cabinet, 2 door, 30" wide | 1.000 | Ea. | 1.000 | 204.00 | 35.50 | 239.50 |
| Vanity top, plastic laminated square edge | 2.670 | L.F. | .712 | 58.74 | 25.37 | 84.11 |
| TOTAL | | | 22.586 | 1315.38 | 811.79 | 2127.17 |
| **BATHROOM WITH WALL HUNG LAVATORY** | | | | | | |
| Water closet, floor mounted, 2 piece, close coupled, white | 1.000 | Ea. | 3.019 | 154.00 | 106.00 | 260.00 |
| Rough-in, vent, 2" diameter DWV piping | 1.000 | Ea. | .955 | 25.60 | 33.60 | 59.20 |
| Waste, 4" diameter DWV piping | 1.000 | Ea. | .828 | 31.20 | 29.10 | 60.30 |
| Supply, 1/2" diameter type "L" copper supply piping | 1.000 | Ea. | .593 | 10.14 | 23.22 | 33.36 |
| Lavatory, 20" x 18" P.E. cast iron, wall hung, white | 1.000 | Ea. | 2.000 | 218.00 | 70.50 | 288.50 |
| Rough-in, waste, 1-1/2" diameter DWV piping | 1.000 | Ea. | 1.803 | 50.00 | 63.60 | 113.60 |
| Supply, 1/2" diameter type "L" copper supply piping | 1.000 | Ea. | .988 | 16.90 | 38.70 | 55.60 |
| Bathtub, P.E. cast iron, 5' long with accessories, white | 1.000 | Ea. | 3.636 | 410.00 | 128.00 | 538.00 |
| Rough-in, waste, 4" diameter DWV piping | 1.000 | Ea. | .828 | 31.20 | 29.10 | 60.30 |
| Supply, 1/2" diameter type "L" copper supply piping | 1.000 | Ea. | .988 | 16.90 | 38.70 | 55.60 |
| Vent, 1-1/2" diameter DWV piping | 1.000 | Ea. | .593 | 15.28 | 23.20 | 38.48 |
| Piping, supply, 1/2" diameter type "L" copper supply piping | 10.000 | L.F. | .988 | 16.90 | 38.70 | 55.60 |
| Waste, 4" diameter DWV piping | 6.000 | L.F. | 1.655 | 62.40 | 58.20 | 120.60 |
| Vent, 2" diameter DWV piping | 6.000 | L.F. | 1.500 | 27.12 | 52.80 | 79.92 |
| Carrier, steel, for studs, no arms | 1.000 | Ea. | 1.143 | 29.00 | 44.50 | 73.50 |
| TOTAL | | | 21.517 | 1114.64 | 777.92 | 1892.56 |

The costs in this system are on a cost each basis. All necessary piping is included.

| Description | QUAN. | UNIT | LABOR HOURS | COST EACH | | |
|---|---|---|---|---|---|---|
| | | | | MAT. | INST. | TOTAL |
| | | | | | | |

**Important: See the Reference Section for critical supporting data - Reference Nos., Crews & Location Factors**

| Three Fixture Bathroom Price Sheet | QUAN. | UNIT | LABOR HOURS | COST EACH | | |
|---|---|---|---|---|---|---|
| | | | | MAT. | INST. | TOTAL |
| Water closet, close coupled standard 2 piece, white | 1.000 | Ea. | 3.019 | 154.00 | 106.00 | 260.00 |
| Color | 1.000 | Ea. | 3.019 | 173.00 | 106.00 | 279.00 |
| One piece elongated bowl, white | 1.000 | Ea. | 3.019 | 545.00 | 106.00 | 651.00 |
| Color | 1.000 | Ea. | 3.019 | 755.00 | 106.00 | 861.00 |
| Low profile, one piece elongated bowl, white | 1.000 | Ea. | 3.019 | 695.00 | 106.00 | 801.00 |
| Color | 1.000 | Ea. | 3.019 | 905.00 | 106.00 | 1011.00 |
| Rough-in for water closet | | | | | | |
| 1/2" copper supply, 4" cast iron waste, 2" cast iron vent | 1.000 | Ea. | 2.376 | 67.00 | 86.00 | 153.00 |
| 4" PVC/DWV waste, 2" PVC vent | 1.000 | Ea. | 2.678 | 38.00 | 96.50 | 134.50 |
| 4" carrier waste, 2" copper vent | 1.000 | Ea. | 2.520 | 73.50 | 94.00 | 167.50 |
| 3" cast iron waste, 1-1/2" cast iron vent | 1.000 | Ea. | 2.244 | 60.00 | 81.50 | 141.50 |
| 3" PVC waste, 1-1/2" PVC vent | 1.000 | Ea. | 2.388 | 34.00 | 90.00 | 124.00 |
| 3" copper waste, 1-1/2" copper vent | 1.000 | Ea. | 2.014 | 50.50 | 75.50 | 126.00 |
| 1/2" PVC supply, 4" PVC waste, 2" PVC vent | 1.000 | Ea. | 2.974 | 46.50 | 108.00 | 154.50 |
| 3" PVC waste, 1-1/2" PVC supply | 1.000 | Ea. | 2.684 | 43.00 | 102.00 | 145.00 |
| 1/2" steel supply, 4" cast iron waste, 2" cast iron vent | 1.000 | Ea. | 2.545 | 67.00 | 92.50 | 159.50 |
| 4" cast iron waste, 2" steel vent | 1.000 | Ea. | 2.590 | 59.00 | 94.00 | 153.00 |
| 4" PVC waste, 2" PVC vent | 1.000 | Ea. | 2.847 | 37.50 | 103.00 | 140.50 |
| Lavatory, wall hung, PE cast iron 20" x 18", white | 1.000 | Ea. | 2.000 | 218.00 | 70.50 | 288.50 |
| Color | 1.000 | Ea. | 2.000 | 255.00 | 70.50 | 325.50 |
| Vitreous china 19" x 17", white | 1.000 | Ea. | 2.286 | 141.00 | 80.50 | 221.50 |
| Color | 1.000 | Ea. | 2.286 | 162.00 | 80.50 | 242.50 |
| Lavatory, for vanity top, PE cast iron 20" x 18", white | 1.000 | Ea. | 2.500 | 185.00 | 88.00 | 273.00 |
| Color | 1.000 | Ea. | 2.500 | 213.00 | 88.00 | 301.00 |
| Steel enameled 20" x 17", white | 1.000 | Ea. | 2.759 | 99.00 | 97.00 | 196.00 |
| Color | 1.000 | Ea. | 2.500 | 105.00 | 88.00 | 193.00 |
| Vitreous china 20" x 16", white | 1.000 | Ea. | 2.963 | 201.00 | 104.00 | 305.00 |
| Color | 1.000 | Ea. | 2.963 | 201.00 | 104.00 | 305.00 |
| Rough-in for lavatory | | | | | | |
| 1/2" copper supply, 1-1/2" cast iron waste, 1-1/2" cast iron vent | 1.000 | Ea. | 2.791 | 67.00 | 102.00 | 169.00 |
| 1-1/2" PVC waste, 1-1/4" PVC vent | 1.000 | Ea. | 2.639 | 38.00 | 103.00 | 141.00 |
| 1/2" steel supply, 1-1/4" cast iron waste, 1-1/4" steel vent | 1.000 | Ea. | 2.890 | 54.00 | 107.00 | 161.00 |
| 1-1/4" PVC waste, 1-1/4" PVC vent | 1.000 | Ea. | 2.794 | 37.50 | 109.00 | 146.50 |
| 1/2" PVC supply, 1-1/2" PVC waste, 1-1/2" PVC vent | 1.000 | Ea. | 3.260 | 53.50 | 128.00 | 181.50 |
| Bathtub, PE cast iron, 5' long corner with fittings, white | 1.000 | Ea. | 3.636 | 410.00 | 128.00 | 538.00 |
| Color | 1.000 | Ea. | 3.636 | 470.00 | 128.00 | 598.00 |
| Rough-in for bathtub | | | | | | |
| 1/2" copper supply, 4" cast iron waste, 1-1/2" copper vent | 1.000 | Ea. | 2.409 | 63.50 | 91.00 | 154.50 |
| 4" PVC waste, 1/2" PVC vent | 1.000 | Ea. | 2.877 | 44.00 | 109.00 | 153.00 |
| 1/2" steel supply, 4" cast iron waste, 1-1/2" steel vent | 1.000 | Ea. | 2.898 | 61.50 | 107.00 | 168.50 |
| 4" PVC waste, 1-1/2" PVC vent | 1.000 | Ea. | 3.159 | 43.50 | 120.00 | 163.50 |
| 1/2" PVC supply, 4" PVC waste, 1-1/2" PVC vent | 1.000 | Ea. | 3.371 | 58.50 | 128.00 | 186.50 |
| Piping supply, 1/2" copper | 10.000 | L.F. | .988 | 16.90 | 38.50 | 55.40 |
| 1/2" steel | 10.000 | L.F. | 1.270 | 16.60 | 49.50 | 66.10 |
| 1/2" PVC | 10.000 | L.F. | 1.482 | 31.50 | 58.00 | 89.50 |
| Piping waste, 4" cast iron no hub | 6.000 | L.F. | 1.655 | 62.50 | 58.00 | 120.50 |
| 4" PVC/DWV | 6.000 | L.F. | 2.000 | 32.00 | 70.50 | 102.50 |
| 4" copper/DWV | 6.000 | L.F. | 2.400 | 87.50 | 84.50 | 172.00 |
| Piping vent 2" cast iron no hub | 6.000 | L.F. | 1.433 | 38.50 | 50.50 | 89.00 |
| 2" copper/DWV | 6.000 | L.F. | 1.091 | 29.50 | 42.50 | 72.00 |
| 2" PVC/DWV | 6.000 | L.F. | 1.627 | 17.45 | 57.50 | 74.95 |
| 2" steel, galvanized | 6.000 | L.F. | 1.500 | 27.00 | 53.00 | 80.00 |
| Vanity base cabinet, 2 door, 24" x 30" | 1.000 | Ea. | 1.000 | 204.00 | 35.50 | 239.50 |
| 24" x 36" | 1.000 | Ea. | 1.200 | 272.00 | 42.50 | 314.50 |
| Vanity top, laminated plastic square edge 25" x 32" | 2.670 | L.F. | .712 | 58.50 | 25.50 | 84.00 |
| 25" x 38" | 3.160 | L.F. | .843 | 69.50 | 30.00 | 99.50 |
| Cultured marble, 25" x 32", with bowl | 1.000 | Ea. | 2.500 | 153.00 | 88.00 | 241.00 |
| 25" x 38", with bowl | 1.000 | Ea. | 2.500 | 187.00 | 88.00 | 275.00 |
| Carrier, for lavatory, steel for studs, no arms | 1.000 | Ea. | 1.143 | 29.00 | 44.50 | 73.50 |
| Wood, 2" x 8" blocking | 1.300 | L.F. | .052 | 1.24 | 1.85 | 3.09 |

MECHANICAL

8

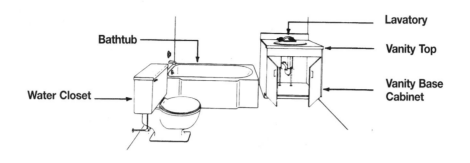

| System Description | QUAN. | UNIT | LABOR HOURS | COST EACH | | |
|---|---|---|---|---|---|---|
| | | | | MAT. | INST. | TOTAL |
| **BATHROOM WITH LAVATORY INSTALLED IN VANITY** | | | | | | |
| Water closet, floor mounted, 2 piece, close coupled, white | 1.000 | Ea. | 3.019 | 154.00 | 106.00 | 260.00 |
| Rough-in, vent, 2" diameter DWV piping | 1.000 | Ea. | .955 | 25.60 | 33.60 | 59.20 |
| Waste, 4" diameter DWV piping | 1.000 | Ea. | .828 | 31.20 | 29.10 | 60.30 |
| Supply, 1/2" diameter type "L" copper supply piping | 1.000 | Ea. | .593 | 10.14 | 23.22 | 33.36 |
| Lavatory, 20" x 18", PE cast iron with accessories, white | 1.000 | Ea. | 2.500 | 185.00 | 88.00 | 273.00 |
| Rough-in, vent, 1-1/2" diameter DWV piping | 1.000 | Ea. | 1.803 | 50.00 | 63.60 | 113.60 |
| Supply, 1/2" diameter type "L" copper supply piping | 1.000 | Ea. | .988 | 16.90 | 38.70 | 55.60 |
| Bathtub, P.E. cast iron, 5' long with accessories, white | 1.000 | Ea. | 3.636 | 410.00 | 128.00 | 538.00 |
| Rough-in, waste, 4" diameter DWV piping | 1.000 | Ea. | .828 | 31.20 | 29.10 | 60.30 |
| Supply, 1/2" diameter type "L" copper supply piping | 1.000 | Ea. | .988 | 16.90 | 38.70 | 55.60 |
| Vent, 1-1/2" diameter DWV piping | 1.000 | Ea. | .593 | 15.28 | 23.20 | 38.48 |
| Piping, supply, 1/2" diameter type "L" copper supply piping | 32.000 | L.F. | 3.161 | 54.08 | 123.84 | 177.92 |
| Waste, 4" diameter DWV piping | 12.000 | L.F. | 3.310 | 124.80 | 116.40 | 241.20 |
| Vent, 2" diameter DWV piping | 6.000 | L.F. | 1.500 | 27.12 | 52.80 | 79.92 |
| Vanity base cabinet, 2 door, 30" wide | 1.000 | Ea. | 1.000 | 204.00 | 35.50 | 239.50 |
| Vanity top, plastic laminated square edge | 2.670 | L.F. | .712 | 58.74 | 25.37 | 84.11 |
| TOTAL | | | 26.414 | 1414.96 | 955.13 | 2370.09 |
| **BATHROOM WITH WALL HUNG LAVATORY** | | | | | | |
| Water closet, floor mounted, 2 piece, close coupled, white | 1.000 | Ea. | 3.019 | 154.00 | 106.00 | 260.00 |
| Rough-in, vent, 2" diameter DWV piping | 1.000 | Ea. | .955 | 25.60 | 33.60 | 59.20 |
| Waste, 4" diameter DWV piping | 1.000 | Ea. | .828 | 31.20 | 29.10 | 60.30 |
| Supply, 1/2" diameter type "L" copper supply piping | 1.000 | Ea. | .593 | 10.14 | 23.22 | 33.36 |
| Lavatory, 20" x 18" P.E. cast iron, wall hung, white | 1.000 | Ea. | 2.000 | 218.00 | 70.50 | 288.50 |
| Rough-in, waste, 1-1/2" diameter DWV piping | 1.000 | Ea. | 1.803 | 50.00 | 63.60 | 113.60 |
| Supply, 1/2" diameter type "L" copper supply piping | 1.000 | Ea. | .988 | 16.90 | 38.70 | 55.60 |
| Bathtub, P.E. cast iron, 5' long with accessories, white | 1.000 | Ea. | 3.636 | 410.00 | 128.00 | 538.00 |
| Rough-in, waste, 4" diameter DWV piping | 1.000 | Ea. | .828 | 31.20 | 29.10 | 60.30 |
| Supply, 1/2" diameter type "L" copper supply piping | 1.000 | Ea. | .988 | 16.90 | 38.70 | 55.60 |
| Vent, 1-1/2" diameter DWV piping | 1.000 | Ea. | .593 | 15.28 | 23.20 | 38.48 |
| Piping, supply, 1/2" diameter type "L" copper supply piping | 32.000 | L.F. | 3.161 | 54.08 | 123.84 | 177.92 |
| Waste, 4" diameter DWV piping | 12.000 | L.F. | 3.310 | 124.80 | 116.40 | 241.20 |
| Vent, 2" diameter DWV piping | 6.000 | L.F. | 1.500 | 27.12 | 52.80 | 79.92 |
| Carrier steel, for studs, no arms | 1.000 | Ea. | 1.143 | 29.00 | 44.50 | 73.50 |
| TOTAL | | | 25.345 | 1214.22 | 921.26 | 2135.48 |

The costs in this system are on a cost each basis. All necessary piping is included.

| Description | QUAN. | UNIT | LABOR HOURS | COST EACH | | |
|---|---|---|---|---|---|---|
| | | | | MAT. | INST. | TOTAL |
| | | | | | | |

**Important: See the Reference Section for critical supporting data - Reference Nos., Crews & Location Factors**

| Three Fixture Bathroom Price Sheet | QUAN. | UNIT | LABOR HOURS | COST EACH | | |
|---|---|---|---|---|---|---|
| | | | | MAT. | INST. | TOTAL |
| Water closet, close coupled, standard 2 piece, white | 1.000 | Ea. | 3.019 | 154.00 | 106.00 | 260.00 |
| Color | 1.000 | Ea. | 3.019 | 173.00 | 106.00 | 279.00 |
| One piece, elongated bowl, white | 1.000 | Ea. | 3.019 | 545.00 | 106.00 | 651.00 |
| Color | 1.000 | Ea. | 3.019 | 755.00 | 106.00 | 861.00 |
| Low profile, one piece, elongated bowl, white | 1.000 | Ea. | 3.019 | 695.00 | 106.00 | 801.00 |
| Color | 1.000 | Ea. | 3.019 | 905.00 | 106.00 | 1011.00 |
| Rough-in, for water closet | | | | | | |
| 1/2" copper supply, 4" cast iron waste, 2" cast iron vent | 1.000 | Ea. | 2.376 | 67.00 | 86.00 | 153.00 |
| 4" PVC/DWV waste, 2" PVC vent | 1.000 | Ea. | 2.678 | 38.00 | 96.50 | 134.50 |
| 4" copper waste, 2" copper vent | 1.000 | Ea. | 2.520 | 73.50 | 94.00 | 167.50 |
| 3" cast iron waste, 1-1/2" cast iron vent | 1.000 | Ea. | 2.244 | 60.00 | 81.50 | 141.50 |
| 3" PVC waste, 1-1/2" PVC vent | 1.000 | Ea. | 2.388 | 34.00 | 90.00 | 124.00 |
| 3" copper waste, 1-1/2" copper vent | 1.000 | Ea. | 2.014 | 50.50 | 75.50 | 126.00 |
| 1/2" PVC supply, 4" PVC waste, 2" PVC vent | 1.000 | Ea. | 2.974 | 46.50 | 108.00 | 154.50 |
| 3" PVC waste, 1-1/2" PVC supply | 1.000 | Ea. | 2.684 | 43.00 | 102.00 | 145.00 |
| 1/2" steel supply, 4" cast iron waste, 2" cast iron vent | 1.000 | Ea. | 2.545 | 67.00 | 92.50 | 159.50 |
| 4" cast iron waste, 2" steel vent | 1.000 | Ea. | 2.590 | 59.00 | 94.00 | 153.00 |
| 4" PVC waste, 2" PVC vent | 1.000 | Ea. | 2.847 | 37.50 | 103.00 | 140.50 |
| Lavatory wall hung, P.E. cast iron, 20" x 18", white | 1.000 | Ea. | 2.000 | 218.00 | 70.50 | 288.50 |
| Color | 1.000 | Ea. | 2.000 | 255.00 | 70.50 | 325.50 |
| Vitreous china, 19" x 17", white | 1.000 | Ea. | 2.286 | 141.00 | 80.50 | 221.50 |
| Color | 1.000 | Ea. | 2.286 | 162.00 | 80.50 | 242.50 |
| Lavatory, for vanity top, P.E., cast iron, 20" x 18", white | 1.000 | Ea. | 2.500 | 185.00 | 88.00 | 273.00 |
| Color | 1.000 | Ea. | 2.500 | 213.00 | 88.00 | 301.00 |
| Steel, enameled, 20" x 17", white | 1.000 | Ea. | 2.759 | 99.00 | 97.00 | 196.00 |
| Color | 1.000 | Ea. | 2.500 | 105.00 | 88.00 | 193.00 |
| Vitreous china, 20" x 16", white | 1.000 | Ea. | 2.963 | 201.00 | 104.00 | 305.00 |
| Color | 1.000 | Ea. | 2.963 | 201.00 | 104.00 | 305.00 |
| Rough-in, for lavatory | | | | | | |
| 1/2" copper supply, 1-1/2" C.I. waste, 1-1/2" C.I. vent | 1.000 | Ea. | 2.791 | 67.00 | 102.00 | 169.00 |
| 1-1/2" PVC waste, 1-1/4" PVC vent | 1.000 | Ea. | 2.639 | 38.00 | 103.00 | 141.00 |
| 1/2" steel supply, 1-1/4" cast iron waste, 1-1/4" steel vent | 1.000 | Ea. | 2.890 | 54.00 | 107.00 | 161.00 |
| 1-1/4" PVC waste, 1-1/4" PVC vent | 1.000 | Ea. | 2.794 | 37.50 | 109.00 | 146.50 |
| 1/2" PVC supply, 1-1/2" PVC waste, 1-1/2" PVC vent | 1.000 | Ea. | 3.260 | 53.50 | 128.00 | 181.50 |
| Bathtub, P.E. cast iron, 5' long corner with fittings, white | 1.000 | Ea. | 3.636 | 410.00 | 128.00 | 538.00 |
| Color | 1.000 | Ea. | 3.636 | 470.00 | 128.00 | 598.00 |
| Rough-in, for bathtub | | | | | | |
| 1/2" copper supply, 4" cast iron waste, 1-1/2" copper vent | 1.000 | Ea. | 2.409 | 63.50 | 91.00 | 154.50 |
| 4" PVC waste, 1/2" PVC vent | 1.000 | Ea. | 2.877 | 44.00 | 109.00 | 153.00 |
| 1/2" steel supply, 4" cast iron waste, 1-1/2" steel vent | 1.000 | Ea. | 2.898 | 61.50 | 107.00 | 168.50 |
| 4" PVC waste, 1-1/2" PVC vent | 1.000 | Ea. | 3.159 | 43.50 | 120.00 | 163.50 |
| 1/2" PVC supply, 4" PVC waste, 1-1/2" PVC vent | 1.000 | Ea. | 3.371 | 58.50 | 128.00 | 186.50 |
| Piping, supply, 1/2" copper | 32.000 | L.F. | 3.161 | 54.00 | 124.00 | 178.00 |
| 1/2" steel | 32.000 | L.F. | 4.063 | 53.00 | 159.00 | 212.00 |
| 1/2" PVC | 32.000 | L.F. | 4.741 | 101.00 | 186.00 | 287.00 |
| Piping, waste, 4" cast iron no hub | 12.000 | L.F. | 3.310 | 125.00 | 116.00 | 241.00 |
| 4" PVC/DWV | 12.000 | L.F. | 4.000 | 64.00 | 141.00 | 205.00 |
| 4" copper/DWV | 12.000 | L.F. | 4.800 | 175.00 | 169.00 | 344.00 |
| Piping, vent, 2" cast iron no hub | 6.000 | L.F. | 1.433 | 38.50 | 50.50 | 89.00 |
| 2" copper/DWV | 6.000 | L.F. | 1.091 | 29.50 | 42.50 | 72.00 |
| 2" PVC/DWV | 6.000 | L.F. | 1.627 | 17.45 | 57.50 | 74.95 |
| 2" steel, galvanized | 6.000 | L.F. | 1.500 | 27.00 | 53.00 | 80.00 |
| Vanity base cabinet, 2 door, 24" x 30" | 1.000 | Ea. | 1.000 | 204.00 | 35.50 | 239.50 |
| 24" x 36" | 1.000 | Ea. | 1.200 | 272.00 | 42.50 | 314.50 |
| Vanity top, laminated plastic square edge, 25" x 32" | 2.670 | L.F. | .712 | 58.50 | 25.50 | 84.00 |
| 25" x 38" | 3.160 | L.F. | .843 | 69.50 | 30.00 | 99.50 |
| Cultured marble, 25" x 32", with bowl | 1.000 | Ea. | 2.500 | 153.00 | 88.00 | 241.00 |
| 25" x 38", with bowl | 1.000 | Ea. | 2.500 | 187.00 | 88.00 | 275.00 |
| Carrier, for lavatory, steel for studs, no arms | 1.000 | Ea. | 1.143 | 29.00 | 44.50 | 73.50 |
| Wood, 2" x 8" blocking | 1.300 | L.F. | .052 | 1.24 | 1.85 | 3.09 |

**MECHANICAL**

**8**

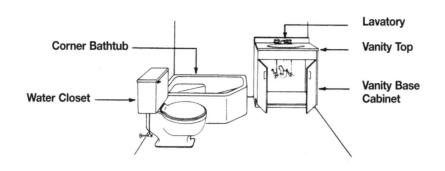

Corner Bathtub

Water Closet

Lavatory

Vanity Top

Vanity Base Cabinet

| System Description | QUAN. | UNIT | LABOR HOURS | COST EACH | | |
|---|---|---|---|---|---|---|
| | | | | MAT. | INST. | TOTAL |
| **BATHROOM WITH LAVATORY INSTALLED IN VANITY** | | | | | | |
| Water closet, floor mounted, 2 piece, close coupled, white | 1.000 | Ea. | 3.019 | 154.00 | 106.00 | 260.00 |
| Rough-in, vent, 2" diameter DWV piping | 1.000 | Ea. | .955 | 25.60 | 33.60 | 59.20 |
| Waste, 4" diameter DWV piping | 1.000 | Ea. | .828 | 31.20 | 29.10 | 60.30 |
| Supply, 1/2" diameter type "L" copper supply piping | 1.000 | Ea. | .593 | 10.14 | 23.22 | 33.36 |
| Lavatory, 20" x 18", P.E. cast iron with fittings, white | 1.000 | Ea. | 2.500 | 185.00 | 88.00 | 273.00 |
| Rough-in, waste, 1-1/2" diameter DWV piping | 1.000 | Ea. | 1.803 | 50.00 | 63.60 | 113.60 |
| Supply, 1/2" diameter type "L" copper supply piping | 1.000 | Ea. | .988 | 16.90 | 38.70 | 55.60 |
| Bathtub, P.E. cast iron, corner with fittings, white | 1.000 | Ea. | 3.636 | 1425.00 | 128.00 | 1553.00 |
| Rough-in, waste, 4" diameter DWV piping | 1.000 | Ea. | .828 | 31.20 | 29.10 | 60.30 |
| Supply, 1/2" diameter type "L" copper supply piping | 1.000 | Ea. | .988 | 16.90 | 38.70 | 55.60 |
| Vent, 1-1/2" diameter DWV piping | 1.000 | Ea. | .593 | 15.28 | 23.20 | 38.48 |
| Piping, supply, 1/2" diameter type "L" copper supply piping | 32.000 | L.F. | 3.161 | 54.08 | 123.84 | 177.92 |
| Waste, 4" diameter DWV piping | 12.000 | L.F. | 3.310 | 124.80 | 116.40 | 241.20 |
| Vent, 2" diameter DWV piping | 6.000 | L.F. | 1.500 | 27.12 | 52.80 | 79.92 |
| Vanity base cabinet, 2 door, 30" wide | 1.000 | Ea. | 1.000 | 204.00 | 35.50 | 239.50 |
| Vanity top, plastic laminated, square edge | 2.670 | L.F. | .712 | 76.10 | 25.37 | 101.47 |
| **TOTAL** | | | 26.414 | 2447.32 | 955.13 | 3402.45 |
| **BATHROOM WITH WALL HUNG LAVATORY** | | | | | | |
| Water closet, floor mounted, 2 piece, close coupled, white | 1.000 | Ea. | 3.019 | 154.00 | 106.00 | 260.00 |
| Rough-in, vent, 2" diameter DWV piping | 1.000 | Ea. | .955 | 25.60 | 33.60 | 59.20 |
| Waste, 4" diameter DWV piping | 1.000 | Ea. | .828 | 31.20 | 29.10 | 60.30 |
| Supply, 1/2" diameter type "L" copper supply piping | 1.000 | Ea. | .593 | 10.14 | 23.22 | 33.36 |
| Lavatory, 20" x 18", P.E. cast iron, with fittings, white | 1.000 | Ea. | 2.000 | 218.00 | 70.50 | 288.50 |
| Rough-in, waste, 1-1/2" diameter DWV piping | 1.000 | Ea. | 1.803 | 50.00 | 63.60 | 113.60 |
| Supply, 1/2" diameter type "L" copper supply piping | 1.000 | Ea. | .988 | 16.90 | 38.70 | 55.60 |
| Bathtub, P.E. cast iron, corner, with fittings, white | 1.000 | Ea. | 3.636 | 1425.00 | 128.00 | 1553.00 |
| Rough-in, waste, 4" diameter DWV piping | 1.000 | Ea. | .828 | 31.20 | 29.10 | 60.30 |
| Supply, 1/2" diameter type "L" copper supply piping | 1.000 | Ea. | .988 | 16.90 | 38.70 | 55.60 |
| Vent, 1-1/2" diameter DWV piping | 1.000 | Ea. | .593 | 15.28 | 23.20 | 38.48 |
| Piping, supply, 1/2" diameter type "L" copper supply piping | 32.000 | L.F. | 3.161 | 54.08 | 123.84 | 177.92 |
| Waste, 4" diameter DWV piping | 12.000 | L.F. | 3.310 | 124.80 | 116.40 | 241.20 |
| Vent, 2" diameter DWV piping | 6.000 | L.F. | 1.500 | 27.12 | 52.80 | 79.92 |
| Carrier, steel, for studs, no arms | 1.000 | Ea. | 1.143 | 29.00 | 44.50 | 73.50 |
| **TOTAL** | | | 25.345 | 2229.22 | 921.26 | 3150.48 |

The costs in this system are on a cost each basis. All necessary piping is included.

MECHANICAL 8

| Three Fixture Bathroom Price Sheet | QUAN. | UNIT | LABOR HOURS | COST EACH | | |
|---|---|---|---|---|---|---|
| | | | | MAT. | INST. | TOTAL |
| Water closet, close coupled, standard 2 piece, white | 1.000 | Ea. | 3.019 | 154.00 | 106.00 | 260.00 |
| Color | 1.000 | Ea. | 3.019 | 173.00 | 106.00 | 279.00 |
| One piece elongated bowl, white | 1.000 | Ea. | 3.019 | 545.00 | 106.00 | 651.00 |
| Color | 1.000 | Ea. | 3.019 | 755.00 | 106.00 | 861.00 |
| Low profile, one piece elongated bowl, white | 1.000 | Ea. | 3.019 | 695.00 | 106.00 | 801.00 |
| Color | 1.000 | Ea. | 3.019 | 905.00 | 106.00 | 1011.00 |
| Rough-in, for water closet | | | | | | |
| 1/2" copper supply, 4" cast iron waste, 2" cast iron vent | 1.000 | Ea. | 2.376 | 67.00 | 86.00 | 153.00 |
| 4" PVC/DWV waste, 2" PVC vent | 1.000 | Ea. | 2.678 | 38.00 | 96.50 | 134.50 |
| 4" copper waste, 2" copper vent | 1.000 | Ea. | 2.520 | 73.50 | 94.00 | 167.50 |
| 3" cast iron waste, 1-1/2" cast iron vent | 1.000 | Ea. | 2.244 | 60.00 | 81.50 | 141.50 |
| 3" PVC waste, 1-1/2" PVC vent | 1.000 | Ea. | 2.388 | 34.00 | 90.00 | 124.00 |
| 3" copper waste, 1-1/2" copper vent | 1.000 | Ea. | 2.014 | 50.50 | 75.50 | 126.00 |
| 1/2" PVC supply, 4" PVC waste, 2" PVC vent | 1.000 | Ea. | 2.974 | 46.50 | 108.00 | 154.50 |
| 3" PVC waste, 1-1/2" PVC supply | 1.000 | Ea. | 2.684 | 43.00 | 102.00 | 145.00 |
| 1/2" steel supply, 4" cast iron waste, 2" cast iron vent | 1.000 | Ea. | 2.545 | 67.00 | 92.50 | 159.50 |
| 4" cast iron waste, 2" steel vent | 1.000 | Ea. | 2.590 | 59.00 | 94.00 | 153.00 |
| 4" PVC waste, 2" PVC vent | 1.000 | Ea. | 2.847 | 37.50 | 103.00 | 140.50 |
| Lavatory, wall hung P.E. cast iron 20" x 18", white | 1.000 | Ea. | 2.000 | 218.00 | 70.50 | 288.50 |
| Color | 1.000 | Ea. | 2.000 | 255.00 | 70.50 | 325.50 |
| Vitreous china 19" x 17", white | 1.000 | Ea. | 2.286 | 141.00 | 80.50 | 221.50 |
| Color | 1.000 | Ea. | 2.286 | 162.00 | 80.50 | 242.50 |
| Lavatory, for vanity top, P.E., cast iron, 20" x 18", white | 1.000 | Ea. | 2.500 | 185.00 | 88.00 | 273.00 |
| Color | 1.000 | Ea. | 2.500 | 213.00 | 88.00 | 301.00 |
| Steel enameled 20" x 17", white | 1.000 | Ea. | 2.759 | 99.00 | 97.00 | 196.00 |
| Color | 1.000 | Ea. | 2.500 | 105.00 | 88.00 | 193.00 |
| Vitreous china 20" x 16", white | 1.000 | Ea. | 2.963 | 201.00 | 104.00 | 305.00 |
| Color | 1.000 | Ea. | 2.963 | 201.00 | 104.00 | 305.00 |
| Rough-in, for lavatory | | | | | | |
| 1/2" copper supply, 1-1/2" cast iron waste, 1-1/2" cast iron vent | 1.000 | Ea. | 2.791 | 67.00 | 102.00 | 169.00 |
| 1-1/2" PVC waste, 1-1/4" PVC vent | 1.000 | Ea. | 2.639 | 38.00 | 103.00 | 141.00 |
| 1/2" steel supply, 1-1/4" cast iron waste, 1-1/4" steel vent | 1.000 | Ea. | 2.890 | 54.00 | 107.00 | 161.00 |
| 1-1/4" PVC waste, 1-1/4" PVC vent | 1.000 | Ea. | 2.794 | 37.50 | 109.00 | 146.50 |
| 1/2" PVC supply, 1-1/2" PVC waste, 1-1/2" PVC vent | 1.000 | Ea. | 3.260 | 53.50 | 128.00 | 181.50 |
| Bathtub, P.E. cast iron, corner with fittings, white | 1.000 | Ea. | 3.636 | 1425.00 | 128.00 | 1553.00 |
| Color | 1.000 | Ea. | 4.000 | 1575.00 | 141.00 | 1716.00 |
| Rough-in, for bathtub | | | | | | |
| 1/2" copper supply, 4" cast iron waste, 1-1/2" copper vent | 1.000 | Ea. | 2.409 | 63.50 | 91.00 | 154.50 |
| 4" PVC waste, 1-1/2" PVC vent | 1.000 | Ea. | 2.877 | 44.00 | 109.00 | 153.00 |
| 1/2" steel supply, 4" cast iron waste, 1-1/2" steel vent | 1.000 | Ea. | 2.898 | 61.50 | 107.00 | 168.50 |
| 4" PVC waste, 1-1/2" PVC vent | 1.000 | Ea. | 3.159 | 43.50 | 120.00 | 163.50 |
| 1/2" PVC supply, 4" PVC waste, 1-1/2" PVC vent | 1.000 | Ea. | 3.371 | 58.50 | 128.00 | 186.50 |
| Piping, supply, 1/2" copper | 32.000 | L.F. | 3.161 | 54.00 | 124.00 | 178.00 |
| 1/2" steel | 32.000 | L.F. | 4.063 | 53.00 | 159.00 | 212.00 |
| 1/2" PVC | 32.000 | L.F. | 4.741 | 101.00 | 186.00 | 287.00 |
| Piping, waste, 4" cast iron, no hub | 12.000 | L.F. | 3.310 | 125.00 | 116.00 | 241.00 |
| 4" PVC/DWV | 12.000 | L.F. | 4.000 | 64.00 | 141.00 | 205.00 |
| 4" copper/DWV | 12.000 | L.F. | 4.800 | 175.00 | 169.00 | 344.00 |
| Piping, vent 2" cast iron, no hub | 6.000 | L.F. | 1.433 | 38.50 | 50.50 | 89.00 |
| 2" copper/DWV | 6.000 | L.F. | 1.091 | 29.50 | 42.50 | 72.00 |
| 2" PVC/DWV | 6.000 | L.F. | 1.627 | 17.45 | 57.50 | 74.95 |
| 2" steel, galvanized | 6.000 | L.F. | 1.500 | 27.00 | 53.00 | 80.00 |
| Vanity base cabinet, 2 door, 24" x 30" | 1.000 | Ea. | 1.000 | 204.00 | 35.50 | 239.50 |
| 24" x 36" | 1.000 | Ea. | 1.200 | 272.00 | 42.50 | 314.50 |
| Vanity top, laminated plastic square edge 25" x 32" | 2.670 | L.F. | .712 | 76.00 | 25.50 | 101.50 |
| 25" x 38" | 3.160 | L.F. | .843 | 90.00 | 30.00 | 120.00 |
| Cultured marble, 25" x 32", with bowl | 1.000 | Ea. | 2.500 | 153.00 | 88.00 | 241.00 |
| 25" x 38", with bowl | 1.000 | Ea. | 2.500 | 187.00 | 88.00 | 275.00 |
| Carrier, for lavatory, steel for studs, no arms | 1.000 | Ea. | 1.143 | 29.00 | 44.50 | 73.50 |
| Wood, 2" x 8" blocking | 1.300 | L.F. | .053 | 1.26 | 1.89 | 3.15 |

**MECHANICAL**

**8**

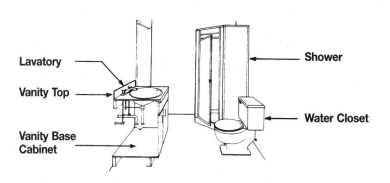

Lavatory
Vanity Top
Vanity Base Cabinet
Shower
Water Closet

| System Description | QUAN. | UNIT | LABOR HOURS | COST EACH | | |
|---|---|---|---|---|---|---|
| | | | | MAT. | INST. | TOTAL |
| **BATHROOM WITH SHOWER, LAVATORY INSTALLED IN VANITY** | | | | | | |
| Water closet, floor mounted, 2 piece, close coupled, white | 1.000 | Ea. | 3.019 | 154.00 | 106.00 | 260.00 |
| Rough-in, vent, 2" diameter DWV piping | 1.000 | Ea. | .955 | 25.60 | 33.60 | 59.20 |
| Waste, 4" diameter DWV piping | 1.000 | Ea. | .828 | 31.20 | 29.10 | 60.30 |
| Supply, 1/2" diameter type "L" copper supply piping | 1.000 | Ea. | .593 | 10.14 | 23.22 | 33.36 |
| Lavatory, 20" x 18" P.E. cast iron with fittings, white | 1.000 | Ea. | 2.500 | 185.00 | 88.00 | 273.00 |
| Rough-in, waste, 1-1/2" diameter DWV piping | 1.000 | Ea. | 1.803 | 50.00 | 63.60 | 113.60 |
| Supply, 1/2" diameter type "L" copper supply piping | 1.000 | Ea. | .988 | 16.90 | 38.70 | 55.60 |
| Shower, steel enameled, stone base, corner, white | 1.000 | Ea. | 8.000 | 360.00 | 282.00 | 642.00 |
| Rough-in, vent, 1-1/2" diameter DWV piping | 1.000 | Ea. | .225 | 6.25 | 7.95 | 14.20 |
| Waste, 2" diameter DWV piping | 1.000 | Ea. | 1.433 | 38.40 | 50.40 | 88.80 |
| Supply, 1/2" diameter type "L" copper supply piping | 1.000 | Ea. | 1.580 | 27.04 | 61.92 | 88.96 |
| Piping, supply, 1/2" diameter type "L" copper supply piping | 36.000 | L.F. | 4.148 | 70.98 | 162.54 | 233.52 |
| Waste, 4" diameter DWV piping | 7.000 | L.F. | 2.759 | 104.00 | 97.00 | 201.00 |
| Vent, 2" diameter DWV piping | 6.000 | L.F. | 2.250 | 40.68 | 79.20 | 119.88 |
| Vanity base 2 door, 30" wide | 1.000 | Ea. | 1.000 | 204.00 | 35.50 | 239.50 |
| Vanity top, plastic laminated, square edge | 2.170 | L.F. | .712 | 60.08 | 25.37 | 85.45 |
| TOTAL | | | 32.793 | 1384.27 | 1184.10 | 2568.37 |
| **BATHROOM WITH SHOWER, WALL HUNG LAVATORY** | | | | | | |
| Water closet, floor mounted, close coupled | 1.000 | Ea. | 3.019 | 154.00 | 106.00 | 260.00 |
| Rough-in, vent, 2" diameter DWV piping | 1.000 | Ea. | .955 | 25.60 | 33.60 | 59.20 |
| Waste, 4" diameter DWV piping | 1.000 | Ea. | .828 | 31.20 | 29.10 | 60.30 |
| Supply, 1/2" diameter type "L" copper supply piping | 1.000 | Ea. | .593 | 10.14 | 23.22 | 33.36 |
| Lavatory, 20" x 18" P.E. cast iron with fittings, white | 1.000 | Ea. | 2.000 | 218.00 | 70.50 | 288.50 |
| Rough-in, waste, 1-1/2" diameter DWV piping | 1.000 | Ea. | 1.803 | 50.00 | 63.60 | 113.60 |
| Supply, 1/2" diameter type "L" copper supply piping | 1.000 | Ea. | .988 | 16.90 | 38.70 | 55.60 |
| Shower, steel enameled, stone base, white | 1.000 | Ea. | 8.000 | 360.00 | 282.00 | 642.00 |
| Rough-in, vent, 1-1/2" diameter DWV piping | 1.000 | Ea. | .225 | 6.25 | 7.95 | 14.20 |
| Waste, 2" diameter DWV piping | 1.000 | Ea. | 1.433 | 38.40 | 50.40 | 88.80 |
| Supply, 1/2" diameter type "L" copper supply piping | 1.000 | Ea. | 1.580 | 27.04 | 61.92 | 88.96 |
| Piping, supply, 1/2" diameter type "L" copper supply piping | 36.000 | L.F. | 4.148 | 70.98 | 162.54 | 233.52 |
| Waste, 4" diameter DWV piping | 7.000 | L.F. | 2.759 | 104.00 | 97.00 | 201.00 |
| Vent, 2" diameter DWV piping | 6.000 | L.F. | 2.250 | 40.68 | 79.20 | 119.88 |
| Carrier, steel, for studs, no arms | 1.000 | Ea. | 1.143 | 29.00 | 44.50 | 73.50 |
| TOTAL | | | 31.724 | 1182.19 | 1150.23 | 2332.42 |

The costs in this system are on a cost each basis. All necessary piping is included.

| Description | QUAN. | UNIT | LABOR HOURS | COST EACH | | |
|---|---|---|---|---|---|---|
| | | | | MAT. | INST. | TOTAL |
| | | | | | | |

**Important: See the Reference Section for critical supporting data - Reference Nos., Crews & Location Factors**

| Three Fixture Bathroom Price Sheet | QUAN. | UNIT | LABOR HOURS | COST EACH MAT. | COST EACH INST. | COST EACH TOTAL |
|---|---|---|---|---|---|---|
| Water closet, close coupled, standard 2 piece, white | 1.000 | Ea. | 3.019 | 154.00 | 106.00 | 260.00 |
| Color | 1.000 | Ea. | 3.019 | 173.00 | 106.00 | 279.00 |
| One piece elongated bowl, white | 1.000 | Ea. | 3.019 | 545.00 | 106.00 | 651.00 |
| Color | 1.000 | Ea. | 3.019 | 755.00 | 106.00 | 861.00 |
| Low profile, one piece elongated bowl, white | 1.000 | Ea. | 3.019 | 695.00 | 106.00 | 801.00 |
| Color | 1.000 | Ea. | 3.019 | 905.00 | 106.00 | 1011.00 |
| Rough-in, for water closet | | | | | | |
| 1/2" copper supply, 4" cast iron waste, 2" cast iron vent | 1.000 | Ea. | 2.376 | 67.00 | 86.00 | 153.00 |
| 4" PVC/DWV waste, 2" PVC vent | 1.000 | Ea. | 2.678 | 38.00 | 96.50 | 134.50 |
| 4" copper waste, 2" copper vent | 1.000 | Ea. | 2.520 | 73.50 | 94.00 | 167.50 |
| 3" cast iron waste, 1-1/2" cast iron vent | 1.000 | Ea. | 2.244 | 60.00 | 81.50 | 141.50 |
| 3" PVC waste, 1-1/2" PVC vent | 1.000 | Ea. | 2.388 | 34.00 | 90.00 | 124.00 |
| 3" copper waste, 1-1/2" copper vent | 1.000 | Ea. | 2.014 | 50.50 | 75.50 | 126.00 |
| 1/2" PVC supply, 4" PVC waste, 2" PVC vent | 1.000 | Ea. | 2.974 | 46.50 | 108.00 | 154.50 |
| 3" PVC waste, 1-1/2" PVC supply | 1.000 | Ea. | 2.684 | 43.00 | 102.00 | 145.00 |
| 1/2" steel supply, 4" cast iron waste, 2" cast iron vent | 1.000 | Ea. | 2.545 | 67.00 | 92.50 | 159.50 |
| 4" cast iron waste, 2" steel vent | 1.000 | Ea. | 2.590 | 59.00 | 94.00 | 153.00 |
| 4" PVC waste, 2" PVC vent | 1.000 | Ea. | 2.847 | 37.50 | 103.00 | 140.50 |
| Lavatory, wall hung, P.E. cast iron 20" x 18", white | 1.000 | Ea. | 2.000 | 218.00 | 70.50 | 288.50 |
| Color | 1.000 | Ea. | 2.000 | 255.00 | 70.50 | 325.50 |
| Vitreous china 19" x 17", white | 1.000 | Ea. | 2.286 | 141.00 | 80.50 | 221.50 |
| Color | 1.000 | Ea. | 2.286 | 162.00 | 80.50 | 242.50 |
| Lavatory, for vanity top, P.E. cast iron 20" x 18", white | 1.000 | Ea. | 2.500 | 185.00 | 88.00 | 273.00 |
| Color | 1.000 | Ea. | 2.500 | 213.00 | 88.00 | 301.00 |
| Steel enameled 20" x 17", white | 1.000 | Ea. | 2.759 | 99.00 | 97.00 | 196.00 |
| Color | 1.000 | Ea. | 2.500 | 105.00 | 88.00 | 193.00 |
| Vitreous china 20" x 16", white | 1.000 | Ea. | 2.963 | 201.00 | 104.00 | 305.00 |
| Color | 1.000 | Ea. | 2.963 | 201.00 | 104.00 | 305.00 |
| Rough-in, for lavatory | | | | | | |
| 1/2" copper supply, 1-1/2" cast iron waste, 1-1/2" cast iron vent | 1.000 | Ea. | 2.791 | 67.00 | 102.00 | 169.00 |
| 1-1/2" PVC waste, 1-1/2" PVC vent | 1.000 | Ea. | 2.639 | 38.00 | 103.00 | 141.00 |
| 1/2" steel supply, 1-1/4" cast iron waste, 1-1/4" steel vent | 1.000 | Ea. | 2.890 | 54.00 | 107.00 | 161.00 |
| 1-1/4" PVC waste, 1-1/4" PVC vent | 1.000 | Ea. | 2.921 | 38.00 | 114.00 | 152.00 |
| 1/2" PVC supply, 1-1/2" PVC waste, 1-1/2" PVC vent | 1.000 | Ea. | 3.260 | 53.50 | 128.00 | 181.50 |
| Shower, steel enameled stone base, 32" x 32", white | 1.000 | Ea. | 8.000 | 360.00 | 282.00 | 642.00 |
| Color | 1.000 | Ea. | 7.822 | 840.00 | 277.00 | 1117.00 |
| 36" x 36" white | 1.000 | Ea. | 8.889 | 905.00 | 315.00 | 1220.00 |
| Color | 1.000 | Ea. | 8.889 | 955.00 | 315.00 | 1270.00 |
| Rough-in, for shower | | | | | | |
| 1/2" copper supply, 4" cast iron waste, 1-1/2" copper vent | 1.000 | Ea. | 3.238 | 71.50 | 120.00 | 191.50 |
| 4" PVC waste, 1-1/2" PVC vent | 1.000 | Ea. | 3.429 | 47.00 | 128.00 | 175.00 |
| 1/2" steel supply, 4" cast iron waste, 1-1/2" steel vent | 1.000 | Ea. | 3.665 | 68.50 | 137.00 | 205.50 |
| 4" PVC waste, 1-1/2" PVC vent | 1.000 | Ea. | 3.881 | 47.00 | 146.00 | 193.00 |
| 1/2" PVC supply, 4" PVC waste, 1-1/2" PVC vent | 1.000 | Ea. | 4.219 | 70.50 | 159.00 | 229.50 |
| Piping, supply, 1/2" copper | 36.000 | L.F. | 4.148 | 71.00 | 163.00 | 234.00 |
| 1/2" steel | 36.000 | L.F. | 5.333 | 69.50 | 209.00 | 278.50 |
| 1/2" PVC | 36.000 | L.F. | 6.222 | 132.00 | 244.00 | 376.00 |
| Piping, waste, 4" cast iron no hub | 7.000 | L.F. | 2.759 | 104.00 | 97.00 | 201.00 |
| 4" PVC/DWV | 7.000 | L.F. | 3.333 | 53.50 | 118.00 | 171.50 |
| 4" copper/DWV | 7.000 | L.F. | 4.000 | 146.00 | 141.00 | 287.00 |
| Piping, vent, 2" cast iron no hub | 6.000 | L.F. | 2.149 | 57.50 | 75.50 | 133.00 |
| 2" copper/DWV | 6.000 | L.F. | 1.636 | 44.00 | 64.00 | 108.00 |
| 2" PVC/DWV | 6.000 | L.F. | 2.441 | 26.00 | 86.00 | 112.00 |
| 2" steel, galvanized | 6.000 | L.F. | 2.250 | 40.50 | 79.00 | 119.50 |
| Vanity base cabinet, 2 door, 24" x 30" | 1.000 | Ea. | 1.000 | 204.00 | 35.50 | 239.50 |
| 24" x 36" | 1.000 | Ea. | 1.200 | 272.00 | 42.50 | 314.50 |
| Vanity top, laminated plastic square edge, 25" x 32" | 2.170 | L.F. | .712 | 60.00 | 25.50 | 85.50 |
| 25" x 38" | 2.670 | L.F. | .845 | 71.50 | 30.00 | 101.50 |
| Carrier, for lavatory, steel for studs, no arms | 1.000 | Ea. | 1.143 | 29.00 | 44.50 | 73.50 |
| Wood, 2" x 8" blocking | 1.300 | L.F. | .052 | 1.24 | 1.85 | 3.09 |

MECHANICAL

8

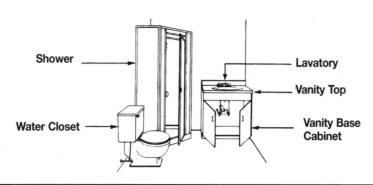

Shower
Lavatory
Vanity Top
Water Closet
Vanity Base Cabinet

| System Description | QUAN. | UNIT | LABOR HOURS | COST EACH | | |
|---|---|---|---|---|---|---|
| | | | | MAT. | INST. | TOTAL |
| **BATHROOM WITH LAVATORY INSTALLED IN VANITY** | | | | | | |
| Water closet, floor mounted, 2 piece, close coupled, white | 1.000 | Ea. | 3.019 | 154.00 | 106.00 | 260.00 |
| Rough-in, vent, 2" diameter DWV piping | 1.000 | Ea. | .955 | 25.60 | 33.60 | 59.20 |
| Waste, 4" diameter DWV piping | 1.000 | Ea. | .828 | 31.20 | 29.10 | 60.30 |
| Supply, 1/2" diameter type "L" copper supply piping | 1.000 | Ea. | .593 | 10.14 | 23.22 | 33.36 |
| Lavatory, 20" x 18", P.E. cast iron with fittings, white | 1.000 | Ea. | 2.500 | 185.00 | 88.00 | 273.00 |
| Rough-in, waste, 1-1/2" diameter DWV piping | 1.000 | Ea. | 1.803 | 50.00 | 63.60 | 113.60 |
| Supply, 1/2" diameter type "L" copper supply piping | 1.000 | Ea. | .988 | 16.90 | 38.70 | 55.60 |
| Shower, steel enameled, stone base, corner, white | 1.000 | Ea. | 8.000 | 360.00 | 282.00 | 642.00 |
| Rough-in, vent, 1-1/2" diameter DWV piping | 1.000 | Ea. | .225 | 6.25 | 7.95 | 14.20 |
| Waste, 2" diameter DWV piping | 1.000 | Ea. | 1.433 | 38.40 | 50.40 | 88.80 |
| Supply, 1/2" diameter type "L" copper supply piping | 1.000 | Ea. | 1.580 | 27.04 | 61.92 | 88.96 |
| Piping, supply, 1/2" diameter type "L" copper supply piping | 36.000 | L.F. | 3.556 | 60.84 | 139.32 | 200.16 |
| Waste, 4" diameter DWV piping | 7.000 | L.F. | 1.931 | 72.80 | 67.90 | 140.70 |
| Vent, 2" diameter DWV piping | 6.000 | L.F. | 1.500 | 27.12 | 52.80 | 79.92 |
| Vanity base, 2 door, 30" wide | 1.000 | Ea. | 1.000 | 204.00 | 35.50 | 239.50 |
| Vanity top, plastic laminated, square edge | 2.670 | L.F. | .712 | 58.74 | 25.37 | 84.11 |
| TOTAL | | | 30.623 | 1328.03 | 1105.38 | 2433.41 |
| | | | | | | |
| **BATHROOM, WITH WALL HUNG LAVATORY** | | | | | | |
| Water closet, floor mounted, 2 piece, close coupled, white | 1.000 | Ea. | 3.019 | 154.00 | 106.00 | 260.00 |
| Rough-in, vent, 2" diameter DWV piping | 1.000 | Ea. | .955 | 25.60 | 33.60 | 59.20 |
| Waste, 4" diameter DWV piping | 1.000 | Ea. | .828 | 31.20 | 29.10 | 60.30 |
| Supply, 1/2" diameter type "L" copper supply piping | 1.000 | Ea. | .593 | 10.14 | 23.22 | 33.36 |
| Lavatory, wall hung, 20" x 18" P.E. cast iron with fittings, white | 1.000 | Ea. | 2.000 | 218.00 | 70.50 | 288.50 |
| Rough-in, waste, 1-1/2" diameter DWV piping | 1.000 | Ea. | 1.803 | 50.00 | 63.60 | 113.60 |
| Supply, 1/2" diameter type "L" copper supply piping | 1.000 | Ea. | .988 | 16.90 | 38.70 | 55.60 |
| Shower, steel enameled, stone base, corner, white | 1.000 | Ea. | 8.000 | 360.00 | 282.00 | 642.00 |
| Rough-in, waste, 1-1/2" diameter DWV piping | 1.000 | Ea. | .225 | 6.25 | 7.95 | 14.20 |
| Waste, 2" diameter DWV piping | 1.000 | Ea. | 1.433 | 38.40 | 50.40 | 88.80 |
| Supply, 1/2" diameter type "L" copper supply piping | 1.000 | Ea. | 1.580 | 27.04 | 61.92 | 88.96 |
| Piping, supply, 1/2" diameter type "L" copper supply piping | 36.000 | L.F. | 3.556 | 60.84 | 139.32 | 200.16 |
| Waste, 4" diameter DWV piping | 7.000 | L.F. | 1.931 | 72.80 | 67.90 | 140.70 |
| Vent, 2" diameter DWV piping | 6.000 | L.F. | 1.500 | 27.12 | 52.80 | 79.92 |
| Carrier, steel, for studs, no arms | 1.000 | Ea. | 1.143 | 29.00 | 44.50 | 73.50 |
| TOTAL | | | 29.554 | 1127.29 | 1071.51 | 2198.80 |

The costs in this system are on a cost each basis. All necessary piping is included.

| Description | QUAN. | UNIT | LABOR HOURS | COST EACH | | |
|---|---|---|---|---|---|---|
| | | | | MAT. | INST. | TOTAL |
| | | | | | | |

**Important: See the Reference Section for critical supporting data - Reference Nos., Crews & Location Factors**

| Three Fixture Bathroom Price Sheet | QUAN. | UNIT | LABOR HOURS | COST EACH | | |
|---|---|---|---|---|---|---|
| | | | | MAT. | INST. | TOTAL |
| Water closet, close coupled, standard 2 piece, white | 1.000 | Ea. | 3.019 | 154.00 | 106.00 | 260.00 |
| Color | 1.000 | Ea. | 3.019 | 173.00 | 106.00 | 279.00 |
| One piece elongated bowl, white | 1.000 | Ea. | 3.019 | 545.00 | 106.00 | 651.00 |
| Color | 1.000 | Ea. | 3.019 | 755.00 | 106.00 | 861.00 |
| Low profile one piece elongated bowl, white | 1.000 | Ea. | 3.019 | 695.00 | 106.00 | 801.00 |
| Color | 1.000 | Ea. | 3.623 | 1075.00 | 127.00 | 1202.00 |
| Rough-in, for water closet | | | | | | |
| 1/2" copper supply, 4" cast iron waste, 2" cast iron vent | 1.000 | Ea. | 2.376 | 67.00 | 86.00 | 153.00 |
| 4" P.V.C./DWV waste, 2" PVC vent | 1.000 | Ea. | 2.678 | 38.00 | 96.50 | 134.50 |
| 4" copper waste, 2" copper vent | 1.000 | Ea. | 2.520 | 73.50 | 94.00 | 167.50 |
| 3" cast iron waste, 1-1/2" cast iron vent | 1.000 | Ea. | 2.244 | 60.00 | 81.50 | 141.50 |
| 3" PVC waste, 1-1/2" PVC vent | 1.000 | Ea. | 2.388 | 34.00 | 90.00 | 124.00 |
| 3" copper waste, 1-1/2" copper vent | 1.000 | Ea. | 2.014 | 50.50 | 75.50 | 126.00 |
| 1/2" P.V.C. supply, 4" P.V.C. waste, 2" P.V.C. vent | 1.000 | Ea. | 2.974 | 46.50 | 108.00 | 154.50 |
| 3" P.V.C. waste, 1-1/2" P.V.C. vent | 1.000 | Ea. | 2.684 | 43.00 | 102.00 | 145.00 |
| 1/2" steel supply, 4" cast iron waste, 2" cast iron vent | 1.000 | Ea. | 2.545 | 67.00 | 92.50 | 159.50 |
| 4" cast iron waste, 2" steel vent | 1.000 | Ea. | 2.590 | 59.00 | 94.00 | 153.00 |
| 4" P.V.C. waste, 2" P.V.C. vent | 1.000 | Ea. | 2.847 | 37.50 | 103.00 | 140.50 |
| Lavatory, wall hung P.E. cast iron 20" x 18", white | 1.000 | Ea. | 2.000 | 218.00 | 70.50 | 288.50 |
| Color | 1.000 | Ea. | 2.000 | 255.00 | 70.50 | 325.50 |
| Vitreous china 19" x 17", white | 1.000 | Ea. | 2.286 | 141.00 | 80.50 | 221.50 |
| Color | 1.000 | Ea. | 2.286 | 162.00 | 80.50 | 242.50 |
| Lavatory, for vanity top P.E. cast iron 20" x 18", white | 1.000 | Ea. | 2.500 | 185.00 | 88.00 | 273.00 |
| Color | 1.000 | Ea. | 2.500 | 213.00 | 88.00 | 301.00 |
| Steel enameled 20" x 17", white | 1.000 | Ea. | 2.759 | 99.00 | 97.00 | 196.00 |
| Color | 1.000 | Ea. | 2.500 | 105.00 | 88.00 | 193.00 |
| Vitreous china 20" x 16", white | 1.000 | Ea. | 2.963 | 201.00 | 104.00 | 305.00 |
| Color | 1.000 | Ea. | 2.963 | 201.00 | 104.00 | 305.00 |
| Rough-in, for lavatory | | | | | | |
| 1/2" copper supply, 1-1/2" cast iron waste, 1-1/2" cast iron vent | 1.000 | Ea. | 2.791 | 67.00 | 102.00 | 169.00 |
| 1-1/2" P.V.C. waste, 1-1/2" P.V.C. vent | 1.000 | Ea. | 2.639 | 38.00 | 103.00 | 141.00 |
| 1/2" steel supply, 1-1/2" cast iron waste, 1-1/4" steel vent | 1.000 | Ea. | 2.890 | 54.00 | 107.00 | 161.00 |
| 1-1/2" P.V.C. waste, 1-1/4" P.V.C. vent | 1.000 | Ea. | 2.921 | 38.00 | 114.00 | 152.00 |
| 1/2" P.V.C. supply, 1-1/2" P.V.C. waste, 1-1/2" P.V.C. vent | 1.000 | Ea. | 3.260 | 53.50 | 128.00 | 181.50 |
| Shower, steel enameled stone base, 32" x 32", white | 1.000 | Ea. | 8.000 | 360.00 | 282.00 | 642.00 |
| Color | 1.000 | Ea. | 7.822 | 840.00 | 277.00 | 1117.00 |
| 36" x 36", white | 1.000 | Ea. | 8.889 | 905.00 | 315.00 | 1220.00 |
| Color | 1.000 | Ea. | 8.889 | 955.00 | 315.00 | 1270.00 |
| Rough-in, for shower | | | | | | |
| 1/2" copper supply, 2" cast iron waste, 1-1/2" copper vent | 1.000 | Ea. | 3.161 | 69.50 | 118.00 | 187.50 |
| 2" P.V.C. waste, 1-1/2" P.V.C. vent | 1.000 | Ea. | 3.429 | 47.00 | 128.00 | 175.00 |
| 1/2" steel supply, 2" cast iron waste, 1-1/2" steel vent | 1.000 | Ea. | 3.887 | 92.50 | 145.00 | 237.50 |
| 2" P.V.C. waste, 1-1/2" P.V.C. vent | 1.000 | Ea. | 3.881 | 47.00 | 146.00 | 193.00 |
| 1/2" P.V.C. supply, 2" P.V.C. waste, 1-1/2" P.V.C. vent | 1.000 | Ea. | 4.219 | 70.50 | 159.00 | 229.50 |
| Piping, supply, 1/2" copper | 36.000 | L.F. | 3.556 | 61.00 | 139.00 | 200.00 |
| 1/2" steel | 36.000 | L.F. | 4.571 | 60.00 | 179.00 | 239.00 |
| 1/2" P.V.C. | 36.000 | L.F. | 5.333 | 113.00 | 209.00 | 322.00 |
| Waste, 4" cast iron, no hub | 7.000 | L.F. | 1.931 | 73.00 | 68.00 | 141.00 |
| 4" P.V.C./DWV | 7.000 | L.F. | 2.333 | 37.50 | 82.50 | 120.00 |
| 4" copper/DWV | 7.000 | L.F. | 2.800 | 102.00 | 98.50 | 200.50 |
| Vent, 2" cast iron, no hub | 6.000 | L.F. | 1.091 | 29.50 | 42.50 | 72.00 |
| 2" copper/DWV | 6.000 | L.F. | 1.091 | 29.50 | 42.50 | 72.00 |
| 2" P.V.C./DWV | 6.000 | L.F. | 1.627 | 17.45 | 57.50 | 74.95 |
| 2" steel, galvanized | 6.000 | L.F. | 1.500 | 27.00 | 53.00 | 80.00 |
| Vanity base cabinet, 2 door, 24" x 30" | 1.000 | Ea. | 1.000 | 204.00 | 35.50 | 239.50 |
| 24" x 36" | 1.000 | Ea. | 1.200 | 272.00 | 42.50 | 314.50 |
| Vanity top, laminated plastic square edge, 25" x 32" | 2.670 | L.F. | .712 | 58.50 | 25.50 | 84.00 |
| 25" x 38" | 3.170 | L.F. | .845 | 69.50 | 30.00 | 99.50 |
| Carrier , for lavatory, steel, for studs, no arms | 1.000 | Ea. | 1.143 | 29.00 | 44.50 | 73.50 |
| Wood, 2" x 8" blocking | 1.300 | L.F. | .052 | 1.24 | 1.85 | 3.09 |

MECHANICAL

8

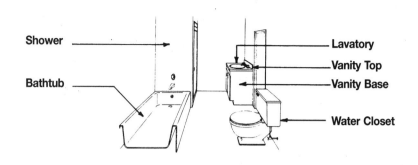

| System Description | QUAN. | UNIT | LABOR HOURS | COST EACH | | |
|---|---|---|---|---|---|---|
| | | | | MAT. | INST. | TOTAL |
| **BATHROOM WITH LAVATORY INSTALLED IN VANITY** | | | | | | |
| Water closet, floor mounted, 2 piece, close coupled, white | 1.000 | Ea. | 3.019 | 154.00 | 106.00 | 260.00 |
| Rough-in, vent, 2" diameter DWV piping | 1.000 | Ea. | .955 | 25.60 | 33.60 | 59.20 |
| Waste, 4" diameter DWV piping | 1.000 | Ea. | .828 | 31.20 | 29.10 | 60.30 |
| Supply, 1/2" diameter type "L" copper supply piping | 1.000 | Ea. | .593 | 10.14 | 23.22 | 33.36 |
| Lavatory, 20" x 18" P.E. cast iron with fittings, white | 1.000 | Ea. | 2.500 | 185.00 | 88.00 | 273.00 |
| Shower, steel, enameled, stone base, corner, white | 1.000 | Ea. | 8.889 | 825.00 | 315.00 | 1140.00 |
| Rough-in, waste, 1-1/2" diameter DWV piping | 2.000 | Ea. | 4.507 | 125.00 | 159.00 | 284.00 |
| Supply, 1/2" diameter type "L" copper supply piping | 2.000 | Ea. | 3.161 | 54.08 | 123.84 | 177.92 |
| Bathtub, P.E. cast iron, 5' long with fittings, white | 1.000 | Ea. | 3.636 | 410.00 | 128.00 | 538.00 |
| Rough-in, waste, 4" diameter DWV piping | 1.000 | Ea. | .828 | 31.20 | 29.10 | 60.30 |
| Supply, 1/2" diameter type "L" copper supply piping | 1.000 | Ea. | .988 | 16.90 | 38.70 | 55.60 |
| Vent, 1-1/2" diameter DWV piping | 1.000 | Ea. | .593 | 15.28 | 23.20 | 38.48 |
| Piping, supply, 1/2" diameter type "L" copper supply piping | 42.000 | L.F. | 4.148 | 70.98 | 162.54 | 233.52 |
| Waste, 4" diameter DWV piping | 10.000 | L.F. | 2.759 | 104.00 | 97.00 | 201.00 |
| Vent, 2" diameter DWV piping | 13.000 | L.F. | 3.250 | 58.76 | 114.40 | 173.16 |
| Vanity base, 2 doors, 30" wide | 1.000 | Ea. | 1.000 | 204.00 | 35.50 | 239.50 |
| Vanity top, plastic laminated, square edge | 2.670 | L.F. | .712 | 58.74 | 25.37 | 84.11 |
| TOTAL | | | 42.366 | 2379.88 | 1531.57 | 3911.45 |
| **BATHROOM WITH WALL HUNG LAVATORY** | | | | | | |
| Water closet, floor mounted, 2 piece, close coupled, white | 1.000 | Ea. | 3.019 | 154.00 | 106.00 | 260.00 |
| Rough-in, vent, 2" diameter DWV piping | 1.000 | Ea. | .955 | 25.60 | 33.60 | 59.20 |
| Waste, 4" diameter DWV piping | 1.000 | Ea. | .828 | 31.20 | 29.10 | 60.30 |
| Supply, 1/2" diameter type "L" copper supply piping | 1.000 | Ea. | .593 | 10.14 | 23.22 | 33.36 |
| Lavatory, 20" x 18" P.E. cast iron with fittings, white | 1.000 | Ea. | 2.000 | 218.00 | 70.50 | 288.50 |
| Shower, steel enameled, stone base, corner, white | 1.000 | Ea. | 8.889 | 825.00 | 315.00 | 1140.00 |
| Rough-in, waste, 1-1/2" diameter DWV piping | 2.000 | Ea. | 4.507 | 125.00 | 159.00 | 284.00 |
| Supply, 1/2" diameter type "L" copper supply piping | 2.000 | Ea. | 3.161 | 54.08 | 123.84 | 177.92 |
| Bathtub, P.E. cast iron, 5' long with fittings, white | 1.000 | Ea. | 3.636 | 410.00 | 128.00 | 538.00 |
| Rough-in, waste, 4" diameter DWV piping | 1.000 | Ea. | .828 | 31.20 | 29.10 | 60.30 |
| Supply, 1/2" diameter type "L" copper supply piping | 1.000 | Ea. | .988 | 16.90 | 38.70 | 55.60 |
| Vent, 1-1/2" diameter copper DWV piping | 1.000 | Ea. | .593 | 15.28 | 23.20 | 38.48 |
| Piping, supply, 1/2" diameter type "L" copper supply piping | 42.000 | L.F. | 4.148 | 70.98 | 162.54 | 233.52 |
| Waste, 4" diameter DWV piping | 10.000 | L.F. | 2.759 | 104.00 | 97.00 | 201.00 |
| Vent, 2" diameter DWV piping | 13.000 | L.F. | 3.250 | 58.76 | 114.40 | 173.16 |
| Carrier, steel, for studs, no arms | 1.000 | Ea. | 1.143 | 29.00 | 44.50 | 73.50 |
| TOTAL | | | 41.297 | 2179.14 | 1497.70 | 3676.84 |

The costs in this system are on a cost each basis. All necessary piping is included.

**Important: See the Reference Section for critical supporting data - Reference Nos., Crews & Location Factors**

| Four Fixture Bathroom Price Sheet | QUAN. | UNIT | LABOR HOURS | COST EACH | | |
|---|---|---|---|---|---|---|
| | | | | MAT. | INST. | TOTAL |
| Water closet, close coupled, standard 2 piece, white | 1.000 | Ea. | 3.019 | 154.00 | 106.00 | 260.00 |
| Color | 1.000 | Ea. | 3.019 | 173.00 | 106.00 | 279.00 |
| One piece elongated bowl, white | 1.000 | Ea. | 3.019 | 545.00 | 106.00 | 651.00 |
| Color | 1.000 | Ea. | 3.019 | 755.00 | 106.00 | 861.00 |
| Low profile, one piece elongated bowl, white | 1.000 | Ea. | 3.019 | 695.00 | 106.00 | 801.00 |
| Color | 1.000 | Ea. | 3.019 | 905.00 | 106.00 | 1011.00 |
| Rough-in, for water closet | | | | | | |
| 1/2" copper supply, 4" cast iron waste, 2" cast iron vent | 1.000 | Ea. | 2.376 | 67.00 | 86.00 | 153.00 |
| 4" PVC/DWV waste, 2" PVC vent | 1.000 | Ea. | 2.678 | 38.00 | 96.50 | 134.50 |
| 4" copper waste, 2" copper vent | 1.000 | Ea. | 2.520 | 73.50 | 94.00 | 167.50 |
| 3" cast iron waste, 1-1/2" cast iron vent | 1.000 | Ea. | 2.244 | 60.00 | 81.50 | 141.50 |
| 3" P.V.C. waste, 1-1/2" P.V.C. vent | 1.000 | Ea. | 2.388 | 34.00 | 90.00 | 124.00 |
| 3" copper waste, 1-1/2" copper vent | 1.000 | Ea. | 2.014 | 50.50 | 75.50 | 126.00 |
| 1/2" P.V.C. supply, 4" P.V.C. waste, 2" P.V.C. vent | 1.000 | Ea. | 2.974 | 46.50 | 108.00 | 154.50 |
| 3" P.V.C. waste, 1-1/2" P.V.C. vent | 1.000 | Ea. | 2.684 | 43.00 | 102.00 | 145.00 |
| 1/2" steel supply, 4" cast iron waste, 2" cast iron vent | 1.000 | Ea. | 2.545 | 67.00 | 92.50 | 159.50 |
| 4" cast iron waste, 2" steel vent | 1.000 | Ea. | 2.590 | 59.00 | 94.00 | 153.00 |
| 4" P.V.C. waste, 2" P.V.C. vent | 1.000 | Ea. | 2.847 | 37.50 | 103.00 | 140.50 |
| Lavatory, wall hung P.E. cast iron 20" x 18", white | 1.000 | Ea. | 2.000 | 218.00 | 70.50 | 288.50 |
| Color | 1.000 | Ea. | 2.000 | 255.00 | 70.50 | 325.50 |
| Vitreous china 19" x 17", white | 1.000 | Ea. | 2.286 | 141.00 | 80.50 | 221.50 |
| Color | 1.000 | Ea. | 2.286 | 162.00 | 80.50 | 242.50 |
| Lavatory for vanity top, P.E. cast iron 20" x 18", white | 1.000 | Ea. | 2.500 | 185.00 | 88.00 | 273.00 |
| Color | 1.000 | Ea. | 2.500 | 213.00 | 88.00 | 301.00 |
| Steel enameled, 20" x 17", white | 1.000 | Ea. | 2.759 | 99.00 | 97.00 | 196.00 |
| Color | 1.000 | Ea. | 2.500 | 105.00 | 88.00 | 193.00 |
| Vitreous china 20" x 16", white | 1.000 | Ea. | 2.963 | 201.00 | 104.00 | 305.00 |
| Color | 1.000 | Ea. | 2.963 | 201.00 | 104.00 | 305.00 |
| Shower, steel enameled stone base, 36" square, white | 1.000 | Ea. | 8.889 | 825.00 | 315.00 | 1140.00 |
| Color | 1.000 | Ea. | 8.889 | 890.00 | 315.00 | 1205.00 |
| Rough-in, for lavatory or shower | | | | | | |
| 1/2" copper supply, 1-1/2" cast iron waste, 1-1/2" cast iron vent | 1.000 | Ea. | 3.834 | 89.50 | 141.00 | 230.50 |
| 1-1/2" P.V.C. waste, 1-1/4" P.V.C. vent | 1.000 | Ea. | 3.675 | 54.00 | 144.00 | 198.00 |
| 1/2" steel supply, 1-1/4" cast iron waste, 1-1/4" steel vent | 1.000 | Ea. | 4.103 | 76.50 | 153.00 | 229.50 |
| 1-1/4" P.V.C. waste, 1-1/4" P.V.C. vent | 1.000 | Ea. | 3.937 | 52.50 | 154.00 | 206.50 |
| 1/2" P.V.C. supply, 1-1/2" P.V.C. waste, 1-1/2" P.V.C. vent | 1.000 | Ea. | 4.592 | 77.50 | 180.00 | 257.50 |
| Bathtub, P.E. cast iron, 5' long with fittings, white | 1.000 | Ea. | 3.636 | 410.00 | 128.00 | 538.00 |
| Color | 1.000 | Ea. | 3.636 | 470.00 | 128.00 | 598.00 |
| Steel, enameled 5' long with fittings, white | 1.000 | Ea. | 2.909 | 320.00 | 102.00 | 422.00 |
| Color | 1.000 | Ea. | 2.909 | 320.00 | 102.00 | 422.00 |
| Rough-in, for bathtub | | | | | | |
| 1/2" copper supply, 4" cast iron waste, 1-1/2" copper vent | 1.000 | Ea. | 2.409 | 63.50 | 91.00 | 154.50 |
| 4" P.V.C. waste, 1-1/2" P.V.C. vent | 1.000 | Ea. | 2.877 | 44.00 | 109.00 | 153.00 |
| 1/2" steel supply, 4" cast iron waste, 1-1/2" steel vent | 1.000 | Ea. | 2.898 | 61.50 | 107.00 | 168.50 |
| 4" P.V.C. waste, 1-1/2" P.V.C. vent | 1.000 | Ea. | 3.159 | 43.50 | 120.00 | 163.50 |
| 1/2" P.V.C. supply, 4" P.V.C. waste, 1-1/2" P.V.C. vent | 1.000 | Ea. | 3.371 | 58.50 | 128.00 | 186.50 |
| Piping, supply, 1/2" copper | 42.000 | L.F. | 4.148 | 71.00 | 163.00 | 234.00 |
| 1/2" steel | 42.000 | L.F. | 5.333 | 69.50 | 209.00 | 278.50 |
| 1/2" P.V.C. | 42.000 | L.F. | 6.222 | 132.00 | 244.00 | 376.00 |
| Waste, 4" cast iron, no hub | 10.000 | L.F. | 2.759 | 104.00 | 97.00 | 201.00 |
| 4" P.V.C./DWV | 10.000 | L.F. | 3.333 | 53.50 | 118.00 | 171.50 |
| 4" copper/DWV | 10.000 | Ea. | 4.000 | 146.00 | 141.00 | 287.00 |
| Vent 2" cast iron, no hub | 13.000 | L.F. | 3.105 | 83.00 | 109.00 | 192.00 |
| 2" copper/DWV | 13.000 | L.F. | 2.364 | 64.00 | 92.50 | 156.50 |
| 2" P.V.C./DWV | 13.000 | L.F. | 3.525 | 38.00 | 124.00 | 162.00 |
| 2" steel, galvanized | 13.000 | L.F. | 3.250 | 59.00 | 114.00 | 173.00 |
| Vanity base cabinet, 2 doors, 30" wide | 1.000 | Ea. | 1.000 | 204.00 | 35.50 | 239.50 |
| Vanity top, plastic laminated, square edge | 2.670 | L.F. | .712 | 58.50 | 25.50 | 84.00 |
| Carrier, steel for studs, no arms | 1.000 | Ea. | 1.143 | 29.00 | 44.50 | 73.50 |
| Wood, 2" x 8" blocking | 1.300 | L.F. | .052 | 1.24 | 1.85 | 3.09 |

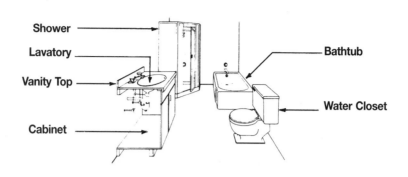

| System Description | QUAN. | UNIT | LABOR HOURS | COST EACH | | |
|---|---|---|---|---|---|---|
| | | | | MAT. | INST. | TOTAL |
| **BATHROOM WITH LAVATORY INSTALLED IN VANITY** | | | | | | |
| Water closet, floor mounted, 2 piece, close coupled, white | 1.000 | Ea. | 3.019 | 154.00 | 106.00 | 260.00 |
| Rough-in, vent, 2" diameter DWV piping | 1.000 | Ea. | .955 | 25.60 | 33.60 | 59.20 |
| Waste, 4" diameter DWV piping | 1.000 | Ea. | .828 | 31.20 | 29.10 | 60.30 |
| Supply, 1/2" diameter type "L" copper supply piping | 1.000 | Ea. | .593 | 10.14 | 23.22 | 33.36 |
| Lavatory, 20" x 18" P.E. cast iron with fittings, white | 1.000 | Ea. | 2.500 | 185.00 | 88.00 | 273.00 |
| Shower, steel, enameled, stone base, corner, white | 1.000 | Ea. | 8.889 | 825.00 | 315.00 | 1140.00 |
| Rough-in, waste, 1-1/2" diameter DWV piping | 2.000 | Ea. | 4.507 | 125.00 | 159.00 | 284.00 |
| Supply, 1/2" diameter type "L" copper supply piping | 2.000 | Ea. | 3.161 | 54.08 | 123.84 | 177.92 |
| Bathtub, P.E. cast iron, 5' long with fittings, white | 1.000 | Ea. | 3.636 | 410.00 | 128.00 | 538.00 |
| Rough-in, waste, 4" diameter DWV piping | 1.000 | Ea. | .828 | 31.20 | 29.10 | 60.30 |
| Supply, 1/2" diameter type "L" copper supply piping | 1.000 | Ea. | .988 | 16.90 | 38.70 | 55.60 |
| Vent, 1-1/2" diameter DWV piping | 1.000 | Ea. | .593 | 15.28 | 23.20 | 38.48 |
| Piping, supply, 1/2" diameter type "L" copper supply piping | 42.000 | L.F. | 4.939 | 84.50 | 193.50 | 278.00 |
| Waste, 4" diameter DWV piping | 10.000 | L.F. | 4.138 | 156.00 | 145.50 | 301.50 |
| Vent, 2" diameter DWV piping | 13.000 | L.F. | 4.500 | 81.36 | 158.40 | 239.76 |
| Vanity base, 2 doors, 30" wide | 1.000 | Ea. | 1.000 | 204.00 | 35.50 | 239.50 |
| Vanity top, plastic laminated, square edge | 2.670 | L.F. | .712 | 60.08 | 25.37 | 85.45 |
| TOTAL | | | 45.786 | 2469.34 | 1655.03 | 4124.37 |
| **BATHROOM WITH WALL HUNG LAVATORY** | | | | | | |
| Water closet, floor mounted, 2 piece, close coupled, white | 1.000 | Ea. | 3.019 | 154.00 | 106.00 | 260.00 |
| Rough-in, vent, 2" diameter DWV piping | 1.000 | Ea. | .955 | 25.60 | 33.60 | 59.20 |
| Waste, 4" diameter DWV piping | 1.000 | Ea. | .828 | 31.20 | 29.10 | 60.30 |
| Supply, 1/2" diameter type "L" copper supply piping | 1.000 | Ea. | .593 | 10.14 | 23.22 | 33.36 |
| Lavatory, 20" x 18" P.E. cast iron with fittings, white | 1.000 | Ea. | 2.000 | 218.00 | 70.50 | 288.50 |
| Shower, steel enameled, stone base, corner, white | 1.000 | Ea. | 8.889 | 825.00 | 315.00 | 1140.00 |
| Rough-in, waste, 1-1/2" diameter DWV piping | 2.000 | Ea. | 4.507 | 125.00 | 159.00 | 284.00 |
| Supply, 1/2" diameter type "L" copper supply piping | 2.000 | Ea. | 3.161 | 54.08 | 123.84 | 177.92 |
| Bathtub, P.E. cast iron, 5' long with fittings, white | 1.000 | Ea. | 3.636 | 410.00 | 128.00 | 538.00 |
| Rough-in, waste, 4" diameter DWV piping | 1.000 | Ea. | .828 | 31.20 | 29.10 | 60.30 |
| Supply, 1/2" diameter type "L" copper supply piping | 1.000 | Ea. | .988 | 16.90 | 38.70 | 55.60 |
| Vent, 1-1/2" diameter DWV piping | 1.000 | Ea. | .593 | 15.28 | 23.20 | 38.48 |
| Piping, supply, 1/2" diameter type "L" copper supply piping | 42.000 | L.F. | 4.939 | 84.50 | 193.50 | 278.00 |
| Waste, 4" diameter DWV piping | 10.000 | L.F. | 4.138 | 156.00 | 145.50 | 301.50 |
| Vent, 2" diameter DWV piping | 13.000 | L.F. | 4.500 | 81.36 | 158.40 | 239.76 |
| Carrier, steel for studs, no arms | 1.000 | Ea. | 1.143 | 29.00 | 44.50 | 73.50 |
| TOTAL | | | 44.717 | 2267.26 | 1621.16 | 3888.42 |

The costs in this system are on a cost each basis. All necessary piping is included

| Four Fixture Bathroom Price Sheet | QUAN. | UNIT | LABOR HOURS | COST EACH | | |
|---|---|---|---|---|---|---|
| | | | | MAT. | INST. | TOTAL |
| Water closet, close coupled, standard 2 piece, white | 1.000 | Ea. | 3.019 | 154.00 | 106.00 | 260.00 |
| Color | 1.000 | Ea. | 3.019 | 173.00 | 106.00 | 279.00 |
| One piece, elongated bowl, white | 1.000 | Ea. | 3.019 | 545.00 | 106.00 | 651.00 |
| Color | 1.000 | Ea. | 3.019 | 755.00 | 106.00 | 861.00 |
| Low profile, one piece elongated bowl, white | 1.000 | Ea. | 3.019 | 695.00 | 106.00 | 801.00 |
| Color | 1.000 | Ea. | 3.019 | 905.00 | 106.00 | 1011.00 |
| Rough-in, for water closet | | | | | | |
| 1/2" copper supply, 4" cast iron waste, 2" cast iron vent | 1.000 | Ea. | 2.376 | 67.00 | 86.00 | 153.00 |
| 4" PVC/DWV waste, 2" PVC vent | 1.000 | Ea. | 2.678 | 38.00 | 96.50 | 134.50 |
| 4" copper waste, 2" copper vent | 1.000 | Ea. | 2.520 | 73.50 | 94.00 | 167.50 |
| 3" cast iron waste, 1-1/2" cast iron vent | 1.000 | Ea. | 2.244 | 60.00 | 81.50 | 141.50 |
| 3" PVC waste, 1-1/2" PVC vent | 1.000 | Ea. | 2.388 | 34.00 | 90.00 | 124.00 |
| 3" PVC waste, 1-1/2" PVC vent | 1.000 | Ea. | 2.014 | 50.50 | 75.50 | 126.00 |
| 1/2" PVC supply, 4" PVC waste, 2" PVC vent | 1.000 | Ea. | 2.974 | 46.50 | 108.00 | 154.50 |
| 3" PVC waste, 1-1/2" PVC vent | 1.000 | Ea. | 2.684 | 43.00 | 102.00 | 145.00 |
| 1/2" steel supply, 4" cast iron waste, 2" cast iron vent | 1.000 | Ea. | 2.545 | 67.00 | 92.50 | 159.50 |
| 4" cast iron waste, 2" steel vent | 1.000 | Ea. | 2.590 | 59.00 | 94.00 | 153.00 |
| 4" PVC waste, 2" PVC vent | 1.000 | Ea. | 2.847 | 37.50 | 103.00 | 140.50 |
| Lavatory wall hung, P.E. cast iron 20" x 18", white | 1.000 | Ea. | 2.000 | 218.00 | 70.50 | 288.50 |
| Color | 1.000 | Ea. | 2.000 | 255.00 | 70.50 | 325.50 |
| Vitreous china 19" x 17", white | 1.000 | Ea. | 2.286 | 141.00 | 80.50 | 221.50 |
| Color | 1.000 | Ea. | 2.286 | 162.00 | 80.50 | 242.50 |
| Lavatory for vanity top, P.E. cast iron, 20" x 18", white | 1.000 | Ea. | 2.500 | 185.00 | 88.00 | 273.00 |
| Color | 1.000 | Ea. | 2.500 | 213.00 | 88.00 | 301.00 |
| Steel, enameled 20" x 17", white | 1.000 | Ea. | 2.759 | 99.00 | 97.00 | 196.00 |
| Color | 1.000 | Ea. | 2.500 | 105.00 | 88.00 | 193.00 |
| Vitreous china 20" x 16", white | 1.000 | Ea. | 2.963 | 201.00 | 104.00 | 305.00 |
| Color | 1.000 | Ea. | 2.963 | 201.00 | 104.00 | 305.00 |
| Shower, steel enameled, stone base 36" square, white | 1.000 | Ea. | 8.889 | 825.00 | 315.00 | 1140.00 |
| Color | 1.000 | Ea. | 8.889 | 890.00 | 315.00 | 1205.00 |
| Rough-in, for lavatory and shower | | | | | | |
| 1/2" copper supply, 1-1/2" cast iron waste, 1-1/2" cast iron vent | 1.000 | Ea. | 7.668 | 179.00 | 283.00 | 462.00 |
| 1-1/2" PVC waste, 1-1/4" PVC vent | 1.000 | Ea. | 7.352 | 108.00 | 288.00 | 396.00 |
| 1/2" steel supply, 1-1/4" cast iron waste, 1-1/4" steel vent | 1.000 | Ea. | 8.205 | 153.00 | 305.00 | 458.00 |
| 1-1/4" PVC waste, 1-1/4" PVC vent | 1.000 | Ea. | 7.873 | 105.00 | 310.00 | 415.00 |
| 1/2" PVC supply, 1-1/2" PVC waste, 1-1/2" PVC vent | 1.000 | Ea. | 9.185 | 155.00 | 360.00 | 515.00 |
| Bathtub, P.E. cast iron, 5' long with fittings, white | 1.000 | Ea. | 3.636 | 410.00 | 128.00 | 538.00 |
| Color | 1.000 | Ea. | 3.636 | 470.00 | 128.00 | 598.00 |
| Steel enameled, 5' long with fittings, white | 1.000 | Ea. | 2.909 | 320.00 | 102.00 | 422.00 |
| Color | 1.000 | Ea. | 2.909 | 320.00 | 102.00 | 422.00 |
| Rough-in, for bathtub | | | | | | |
| 1/2" copper supply, 4" cast iron waste, 1-1/2" copper vent | 1.000 | Ea. | 2.409 | 63.50 | 91.00 | 154.50 |
| 4" PVC waste, 1-1/2" PVC vent | 1.000 | Ea. | 2.877 | 44.00 | 109.00 | 153.00 |
| 1/2" steel supply, 4" cast iron waste, 1-1/2" steel vent | 1.000 | Ea. | 2.898 | 61.50 | 107.00 | 168.50 |
| 4" PVC waste, 1-1/2" PVC vent | 1.000 | Ea. | 3.159 | 43.50 | 120.00 | 163.50 |
| 1/2" PVC supply, 4" PVC waste, 1-1/2" PVC vent | 1.000 | Ea. | 3.371 | 58.50 | 128.00 | 186.50 |
| Piping supply, 1/2" copper | 42.000 | L.F. | 4.148 | 71.00 | 163.00 | 234.00 |
| 1/2" steel | 42.000 | L.F. | 5.333 | 69.50 | 209.00 | 278.50 |
| 1/2" PVC | 42.000 | L.F. | 6.222 | 132.00 | 244.00 | 376.00 |
| Piping, waste, 4" cast iron, no hub | 10.000 | L.F. | 3.586 | 135.00 | 126.00 | 261.00 |
| 4" PVC/DWV | 10.000 | L.F. | 4.333 | 69.50 | 153.00 | 222.50 |
| 4" copper/DWV | 10.000 | L.F. | 5.200 | 189.00 | 183.00 | 372.00 |
| Piping, vent, 2" cast iron, no hub | 13.000 | L.F. | 3.105 | 83.00 | 109.00 | 192.00 |
| 2" copper/DWV | 13.000 | L.F. | 2.364 | 64.00 | 92.50 | 156.50 |
| 2" PVC/DWV | 13.000 | L.F. | 3.525 | 38.00 | 124.00 | 162.00 |
| 2" steel, galvanized | 13.000 | L.F. | 3.250 | 59.00 | 114.00 | 173.00 |
| Vanity base cabinet, 2 doors, 30" wide | 1.000 | Ea. | 1.000 | 204.00 | 35.50 | 239.50 |
| Vanity top, plastic laminated, square edge | 3.160 | L.F. | .843 | 69.50 | 30.00 | 99.50 |
| Carrier, steel, for studs, no arms | 1.000 | Ea. | 1.143 | 29.00 | 44.50 | 73.50 |
| Wood, 2" x 8" blocking | 1.300 | L.F. | .052 | 1.24 | 1.85 | 3.09 |

MECHANICAL

8

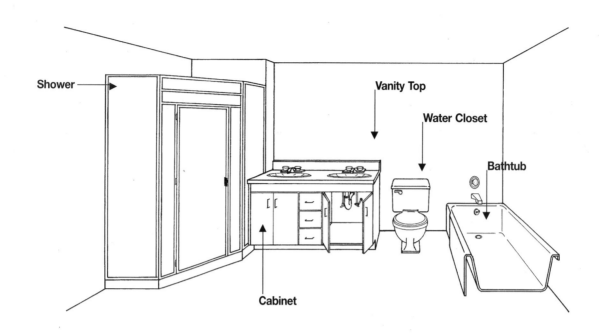

Shower · Vanity Top · Water Closet · Bathtub · Cabinet

| System Description | QUAN. | UNIT | LABOR HOURS | COST EACH | | |
|---|---|---|---|---|---|---|
| | | | | MAT. | INST. | TOTAL |
| **BATHROOM WITH SHOWER, BATHTUB, LAVATORIES IN VANITY** | | | | | | |
| Water closet, floor mounted, 1 piece combination, white | 1.000 | Ea. | 3.019 | 695.00 | 106.00 | 801.00 |
| Rough-in, vent, 2" diameter DWV piping | 1.000 | Ea. | .955 | 25.60 | 33.60 | 59.20 |
| Waste, 4" diameter DWV piping | 1.000 | Ea. | .828 | 31.20 | 29.10 | 60.30 |
| Supply, 1/2" diameter type "L" copper supply piping | 1.000 | Ea. | .593 | 10.14 | 23.22 | 33.36 |
| Lavatory, 20" x 16", vitreous china oval, with fittings, white | 2.000 | Ea. | 5.926 | 402.00 | 208.00 | 610.00 |
| Shower, steel enameled, stone base, corner, white | 1.000 | Ea. | 8.889 | 825.00 | 315.00 | 1140.00 |
| Rough-in, waste, 1-1/2" diameter DWV piping | 3.000 | Ea. | 5.408 | 150.00 | 190.80 | 340.80 |
| Supply, 1/2" diameter type "L" copper supply piping | 3.000 | Ea. | 2.963 | 50.70 | 116.10 | 166.80 |
| Bathtub, P.E. cast iron, 5' long with fittings, white | 1.000 | Ea. | 3.636 | 410.00 | 128.00 | 538.00 |
| Rough-in, waste, 4" diameter DWV piping | 1.000 | Ea. | 1.103 | 41.60 | 38.80 | 80.40 |
| Supply, 1/2" diameter type "L" copper supply piping | 1.000 | Ea. | .988 | 16.90 | 38.70 | 55.60 |
| Vent, 1-1/2" diameter copper DWV piping | 1.000 | Ea. | .593 | 15.28 | 23.20 | 38.48 |
| Piping, supply, 1/2" diameter type "L" copper supply piping | 42.000 | L.F. | 4.148 | 70.98 | 162.54 | 233.52 |
| Waste, 4" diameter DWV piping | 10.000 | L.F. | 2.759 | 104.00 | 97.00 | 201.00 |
| Vent, 2" diameter DWV piping | 13.000 | L.F. | 3.250 | 58.76 | 114.40 | 173.16 |
| Vanity base, 2 door, 24" x 48" | 1.000 | Ea. | 1.400 | 325.00 | 50.00 | 375.00 |
| Vanity top, plastic laminated, square edge | 4.170 | L.F. | 1.112 | 91.74 | 39.62 | 131.36 |
| | | | | | | |
| TOTAL | | | 47.570 | 3323.90 | 1714.08 | 5037.98 |

The costs in this system are on a cost each basis. All necessary piping is included

| Description | QUAN. | UNIT | LABOR HOURS | COST EACH | | |
|---|---|---|---|---|---|---|
| | | | | MAT. | INST. | TOTAL |
| | | | | | | |
| | | | | | | |
| | | | | | | |
| | | | | | | |

| Five Fixture Bathroom Price Sheet | QUAN. | UNIT | LABOR HOURS | COST EACH | | |
|---|---|---|---|---|---|---|
| | | | | MAT. | INST. | TOTAL |
| Water closet, close coupled, standard 2 piece, white | 1.000 | Ea. | 3.019 | 154.00 | 106.00 | 260.00 |
| Color | 1.000 | Ea. | 3.019 | 173.00 | 106.00 | 279.00 |
| One piece elongated bowl, white | 1.000 | Ea. | 3.019 | 545.00 | 106.00 | 651.00 |
| Color | 1.000 | Ea. | 3.019 | 755.00 | 106.00 | 861.00 |
| Low profile, one piece elongated bowl, white | 1.000 | Ea. | 3.019 | 695.00 | 106.00 | 801.00 |
| Color | 1.000 | Ea. | 3.019 | 905.00 | 106.00 | 1011.00 |
| Rough-in, supply, waste and vent for water closet | | | | | | |
| 1/2″ copper supply, 4″ cast iron waste, 2″ cast iron vent | 1.000 | Ea. | 2.376 | 67.00 | 86.00 | 153.00 |
| 4″ P.V.C./DWV waste, 2″ P.V.C. vent | 1.000 | Ea. | 2.678 | 38.00 | 96.50 | 134.50 |
| 4″ copper waste, 2″ copper vent | 1.000 | Ea. | 2.520 | 73.50 | 94.00 | 167.50 |
| 3″ cast iron waste, 1-1/2″ cast iron vent | 1.000 | Ea. | 2.244 | 60.00 | 81.50 | 141.50 |
| 3″ P.V.C. waste, 1-1/2″ P.V.C. vent | 1.000 | Ea. | 2.388 | 34.00 | 90.00 | 124.00 |
| 3″ copper waste, 1-1/2″ copper vent | 1.000 | Ea. | 2.014 | 50.50 | 75.50 | 126.00 |
| 1/2″ P.V.C. supply, 4″ P.V.C. waste, 2″ P.V.C. vent | 1.000 | Ea. | 2.974 | 46.50 | 108.00 | 154.50 |
| 3″ P.V.C. waste, 1-1/2″ P.V.C. supply | 1.000 | Ea. | 2.684 | 43.00 | 102.00 | 145.00 |
| 1/2″ steel supply, 4″ cast iron waste, 2″ cast iron vent | 1.000 | Ea. | 2.545 | 67.00 | 92.50 | 159.50 |
| 4″ cast iron waste, 2″ steel vent | 1.000 | Ea. | 2.590 | 59.00 | 94.00 | 153.00 |
| 4″ P.V.C. waste, 2″ P.V.C. vent | 1.000 | Ea. | 2.847 | 37.50 | 103.00 | 140.50 |
| Lavatory, wall hung, P.E. cast iron 20″ x 18″, white | 2.000 | Ea. | 4.000 | 435.00 | 141.00 | 576.00 |
| Color | 2.000 | Ea. | 4.000 | 510.00 | 141.00 | 651.00 |
| Vitreous china, 19″ x 17″, white | 2.000 | Ea. | 4.571 | 282.00 | 161.00 | 443.00 |
| Color | 2.000 | Ea. | 4.571 | 325.00 | 161.00 | 486.00 |
| Lavatory, for vanity top, P.E. cast iron, 20″ x 18″, white | 2.000 | Ea. | 5.000 | 370.00 | 176.00 | 546.00 |
| Color | 2.000 | Ea. | 5.000 | 425.00 | 176.00 | 601.00 |
| Steel enameled 20″ x 17″, white | 2.000 | Ea. | 5.517 | 198.00 | 194.00 | 392.00 |
| Color | 2.000 | Ea. | 5.000 | 210.00 | 176.00 | 386.00 |
| Vitreous china 20″ x 16″, white | 2.000 | Ea. | 5.926 | 400.00 | 208.00 | 608.00 |
| Color | 2.000 | Ea. | 5.926 | 400.00 | 208.00 | 608.00 |
| Shower, steel enameled, stone base 36″ square, white | 1.000 | Ea. | 8.889 | 825.00 | 315.00 | 1140.00 |
| Color | 1.000 | Ea. | 8.889 | 890.00 | 315.00 | 1205.00 |
| Rough-in, for lavatory or shower | | | | | | |
| 1/2″ copper supply, 1-1/2″ cast iron waste, 1-1/2″ cast iron vent | 3.000 | Ea. | 8.371 | 201.00 | 305.00 | 506.00 |
| 1-1/2″ P.V.C. waste, 1-1/4″ P.V.C. vent | 3.000 | Ea. | 7.916 | 115.00 | 310.00 | 425.00 |
| 1/2″ steel supply, 1-1/4″ cast iron waste, 1-1/4″ steel vent | 3.000 | Ea. | 8.670 | 162.00 | 320.00 | 482.00 |
| 1-1/4″ P.V.C. waste, 1-1/4″ P.V.C. vent | 3.000 | Ea. | 8.381 | 112.00 | 330.00 | 442.00 |
| 1/2″ P.V.C. supply, 1-1/2″ P.V.C. waste, 1-1/2″ P.V.C. vent | 3.000 | Ea. | 9.778 | 160.00 | 385.00 | 545.00 |
| Bathtub, P.E. cast iron 5′ long with fittings, white | 1.000 | Ea. | 3.636 | 410.00 | 128.00 | 538.00 |
| Color | 1.000 | Ea. | 3.636 | 470.00 | 128.00 | 598.00 |
| Steel, enameled 5′ long with fittings, white | 1.000 | Ea. | 2.909 | 320.00 | 102.00 | 422.00 |
| Color | 1.000 | Ea. | 2.909 | 320.00 | 102.00 | 422.00 |
| Rough-in, for bathtub | | | | | | |
| 1/2″ copper supply, 4″ cast iron waste, 1-1/2″ copper vent | 1.000 | Ea. | 2.684 | 74.00 | 101.00 | 175.00 |
| 4″ P.V.C. waste, 1-1/2″ P.V.C. vent | 1.000 | Ea. | 3.210 | 49.00 | 121.00 | 170.00 |
| 1/2″ steel supply, 4″ cast iron waste, 1-1/2″ steel vent | 1.000 | Ea. | 3.173 | 71.50 | 117.00 | 188.50 |
| 4″ P.V.C. waste, 1-1/2″ P.V.C. vent | 1.000 | Ea. | 3.492 | 49.00 | 132.00 | 181.00 |
| 1/2″ P.V.C. supply, 4″ P.V.C. waste, 1-1/2″ P.V.C. vent | 1.000 | Ea. | 3.704 | 64.00 | 140.00 | 204.00 |
| Piping, supply, 1/2″ copper | 42.000 | L.F. | 4.148 | 71.00 | 163.00 | 234.00 |
| 1/2″ steel | 42.000 | L.F. | 5.333 | 69.50 | 209.00 | 278.50 |
| 1/2″ P.V.C. | 42.000 | L.F. | 6.222 | 132.00 | 244.00 | 376.00 |
| Piping, waste, 4″ cast iron, no hub | 10.000 | L.F. | 2.759 | 104.00 | 97.00 | 201.00 |
| 4″ P.V.C./DWV | 10.000 | L.F. | 3.333 | 53.50 | 118.00 | 171.50 |
| 4″ copper/DWV | 10.000 | L.F. | 4.000 | 146.00 | 141.00 | 287.00 |
| Piping, vent, 2″ cast iron, no hub | 13.000 | L.F. | 3.105 | 83.00 | 109.00 | 192.00 |
| 2″ copper/DWV | 13.000 | L.F. | 2.364 | 64.00 | 92.50 | 156.50 |
| 2″ P.V.C./DWV | 13.000 | L.F. | 3.525 | 38.00 | 124.00 | 162.00 |
| 2″ steel, galvanized | 13.000 | L.F. | 3.250 | 59.00 | 114.00 | 173.00 |
| Vanity base cabinet, 2 doors, 24″ x 48″ | 1.000 | Ea. | 1.400 | 325.00 | 50.00 | 375.00 |
| Vanity top, plastic laminated, square edge | 4.170 | L.F. | 1.112 | 91.50 | 39.50 | 131.00 |
| Carrier, steel, for studs, no arms | 1.000 | Ea. | 1.143 | 29.00 | 44.50 | 73.50 |
| Wood, 2″ x 8″ blocking | 1.300 | L.F. | .052 | 1.24 | 1.85 | 3.09 |

**MECHANICAL**

**8**

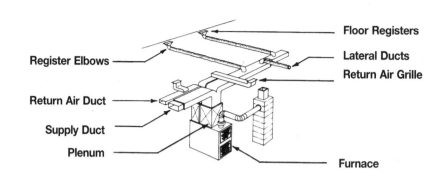

| System Description | QUAN. | UNIT | LABOR HOURS | COST PER SYSTEM | | |
|---|---|---|---|---|---|---|
| | | | | MAT. | INST. | TOTAL |
| **HEATING ONLY, GAS FIRED HOT AIR, ONE ZONE, 1200 S.F. BUILDING** | | | | | | |
| Furnace, gas, up flow | 1.000 | Ea. | 5.000 | 720.00 | 175.00 | 895.00 |
| Intermittent pilot | 1.000 | Ea. | | 145.00 | | 145.00 |
| Supply duct, rigid fiberglass | 176.000 | S.F. | 12.068 | 197.12 | 438.24 | 635.36 |
| Return duct, sheet metal, galvanized | 158.000 | Lb. | 16.137 | 132.72 | 586.18 | 718.90 |
| Lateral ducts, 6" flexible fiberglass | 144.000 | L.F. | 8.862 | 233.28 | 311.04 | 544.32 |
| Register, elbows | 12.000 | Ea. | 3.200 | 318.00 | 112.20 | 430.20 |
| Floor registers, enameled steel | 12.000 | Ea. | 3.000 | 220.20 | 117.00 | 337.20 |
| Floor grille, return air | 2.000 | Ea. | .727 | 49.00 | 28.30 | 77.30 |
| Thermostat | 1.000 | Ea. | 1.000 | 30.00 | 39.00 | 69.00 |
| Plenum | 1.000 | Ea. | 1.000 | 67.50 | 35.00 | 102.50 |
| TOTAL | | | 50.994 | 2112.82 | 1841.96 | 3954.78 |
| **HEATING/COOLING, GAS FIRED FORCED AIR, ONE ZONE, 1200 S.F. BUILDING** | | | | | | |
| Furnace, including plenum, compressor, coil | 1.000 | Ea. | 14.720 | 3404.00 | 515.20 | 3919.20 |
| Intermittent pilot | 1.000 | Ea. | | 145.00 | | 145.00 |
| Supply duct, rigid fiberglass | 176.000 | S.F. | 12.068 | 197.12 | 438.24 | 635.36 |
| Return duct, sheet metal, galvanized | 158.000 | Lb. | 16.137 | 132.72 | 586.18 | 718.90 |
| Lateral duct, 6" flexible fiberglass | 144.000 | L.F. | 8.862 | 233.28 | 311.04 | 544.32 |
| Register elbows | 12.000 | Ea. | 3.200 | 318.00 | 112.20 | 430.20 |
| Floor registers, enameled steel | 12.000 | Ea. | 3.000 | 220.20 | 117.00 | 337.20 |
| Floor grille return air | 2.000 | Ea. | .727 | 49.00 | 28.30 | 77.30 |
| Thermostat | 1.000 | Ea. | 1.000 | 30.00 | 39.00 | 69.00 |
| Refrigeration piping, 25 ft. (pre-charged) | 1.000 | Ea. | | 175.00 | | 175.00 |
| TOTAL | | | 59.714 | 4904.32 | 2147.16 | 7051.48 |

The costs in these systems are based on complete system basis. For larger buildings use the price sheet on the opposite page.

| Description | QUAN. | UNIT | LABOR HOURS | COST PER SYSTEM | | |
|---|---|---|---|---|---|---|
| | | | | MAT. | INST. | TOTAL |
| | | | | | | |
| | | | | | | |
| | | | | | | |
| | | | | | | |
| | | | | | | |

MECHANICAL 8

# Gas Heating/Cooling Price Sheet

| Gas Heating/Cooling Price Sheet | QUAN. | UNIT | LABOR HOURS | COST EACH MAT. | COST EACH INST. | COST EACH TOTAL |
|---|---|---|---|---|---|---|
| Furnace, heating only, 100 MBH, area to 1200 S.F. | 1.000 | Ea. | 5.000 | 720.00 | 175.00 | 895.00 |
| 120 MBH, area to 1500 S.F. | 1.000 | Ea. | 5.000 | 720.00 | 175.00 | 895.00 |
| 160 MBH, area to 2000 S.F. | 1.000 | Ea. | 5.714 | 965.00 | 200.00 | 1165.00 |
| 200 MBH, area to 2400 S.F. | 1.000 | Ea. | 6.154 | 1775.00 | 216.00 | 1991.00 |
| Heating/cooling, 100 MBH heat, 36 MBH cool, to 1200 S.F. | 1.000 | Ea. | 16.000 | 3700.00 | 560.00 | 4260.00 |
| 120 MBH heat, 42 MBH cool, to 1500 S.F. | 1.000 | Ea. | 18.462 | 3950.00 | 670.00 | 4620.00 |
| 144 MBH heat, 47 MBH cool, to 2000 S.F. | 1.000 | Ea. | 20.000 | 4550.00 | 725.00 | 5275.00 |
| 200 MBH heat, 60 MBH cool, to 2400 S.F. | 1.000 | Ea. | 34.286 | 4800.00 | 1250.00 | 6050.00 |
| Intermittent pilot, 100 MBH furnace | 1.000 | Ea. | | 145.00 | | 145.00 |
| 200 MBH furnace | 1.000 | Ea. | | 145.00 | | 145.00 |
| Supply duct, rectangular, area to 1200 S.F., rigid fiberglass | 176.000 | S.F. | 12.068 | 197.00 | 440.00 | 637.00 |
| Sheet metal insulated | 228.000 | Lb. | 31.331 | 274.00 | 1125.00 | 1399.00 |
| Area to 1500 S.F., rigid fiberglass | 176.000 | S.F. | 12.068 | 197.00 | 440.00 | 637.00 |
| Sheet metal insulated | 228.000 | Lb. | 31.331 | 274.00 | 1125.00 | 1399.00 |
| Area to 2400 S.F., rigid fiberglass | 205.000 | S.F. | 14.057 | 230.00 | 510.00 | 740.00 |
| Sheet metal insulated | 271.000 | Lb. | 37.048 | 325.00 | 1325.00 | 1650.00 |
| Round flexible, insulated 6" diameter, to 1200 S.F. | 156.000 | L.F. | 9.600 | 253.00 | 335.00 | 588.00 |
| To 1500 S.F. | 184.000 | L.F. | 11.323 | 298.00 | 395.00 | 693.00 |
| 8" diameter, to 2000 S.F. | 269.000 | L.F. | 23.911 | 545.00 | 835.00 | 1380.00 |
| To 2400 S.F. | 248.000 | L.F. | 22.045 | 500.00 | 770.00 | 1270.00 |
| Return duct, sheet metal galvanized, to 1500 S.F. | 158.000 | Lb. | 16.137 | 133.00 | 585.00 | 718.00 |
| To 2400 S.F. | 191.000 | Lb. | 19.507 | 160.00 | 710.00 | 870.00 |
| Lateral ducts, flexible round 6" insulated, to 1200 S.F. | 144.000 | L.F. | 8.862 | 233.00 | 310.00 | 543.00 |
| To 1500 S.F. | 172.000 | L.F. | 10.585 | 279.00 | 370.00 | 649.00 |
| To 2000 S.F. | 261.000 | L.F. | 16.062 | 425.00 | 565.00 | 990.00 |
| To 2400 S.F. | 300.000 | L.F. | 18.462 | 485.00 | 650.00 | 1135.00 |
| Spiral steel insulated, to 1200 S.F. | 144.000 | L.F. | 20.067 | 315.00 | 695.00 | 1010.00 |
| To 1500 S.F. | 172.000 | L.F. | 23.952 | 375.00 | 830.00 | 1205.00 |
| To 2000 S.F. | 261.000 | L.F. | 36.352 | 575.00 | 1250.00 | 1825.00 |
| To 2400 S.F. | 300.000 | L.F. | 41.825 | 660.00 | 1450.00 | 2110.00 |
| Rectangular sheet metal galvanized insulated, to 1200 S.F. | 228.000 | Lb. | 39.056 | 355.00 | 1400.00 | 1755.00 |
| To 1500 S.F. | 344.000 | Lb. | 53.966 | 485.00 | 1925.00 | 2410.00 |
| To 2000 S.F. | 522.000 | Lb. | 81.926 | 735.00 | 2925.00 | 3660.00 |
| To 2400 S.F. | 600.000 | Lb. | 94.189 | 840.00 | 3350.00 | 4190.00 |
| Register elbows, to 1500 S.F. | 12.000 | Ea. | 3.200 | 320.00 | 112.00 | 432.00 |
| To 2400 S.F. | 14.000 | Ea. | 3.733 | 370.00 | 131.00 | 501.00 |
| Floor registers, enameled steel w/damper, to 1500 S.F. | 12.000 | Ea. | 3.000 | 220.00 | 117.00 | 337.00 |
| To 2400 S.F. | 14.000 | Ea. | 4.308 | 300.00 | 168.00 | 468.00 |
| Return air grille, area to 1500 S.F. 12" x 12" | 2.000 | Ea. | .727 | 49.00 | 28.50 | 77.50 |
| Area to 2400 S.F. 8" x 16" | 2.000 | Ea. | .444 | 43.50 | 17.30 | 60.80 |
| Area to 2400 S.F. 8" x 16" | 2.000 | Ea. | .727 | 49.00 | 28.50 | 77.50 |
| 16" x 16" | 1.000 | Ea. | .364 | 35.00 | 14.15 | 49.15 |
| Thermostat, manual, 1 set back | 1.000 | Ea. | 1.000 | 30.00 | 39.00 | 69.00 |
| Electric, timed, 1 set back | 1.000 | Ea. | 1.000 | 89.00 | 39.00 | 128.00 |
| 2 set back | 1.000 | Ea. | 1.000 | 122.00 | 39.00 | 161.00 |
| Plenum, heating only, 100 M.B.H. | 1.000 | Ea. | 1.000 | 67.50 | 35.00 | 102.50 |
| 120 MBH | 1.000 | Ea. | 1.000 | 67.50 | 35.00 | 102.50 |
| 160 MBH | 1.000 | Ea. | 1.000 | 67.50 | 35.00 | 102.50 |
| 200 MBH | 1.000 | Ea. | 1.000 | 67.50 | 35.00 | 102.50 |
| Refrigeration piping, 3/8" | 25.000 | L.F. | | 15.75 | | 15.75 |
| 3/4" | 25.000 | L.F. | | 31.50 | | 31.50 |
| 7/8" | 25.000 | L.F. | | 36.50 | | 36.50 |
| Refrigerant piping, 25 ft. (precharged) | 1.000 | Ea. | | 175.00 | | 175.00 |
| Diffusers, ceiling, 6" diameter, to 1500 S.F. | 10.000 | Ea. | 4.444 | 168.00 | 173.00 | 341.00 |
| To 2400 S.F. | 12.000 | Ea. | 6.000 | 217.00 | 234.00 | 451.00 |
| Floor, aluminum, adjustable, 2-1/4" x 12" to 1500 S.F. | 12.000 | Ea. | 3.000 | 160.00 | 117.00 | 277.00 |
| To 2400 S.F. | 14.000 | Ea. | 3.500 | 186.00 | 137.00 | 323.00 |
| Side wall, aluminum, adjustable, 8" x 4", to 1500 S.F. | 12.000 | Ea. | 3.000 | 350.00 | 117.00 | 467.00 |
| 5" x 10" to 2400 S.F. | 12.000 | Ea. | 3.692 | 450.00 | 144.00 | 594.00 |

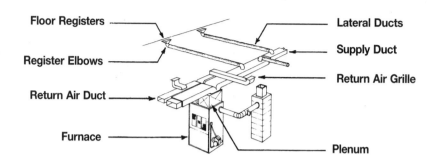

Floor Registers · Lateral Ducts · Register Elbows · Supply Duct · Return Air Duct · Return Air Grille · Furnace · Plenum

| System Description | QUAN. | UNIT | LABOR HOURS | COST PER SYSTEM | | |
|---|---|---|---|---|---|---|
| | | | | MAT. | INST. | TOTAL |
| **HEATING ONLY, OIL FIRED HOT AIR, ONE ZONE, 1200 S.F. BUILDING** | | | | | | |
| Furnace, oil fired, atomizing gun type burner | 1.000 | Ea. | 4.571 | 850.00 | 160.00 | 1010.00 |
| 3/8" diameter copper supply pipe | 1.000 | Ea. | 2.759 | 39.60 | 108.00 | 147.60 |
| Shut off valve | 1.000 | Ea. | .333 | 7.40 | 13.05 | 20.45 |
| Oil tank, 275 gallon, on legs | 1.000 | Ea. | 3.200 | 265.00 | 113.00 | 378.00 |
| Supply duct, rigid fiberglass | 176.000 | S.F. | 12.068 | 197.12 | 438.24 | 635.36 |
| Return duct, sheet metal, galvanized | 158.000 | Lb. | 16.137 | 132.72 | 586.18 | 718.90 |
| Lateral ducts, 6" flexible fiberglass | 144.000 | L.F. | 8.862 | 233.28 | 311.04 | 544.32 |
| Register elbows | 12.000 | Ea. | 3.200 | 318.00 | 112.20 | 430.20 |
| Floor register, enameled steel | 12.000 | Ea. | 3.000 | 220.20 | 117.00 | 337.20 |
| Floor grille, return air | 2.000 | Ea. | .727 | 49.00 | 28.30 | 77.30 |
| Thermostat | 1.000 | Ea. | 1.000 | 30.00 | 39.00 | 69.00 |
| TOTAL | | | 55.857 | 2342.32 | 2026.01 | 4368.33 |
| **HEATING/COOLING, OIL FIRED, FORCED AIR, ONE ZONE, 1200 S.F. BUILDING** | | | | | | |
| Furnace, including plenum, compressor, coil | 1.000 | Ea. | 16.000 | 3950.00 | 560.00 | 4510.00 |
| 3/8" diameter copper supply pipe | 1.000 | Ea. | 2.759 | 39.60 | 108.00 | 147.60 |
| Shut off valve | 1.000 | Ea. | .333 | 7.40 | 13.05 | 20.45 |
| Oil tank, 275 gallon on legs | 1.000 | Ea. | 3.200 | 265.00 | 113.00 | 378.00 |
| Supply duct, rigid fiberglass | 176.000 | S.F. | 12.068 | 197.12 | 438.24 | 635.36 |
| Return duct, sheet metal, galvanized | 158.000 | Lb. | 16.137 | 132.72 | 586.18 | 718.90 |
| Lateral ducts, 6" flexible fiberglass | 144.000 | L.F. | 8.862 | 233.28 | 311.04 | 544.32 |
| Register elbows | 12.000 | Ea. | 3.200 | 318.00 | 112.20 | 430.20 |
| Floor registers, enameled steel | 12.000 | Ea. | 3.000 | 220.20 | 117.00 | 337.20 |
| Floor grille, return air | 2.000 | Ea. | .727 | 49.00 | 28.30 | 77.30 |
| Refrigeration piping (precharged) | 25.000 | L.F. | | 175.00 | | 175.00 |
| TOTAL | | | 66.286 | 5587.32 | 2387.01 | 7974.33 |

| Description | QUAN. | UNIT | LABOR HOURS | COST EACH | | |
|---|---|---|---|---|---|---|
| | | | | MAT. | INST. | TOTAL |
| | | | | | | |
| | | | | | | |
| | | | | | | |
| | | | | | | |
| | | | | | | |

**Important: See the Reference Section for critical supporting data - Reference Nos., Crews & Location Factors**

| Oil Fired Heating/Cooling | QUAN. | UNIT | LABOR HOURS | COST EACH | | |
|---|---|---|---|---|---|---|
| | | | | MAT. | INST. | TOTAL |
| Furnace, heating, 95.2 MBH, area to 1200 S.F. | 1.000 | Ea. | 4.706 | 870.00 | 165.00 | 1035.00 |
| 123.2 MBH, area to 1500 S.F. | 1.000 | Ea. | 5.000 | 1200.00 | 175.00 | 1375.00 |
| 151.2 MBH, area to 2000 S.F. | 1.000 | Ea. | 5.333 | 1325.00 | 187.00 | 1512.00 |
| 200 MBH, area to 2400 S.F. | 1.000 | Ea. | 6.154 | 1825.00 | 216.00 | 2041.00 |
| Heating/cooling, 95.2 MBH heat, 36 MBH cool, to 1200 S.F. | 1.000 | Ea. | 16.000 | 3950.00 | 560.00 | 4510.00 |
| 112 MBH heat, 42 MBH cool, to 1500 S.F. | 1.000 | Ea. | 24.000 | 5925.00 | 840.00 | 6765.00 |
| 151 MBH heat, 47 MBH cool, to 2000 S.F. | 1.000 | Ea. | 20.800 | 5125.00 | 730.00 | 5855.00 |
| 184.8 MBH heat, 60 MBH cool, to 2400 S.F. | 1.000 | Ea. | 24.000 | 5425.00 | 870.00 | 6295.00 |
| Oil piping to furnace, 3/8" dia., copper | 1.000 | Ea. | 3.412 | 126.00 | 132.00 | 258.00 |
| Oil tank, on legs above ground, 275 gallons | 1.000 | Ea. | 3.200 | 265.00 | 113.00 | 378.00 |
| 550 gallons | 1.000 | Ea. | 5.926 | 1250.00 | 210.00 | 1460.00 |
| Below ground, 275 gallons | 1.000 | Ea. | 3.200 | 265.00 | 113.00 | 378.00 |
| 550 gallons | 1.000 | Ea. | 5.926 | 1250.00 | 210.00 | 1460.00 |
| 1000 gallons | 1.000 | Ea. | 6.400 | 1975.00 | 227.00 | 2202.00 |
| Supply duct, rectangular, area to 1200 S.F., rigid fiberglass | 176.000 | S.F. | 12.068 | 197.00 | 440.00 | 637.00 |
| Sheet metal, insulated | 228.000 | Lb. | 31.331 | 274.00 | 1125.00 | 1399.00 |
| Area to 1500 S.F., rigid fiberglass | 176.000 | S.F. | 12.068 | 197.00 | 440.00 | 637.00 |
| Sheet metal, insulated | 228.000 | Lb. | 31.331 | 274.00 | 1125.00 | 1399.00 |
| Area to 2400 S.F., rigid fiberglass | 205.000 | S.F. | 14.057 | 230.00 | 510.00 | 740.00 |
| Sheet metal, insulated | 271.000 | Lb. | 37.048 | 325.00 | 1325.00 | 1650.00 |
| Round flexible, insulated, 6" diameter to 1200 S.F. | 156.000 | L.F. | 9.600 | 253.00 | 335.00 | 588.00 |
| To 1500 S.F. | 184.000 | L.F. | 11.323 | 298.00 | 395.00 | 693.00 |
| 8" diameter to 2000 S.F. | 269.000 | L.F. | 23.911 | 545.00 | 835.00 | 1380.00 |
| To 2400 S.F. | 269.000 | L.F. | 22.045 | 500.00 | 770.00 | 1270.00 |
| Return duct, sheet metal galvanized, to 1500 S.F. | 158.000 | Lb. | 16.137 | 133.00 | 585.00 | 718.00 |
| To 2400 S.F. | 191.000 | Lb. | 19.507 | 160.00 | 710.00 | 870.00 |
| Lateral ducts, flexible round, 6", insulated to 1200 S.F. | 144.000 | L.F. | 8.862 | 233.00 | 310.00 | 543.00 |
| To 1500 S.F. | 172.000 | L.F. | 10.585 | 279.00 | 370.00 | 649.00 |
| To 2000 S.F. | 261.000 | L.F. | 16.062 | 425.00 | 565.00 | 990.00 |
| To 2400 S.F. | 300.000 | L.F. | 18.462 | 485.00 | 650.00 | 1135.00 |
| Spiral steel, insulated to 1200 S.F. | 144.000 | L.F. | 20.067 | 315.00 | 695.00 | 1010.00 |
| To 1500 S.F. | 172.000 | L.F. | 23.952 | 375.00 | 830.00 | 1205.00 |
| To 2000 S.F. | 261.000 | L.F. | 36.352 | 575.00 | 1250.00 | 1825.00 |
| To 2400 S.F. | 300.000 | L.F. | 41.825 | 660.00 | 1450.00 | 2110.00 |
| Rectangular sheet metal galvanized insulated, to 1200 S.F. | 288.000 | Lb. | 45.183 | 405.00 | 1600.00 | 2005.00 |
| To 1500 S.F. | 344.000 | Lb. | 53.966 | 485.00 | 1925.00 | 2410.00 |
| To 2000 S.F. | 522.000 | Lb. | 81.926 | 735.00 | 2925.00 | 3660.00 |
| To 2400 S.F. | 600.000 | Lb. | 94.189 | 840.00 | 3350.00 | 4190.00 |
| Register elbows, to 1500 S.F. | 12.000 | Ea. | 3.200 | 320.00 | 112.00 | 432.00 |
| To 2400 S.F. | 14.000 | Ea. | 3.733 | 370.00 | 131.00 | 501.00 |
| Floor registers, enameled steel w/damper, to 1500 S.F. | 12.000 | Ea. | 3.000 | 220.00 | 117.00 | 337.00 |
| To 2400 S.F. | 14.000 | Ea. | 4.308 | 300.00 | 168.00 | 468.00 |
| Return air grille, area to 1500 S.F., 12" x 12" | 2.000 | Ea. | .727 | 49.00 | 28.50 | 77.50 |
| 12" x 24" | 1.000 | Ea. | .444 | 43.50 | 17.30 | 60.80 |
| Area to 2400 S.F., 8" x 16" | 2.000 | Ea. | .727 | 49.00 | 28.50 | 77.50 |
| 16" x 16" | 1.000 | Ea. | .364 | 35.00 | 14.15 | 49.15 |
| Thermostat, manual, 1 set back | 1.000 | Ea. | 1.000 | 30.00 | 39.00 | 69.00 |
| Electric, timed, 1 set back | 1.000 | Ea. | 1.000 | 89.00 | 39.00 | 128.00 |
| 2 set back | 1.000 | Ea. | 1.000 | 122.00 | 39.00 | 161.00 |
| Refrigeration piping, 3/8" | 25.000 | L.F. | | 15.75 | | 15.75 |
| 3/4" | 25.000 | L.F. | | 31.50 | | 31.50 |
| Diffusers, ceiling, 6" diameter, to 1500 S.F. | 10.000 | Ea. | 4.444 | 168.00 | 173.00 | 341.00 |
| To 2400 S.F. | 12.000 | Ea. | 6.000 | 217.00 | 234.00 | 451.00 |
| Floor, aluminum, adjustable, 2-1/4" x 12" to 1500 S.F. | 12.000 | Ea. | 3.000 | 160.00 | 117.00 | 277.00 |
| To 2400 S.F. | 14.000 | Ea. | 3.500 | 186.00 | 137.00 | 323.00 |
| Side wall, aluminum, adjustable, 8" x 4", to 1500 S.F. | 12.000 | Ea. | 3.000 | 350.00 | 117.00 | 467.00 |
| 5" x 10" to 2400 S.F. | 12.000 | Ea. | 3.692 | 450.00 | 144.00 | 594.00 |

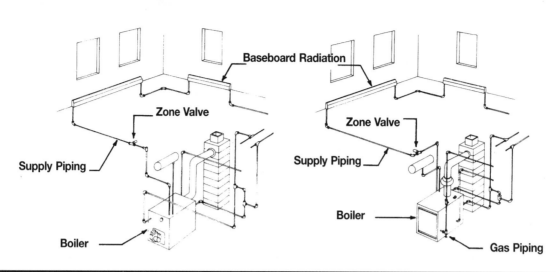

| System Description | QUAN. | UNIT | LABOR HOURS | COST EACH | | |
|---|---|---|---|---|---|---|
| | | | | MAT. | INST. | TOTAL |
| **OIL FIRED HOT WATER HEATING SYSTEM, AREA TO 1200 S.F.** | | | | | | |
| Boiler package, oil fired, 97 MBH, area to 1200 S.F. building | 1.000 | Ea. | 15.000 | 1400.00 | 510.00 | 1910.00 |
| 3/8" diameter copper supply pipe | 1.000 | Ea. | 2.759 | 39.60 | 108.00 | 147.60 |
| Shut off valve | 1.000 | Ea. | .333 | 7.40 | 13.05 | 20.45 |
| Oil tank, 275 gallon, with black iron filler pipe | 1.000 | Ea. | 3.200 | 265.00 | 113.00 | 378.00 |
| Supply piping, 3/4" copper tubing | 176.000 | L.F. | 18.526 | 399.52 | 725.12 | 1124.64 |
| Supply fittings, copper 3/4" | 36.000 | Ea. | 15.158 | 32.04 | 594.00 | 626.04 |
| Supply valves, 3/4" | 2.000 | Ea. | .800 | 99.00 | 31.30 | 130.30 |
| Baseboard radiation, 3/4" | 106.000 | L.F. | 35.333 | 361.46 | 1250.80 | 1612.26 |
| Zone valve | 1.000 | Ea. | .400 | 74.50 | 15.75 | 90.25 |
| TOTAL | | | 91.509 | 2678.52 | 3361.02 | 6039.54 |
| **OIL FIRED HOT WATER HEATING SYSTEM, AREA TO 2400 S.F.** | | | | | | |
| Boiler package, oil fired, 225 MBH, area to 2400 S.F. building | 1.000 | Ea. | 19.704 | 3125.00 | 675.00 | 3800.00 |
| 3/8" diameter copper supply pipe | 1.000 | Ea. | 2.759 | 39.60 | 108.00 | 147.60 |
| Shut off valve | 1.000 | Ea. | .333 | 7.40 | 13.05 | 20.45 |
| Oil tank, 550 gallon, with black iron pipe filler pipe | 1.000 | Ea. | 5.926 | 1250.00 | 210.00 | 1460.00 |
| Supply piping, 3/4" copper tubing | 228.000 | L.F. | 23.999 | 517.56 | 939.36 | 1456.92 |
| Supply fittings, copper | 46.000 | Ea. | 19.368 | 40.94 | 759.00 | 799.94 |
| Supply valves | 2.000 | Ea. | .800 | 99.00 | 31.30 | 130.30 |
| Baseboard radiation | 212.000 | L.F. | 70.666 | 722.92 | 2501.60 | 3224.52 |
| Zone valve | 1.000 | Ea. | .400 | 74.50 | 15.75 | 90.25 |
| TOTAL | | | 143.955 | 5876.92 | 5253.06 | 11129.98 |

The costs in this system are on a cost each basis. the costs represent total cost for the system based on a gross square foot of plan area.

| Description | QUAN. | UNIT | LABOR HOURS | COST EACH | | |
|---|---|---|---|---|---|---|
| | | | | MAT. | INST. | TOTAL |
| | | | | | | |
| | | | | | | |
| | | | | | | |

**Important: See the Reference Section for critical supporting data - Reference Nos., Crews & Location Factors**

# Hot Water Heating Price Sheet

| | QUAN. | UNIT | LABOR HOURS | COST EACH MAT. | COST EACH INST. | COST EACH TOTAL |
|---|---|---|---|---|---|---|
| Boiler, oil fired, 97 MBH, area to 1200 S.F. | 1.000 | Ea. | 15.000 | 1400.00 | 510.00 | 1910.00 |
| 118 MBH, area to 1500 S.F. | 1.000 | Ea. | 16.506 | 2425.00 | 565.00 | 2990.00 |
| 161 MBH, area to 2000 S.F. | 1.000 | Ea. | 18.405 | 3050.00 | 630.00 | 3680.00 |
| 215 MBH, area to 2400 S.F. | 1.000 | Ea. | 19.704 | 3125.00 | 675.00 | 3800.00 |
| Oil piping, (valve & filter), 3/8" copper | 1.000 | Ea. | 3.289 | 78.50 | 128.00 | 206.50 |
| 1/4" copper | 1.000 | Ea. | 3.242 | 61.00 | 126.00 | 187.00 |
| Oil tank, filler pipe and cap on legs, 275 gallon | 1.000 | Ea. | 3.200 | 265.00 | 113.00 | 378.00 |
| 550 gallon | 1.000 | Ea. | 5.926 | 1250.00 | 210.00 | 1460.00 |
| Buried underground, 275 gallon | 1.000 | Ea. | 3.200 | 265.00 | 113.00 | 378.00 |
| 550 gallon | 1.000 | Ea. | 5.926 | 1250.00 | 210.00 | 1460.00 |
| 1000 gallon | 1.000 | Ea. | 6.400 | 1975.00 | 227.00 | 2202.00 |
| Supply piping copper, area to 1200 S.F., 1/2" tubing | 176.000 | L.F. | 17.384 | 297.00 | 680.00 | 977.00 |
| 3/4" tubing | 176.000 | L.F. | 18.526 | 400.00 | 725.00 | 1125.00 |
| Area to 1500 S.F., 1/2" tubing | 186.000 | L.F. | 18.371 | 315.00 | 720.00 | 1035.00 |
| 3/4" tubing | 186.000 | L.F. | 19.578 | 420.00 | 765.00 | 1185.00 |
| Area to 2000 S.F., 1/2" tubing | 204.000 | L.F. | 20.149 | 345.00 | 790.00 | 1135.00 |
| 3/4" tubing | 204.000 | L.F. | 21.473 | 465.00 | 840.00 | 1305.00 |
| Area to 2400 S.F., 1/2" tubing | 228.000 | L.F. | 22.520 | 385.00 | 880.00 | 1265.00 |
| 3/4" tubing | 228.000 | L.F. | 23.999 | 520.00 | 940.00 | 1460.00 |
| Supply pipe fittings copper, area to 1200 S.F., 1/2" | 36.000 | Ea. | 14.400 | 14.75 | 565.00 | 579.75 |
| 3/4" | 36.000 | Ea. | 15.158 | 32.00 | 595.00 | 627.00 |
| Area to 1500 S.F., 1/2" | 40.000 | Ea. | 16.000 | 16.40 | 625.00 | 641.40 |
| 3/4" | 40.000 | Ea. | 16.842 | 35.50 | 660.00 | 695.50 |
| Area to 2000 S.F., 1/2" | 44.000 | Ea. | 17.600 | 18.05 | 690.00 | 708.05 |
| 3/4" | 44.000 | Ea. | 18.526 | 39.00 | 725.00 | 764.00 |
| Area to 2400, S.F., 1/2" | 46.000 | Ea. | 18.400 | 18.85 | 720.00 | 738.85 |
| 3/4" | 46.000 | Ea. | 19.368 | 41.00 | 760.00 | 801.00 |
| Supply valves, 1/2" pipe size | 2.000 | Ea. | .667 | 72.00 | 26.00 | 98.00 |
| 3/4" | 2.000 | Ea. | .800 | 99.00 | 31.50 | 130.50 |
| Baseboard radiation, area to 1200 S.F., 1/2" tubing | 106.000 | L.F. | 28.267 | 620.00 | 1000.00 | 1620.00 |
| 3/4" tubing | 106.000 | L.F. | 35.333 | 360.00 | 1250.00 | 1610.00 |
| Area to 1500 S.F., 1/2" tubing | 134.000 | L.F. | 35.734 | 785.00 | 1275.00 | 2060.00 |
| 3/4" tubing | 134.000 | L.F. | 44.666 | 455.00 | 1575.00 | 2030.00 |
| Area to 2000 S.F., 1/2" tubing | 178.000 | L.F. | 47.467 | 1050.00 | 1675.00 | 2725.00 |
| 3/4" tubing | 178.000 | L.F. | 59.333 | 605.00 | 2100.00 | 2705.00 |
| Area to 2400 S.F., 1/2" tubing | 212.000 | L.F. | 56.534 | 1250.00 | 2000.00 | 3250.00 |
| 3/4" tubing | 212.000 | L.F. | 70.666 | 725.00 | 2500.00 | 3225.00 |
| Zone valves, 1/2" tubing | 1.000 | Ea. | .400 | 74.50 | 15.75 | 90.25 |
| 3/4" tubing | 1.000 | Ea. | .400 | 79.50 | 15.75 | 95.25 |

**MECHANICAL**

**8**

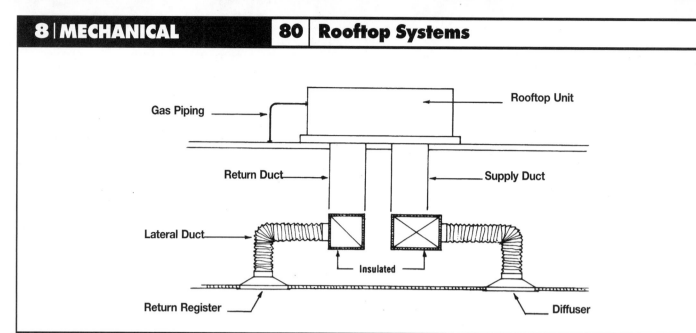

| System Description | QUAN. | UNIT | LABOR HOURS | COST EACH | | |
|---|---|---|---|---|---|---|
| | | | | MAT. | INST. | TOTAL |
| **ROOFTOP HEATING/COOLING UNIT, AREA TO 2000 S.F.** | | | | | | |
| Rooftop unit, single zone, electric cool, gas heat, to 2000 s.f. | 1.000 | Ea. | 28.521 | 6400.00 | 1000.00 | 7400.00 |
| Gas piping | 34.500 | L.F. | 5.207 | 71.42 | 203.55 | 274.97 |
| Duct, supply and return, galvanized steel | 38.000 | Lb. | 3.881 | 31.92 | 140.98 | 172.90 |
| Insulation, ductwork | 33.000 | S.F. | 1.508 | 15.51 | 51.81 | 67.32 |
| Lateral duct, flexible duct 12″ diameter, insulated | 72.000 | L.F. | 11.520 | 220.32 | 403.20 | 623.52 |
| Diffusers | 4.000 | Ea. | 4.571 | 1064.00 | 178.00 | 1242.00 |
| Return registers | 1.000 | Ea. | .727 | 117.00 | 28.50 | 145.50 |
| | | | | | | |
| TOTAL | | | 55.935 | 7920.17 | 2006.04 | 9926.21 |
| **ROOFTOP HEATING/COOLING UNIT, AREA TO 5000 S.F.** | | | | | | |
| Rooftop unit, single zone, electric cool, gas heat, to 5000 s.f. | 1.000 | Ea. | 42.032 | 15000.00 | 1425.00 | 16425.00 |
| Gas piping | 86.250 | L.F. | 13.019 | 178.54 | 508.88 | 687.42 |
| Duct supply and return, galvanized steel | 95.000 | Lb. | 9.702 | 79.80 | 352.45 | 432.25 |
| Insulation, ductwork | 82.000 | S.F. | 3.748 | 38.54 | 128.74 | 167.28 |
| Lateral duct, flexible duct, 12″ diameter, insulated | 180.000 | L.F. | 28.800 | 550.80 | 1008.00 | 1558.80 |
| Diffusers | 10.000 | Ea. | 11.429 | 2660.00 | 445.00 | 3105.00 |
| Return registers | 3.000 | Ea. | 2.182 | 351.00 | 85.50 | 436.50 |
| | | | | | | |
| TOTAL | | | 110.912 | 18858.68 | 3953.57 | 22812.25 |

| Description | QUAN. | UNIT | LABOR HOURS | COST EACH | | |
|---|---|---|---|---|---|---|
| | | | | MAT. | INST. | TOTAL |
| | | | | | | |
| | | | | | | |
| | | | | | | |
| | | | | | | |
| | | | | | | |

**Important: See the Reference Section for critical supporting data - Reference Nos., Crews & Location Factors**

| Rooftop Price Sheet | QUAN. | UNIT | LABOR HOURS | COST EACH | | |
|---|---|---|---|---|---|---|
| | | | | MAT. | INST. | TOTAL |
| Rooftop unit, single zone, electric cool, gas heat to 2000 S.F. | 1.000 | Ea. | 28.521 | 6400.00 | 1000.00 | 7400.00 |
| Area to 3000 S.F. | 1.000 | Ea. | 35.982 | 13700.00 | 1225.00 | 14925.00 |
| Area to 5000 S.F. | 1.000 | Ea. | 42.032 | 15000.00 | 1425.00 | 16425.00 |
| Area to 10000 S.F. | 1.000 | Ea. | 68.376 | 35900.00 | 2425.00 | 38325.00 |
| Gas piping, area 2000 through 4000 S.F. | 34.500 | L.F. | 5.207 | 71.50 | 204.00 | 275.50 |
| Area 5000 to 10000 S.F. | 86.250 | L.F. | 13.019 | 179.00 | 510.00 | 689.00 |
| Duct, supply and return, galvanized steel, to 2000 S.F. | 38.000 | Lb. | 3.881 | 32.00 | 141.00 | 173.00 |
| Area to 3000 S.F. | 57.000 | Lb. | 5.821 | 48.00 | 211.00 | 259.00 |
| Area to 5000 S.F. | 95.000 | Lb. | 9.702 | 80.00 | 350.00 | 430.00 |
| Area to 10000 S.F. | 190.000 | Lb. | 19.405 | 160.00 | 705.00 | 865.00 |
| Rigid fiberglass, area to 2000 S.F. | 33.000 | S.F. | 2.263 | 37.00 | 82.00 | 119.00 |
| Area to 3000 S.F. | 49.000 | S.F. | 3.360 | 55.00 | 122.00 | 177.00 |
| Area to 5000 S.F. | 82.000 | S.F. | 5.623 | 92.00 | 204.00 | 296.00 |
| Area to 10000 S.F. | 164.000 | S.F. | 11.245 | 184.00 | 410.00 | 594.00 |
| Insulation, supply and return, blanket type, area to 2000 S.F. | 33.000 | S.F. | 1.508 | 15.50 | 52.00 | 67.50 |
| Area to 3000 S.F. | 49.000 | S.F. | 2.240 | 23.00 | 77.00 | 100.00 |
| Area to 5000 S.F. | 82.000 | S.F. | 3.748 | 38.50 | 129.00 | 167.50 |
| Area to 10000 S.F. | 164.000 | S.F. | 7.496 | 77.00 | 257.00 | 334.00 |
| Lateral ducts, flexible round, 12" insulated, to 2000 S.F. | 72.000 | L.F. | 11.520 | 220.00 | 405.00 | 625.00 |
| Area to 3000 S.F. | 108.000 | L.F. | 17.280 | 330.00 | 605.00 | 935.00 |
| Area to 5000 S.F. | 180.000 | L.F. | 28.800 | 550.00 | 1000.00 | 1550.00 |
| Area to 10000 S.F. | 360.000 | L.F. | 57.600 | 1100.00 | 2025.00 | 3125.00 |
| Rectangular, galvanized steel, to 2000 S.F. | 239.000 | Lb. | 24.409 | 201.00 | 885.00 | 1086.00 |
| Area to 3000 S.F. | 360.000 | Lb. | 36.767 | 300.00 | 1325.00 | 1625.00 |
| Area to 5000 S.F. | 599.000 | Lb. | 61.176 | 505.00 | 2225.00 | 2730.00 |
| Area to 10000 S.F. | 998.000 | Lb. | 101.926 | 840.00 | 3700.00 | 4540.00 |
| Diffusers, ceiling, 1 to 4 way blow, 24" x 24", to 2000 S.F. | 4.000 | Ea. | 4.571 | 1075.00 | 178.00 | 1253.00 |
| Area to 3000 S.F. | 6.000 | Ea. | 6.857 | 1600.00 | 267.00 | 1867.00 |
| Area to 5000 S.F. | 10.000 | Ea. | 11.429 | 2650.00 | 445.00 | 3095.00 |
| Area to 10000 S.F. | 20.000 | Ea. | 22.857 | 5325.00 | 890.00 | 6215.00 |
| Return grilles, 24" x 24", to 2000 S.F. | 1.000 | Ea. | .727 | 117.00 | 28.50 | 145.50 |
| Area to 3000 S.F. | 2.000 | Ea. | 1.455 | 234.00 | 57.00 | 291.00 |
| Area to 5000 S.F. | 3.000 | Ea. | 2.182 | 350.00 | 85.50 | 435.50 |
| Area to 10000 S.F. | 5.000 | Ea. | 3.636 | 585.00 | 143.00 | 728.00 |
| | | | | | | |
| | | | | | | |
| | | | | | | |
| | | | | | | |
| | | | | | | |
| | | | | | | |
| | | | | | | |
| | | | | | | |
| | | | | | | |
| | | | | | | |
| | | | | | | |

MECHANICAL

8

For information about Means Estimating Seminars, see yellow pages 11 and 12 in back of book

# Division 9
# Electrical

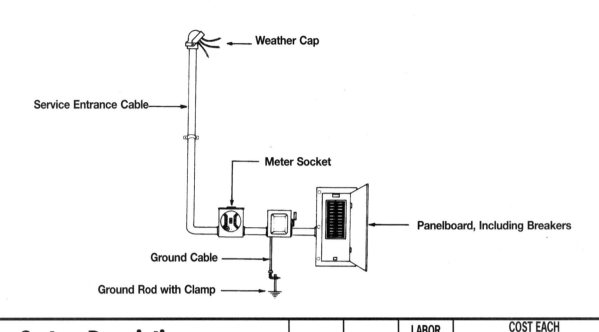

- Weather Cap
- Service Entrance Cable
- Meter Socket
- Panelboard, Including Breakers
- Ground Cable
- Ground Rod with Clamp

| System Description | QUAN. | UNIT | LABOR HOURS | COST EACH | | |
|---|---|---|---|---|---|---|
| | | | | MAT. | INST. | TOTAL |
| **100 AMP SERVICE** | | | | | | |
| Weather cap | 1.000 | Ea. | .667 | 9.15 | 26.00 | 35.15 |
| Service entrance cable | 10.000 | L.F. | .762 | 23.60 | 29.50 | 53.10 |
| Meter socket | 1.000 | Ea. | 2.500 | 32.00 | 97.00 | 129.00 |
| Ground rod with clamp | 1.000 | Ea. | 1.455 | 18.05 | 56.50 | 74.55 |
| Ground cable | 5.000 | L.F. | .250 | 6.70 | 9.70 | 16.40 |
| Panel board, 12 circuit | 1.000 | Ea. | 6.667 | 182.00 | 258.00 | 440.00 |
| TOTAL | | | 12.301 | 271.50 | 476.70 | 748.20 |
| **200 AMP SERVICE** | | | | | | |
| Weather cap | 1.000 | Ea. | 1.000 | 20.00 | 39.00 | 59.00 |
| Service entrance cable | 10.000 | L.F. | 1.143 | 57.50 | 44.30 | 101.80 |
| Meter socket | 1.000 | Ea. | 4.211 | 46.00 | 163.00 | 209.00 |
| Ground rod with clamp | 1.000 | Ea. | 1.818 | 32.50 | 70.50 | 103.00 |
| Ground cable | 10.000 | L.F. | .500 | 13.40 | 19.40 | 32.80 |
| 3/4" EMT | 5.000 | L.F. | .308 | 3.10 | 11.90 | 15.00 |
| Panel board, 24 circuit | 1.000 | Ea. | 12.308 | 490.00 | 400.00 | 890.00 |
| TOTAL | | | 21.288 | 662.50 | 748.10 | 1410.60 |
| **400 AMP SERVICE** | | | | | | |
| Weather cap | 1.000 | Ea. | 2.963 | 345.00 | 115.00 | 460.00 |
| Service entrance cable | 180.000 | L.F. | 5.760 | 275.40 | 223.20 | 498.60 |
| Meter socket | 1.000 | Ea. | 4.211 | 46.00 | 163.00 | 209.00 |
| Ground rod with clamp | 1.000 | Ea. | 2.000 | 87.00 | 77.50 | 164.50 |
| Ground cable | 20.000 | L.F. | .485 | 19.80 | 18.80 | 38.60 |
| 3/4" greenfield | 20.000 | L.F. | 1.000 | 9.40 | 38.80 | 48.20 |
| Current transformer cabinet | 1.000 | Ea. | 6.154 | 132.00 | 238.00 | 370.00 |
| Panel board, 42 circuit | 1.000 | Ea. | 33.333 | 2575.00 | 1300.00 | 3875.00 |
| TOTAL | | | 55.906 | 3489.60 | 2174.30 | 5663.90 |

**ELECTRICAL 9**

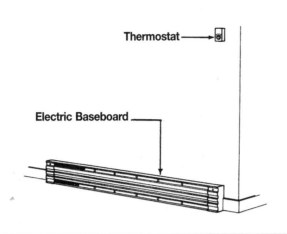

Thermostat

Electric Baseboard

| System Description | QUAN. | UNIT | LABOR HOURS | COST EACH | | |
|---|---|---|---|---|---|---|
| | | | | MAT. | INST. | TOTAL |
| **4' BASEBOARD HEATER** | | | | | | |
| Electric baseboard heater, 4' long | 1.000 | Ea. | 1.194 | 54.00 | 46.50 | 100.50 |
| Thermostat, integral | 1.000 | Ea. | .500 | 28.00 | 19.40 | 47.40 |
| Romex, 12-3 with ground | 40.000 | L.F. | 1.600 | 14.00 | 62.00 | 76.00 |
| Panel board breaker, 20 Amp | 1.000 | Ea. | .300 | 7.65 | 11.70 | 19.35 |
| TOTAL | | | 3.594 | 103.65 | 139.60 | 243.25 |
| **6' BASEBOARD HEATER** | | | | | | |
| Electric baseboard heater, 6' long | 1.000 | Ea. | 1.600 | 73.00 | 62.00 | 135.00 |
| Thermostat, integral | 1.000 | Ea. | .500 | 28.00 | 19.40 | 47.40 |
| Romex, 12-3 with ground | 40.000 | L.F. | 1.600 | 14.00 | 62.00 | 76.00 |
| Panel board breaker, 20 Amp | 1.000 | Ea. | .400 | 10.20 | 15.60 | 25.80 |
| TOTAL | | | 4.100 | 125.20 | 159.00 | 284.20 |
| **8' BASEBOARD HEATER** | | | | | | |
| Electric baseboard heater, 8' long | 1.000 | Ea. | 2.000 | 91.00 | 77.50 | 168.50 |
| Thermostat, integral | 1.000 | Ea. | .500 | 28.00 | 19.40 | 47.40 |
| Romex, 12-3 with ground | 40.000 | L.F. | 1.600 | 14.00 | 62.00 | 76.00 |
| Panel board breaker, 20 Amp | 1.000 | Ea. | .500 | 12.75 | 19.50 | 32.25 |
| TOTAL | | | 4.600 | 145.75 | 178.40 | 324.15 |
| **10' BASEBOARD HEATER** | | | | | | |
| Electric baseboard heater, 10' long | 1.000 | Ea. | 2.424 | 150.00 | 94.00 | 244.00 |
| Thermostat, integral | 1.000 | Ea. | .500 | 28.00 | 19.40 | 47.40 |
| Romex, 12-3 with ground | 40.000 | L.F. | 1.600 | 14.00 | 62.00 | 76.00 |
| Panel board breaker, 20 Amp | 1.000 | Ea. | .750 | 19.13 | 29.25 | 48.38 |
| TOTAL | | | 5.274 | 211.13 | 204.65 | 415.78 |

The costs in this system are on a cost each basis and include all
necessary conduit fittings.

| Description | QUAN. | UNIT | LABOR HOURS | COST EACH | | |
|---|---|---|---|---|---|---|
| | | | | MAT. | INST. | TOTAL |
| | | | | | | |
| | | | | | | |

| System Description | QUAN. | UNIT | LABOR HOURS | COST EACH | | |
|---|---|---|---|---|---|---|
| | | | | MAT. | INST. | TOTAL |
| **Air conditioning receptacles** | | | | | | |
| Using non-metallic sheathed cable | 1.000 | Ea. | .800 | 13.05 | 31.00 | 44.05 |
| Using BX cable | 1.000 | Ea. | .964 | 23.50 | 37.50 | 61.00 |
| Using EMT conduit | 1.000 | Ea. | 1.194 | 26.00 | 46.50 | 72.50 |
| **Disposal wiring** | | | | | | |
| Using non-metallic sheathed cable | 1.000 | Ea. | .889 | 10.45 | 34.50 | 44.95 |
| Using BX cable | 1.000 | Ea. | 1.067 | 20.50 | 41.50 | 62.00 |
| Using EMT conduit | 1.000 | Ea. | 1.333 | 24.00 | 51.50 | 75.50 |
| **Dryer circuit** | | | | | | |
| Using non-metallic sheathed cable | 1.000 | Ea. | 1.455 | 32.00 | 56.50 | 88.50 |
| Using BX cable | 1.000 | Ea. | 1.739 | 42.50 | 67.50 | 110.00 |
| Using EMT conduit | 1.000 | Ea. | 2.162 | 38.50 | 84.00 | 122.50 |
| **Duplex receptacles** | | | | | | |
| Using non-metallic sheathed cable | 1.000 | Ea. | .615 | 13.05 | 24.00 | 37.05 |
| Using BX cable | 1.000 | Ea. | .741 | 23.50 | 28.50 | 52.00 |
| Using EMT conduit | 1.000 | Ea. | .920 | 26.00 | 35.50 | 61.50 |
| **Exhaust fan wiring** | | | | | | |
| Using non-metallic sheathed cable | 1.000 | Ea. | .800 | 12.70 | 31.00 | 43.70 |
| Using BX cable | 1.000 | Ea. | .964 | 23.50 | 37.50 | 61.00 |
| Using EMT conduit | 1.000 | Ea. | 1.194 | 26.00 | 46.50 | 72.50 |
| **Furnace circuit & switch** | | | | | | |
| Using non-metallic sheathed cable | 1.000 | Ea. | 1.333 | 19.60 | 51.50 | 71.10 |
| Using BX cable | 1.000 | Ea. | 1.600 | 31.00 | 62.00 | 93.00 |
| Using EMT conduit | 1.000 | Ea. | 2.000 | 32.50 | 77.50 | 110.00 |
| **Ground fault** | | | | | | |
| Using non-metallic sheathed cable | 1.000 | Ea. | 1.000 | 42.00 | 39.00 | 81.00 |
| Using BX cable | 1.000 | Ea. | 1.212 | 52.50 | 47.00 | 99.50 |
| Using EMT conduit | 1.000 | Ea. | 1.481 | 66.00 | 57.50 | 123.50 |
| **Heater circuits** | | | | | | |
| Using non-metallic sheathed cable | 1.000 | Ea. | 1.000 | 12.30 | 39.00 | 51.30 |
| Using BX cable | 1.000 | Ea. | 1.212 | 20.00 | 47.00 | 67.00 |
| Using EMT conduit | 1.000 | Ea. | 1.481 | 22.50 | 57.50 | 80.00 |
| **Lighting wiring** | | | | | | |
| Using non-metallic sheathed cable | 1.000 | Ea. | .500 | 12.95 | 19.40 | 32.35 |
| Using BX cable | 1.000 | Ea. | .602 | 21.00 | 23.50 | 44.50 |
| Using EMT conduit | 1.000 | Ea. | .748 | 22.00 | 29.00 | 51.00 |
| **Range circuits** | | | | | | |
| Using non-metallic sheathed cable | 1.000 | Ea. | 2.000 | 65.50 | 77.50 | 143.00 |
| Using BX cable | 1.000 | Ea. | 2.424 | 91.00 | 94.00 | 185.00 |
| Using EMT conduit | 1.000 | Ea. | 2.963 | 66.00 | 115.00 | 181.00 |
| **Switches, single pole** | | | | | | |
| Using non-metallic sheathed cable | 1.000 | Ea. | .500 | 12.70 | 19.40 | 32.10 |
| Using BX cable | 1.000 | Ea. | .602 | 23.50 | 23.50 | 47.00 |
| Using EMT conduit | 1.000 | Ea. | .748 | 26.00 | 29.00 | 55.00 |
| **Switches, 3-way** | | | | | | |
| Using non-metallic sheathed cable | 1.000 | Ea. | .667 | 17.05 | 26.00 | 43.05 |
| Using BX cable | 1.000 | Ea. | .800 | 26.50 | 31.00 | 57.50 |
| Using EMT conduit | 1.000 | Ea. | 1.333 | 35.00 | 51.50 | 86.50 |
| **Water heater** | | | | | | |
| Using non-metallic sheathed cable | 1.000 | Ea. | 1.600 | 19.10 | 62.00 | 81.10 |
| Using BX cable | 1.000 | Ea. | 1.905 | 35.00 | 74.00 | 109.00 |
| Using EMT conduit | 1.000 | Ea. | 2.353 | 29.00 | 91.00 | 120.00 |
| **Weatherproof receptacle** | | | | | | |
| Using non-metallic sheathed cable | 1.000 | Ea. | 1.333 | 104.00 | 51.50 | 155.50 |
| Using BX cable | 1.000 | Ea. | 1.600 | 112.00 | 62.00 | 174.00 |
| Using EMT conduit | 1.000 | Ea. | 2.000 | 114.00 | 77.50 | 191.50 |

ELECTRICAL 9

| DESCRIPTION | QUAN. | UNIT | LABOR HOURS | COST EACH | | |
|---|---|---|---|---|---|---|
| | | | | MAT. | INST. | TOTAL |
| Fluorescent strip, 4' long, 1 light, average | 1.000 | Ea. | .941 | 29.00 | 36.50 | 65.50 |
| Deluxe | 1.000 | Ea. | 1.129 | 35.00 | 44.00 | 79.00 |
| 2 lights, average | 1.000 | Ea. | 1.000 | 31.00 | 39.00 | 70.00 |
| Deluxe | 1.000 | Ea. | 1.200 | 37.00 | 47.00 | 84.00 |
| 8' long, 1 light, average | 1.000 | Ea. | 1.194 | 43.50 | 46.50 | 90.00 |
| Deluxe | 1.000 | Ea. | 1.433 | 52.00 | 56.00 | 108.00 |
| 2 lights, average | 1.000 | Ea. | 1.290 | 52.50 | 50.00 | 102.50 |
| Deluxe | 1.000 | Ea. | 1.548 | 63.00 | 60.00 | 123.00 |
| Surface mounted, 4' x 1', economy | 1.000 | Ea. | .914 | 62.50 | 35.50 | 98.00 |
| Average | 1.000 | Ea. | 1.143 | 78.00 | 44.50 | 122.50 |
| Deluxe | 1.000 | Ea. | 1.371 | 93.50 | 53.50 | 147.00 |
| 4' x 2', economy | 1.000 | Ea. | 1.208 | 80.00 | 47.00 | 127.00 |
| Average | 1.000 | Ea. | 1.509 | 100.00 | 58.50 | 158.50 |
| Deluxe | 1.000 | Ea. | 1.811 | 120.00 | 70.00 | 190.00 |
| Recessed, 4'x 1', 2 lamps, economy | 1.000 | Ea. | 1.123 | 41.00 | 43.50 | 84.50 |
| Average | 1.000 | Ea. | 1.404 | 51.00 | 54.50 | 105.50 |
| Deluxe | 1.000 | Ea. | 1.684 | 61.00 | 65.50 | 126.50 |
| 4' x 2', 4' lamps, economy | 1.000 | Ea. | 1.362 | 49.50 | 53.00 | 102.50 |
| Average | 1.000 | Ea. | 1.702 | 62.00 | 66.00 | 128.00 |
| Deluxe | 1.000 | Ea. | 2.043 | 74.50 | 79.00 | 153.50 |
| Incandescent, exterior, 150W, single spot | 1.000 | Ea. | .500 | 18.60 | 19.40 | 38.00 |
| Double spot | 1.000 | Ea. | 1.167 | 75.50 | 45.50 | 121.00 |
| Recessed, 100W, economy | 1.000 | Ea. | .800 | 49.00 | 31.00 | 80.00 |
| Average | 1.000 | Ea. | 1.000 | 61.00 | 39.00 | 100.00 |
| Deluxe | 1.000 | Ea. | 1.200 | 73.00 | 47.00 | 120.00 |
| 150W, economy | 1.000 | Ea. | .800 | 71.00 | 31.00 | 102.00 |
| Average | 1.000 | Ea. | 1.000 | 89.00 | 39.00 | 128.00 |
| Deluxe | 1.000 | Ea. | 1.200 | 107.00 | 47.00 | 154.00 |
| Surface mounted, 60W, economy | 1.000 | Ea. | .800 | 37.50 | 31.00 | 68.50 |
| Average | 1.000 | Ea. | 1.000 | 42.00 | 39.00 | 81.00 |
| Deluxe | 1.000 | Ea. | 1.194 | 60.50 | 46.50 | 107.00 |
| Mercury vapor, recessed, 2' x 2' with 250W DX lamp | 1.000 | Ea. | 2.500 | 289.00 | 97.00 | 386.00 |
| 2' x 2' with 400W DX lamp | 1.000 | Ea. | 2.759 | 295.00 | 107.00 | 402.00 |
| Surface mounted, 2' x 2' with 250W DX lamp | 1.000 | Ea. | 2.963 | 289.00 | 115.00 | 404.00 |
| 2' x 2' with 400W DX lamp | 1.000 | Ea. | 3.333 | 325.00 | 129.00 | 454.00 |
| High bay, single unit, 400W DX lamp | 1.000 | Ea. | 3.478 | 286.00 | 135.00 | 421.00 |
| Twin unit, 400W DX lamp | 1.000 | Ea. | 5.000 | 485.00 | 194.00 | 679.00 |
| Low bay, 250W DX lamp | 1.000 | Ea. | 2.500 | 295.00 | 97.00 | 392.00 |
| Metal halide, recessed 2' x 2' 250W | 1.000 | Ea. | 2.500 | 283.00 | 97.00 | 380.00 |
| 2' x 2', 400W | 1.000 | Ea. | 2.759 | 330.00 | 107.00 | 437.00 |
| Surface mounted, 2' x 2', 250W | 1.000 | Ea. | 2.963 | 283.00 | 115.00 | 398.00 |
| 2' x 2', 400W | 1.000 | Ea. | 3.333 | 335.00 | 129.00 | 464.00 |
| High bay, single, unit, 400W | 1.000 | Ea. | 3.478 | 395.00 | 135.00 | 530.00 |
| Twin unit, 400W | 1.000 | Ea. | 5.000 | 780.00 | 194.00 | 974.00 |
| Low bay, 250W | 1.000 | Ea. | 2.500 | 375.00 | 97.00 | 472.00 |
| | | | | | | |
| | | | | | | |
| | | | | | | |
| | | | | | | |
| | | | | | | |
| | | | | | | |

**ELECTRICAL**

**9**

For information about Means Estimating Seminars, see yellow pages 11 and 12 in back of book

Costs shown in *Means cost data publications* are based on National Averages for materials and installation. To adjust these costs to a specific location, simply multiply the base cost by the factor for that city. The data is arranged alphabetically by state and postal zip code numbers. For a city not listed, use the factor for a nearby city with similar economic characteristics.

| STATE/ZIP | CITY | Residential | Commercial |
|---|---|---|---|
| **ALABAMA** | | | |
| 350-352 | Birmingham | .85 | .86 |
| 354 | Tuscaloosa | .80 | .78 |
| 355 | Jasper | .76 | .77 |
| 356 | Decatur | .79 | .80 |
| 357-358 | Huntsville | .81 | .82 |
| 359 | Gadsden | .80 | .81 |
| 360-361 | Montgomery | .82 | .80 |
| 362 | Anniston | .73 | .74 |
| 363 | Dothan | .79 | .77 |
| 364 | Evergreen | .79 | .77 |
| 365-366 | Mobile | .81 | .82 |
| 367 | Selma | .79 | .77 |
| 368 | Phenix City | .82 | .80 |
| 369 | Butler | .79 | .77 |
| **ALASKA** | | | |
| 995-996 | Anchorage | 1.25 | 1.24 |
| 997 | Fairbanks | 1.25 | 1.24 |
| 998 | Juneau | 1.24 | 1.23 |
| 999 | Ketchikan | 1.30 | 1.29 |
| **ARIZONA** | | | |
| 850,853 | Phoenix | .92 | .89 |
| 852 | Mesa/Tempe | .87 | .84 |
| 855 | Globe | .88 | .85 |
| 856-857 | Tucson | .90 | .87 |
| 859 | Show Low | .89 | .85 |
| 860 | Flagstaff | .92 | .88 |
| 863 | Prescott | .90 | .86 |
| 864 | Kingman | .89 | .85 |
| 865 | Chambers | .88 | .84 |
| **ARKANSAS** | | | |
| 716 | Pine Bluff | .80 | .80 |
| 717 | Camden | .70 | .70 |
| 718 | Texarkana | .74 | .74 |
| 719 | Hot Springs | .69 | .69 |
| 720-722 | Little Rock | .81 | .81 |
| 723 | West Memphis | .79 | .79 |
| 724 | Jonesboro | .79 | .79 |
| 725 | Batesville | .75 | .75 |
| 726 | Harrison | .76 | .76 |
| 727 | Fayetteville | .69 | .66 |
| 728 | Russellville | .77 | .74 |
| 729 | Fort Smith | .83 | .80 |
| **CALIFORNIA** | | | |
| 900-902 | Los Angeles | 1.08 | 1.08 |
| 903-905 | Inglewood | 1.06 | 1.06 |
| 906-908 | Long Beach | 1.07 | 1.07 |
| 910-912 | Pasadena | 1.07 | 1.07 |
| 913-916 | Van Nuys | 1.09 | 1.09 |
| 917-918 | Alhambra | 1.08 | 1.08 |
| 919-921 | San Diego | 1.10 | 1.06 |
| 922 | Palm Springs | 1.09 | 1.05 |
| 923-924 | San Bernardino | 1.08 | 1.04 |
| 925 | Riverside | 1.11 | 1.07 |
| 926-927 | Santa Ana | 1.09 | 1.06 |
| 928 | Anaheim | 1.10 | 1.08 |
| 930 | Oxnard | 1.13 | 1.08 |
| 931 | Santa Barbara | 1.11 | 1.08 |
| 932-933 | Bakersfield | 1.11 | 1.06 |
| 934 | San Luis Obispo | 1.14 | 1.08 |
| 935 | Mojave | 1.09 | 1.05 |
| 936-938 | Fresno | 1.12 | 1.08 |
| 939 | Salinas | 1.12 | 1.12 |
| 940-941 | San Francisco | 1.21 | 1.24 |
| 942,956-958 | Sacramento | 1.11 | 1.10 |
| 943 | Palo Alto | 1.15 | 1.18 |
| 944 | San Mateo | 1.16 | 1.19 |
| 945 | Vallejo | 1.11 | 1.14 |
| 946 | Oakland | 1.16 | 1.19 |
| 947 | Berkeley | 1.16 | 1.18 |
| 948 | Richmond | 1.14 | 1.17 |
| 949 | San Rafael | 1.26 | 1.20 |
| 950 | Santa Cruz | 1.16 | 1.14 |
| 951 | San Jose | 1.22 | 1.20 |
| 952 | Stockton | 1.13 | 1.09 |
| 953 | Modesto | 1.13 | 1.09 |

| STATE/ZIP | CITY | Residential | Commercial |
|---|---|---|---|
| **CALIFORNIA (CONT'D)** | | | |
| 954 | Santa Rosa | 1.14 | 1.17 |
| 955 | Eureka | 1.10 | 1.09 |
| 959 | Marysville | 1.10 | 1.09 |
| 960 | Redding | 1.11 | 1.10 |
| 961 | Susanville | 1.11 | 1.10 |
| **COLORADO** | | | |
| 800-802 | Denver | .99 | .95 |
| 803 | Boulder | .88 | .84 |
| 804 | Golden | .97 | .93 |
| 805 | Fort Collins | .98 | .92 |
| 806 | Greeley | .90 | .84 |
| 807 | Fort Morgan | .97 | .91 |
| 808-809 | Colorado Springs | .94 | .92 |
| 810 | Pueblo | .94 | .92 |
| 811 | Alamosa | .89 | .87 |
| 812 | Salida | .89 | .87 |
| 813 | Durango | .88 | .86 |
| 814 | Montrose | .86 | .84 |
| 815 | Grand Junction | .90 | .85 |
| 816 | Glenwood Springs | .95 | .91 |
| **CONNECTICUT** | | | |
| 060 | New Britain | 1.04 | 1.05 |
| 061 | Hartford | 1.04 | 1.05 |
| 062 | Willimantic | 1.03 | 1.04 |
| 063 | New London | 1.05 | 1.04 |
| 064 | Meriden | 1.03 | 1.04 |
| 065 | New Haven | 1.04 | 1.05 |
| 066 | Bridgeport | 1.02 | 1.05 |
| 067 | Waterbury | 1.05 | 1.05 |
| 068 | Norwalk | 1.01 | 1.05 |
| 069 | Stamford | 1.04 | 1.08 |
| **D.C.** | | | |
| 200-205 | Washington | .93 | .95 |
| **DELAWARE** | | | |
| 197 | Newark | 1.00 | 1.01 |
| 198 | Wilmington | 1.00 | 1.01 |
| 199 | Dover | 1.00 | 1.01 |
| **FLORIDA** | | | |
| 320,322 | Jacksonville | .83 | .82 |
| 321 | Daytona Beach | .87 | .86 |
| 323 | Tallahassee | .75 | .77 |
| 324 | Panama City | .70 | .72 |
| 325 | Pensacola | .84 | .82 |
| 326,344 | Gainesville | .84 | .81 |
| 327-328,347 | Orlando | .86 | .84 |
| 329 | Melbourne | .90 | .89 |
| 330-332,340 | Miami | .83 | .85 |
| 333 | Fort Lauderdale | .83 | .85 |
| 334,349 | West Palm Beach | .86 | .83 |
| 335-336,346 | Tampa | .80 | .82 |
| 337 | St. Petersburg | .81 | .83 |
| 338 | Lakeland | .79 | .81 |
| 339,341 | Fort Myers | .79 | .79 |
| 342 | Sarasota | .78 | .80 |
| **GEORGIA** | | | |
| 300-303,399 | Atlanta | .85 | .90 |
| 304 | Statesboro | .72 | .74 |
| 305 | Gainesville | .76 | .80 |
| 306 | Athens | .77 | .82 |
| 307 | Dalton | .68 | .67 |
| 308-309 | Augusta | .76 | .78 |
| 310-312 | Macon | .81 | .81 |
| 313-314 | Savannah | .80 | .81 |
| 315 | Waycross | .74 | .74 |
| 316 | Valdosta | .76 | .76 |
| 317 | Albany | .77 | .79 |
| 318-319 | Columbus | .78 | .78 |
| **HAWAII** | | | |
| 967 | Hilo | 1.27 | 1.23 |
| 968 | Honolulu | 1.27 | 1.23 |

| STATE/ZIP | CITY | Residential | Commercial |
|---|---|---|---|
| **STATES & POSS.** | | | |
| 969 | Guam | 1.37 | 1.33 |
| **IDAHO** | | | |
| 832 | Pocatello | .94 | .93 |
| 833 | Twin Falls | .79 | .78 |
| 834 | Idaho Falls | .83 | .82 |
| 835 | Lewiston | 1.09 | 1.01 |
| 836-837 | Boise | .94 | .93 |
| 838 | Coeur d'Alene | .95 | .88 |
| **ILLINOIS** | | | |
| 600-603 | North Suburban | 1.11 | 1.09 |
| 604 | Joliet | 1.11 | 1.10 |
| 605 | South Suburban | 1.10 | 1.09 |
| 606 | Chicago | 1.13 | 1.12 |
| 609 | Kankakee | 1.00 | 1.00 |
| 610-611 | Rockford | 1.05 | 1.04 |
| 612 | Rock Island | 1.06 | .97 |
| 613 | La Salle | 1.06 | .98 |
| 614 | Galesburg | 1.09 | 1.01 |
| 615-616 | Peoria | 1.09 | 1.02 |
| 617 | Bloomington | 1.05 | 1.00 |
| 618-619 | Champaign | 1.04 | 1.01 |
| 620-622 | East St. Louis | 1.00 | 1.00 |
| 623 | Quincy | .99 | .97 |
| 624 | Effingham | 1.02 | .99 |
| 625 | Decatur | 1.01 | .98 |
| 626-627 | Springfield | 1.01 | .98 |
| 628 | Centralia | .99 | .99 |
| 629 | Carbondale | .97 | .97 |
| **INDIANA** | | | |
| 460 | Anderson | .95 | .93 |
| 461-462 | Indianapolis | .98 | .96 |
| 463-464 | Gary | 1.04 | 1.02 |
| 465-466 | South Bend | .94 | .92 |
| 467-468 | Fort Wayne | .92 | .93 |
| 469 | Kokomo | .93 | .92 |
| 470 | Lawrenceburg | .93 | .90 |
| 471 | New Albany | .93 | .89 |
| 472 | Columbus | .96 | .93 |
| 473 | Muncie | .94 | .93 |
| 474 | Bloomington | .96 | .93 |
| 475 | Washington | .93 | .93 |
| 476-477 | Evansville | .95 | .95 |
| 478 | Terre Haute | .96 | .95 |
| 479 | Lafayette | .92 | .92 |
| **IOWA** | | | |
| 500-503,509 | Des Moines | .97 | .93 |
| 504 | Mason City | .86 | .81 |
| 505 | Fort Dodge | .84 | .78 |
| 506-507 | Waterloo | .88 | .82 |
| 508 | Creston | .89 | .84 |
| 510-511 | Sioux City | .95 | .88 |
| 512 | Sibley | .80 | .78 |
| 513 | Spencer | .80 | .78 |
| 514 | Carroll | .84 | .80 |
| 515 | Council Bluffs | .96 | .89 |
| 516 | Shenandoah | .82 | .77 |
| 520 | Dubuque | .99 | .88 |
| 521 | Decorah | .88 | .79 |
| 522-524 | Cedar Rapids | 1.01 | .92 |
| 525 | Ottumwa | .94 | .86 |
| 526 | Burlington | .92 | .86 |
| 527-528 | Davenport | .98 | .96 |
| **KANSAS** | | | |
| 660-662 | Kansas City | .96 | .94 |
| 664-666 | Topeka | .86 | .85 |
| 667 | Fort Scott | .85 | .83 |
| 668 | Emporia | .81 | .81 |
| 669 | Belleville | .87 | .81 |
| 670-672 | Wichita | .89 | .86 |
| 673 | Independence | .82 | .79 |
| 674 | Salina | .85 | .81 |
| 675 | Hutchinson | .79 | .76 |
| 676 | Hays | .84 | .80 |
| 677 | Colby | .85 | .81 |
| 678 | Dodge City | .84 | .81 |
| 679 | Liberal | .78 | .75 |
| **KENTUCKY** | | | |
| 400-402 | Louisville | .95 | .92 |
| 403-405 | Lexington | .87 | .84 |

| STATE/ZIP | CITY | Residential | Commercial |
|---|---|---|---|
| **KENTUCKY (CONT'D)** | | | |
| 406 | Frankfort | .92 | .86 |
| 407-409 | Corbin | .78 | .73 |
| 410 | Covington | .98 | .95 |
| 411-412 | Ashland | .96 | .97 |
| 413-414 | Campton | .77 | .73 |
| 415-416 | Pikeville | .82 | .83 |
| 417-418 | Hazard | .76 | .73 |
| 420 | Paducah | .97 | .92 |
| 421-422 | Bowling Green | .96 | .91 |
| 423 | Owensboro | .91 | .89 |
| 424 | Henderson | .94 | .92 |
| 425-426 | Somerset | .75 | .72 |
| 427 | Elizabethtown | .94 | .90 |
| **LOUISIANA** | | | |
| 700-701 | New Orleans | .86 | .85 |
| 703 | Thibodaux | .85 | .85 |
| 704 | Hammond | .84 | .83 |
| 705 | Lafayette | .84 | .81 |
| 706 | Lake Charles | .83 | .83 |
| 707-708 | Baton Rouge | .82 | .81 |
| 710-711 | Shreveport | .81 | .81 |
| 712 | Monroe | .79 | .79 |
| 713-714 | Alexandria | .78 | .78 |
| **MAINE** | | | |
| 039 | Kittery | .86 | .88 |
| 040-041 | Portland | .91 | .93 |
| 042 | Lewiston | .92 | .93 |
| 043 | Augusta | .88 | .88 |
| 044 | Bangor | .93 | .93 |
| 045 | Bath | .89 | .89 |
| 046 | Machias | .87 | .87 |
| 047 | Houlton | .89 | .89 |
| 048 | Rockland | .86 | .86 |
| 049 | Waterville | .86 | .85 |
| **MARYLAND** | | | |
| 206 | Waldorf | .87 | .87 |
| 207-208 | College Park | .90 | .90 |
| 209 | Silver Spring | .89 | .89 |
| 210-212 | Baltimore | .91 | .91 |
| 214 | Annapolis | .89 | .90 |
| 215 | Cumberland | .87 | .88 |
| 216 | Easton | .73 | .73 |
| 217 | Hagerstown | .90 | .88 |
| 218 | Salisbury | .76 | .77 |
| 219 | Elkton | .82 | .83 |
| **MASSACHUSETTS** | | | |
| 010-011 | Springfield | 1.04 | 1.02 |
| 012 | Pittsfield | .99 | .99 |
| 013 | Greenfield | 1.02 | 1.00 |
| 014 | Fitchburg | 1.08 | 1.04 |
| 015-016 | Worcester | 1.10 | 1.06 |
| 017 | Framingham | 1.06 | 1.06 |
| 018 | Lowell | 1.08 | 1.08 |
| 019 | Lawrence | 1.09 | 1.09 |
| 020-022, 024 | Boston | 1.14 | 1.15 |
| 023 | Brockton | 1.06 | 1.08 |
| 025 | Buzzards Bay | 1.02 | 1.04 |
| 026 | Hyannis | 1.04 | 1.05 |
| 027 | New Bedford | 1.06 | 1.07 |
| **MICHIGAN** | | | |
| 480,483 | Royal Oak | 1.03 | 1.02 |
| 481 | Ann Arbor | 1.04 | 1.03 |
| 482 | Detroit | 1.07 | 1.06 |
| 484-485 | Flint | .99 | 1.00 |
| 486 | Saginaw | .97 | .98 |
| 487 | Bay City | .96 | .97 |
| 488-489 | Lansing | 1.01 | .98 |
| 490 | Battle Creek | 1.01 | .95 |
| 491 | Kalamazoo | 1.00 | .94 |
| 492 | Jackson | .99 | .96 |
| 493,495 | Grand Rapids | .88 | .85 |
| 494 | Muskegon | .95 | .92 |
| 496 | Traverse City | .87 | .84 |
| 497 | Gaylord | .87 | .88 |
| 498-499 | Iron Mountain | .98 | .95 |
| **MINNESOTA** | | | |
| 550-551 | Saint Paul | 1.09 | 1.07 |
| 553-555 | Minneapolis | 1.11 | 1.08 |

| STATE/ZIP | CITY | Residential | Commercial |
|---|---|---|---|
| 556-558 | Duluth | 1.04 | 1.05 |
| 559 | Rochester | 1.03 | .99 |
| 560 | Mankato | 1.00 | .99 |
| 561 | Windom | .90 | .89 |
| 562 | Willmar | .93 | .92 |
| 563 | St. Cloud | 1.11 | 1.03 |
| 564 | Brainerd | 1.06 | .99 |
| 565 | Detroit Lakes | .88 | .95 |
| 566 | Bemidji | .91 | .98 |
| 567 | Thief River Falls | .87 | .94 |
| **MISSISSIPPI** | | | |
| 386 | Clarksdale | .70 | .66 |
| 387 | Greenville | .80 | .77 |
| 388 | Tupelo | .71 | .71 |
| 389 | Greenwood | .72 | .68 |
| 390-392 | Jackson | .80 | .76 |
| 393 | Meridian | .76 | .75 |
| 394 | Laurel | .72 | .69 |
| 395 | Biloxi | .84 | .80 |
| 396 | McComb | .72 | .70 |
| 397 | Columbus | .70 | .71 |
| **MISSOURI** | | | |
| 630-631 | St. Louis | 1.00 | 1.03 |
| 633 | Bowling Green | .92 | .94 |
| 634 | Hannibal | .99 | .93 |
| 635 | Kirksville | .86 | .90 |
| 636 | Flat River | .94 | .97 |
| 637 | Cape Girardeau | .93 | .96 |
| 638 | Sikeston | .90 | .92 |
| 639 | Poplar Bluff | .90 | .92 |
| 640-641 | Kansas City | 1.04 | 1.01 |
| 644-645 | St. Joseph | .90 | .94 |
| 646 | Chillicothe | .82 | .86 |
| 647 | Harrisonville | .98 | .96 |
| 648 | Joplin | .84 | .86 |
| 650-651 | Jefferson City | .98 | .92 |
| 652 | Columbia | .99 | .93 |
| 653 | Sedalia | .99 | .92 |
| 654-655 | Rolla | .95 | .89 |
| 656-658 | Springfield | .86 | .88 |
| **MONTANA** | | | |
| 590-591 | Billings | .93 | .91 |
| 592 | Wolf Point | .92 | .90 |
| 593 | Miles City | .91 | .89 |
| 594 | Great Falls | .92 | .91 |
| 595 | Havre | .90 | .89 |
| 596 | Helena | .91 | .90 |
| 597 | Butte | .90 | .89 |
| 598 | Missoula | .89 | .88 |
| 599 | Kalispell | .88 | .87 |
| **NEBRASKA** | | | |
| 680-681 | Omaha | .92 | .91 |
| 683-685 | Lincoln | .88 | .83 |
| 686 | Columbus | .74 | .73 |
| 687 | Norfolk | .84 | .83 |
| 688 | Grand Island | .88 | .83 |
| 689 | Hastings | .82 | .78 |
| 690 | Mccook | .79 | .75 |
| 691 | North Platte | .86 | .82 |
| 692 | Valentine | .77 | .74 |
| 693 | Alliance | .75 | .71 |
| **NEVADA** | | | |
| 889-891 | Las Vegas | 1.05 | 1.04 |
| 893 | Ely | .93 | .94 |
| 894-895 | Reno | .95 | 1.00 |
| 897 | Carson City | .96 | .99 |
| 898 | Elko | .90 | .93 |
| **NEW HAMPSHIRE** | | | |
| 030 | Nashua | .94 | .95 |
| 031 | Manchester | .94 | .95 |
| 032-033 | Concord | .93 | .94 |
| 034 | Keene | .78 | .79 |
| 035 | Littleton | .82 | .83 |
| 036 | Charleston | .77 | .77 |
| 037 | Claremont | .76 | .77 |
| 038 | Portsmouth | .93 | .92 |

| STATE/ZIP | CITY | Residential | Commercial |
|---|---|---|---|
| **NEW JERSEY** | | | |
| 070-071 | Newark | 1.14 | 1.12 |
| 072 | Elizabeth | 1.09 | 1.08 |
| 073 | Jersey City | 1.11 | 1.10 |
| 074-075 | Paterson | 1.12 | 1.12 |
| 076 | Hackensack | 1.10 | 1.10 |
| 077 | Long Branch | 1.10 | 1.08 |
| 078 | Dover | 1.11 | 1.09 |
| 079 | Summit | 1.08 | 1.06 |
| 080,083 | Vineland | 1.11 | 1.08 |
| 081 | Camden | 1.11 | 1.08 |
| 082,084 | Atlantic City | 1.11 | 1.08 |
| 085-086 | Trenton | 1.12 | 1.10 |
| 087 | Point Pleasant | 1.10 | 1.08 |
| 088-089 | New Brunswick | 1.12 | 1.10 |
| **NEW MEXICO** | | | |
| 870-872 | Albuquerque | .88 | .90 |
| 873 | Gallup | .88 | .91 |
| 874 | Farmington | .88 | .91 |
| 875 | Santa Fe | .88 | .90 |
| 877 | Las Vegas | .88 | .90 |
| 878 | Socorro | .87 | .89 |
| 879 | Truth/Consequences | .87 | .87 |
| 880 | Las Cruces | .84 | .84 |
| 881 | Clovis | .89 | .89 |
| 882 | Roswell | .90 | .90 |
| 883 | Carrizozo | .91 | .91 |
| 884 | Tucumcari | .90 | .90 |
| **NEW YORK** | | | |
| 100-102 | New York | 1.35 | 1.35 |
| 103 | Staten Island | 1.31 | 1.31 |
| 104 | Bronx | 1.30 | 1.30 |
| 105 | Mount Vernon | 1.20 | 1.20 |
| 106 | White Plains | 1.19 | 1.19 |
| 107 | Yonkers | 1.22 | 1.22 |
| 108 | New Rochelle | 1.20 | 1.20 |
| 109 | Suffern | 1.14 | 1.14 |
| 110 | Queens | 1.30 | 1.30 |
| 111 | Long Island City | 1.31 | 1.31 |
| 112 | Brooklyn | 1.31 | 1.31 |
| 113 | Flushing | 1.32 | 1.32 |
| 114 | Jamaica | 1.30 | 1.30 |
| 115,117,118 | Hicksville | 1.26 | 1.26 |
| 116 | Far Rockaway | 1.32 | 1.32 |
| 119 | Riverhead | 1.27 | 1.27 |
| 120-122 | Albany | .97 | .97 |
| 123 | Schenectady | .98 | .98 |
| 124 | Kingston | 1.11 | 1.09 |
| 125-126 | Poughkeepsie | 1.13 | 1.11 |
| 127 | Monticello | 1.09 | 1.07 |
| 128 | Glens Falls | .95 | .93 |
| 129 | Plattsburgh | .95 | .93 |
| 130-132 | Syracuse | .99 | .97 |
| 133-135 | Utica | .91 | .94 |
| 136 | Watertown | .92 | .95 |
| 137-139 | Binghamton | .94 | .94 |
| 140-142 | Buffalo | 1.05 | 1.02 |
| 143 | Niagara Falls | 1.07 | 1.03 |
| 144-146 | Rochester | .99 | 1.00 |
| 147 | Jamestown | .98 | .94 |
| 148-149 | Elmira | .95 | .93 |
| **NORTH CAROLINA** | | | |
| 270,272-274 | Greensboro | .75 | .76 |
| 271 | Winston-Salem | .74 | .75 |
| 275-276 | Raleigh | .76 | .76 |
| 277 | Durham | .75 | .76 |
| 278 | Rocky Mount | .68 | .68 |
| 279 | Elizabeth City | .70 | .70 |
| 280 | Gastonia | .74 | .75 |
| 281-282 | Charlotte | .74 | .75 |
| 283 | Fayetteville | .75 | .75 |
| 284 | Wilmington | .73 | .75 |
| 285 | Kinston | .67 | .67 |
| 286 | Hickory | .66 | .67 |
| 287-288 | Asheville | .73 | .75 |
| 289 | Murphy | .66 | .67 |
| **NORTH DAKOTA** | | | |
| 580-581 | Fargo | .79 | .84 |
| 582 | Grand Forks | .77 | .82 |
| 583 | Devils Lake | .76 | .81 |
| 584 | Jamestown | .76 | .81 |
| 585 | Bismarck | .80 | .84 |

# Location Factors

| STATE/ZIP | CITY | Residential | Commercial |
|---|---|---|---|
| **NORTH DAKOTA (CONT'D)** | | | |
| 586 | Dickinson | .81 | .85 |
| 587 | Minot | .82 | .87 |
| 588 | Williston | .76 | .80 |
| **OHIO** | | | |
| 430-432 | Columbus | .98 | .96 |
| 433 | Marion | .92 | .93 |
| 434-436 | Toledo | 1.02 | 1.01 |
| 437-438 | Zanesville | .93 | .92 |
| 439 | Steubenville | .97 | .97 |
| 440 | Lorain | 1.05 | .98 |
| 441 | Cleveland | 1.09 | 1.03 |
| 442-443 | Akron | 1.02 | 1.01 |
| 444-445 | Youngstown | 1.01 | .98 |
| 446-447 | Canton | .97 | .96 |
| 448-449 | Mansfield | .96 | .94 |
| 450 | Hamilton | 1.00 | .94 |
| 451-452 | Cincinnati | 1.00 | .94 |
| 453-454 | Dayton | .94 | .93 |
| 455 | Springfield | .95 | .93 |
| 456 | Chillicothe | 1.02 | .96 |
| 457 | Athens | .92 | .91 |
| 458 | Lima | .96 | .95 |
| **OKLAHOMA** | | | |
| 730-731 | Oklahoma City | .82 | .84 |
| 734 | Ardmore | .83 | .82 |
| 735 | Lawton | .84 | .83 |
| 736 | Clinton | .80 | .82 |
| 737 | Enid | .83 | .82 |
| 738 | Woodward | .82 | .81 |
| 739 | Guymon | .69 | .68 |
| 740-741 | Tulsa | .84 | .81 |
| 743 | Miami | .86 | .83 |
| 744 | Muskogee | .75 | .73 |
| 745 | Mcalester | .76 | .77 |
| 746 | Ponca City | .82 | .81 |
| 747 | Durant | .79 | .81 |
| 748 | Shawnee | .79 | .81 |
| 749 | Poteau | .85 | .81 |
| **OREGON** | | | |
| 970-972 | Portland | 1.08 | 1.06 |
| 973 | Salem | 1.06 | 1.05 |
| 974 | Eugene | 1.05 | 1.04 |
| 975 | Medford | 1.05 | 1.04 |
| 976 | Klamath Falls | 1.05 | 1.04 |
| 977 | Bend | 1.06 | 1.05 |
| 978 | Pendleton | 1.03 | 1.01 |
| 979 | Vale | .98 | .96 |
| **PENNSYLVANIA** | | | |
| 150-152 | Pittsburgh | 1.04 | 1.02 |
| 153 | Washington | 1.02 | 1.00 |
| 154 | Uniontown | 1.01 | .99 |
| 155 | Bedford | 1.03 | .96 |
| 156 | Greensburg | 1.02 | 1.00 |
| 157 | Indiana | 1.05 | .98 |
| 158 | Dubois | 1.04 | .97 |
| 159 | Johnstown | 1.04 | .97 |
| 160 | Butler | 1.01 | .98 |
| 161 | New Castle | 1.01 | .98 |
| 162 | Kittanning | 1.02 | .99 |
| 163 | Oil City | .91 | .95 |
| 164-165 | Erie | .98 | .97 |
| 166 | Altoona | 1.04 | .96 |
| 167 | Bradford | .99 | .97 |
| 168 | State College | .96 | .96 |
| 169 | Wellsboro | .93 | .94 |
| 170-171 | Harrisburg | .98 | .97 |
| 172 | Chambersburg | .96 | .95 |
| 173-174 | York | .97 | .95 |
| 175-176 | Lancaster | .95 | .93 |
| 177 | Williamsport | .91 | .90 |
| 178 | Sunbury | .95 | .94 |
| 179 | Pottsville | .95 | .94 |
| 180 | Lehigh Valley | 1.04 | 1.03 |
| 181 | Allentown | 1.01 | 1.00 |
| 182 | Hazleton | .97 | .96 |
| 183 | Stroudsburg | 1.01 | 1.00 |
| 184-185 | Scranton | .95 | .98 |
| 186-187 | Wilkes-Barre | .93 | .96 |
| 188 | Montrose | .93 | .96 |
| 189 | Doylestown | .94 | 1.06 |

| STATE/ZIP | CITY | Residential | Commercial |
|---|---|---|---|
| **PENNSYLVANIA (CONT'D)** | | | |
| 190-191 | Philadelphia | 1.13 | 1.11 |
| 193 | Westchester | 1.08 | 1.06 |
| 194 | Norristown | 1.09 | 1.07 |
| 195-196 | Reading | .97 | .98 |
| **PUERTO RICO** | | | |
| 009 | San Juan | .86 | .86 |
| **RHODE ISLAND** | | | |
| 028 | Newport | 1.02 | 1.04 |
| 029 | Providence | 1.02 | 1.04 |
| **SOUTH CAROLINA** | | | |
| 290-292 | Columbia | .72 | .75 |
| 293 | Spartanburg | .71 | .73 |
| 294 | Charleston | .73 | .75 |
| 295 | Florence | .71 | .73 |
| 296 | Greenville | .70 | .73 |
| 297 | Rock Hill | .64 | .67 |
| 298 | Aiken | .80 | .83 |
| 299 | Beaufort | .68 | .70 |
| **SOUTH DAKOTA** | | | |
| 570-571 | Sioux Falls | .88 | .81 |
| 572 | Watertown | .84 | .78 |
| 573 | Mitchell | .83 | .77 |
| 574 | Aberdeen | .84 | .78 |
| 575 | Pierre | .84 | .79 |
| 576 | Mobridge | .84 | .77 |
| 577 | Rapid City | .85 | .79 |
| **TENNESSEE** | | | |
| 370-372 | Nashville | .86 | .86 |
| 373-374 | Chattanooga | .82 | .81 |
| 375,380-381 | Memphis | .84 | .84 |
| 376 | Johnson City | .80 | .79 |
| 377-379 | Knoxville | .80 | .80 |
| 382 | Mckenzie | .69 | .69 |
| 383 | Jackson | .68 | .75 |
| 384 | Columbia | .76 | .76 |
| 385 | Cookeville | .68 | .68 |
| **TEXAS** | | | |
| 750 | Mckinney | .88 | .81 |
| 751 | Waxahackie | .82 | .82 |
| 752-753 | Dallas | .89 | .85 |
| 754 | Greenville | .78 | .72 |
| 755 | Texarkana | .87 | .77 |
| 756 | Longview | .84 | .74 |
| 757 | Tyler | .91 | .80 |
| 758 | Palestine | .72 | .72 |
| 759 | Lufkin | .76 | .76 |
| 760-761 | Fort Worth | .83 | .82 |
| 762 | Denton | .87 | .78 |
| 763 | Wichita Falls | .80 | .80 |
| 764 | Eastland | .73 | .72 |
| 765 | Temple | .77 | .76 |
| 766-767 | Waco | .81 | .80 |
| 768 | Brownwood | .72 | .71 |
| 769 | San Angelo | .79 | .76 |
| 770-772 | Houston | .87 | .88 |
| 773 | Huntsville | .73 | .73 |
| 774 | Wharton | .75 | .76 |
| 775 | Galveston | .86 | .87 |
| 776-777 | Beaumont | .82 | .83 |
| 778 | Bryan | .81 | .82 |
| 779 | Victoria | .78 | .78 |
| 780 | Laredo | .76 | .77 |
| 781-782 | San Antonio | .82 | .83 |
| 783-784 | Corpus Christi | .80 | .79 |
| 785 | Mc Allen | .78 | .76 |
| 786-787 | Austin | .78 | .81 |
| 788 | Del Rio | .68 | .68 |
| 789 | Giddings | .72 | .71 |
| 790-791 | Amarillo | .81 | .81 |
| 792 | Childress | .75 | .78 |
| 793-794 | Lubbock | .78 | .80 |
| 795-796 | Abilene | .79 | .79 |
| 797 | Midland | .78 | .79 |
| 798-799,885 | El Paso | .79 | .78 |
| **UTAH** | | | |
| 840-841 | Salt Lake City | .90 | .89 |
| 842,844 | Ogden | .90 | .88 |

# Location Factors

| STATE/ZIP | CITY | Residential | Commercial |
|---|---|---|---|
| **UTAH(CONT'D)** | | | |
| 843 | Logan | .91 | .89 |
| 845 | Price | .81 | .80 |
| 846-847 | Provo | .90 | .89 |
| | | | |
| **VERMONT** | | | |
| 050 | White River Jct. | .73 | .72 |
| 051 | Bellows Falls | .73 | .73 |
| 052 | Bennington | .72 | .71 |
| 053 | Brattleboro | .74 | .73 |
| 054 | Burlington | .85 | .86 |
| 056 | Montpelier | .84 | .85 |
| 057 | Rutland | .87 | .86 |
| 058 | St. Johnsbury | .74 | .75 |
| 059 | Guildhall | .73 | .74 |
| | | | |
| **VIRGINIA** | | | |
| 220-221 | Fairfax | .89 | .90 |
| 222 | Arlington | .89 | .90 |
| 223 | Alexandria | .89 | .90 |
| 224-225 | Fredericksburg | .83 | .84 |
| 226 | Winchester | .78 | .79 |
| 227 | Culpeper | .78 | .79 |
| 228 | Harrisonburg | .75 | .75 |
| 229 | Charlottesville | .83 | .82 |
| 230-232 | Richmond | .86 | .84 |
| 233-235 | Norfolk | .82 | .82 |
| 236 | Newport News | .82 | .81 |
| 237 | Portsmouth | .81 | .81 |
| 238 | Petersburg | .86 | .84 |
| 239 | Farmville | .74 | .72 |
| 240-241 | Roanoke | .76 | .76 |
| 242 | Bristol | .79 | .74 |
| 243 | Pulaski | .73 | .72 |
| 244 | Staunton | .76 | .74 |
| 245 | Lynchburg | .80 | .76 |
| 246 | Grundy | .72 | .72 |
| | | | |
| **WASHINGTON** | | | |
| 980-981,987 | Seattle | 1.00 | 1.05 |
| 982 | Everett | .97 | 1.03 |
| 983-984 | Tacoma | 1.05 | 1.03 |
| 985 | Olympia | 1.05 | 1.03 |
| 986 | Vancouver | 1.11 | 1.04 |
| 988 | Wenatchee | .94 | .97 |
| 989 | Yakima | 1.02 | 1.00 |
| 990-992 | Spokane | .99 | .98 |
| 993 | Richland | 1.00 | .99 |
| 994 | Clarkston | .99 | .98 |
| | | | |
| **WEST VIRGINIA** | | | |
| 247-248 | Bluefield | .89 | .89 |
| 249 | Lewisburg | .90 | .90 |
| 250-253 | Charleston | .93 | .93 |
| 254 | Martinsburg | .75 | .76 |
| 255-257 | Huntington | .93 | .95 |
| 258-259 | Beckley | .90 | .90 |
| 260 | Wheeling | .93 | .95 |
| 261 | Parkersburg | .92 | .94 |
| 262 | Buckhannon | .97 | .94 |
| 263-264 | Clarksburg | .97 | .94 |
| 265 | Morgantown | .97 | .94 |
| 266 | Gassaway | .93 | .93 |
| 267 | Romney | .90 | .90 |
| 268 | Petersburg | .95 | .92 |
| | | | |
| **WISCONSIN** | | | |
| 530,532 | Milwaukee | 1.02 | 1.01 |
| 531 | Kenosha | 1.02 | 1.01 |
| 534 | Racine | 1.06 | 1.00 |
| 535 | Beloit | 1.00 | .98 |
| 537 | Madison | 1.00 | .98 |
| 538 | Lancaster | .91 | .89 |
| 539 | Portage | .98 | .96 |
| 540 | New Richmond | 1.03 | .95 |
| 541-543 | Green Bay | 1.00 | .97 |
| 544 | Wausau | .98 | .94 |
| 545 | Rhinelander | .98 | .94 |
| 546 | La Crosse | .98 | .95 |
| 547 | Eau Claire | 1.04 | .96 |
| 548 | Superior | 1.03 | .97 |
| 549 | Oshkosh | .97 | .94 |

| STATE/ZIP | CITY | Residential | Commercial |
|---|---|---|---|
| **WYOMING** | | | |
| 820 | Cheyenne | .86 | .81 |
| 821 | Yellowstone Nat. Pk. | .80 | .77 |
| 822 | Wheatland | .83 | .78 |
| 823 | Rawlins | .81 | .77 |
| 824 | Worland | .78 | .76 |
| 825 | Riverton | .81 | .78 |
| 826 | Casper | .86 | .82 |
| 827 | Newcastle | .79 | .75 |
| 828 | Sheridan | .83 | .80 |
| 829-831 | Rock Springs | .83 | .78 |
| | | | |
| **CANADIAN FACTORS (reflect Canadian currency)** | | | |
| **ALBERTA** | | | |
| | Calgary | .99 | .96 |
| | Edmonton | .99 | .96 |
| | Fort McMurray | .98 | .95 |
| | Lethbridge | .98 | .95 |
| | Lloydminster | .98 | .95 |
| | Medicine Hat | .98 | .95 |
| | Red Deer | .98 | .95 |
| **BRITISH COLUMBIA** | | | |
| | Kamloops | 1.02 | 1.03 |
| | Prince George | 1.04 | 1.05 |
| | Vancouver | 1.05 | 1.06 |
| | Victoria | 1.04 | 1.05 |
| **MANITOBA** | | | |
| | Brandon | .97 | .96 |
| | Portage la Prairie | .97 | .96 |
| | Winnipeg | .97 | .96 |
| **NEW BRUNSWICK** | | | |
| | Bathurst | .93 | .91 |
| | Dalhousie | .93 | .91 |
| | Fredericton | .95 | .93 |
| | Moncton | .92 | .90 |
| | Newcastle | .93 | .91 |
| | Saint John | .96 | .94 |
| **NEWFOUNDLAND** | | | |
| | Corner Brook | .95 | .94 |
| | St. John's | .94 | .93 |
| **NORTHWEST TERRITORIES** | | | |
| | Yellowknife | .91 | .90 |
| **NOVA SCOTIA** | | | |
| | Dartmouth | .96 | .95 |
| | Halifax | .96 | .95 |
| | New Glasgow | .96 | .95 |
| | Sydney | .94 | .93 |
| | Yarmouth | .96 | .95 |
| **ONTARIO** | | | |
| | Barrie | 1.08 | 1.07 |
| | Brantford | 1.10 | 1.08 |
| | Cornwall | 1.08 | 1.06 |
| | Hamilton | 1.12 | 1.08 |
| | Kingston | 1.08 | 1.07 |
| | Kitchener | 1.05 | 1.03 |
| | London | 1.08 | 1.06 |
| | North Bay | 1.07 | 1.05 |
| | Oshawa | 1.09 | 1.07 |
| | Ottawa | 1.09 | 1.07 |
| | Owen Sound | 1.08 | 1.07 |
| | Peterborough | 1.07 | 1.06 |
| | Sarnia | 1.10 | 1.08 |
| | St. Catharines | 1.04 | 1.02 |
| | Sudbury | 1.04 | 1.02 |
| | Thunder Bay | 1.05 | 1.03 |
| | Toronto | 1.12 | 1.11 |
| | Windsor | 1.06 | 1.04 |
| **PRINCE EDWARD ISLAND** | | | |
| | Charlottetown | .92 | .90 |
| | Summerside | .92 | .90 |

| STATE/ZIP | CITY | Residential | Commercial |
|---|---|---|---|
| **QUEBEC** | | | |
| | Cap-de-la-Madeleine | 1.03 | 1.02 |
| | Charlesbourg | 1.03 | 1.02 |
| | Chicoutimi | 1.02 | 1.01 |
| | Gatineau | 1.01 | 1.00 |
| | Laval | 1.02 | 1.01 |
| | Montreal | 1.08 | 1.01 |
| | Quebec | 1.10 | 1.02 |
| | Sherbrooke | 1.02 | 1.01 |
| | Trois Rivieres | 1.03 | 1.02 |
| **SASKATCHEWAN** | | | |
| | Moose Jaw | .91 | .91 |
| | Prince Albert | .91 | .91 |
| | Regina | .92 | .92 |
| | Saskatoon | .91 | .91 |
| **YUKON** | | | |
| | Whitehorse | .91 | .90 |

|  |  |
|---|---|
| A | Area Square Feet; Ampere |
| ABS | Acrylonitrile Butadiene Stryrene; Asbestos Bonded Steel |
| A.C. | Alternating Current; Air-Conditioning; Asbestos Cement; Plywood Grade A & C |
| A.C.I. | American Concrete Institute |
| AD | Plywood, Grade A & D |
| Addit. | Additional |
| Adj. | Adjustable |
| af | Audio-frequency |
| A.G.A. | American Gas Association |
| Agg. | Aggregate |
| A.H. | Ampere Hours |
| A hr. | Ampere-hour |
| A.H.U. | Air Handling Unit |
| A.I.A. | American Institute of Architects |
| AIC | Ampere Interrupting Capacity |
| Allow. | Allowance |
| alt. | Altitude |
| Alum. | Aluminum |
| a.m. | Ante Meridiem |
| Amp. | Ampere |
| Anod. | Anodized |
| Approx. | Approximate |
| Apt. | Apartment |
| Asb. | Asbestos |
| A.S.B.C. | American Standard Building Code |
| Asbe. | Asbestos Worker |
| A.S.H.R.A.E. | American Society of Heating, Refrig. & AC Engineers |
| A.S.M.E. | American Society of Mechanical Engineers |
| A.S.T.M. | American Society for Testing and Materials |
| Attchmt. | Attachment |
| Avg. | Average |
| A.W.G. | American Wire Gauge |
| AWWA | American Water Works Assoc. |
| Bbl. | Barrel |
| B. & B. | Grade B and Better; Balled & Burlapped |
| B. & S. | Bell and Spigot |
| B. & W. | Black and White |
| b.c.c. | Body-centered Cubic |
| B.C.Y. | Bank Cubic Yards |
| BE | Bevel End |
| B.F. | Board Feet |
| Bg. cem. | Bag of Cement |
| BHP | Boiler Horsepower; Brake Horsepower |
| B.I. | Black Iron |
| Bit.; Bitum. | Bituminous |
| Bk. | Backed |
| Bkrs. | Breakers |
| Bldg. | Building |
| Blk. | Block |
| Bm. | Beam |
| Boil. | Boilermaker |
| B.P.M. | Blows per Minute |
| BR | Bedroom |
| Brg. | Bearing |
| Brhe. | Bricklayer Helper |
| Bric. | Bricklayer |
| Brk. | Brick |
| Brng. | Bearing |
| Brs. | Brass |
| Brz. | Bronze |
| Bsn. | Basin |
| Btr. | Better |
| BTU | British Thermal Unit |
| BTUH | BTU per Hour |
| B.U.R. | Built-up Roofing |
| BX | Interlocked Armored Cable |
| c | Conductivity, Copper Sweat |
| C | Hundred; Centigrade |
| C/C | Center to Center, Cedar on Cedar |
| Cab. | Cabinet |
| Cair. | Air Tool Laborer |
| Calc | Calculated |
| Cap. | Capacity |
| Carp. | Carpenter |
| C.B. | Circuit Breaker |
| C.C.A. | Chromate Copper Arsenate |
| C.C.F. | Hundred Cubic Feet |
| cd | Candela |
| cd/sf | Candela per Square Foot |
| CD | Grade of Plywood Face & Back |
| CDX | Plywood, Grade C & D, exterior glue |
| Cefi. | Cement Finisher |
| Cem. | Cement |
| CF | Hundred Feet |
| C.F. | Cubic Feet |
| CFM | Cubic Feet per Minute |
| c.g. | Center of Gravity |
| CHW | Chilled Water; Commercial Hot Water |
| C.I. | Cast Iron |
| C.I.P. | Cast in Place |
| Circ. | Circuit |
| C.L. | Carload Lot |
| Clab. | Common Laborer |
| C.L.F. | Hundred Linear Feet |
| CLF | Current Limiting Fuse |
| CLP | Cross Linked Polyethylene |
| cm | Centimeter |
| CMP | Corr. Metal Pipe |
| C.M.U. | Concrete Masonry Unit |
| CN | Change Notice |
| Col. | Column |
| $CO_2$ | Carbon Dioxide |
| Comb. | Combination |
| Compr. | Compressor |
| Conc. | Concrete |
| Cont. | Continuous; Continued |
| Corr. | Corrugated |
| Cos | Cosine |
| Cot | Cotangent |
| Cov. | Cover |
| C/P | Cedar on Paneling |
| CPA | Control Point Adjustment |
| Cplg. | Coupling |
| C.P.M. | Critical Path Method |
| CPVC | Chlorinated Polyvinyl Chloride |
| C.Pr. | Hundred Pair |
| CRC | Cold Rolled Channel |
| Creos. | Creosote |
| Crpt. | Carpet & Linoleum Layer |
| CRT | Cathode-ray Tube |
| CS | Carbon Steel, Constant Shear Bar Joist |
| Csc | Cosecant |
| C.S.F. | Hundred Square Feet |
| CSI | Construction Specifications Institute |
| C.T. | Current Transformer |
| CTS | Copper Tube Size |
| Cu | Copper, Cubic |
| Cu. Ft. | Cubic Foot |
| cw | Continuous Wave |
| C.W. | Cool White; Cold Water |
| Cwt. | 100 Pounds |
| C.W.X. | Cool White Deluxe |
| C.Y. | Cubic Yard (27 cubic feet) |
| C.Y./Hr. | Cubic Yard per Hour |
| Cyl. | Cylinder |
| d | Penny (nail size) |
| D | Deep; Depth; Discharge |
| Dis.;Disch. | Discharge |
| Db. | Decibel |
| Dbl. | Double |
| DC | Direct Current |
| DDC | Direct Digital Control |
| Demob. | Demobilization |
| d.f.u. | Drainage Fixture Units |
| D.H. | Double Hung |
| DHW | Domestic Hot Water |
| Diag. | Diagonal |
| Diam. | Diameter |
| Distrib. | Distribution |
| Dk. | Deck |
| D.L. | Dead Load; Diesel |
| DLH | Deep Long Span Bar Joist |
| Do. | Ditto |
| Dp. | Depth |
| D.P.S.T. | Double Pole, Single Throw |
| Dr. | Driver |
| Drink. | Drinking |
| D.S. | Double Strength |
| D.S.A. | Double Strength A Grade |
| D.S.B. | Double Strength B Grade |
| Dty. | Duty |
| DWV | Drain Waste Vent |
| DX | Deluxe White, Direct Expansion |
| dyn | Dyne |
| e | Eccentricity |
| E | Equipment Only; East |
| Ea. | Each |
| E.B. | Encased Burial |
| Econ. | Economy |
| EDP | Electronic Data Processing |
| EIFS | Exterior Insulation Finish System |
| E.D.R. | Equiv. Direct Radiation |
| Eq. | Equation |
| Elec. | Electrician; Electrical |
| Elev. | Elevator; Elevating |
| EMT | Electrical Metallic Conduit; Thin Wall Conduit |
| Eng. | Engine, Engineered |
| EPDM | Ethylene Propylene Diene Monomer |
| EPS | Expanded Polystyrene |
| Eqhv. | Equip. Oper., Heavy |
| Eqlt. | Equip. Oper., Light |
| Eqmd. | Equip. Oper., Medium |
| Eqmm. | Equip. Oper., Master Mechanic |
| Eqol. | Equip. Oper., Oilers |
| Equip. | Equipment |
| ERW | Electric Resistance Welded |
| E.S. | Energy Saver |
| Est. | Estimated |
| esu | Electrostatic Units |
| E.W. | Each Way |
| EWT | Entering Water Temperature |
| Excav. | Excavation |
| Exp. | Expansion, Exposure |
| Ext. | Exterior |
| Extru. | Extrusion |
| f. | Fiber stress |
| F | Fahrenheit; Female; Fill |
| Fab. | Fabricated |
| FBGS | Fiberglass |
| F.C. | Footcandles |
| f.c.c. | Face-centered Cubic |
| f'c. | Compressive Stress in Concrete; Extreme Compressive Stress |
| F.E. | Front End |
| FEP | Fluorinated Ethylene Propylene (Teflon) |
| F.G. | Flat Grain |
| F.H.A. | Federal Housing Administration |
| Fig. | Figure |
| Fin. | Finished |
| Fixt. | Fixture |
| Fl. Oz. | Fluid Ounces |
| Flr. | Floor |
| F.M. | Frequency Modulation; Factory Mutual |
| Fmg. | Framing |
| Fndtn. | Foundation |
| Fori. | Foreman, Inside |
| Foro. | Foreman, Outside |

| | | | | | |
|---|---|---|---|---|---|
| 'ount. | Fountain | J.I.C. | Joint Industrial Council | M.C.P. | Motor Circuit Protector |
| 'PM | Feet per Minute | K | Thousand; Thousand Pounds; Heavy Wall Copper Tubing, Kelvin | MD | Medium Duty |
| 'PT | Female Pipe Thread | | | M.D.O. | Medium Density Overlaid |
| 'r. | Frame | K.A.H. | Thousand Amp. Hours | Med. | Medium |
| .R. | Fire Rating | KCMIL | Thousand Circular Mils | MF | Thousand Feet |
| RK | Foil Reinforced Kraft | KD | Knock Down | M.F.B.M. | Thousand Feet Board Measure |
| RP | Fiberglass Reinforced Plastic | K.D.A.T. | Kiln Dried After Treatment | Mfg. | Manufacturing |
| 'S | Forged Steel | kg | Kilogram | Mfrs. | Manufacturers |
| SC | Cast Body; Cast Switch Box | kG | Kilogauss | mg | Milligram |
| 't. | Foot; Feet | kgf | Kilogram Force | MGD | Million Gallons per Day |
| tng. | Fitting | kHz | Kilohertz | MGPH | Thousand Gallons per Hour |
| tg. | Footing | Kip. | 1000 Pounds | MH, M.H. | Manhole; Metal Halide; Man-Hour |
| 't. Lb. | Foot Pound | KJ | Kiljoule | MHz | Megahertz |
| urn. | Furniture | K.L. | Effective Length Factor | Mi. | Mile |
| VNR | Full Voltage Non-Reversing | K.L.F. | Kips per Linear Foot | MI | Malleable Iron; Mineral Insulated |
| XM | Female by Male | Km | Kilometer | mm | Millimeter |
| y. | Minimum Yield Stress of Steel | K.S.F. | Kips per Square Foot | Mill. | Millwright |
| | Gram | K.S.I. | Kips per Square Inch | Min., min. | Minimum, minute |
| | Gauss | kV | Kilovolt | Misc. | Miscellaneous |
| Ga. | Gauge | kVA | Kilovolt Ampere | ml | Milliliter, Mainline |
| Gal. | Gallon | K.V.A.R. | Kilovar (Reactance) | M.L.F. | Thousand Linear Feet |
| Gal./Min. | Gallon per Minute | KW | Kilowatt | Mo. | Month |
| Galv. | Galvanized | KWh | Kilowatt-hour | Mobil. | Mobilization |
| Gen. | General | L | Labor Only; Length; Long; Medium Wall Copper Tubing | Mog. | Mogul Base |
| G.F.I. | Ground Fault Interrupter | | | MPH | Miles per Hour |
| Glaz. | Glazier | Lab. | Labor | MPT | Male Pipe Thread |
| GPD | Gallons per Day | lat | Latitude | MRT | Mile Round Trip |
| GPH | Gallons per Hour | Lath. | Lather | ms | Millisecond |
| GPM | Gallons per Minute | Lav. | Lavatory | M.S.F. | Thousand Square Feet |
| GR | Grade | lb.; # | Pound | Mstz. | Mosaic & Terrazzo Worker |
| Gran. | Granular | L.B. | Load Bearing; L Conduit Body | M.S.Y. | Thousand Square Yards |
| Grnd. | Ground | L. & E. | Labor & Equipment | Mtd. | Mounted |
| H | High; High Strength Bar Joist; Henry | lb./hr. | Pounds per Hour | Mthe. | Mosaic & Terrazzo Helper |
| | | lb./L.F. | Pounds per Linear Foot | Mtng. | Mounting |
| H.C. | High Capacity | lbf/sq.in. | Pound-force per Square Inch | Mult. | Multi; Multiply |
| H.D. | Heavy Duty; High Density | L.C.L. | Less than Carload Lot | M.V.A. | Million Volt Amperes |
| H.D.O. | High Density Overlaid | Ld. | Load | M.V.A.R. | Million Volt Amperes Reactance |
| Hdr. | Header | LE | Lead Equivalent | MV | Megavolt |
| Hdwe. | Hardware | LED | Light Emitting Diode | MW | Megawatt |
| Help. | Helpers Average | L.F. | Linear Foot | MXM | Male by Male |
| HEPA | High Efficiency Particulate Air Filter | Lg. | Long; Length; Large | MYD | Thousand Yards |
| | | L & H | Light and Heat | N | Natural; North |
| Hg | Mercury | LH | Long Span Bar Joist | nA | Nanoampere |
| HIC | High Interrupting Capacity | L.H. | Labor Hours | NA | Not Available; Not Applicable |
| HM | Hollow Metal | L.L. | Live Load | N.B.C. | National Building Code |
| H.O. | High Output | L.L.D. | Lamp Lumen Depreciation | NC | Normally Closed |
| Horiz. | Horizontal | L-O-L | Lateralolet | N.E.M.A. | National Electrical Manufacturers Assoc. |
| H.P. | Horsepower; High Pressure | lm | Lumen | | |
| H.P.F. | High Power Factor | lm/sf | Lumen per Square Foot | NEHB | Bolted Circuit Breaker to 600V. |
| Hr. | Hour | lm/W | Lumen per Watt | N.L.B. | Non-Load-Bearing |
| Hrs./Day | Hours per Day | L.O.A. | Length Over All | NM | Non-Metallic Cable |
| HSC | High Short Circuit | log | Logarithm | nm | Nanometer |
| Ht. | Height | L.P. | Liquefied Petroleum; Low Pressure | No. | Number |
| Htg. | Heating | L.P.F. | Low Power Factor | NO | Normally Open |
| Htrs. | Heaters | LR | Long Radius | N.O.C. | Not Otherwise Classified |
| HVAC | Heating, Ventilation & Air-Conditioning | L.S. | Lump Sum | Nose. | Nosing |
| | | Lt. | Light | N.P.T. | National Pipe Thread |
| Hvy. | Heavy | Lt. Ga. | Light Gauge | NQOD | Combination Plug-on/Bolt on Circuit Breaker to 240V. |
| HW | Hot Water | L.T.L. | Less than Truckload Lot | | |
| Hyd.;Hydr. | Hydraulic | Lt. Wt. | Lightweight | N.R.C. | Noise Reduction Coefficient |
| Hz. | Hertz (cycles) | L.V. | Low Voltage | N.R.S. | Non Rising Stem |
| I. | Moment of Inertia | M | Thousand; Material; Male; Light Wall Copper Tubing | ns | Nanosecond |
| I.C. | Interrupting Capacity | | | nW | Nanowatt |
| ID | Inside Diameter | M²CA | Meters Squared Contact Area | OB | Opposing Blade |
| I.D. | Inside Dimension; Identification | m/hr; M.H. | Man-hour | OC | On Center |
| I.F. | Inside Frosted | mA | Milliampere | OD | Outside Diameter |
| I.M.C. | Intermediate Metal Conduit | Mach. | Machine | O.D. | Outside Dimension |
| In. | Inch | Mag. Str. | Magnetic Starter | ODS | Overhead Distribution System |
| Incan. | Incandescent | Maint. | Maintenance | O.G. | Ogee |
| Incl. | Included; Including | Marb. | Marble Setter | O.H. | Overhead |
| Int. | Interior | Mat; Mat'l. | Material | O & P | Overhead and Profit |
| Inst. | Installation | Max. | Maximum | Oper. | Operator |
| Insul. | Insulation/Insulated | MBF | Thousand Board Feet | Opng. | Opening |
| I.P. | Iron Pipe | MBH | Thousand BTU's per hr. | Orna. | Ornamental |
| I.P.S. | Iron Pipe Size | MC | Metal Clad Cable | OSB | Oriented Strand Board |
| I.P.T. | Iron Pipe Threaded | M.C.F. | Thousand Cubic Feet | O. S. & Y. | Outside Screw and Yoke |
| I.W. | Indirect Waste | M.C.F.M. | Thousand Cubic Feet per Minute | Ovhd. | Overhead |
| J | Joule | M.C.M. | Thousand Circular Mils | OWG | Oil, Water or Gas |

# Abbreviations

| | | | | | | | |
|---|---|---|---|---|---|
| Oz. | Ounce | SCFM | Standard Cubic Feet per Minute | THW. | Insulated Strand Wire |
| P. | Pole; Applied Load; Projection | Scaf. | Scaffold | THWN; | Nylon Jacketed Wire |
| p. | Page | Sch.; Sched. | Schedule | T.L. | Truckload |
| Pape. | Paperhanger | S.C.R. | Modular Brick | T.M. | Track Mounted |
| P.A.P.R. | Powered Air Purifying Respirator | S.D. | Sound Deadening | Tot. | Total |
| PAR | Parabolic Reflector | S.D.R. | Standard Dimension Ratio | T-O-L | Threadolet |
| Pc., Pcs. | Piece, Pieces | S.E. | Surfaced Edge | T.S. | Trigger Start |
| P.C. | Portland Cement; Power Connector | Sel. | Select | Tr. | Trade |
| P.C.F. | Pounds per Cubic Foot | S.E.R.; | Service Entrance Cable | Transf. | Transformer |
| P.C.M. | Phase Contract Microscopy | S.E.U. | Service Entrance Cable | Trhv. | Truck Driver, Heavy |
| P.E. | Professional Engineer; | S.F. | Square Foot | Trlr. | Trailer |
| | Porcelain Enamel; | S.F.C.A. | Square Foot Contact Area | Trlt. | Truck Driver, Light |
| | Polyethylene; Plain End | S.F.G. | Square Foot of Ground | TV | Television |
| Perf. | Perforated | S.F. Hor. | Square Foot Horizontal | T.W. | Thermoplastic Water Resistant |
| Ph. | Phase | S.F.R. | Square Feet of Radiation | | Wire |
| P.I. | Pressure Injected | S.F. Shlf. | Square Foot of Shelf | UCI | Uniform Construction Index |
| Pile. | Pile Driver | S4S | Surface 4 Sides | UF | Underground Feeder |
| Pkg. | Package | Shee. | Sheet Metal Worker | UGND | Underground Feeder |
| Pl. | Plate | Sin. | Sine | U.H.F. | Ultra High Frequency |
| Plah. | Plasterer Helper | Skwk. | Skilled Worker | U.L. | Underwriters Laboratory |
| Plas. | Plasterer | SL | Saran Lined | Unfin. | Unfinished |
| Pluh. | Plumbers Helper | S.L. | Slimline | URD | Underground Residential |
| Plum. | Plumber | Sldr. | Solder | | Distribution |
| Ply. | Plywood | SLH | Super Long Span Bar Joist | US | United States |
| p.m. | Post Meridiem | S.N. | Solid Neutral | USP | United States Primed |
| Pntd. | Painted | S-O-L | Socketolet | UTP | Unshielded Twisted Pair |
| Pord. | Painter, Ordinary | sp | Standpipe | V | Volt |
| pp | Pages | S.P. | Static Pressure; Single Pole; Self- | V.A. | Volt Amperes |
| PP; PPL | Polypropylene | | Propelled | V.C.T. | Vinyl Composition Tile |
| P.P.M. | Parts per Million | Spri. | Sprinkler Installer | VAV | Variable Air Volume |
| Pr. | Pair | spwg | Static Pressure Water Gauge | VC | Veneer Core |
| P.E.S.B. | Pre-engineered Steel Building | Sq. | Square; 100 Square Feet | Vent. | Ventilation |
| Prefab. | Prefabricated | S.P.D.T. | Single Pole, Double Throw | Vert. | Vertical |
| Prefin. | Prefinished | SPF | Spruce Pine Fir | V.F. | Vinyl Faced |
| Prop. | Propelled | S.P.S.T. | Single Pole, Single Throw | V.G. | Vertical Grain |
| PSF; psf | Pounds per Square Foot | SPT | Standard Pipe Thread | V.H.F. | Very High Frequency |
| PSI; psi | Pounds per Square Inch | Sq. Hd. | Square Head | VHO | Very High Output |
| PSIG | Pounds per Square Inch Gauge | Sq. In. | Square Inch | Vib. | Vibrating |
| PSP | Plastic Sewer Pipe | S.S. | Single Strength; Stainless Steel | V.L.F. | Vertical Linear Foot |
| Pspr. | Painter, Spray | S.S.B. | Single Strength B Grade | Vol. | Volume |
| Psst. | Painter, Structural Steel | sst | Stainless Steel | VRP | Vinyl Reinforced Polyester |
| P.T. | Potential Transformer | Sswk. | Structural Steel Worker | W | Wire; Watt; Wide; West |
| P. & T. | Pressure & Temperature | Sswl. | Structural Steel Welder | w/ | With |
| Ptd. | Painted | St.; Stl. | Steel | W.C. | Water Column; Water Closet |
| Ptns. | Partitions | S.T.C. | Sound Transmission Coefficient | W.F. | Wide Flange |
| Pu | Ultimate Load | Std. | Standard | W.G. | Water Gauge |
| PVC | Polyvinyl Chloride | STK | Select Tight Knot | Wldg. | Welding |
| Pvmt. | Pavement | STP | Standard Temperature & Pressure | W. Mile | Wire Mile |
| Pwr. | Power | Stpi. | Steamfitter, Pipefitter | W-O-L | Weldolet |
| Q | Quantity Heat Flow | Str. | Strength; Starter; Straight | W.R. | Water Resistant |
| Quan.; Qty. | Quantity | Strd. | Stranded | Wrck. | Wrecker |
| Q.C. | Quick Coupling | Struct. | Structural | W.S.P. | Water, Steam, Petroleum |
| r | Radius of Gyration | Sty. | Story | WT., Wt. | Weight |
| R | Resistance | Subj. | Subject | WWF | Welded Wire Fabric |
| R.C.P. | Reinforced Concrete Pipe | Subs. | Subcontractors | XFER | Transfer |
| Rect. | Rectangle | Surf. | Surface | XFMR | Transformer |
| Reg. | Regular | Sw. | Switch | XHD | Extra Heavy Duty |
| Reinf. | Reinforced | Swbd. | Switchboard | XHHW; XLPE | Cross-Linked Polyethylene Wire |
| Req'd. | Required | S.Y. | Square Yard | | Insulation |
| Res. | Resistant | Syn. | Synthetic | XLP | Cross-linked Polyethylene |
| Resi. | Residential | S.Y.P. | Southern Yellow Pine | Y | Wye |
| Rgh. | Rough | Sys. | System | yd | Yard |
| RGS | Rigid Galvanized Steel | t. | Thickness | yr | Year |
| R.H.W. | Rubber, Heat & Water Resistant; | T | Temperature; Ton | Δ | Delta |
| | Residential Hot Water | Tan | Tangent | % | Percent |
| rms | Root Mean Square | T.C. | Terra Cotta | ~ | Approximately |
| Rnd. | Round | T & C | Threaded and Coupled | Ø | Phase |
| Rodm. | Rodman | T.D. | Temperature Difference | @ | At |
| Rofc. | Roofer, Composition | T.E.M. | Transmission Electron Microscopy | # | Pound; Number |
| Rofp. | Roofer, Precast | TFE | Tetrafluoroethylene (Teflon) | < | Less Than |
| Rohe. | Roofer Helpers (Composition) | T. & G. | Tongue & Groove; | > | Greater Than |
| Rots. | Roofer, Tile & Slate | | Tar & Gravel | | |
| R.O.W. | Right of Way | Th.; Thk. | Thick | | |
| RPM | Revolutions per Minute | Thn. | Thin | | |
| R.S. | Rapid Start | Thrded | Threaded | | |
| Rsr | Riser | Tilf. | Tile Layer, Floor | | |
| RT | Round Trip | Tilh. | Tile Layer, Helper | | |
| S. | Suction; Single Entrance; South | THHN | Nylon Jacketed Wire | | |

# Index

# Index

## W

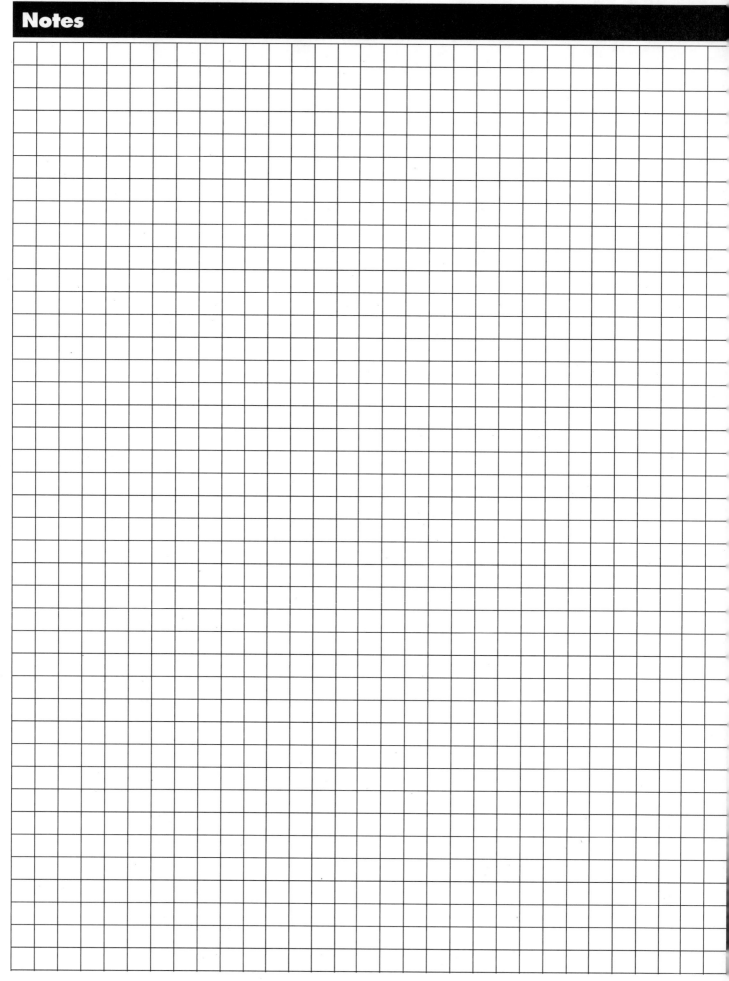

278

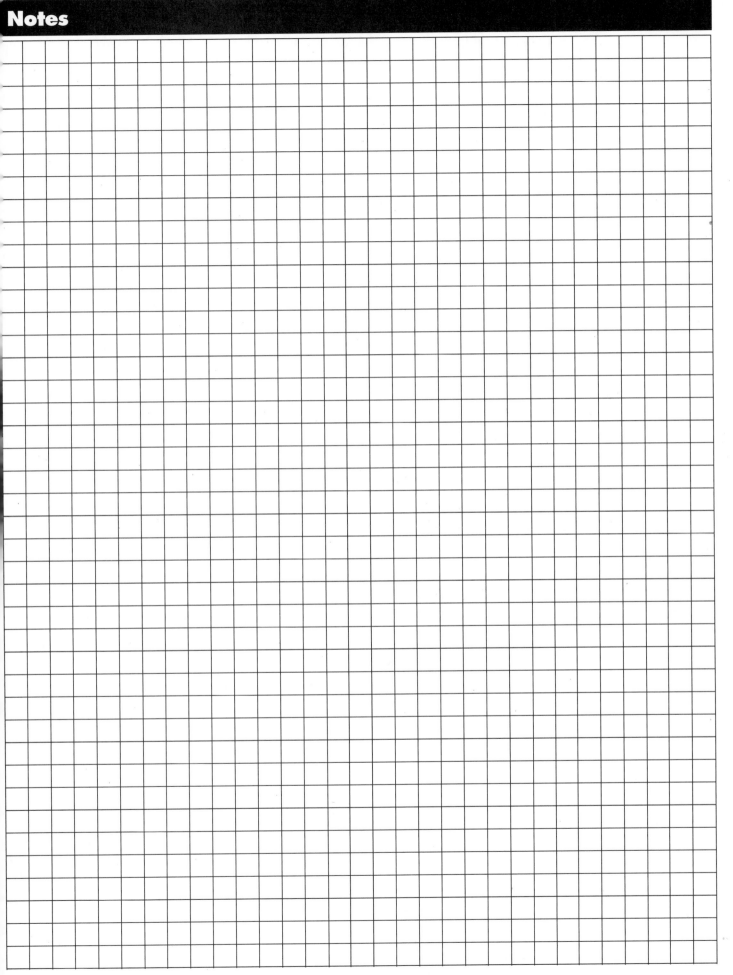

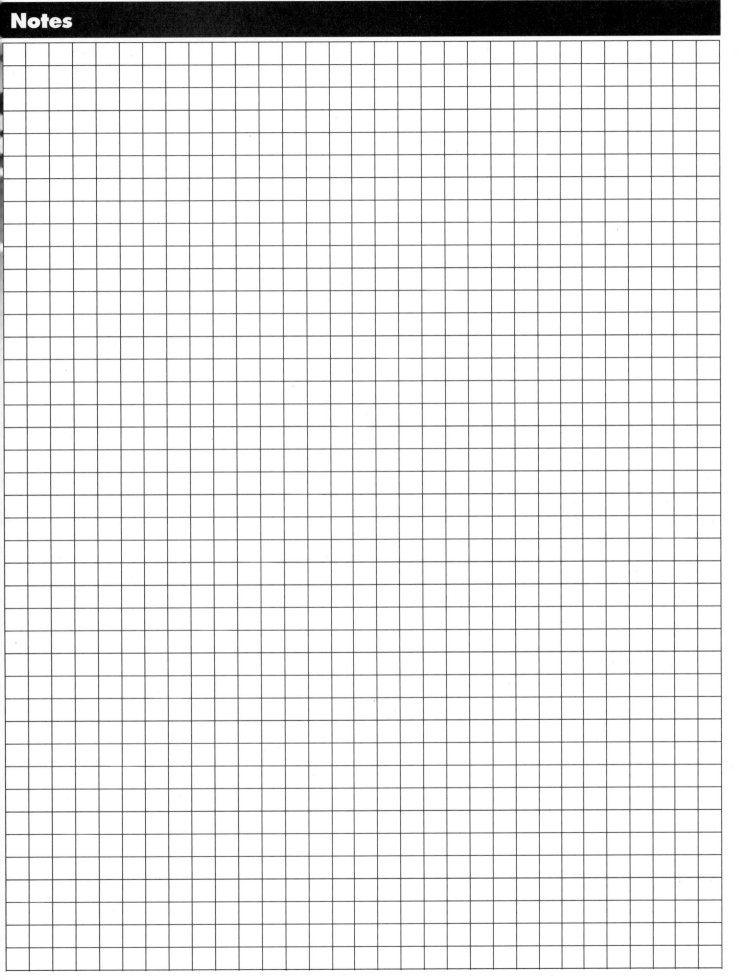

**R.S. Means Company, Inc.,** a CMD company, is the leading provider of construction cost data in North America and supplies comprehensive construction cost guides, related technical publications and education services.

**CMD,** a leading worldwide provider of total construction information solutions, is comprised of three synergistic product groups designed specifically to help construction professionals advance their businesses with timely, accurate and actionable project, product and cost data. CMD is a division of Cahners Business Information, a member of the Reed Elsevier plc group of companies.

The *Project, Product, and Cost & Estimating* divisions offer a variety of innovative products designed for the full spectrum of design, construction and manufacturing professionals. Together with Cahners, CMD created *Buildingteam.com,* a valuable Internet portal of the construction community. Through it's *International* companies, CMD's reputation for quality construction market data is growing worldwide.

## Project Data

CMD provides complete, accurate and relevant project information through all stages of construction. Customers are supplied industry data through leads, project reports, contact lists, plans and specifications surveys, market penetration analyses and sales evaluation reports. Any of these products can pinpoint a county, look at a state, or cover the country. Data is delivered via paper, e-mail, CD-ROM or the Internet.

## Building Product Information

The First Source suite of products is the only integrated building product information system offered to the commercial construction industry for comparing and specifying building products. These print and online resources include *First Source for Products,* SPEC-DATA™, MANU-SPEC™, CADBlocks, First Source Exchange (www.firstsourceexchange.com), and Manufacturer Catalogs. Written by industry professionals and organized using CSI's MasterFormat™, construction professionals use this information to make better design decisions.

## Cost Information

R.S. Means, the undisputed market leader and authority on construction costs, publishes current cost and estimating information in annual cost books and on the CostWorks CD-ROM. R.S. Means furnishes the construction industry with a rich library of complementary reference books and a series of professional seminars that are designed to sharpen professional skills and maximize the effective use of cost estimating and management tools. R.S. Means also provides construction cost consulting for Owners, Manufacturers Designers and Contractors.

## Buildingteam.com

Combining CMD's project, product and cost data with news and information from Cahners' *Building Design & Construction* and *Consulting-Specifying Engineer,* this industry-focused site offers easy and unlimited access to vital information for all construction professionals.

## International

*BIMSA/Mexico* provides construction project news, product information, cost-data, seminars and consulting services to construction professionals in Mexico. Its subsidiary, PRISMA, provides job costing software.

*Byggfakta Scandinavia AB,* founded in 1936, is the parent company for the leaders of customized construction market data for Denmark, Estonia, Finland, Norway and Sweden. Each company fully covers the local construction market and provides information across several platforms including subscription, ad-hoc basis, electronically and on paper.

*CMD Canada* serves the Canadian construction market with reliable and comprehensive project and product information services that cover all facets of construction. Core services include: Buildcore, product selection and specification tools available in print and on the Internet; CMD Building Reports, a national construction project lead service; CanaData, statistical and forecasting information; *Daily Commercial News,* a construction newspaper reporting on news and projects in Ontario; and *Journal of Commerce,* reporting news in British Columbia and Alberta.

*Cordell Building Information Services,* with its complete range of project and cost and estimating services, is Australia's specialist in the construction information industry. Cordell provides in-depth and historical information on all aspects of construction projects and estimation, including several customized reports, construction and sales leads, and detailed cost information among others.

*For more information, please visit our website at www.cmdg.com.*

CMD Corporate Office
30 Technology Parkway South
Norcross, GA 30092-2912
(800) 793-0304
(700) 417-4002 (fax)
info@cmdg.com
www.cmdg.com

**RSMeans**
**CMD**

# Contractor's Pricing Guides

For more information
visit Means Web Site
at www.rsmeans.com

## Means ADA Compliance Pricing Guide

Accurately plan and budget for the ADA modifications you are most likely to need... with the first available cost guide for business owners, facility managers, and all who are involved in building modifications to comply with the Americans With Disabilities Act.

75 major projects—the most frequently needed modifications—include complete estimates with itemized materials and labor, plus contractor's total fees. Over 260 project variations fit almost any site conditions or budget constraints. Location Factors to adjust costs for 927 cities and towns.

A collaboration between Adaptive Environments Center, Inc. and R.S. Means Engineering Staff.

**$59.98 per copy**
**Over 350 pages, illustrated, softcover**
**Catalog No. 67310   ISBN 0-87629-351-8**

## *Contractor's Pricing Guide:* Residential Square Foot Costs 2002

Now available in one concise volume, all you need to know to plan and budget the cost of new homes. If you are looking for a quick reference, the model home section contains costs for over 250 different sizes and types of residences, with hundreds of easily applied modifications. If you need even more detail, the Assemblies Section lets you build your own costs or modify the model costs further. Hundreds of graphics are provided, along with forms and procedures to help you get it right.

**$39.95 per copy**
**Over 250 pages, illustrated, 8-1/2 x 11**
**Catalog No. 60322   ISBN 0-87629-647-9**

## *Contractor's Pricing Guide:* Residential Detailed Costs 2002

Every aspect of residential construction, from overhead costs to residential lighting and wiring, is in here. All the detail you need to accurately estimate the costs of your work with or without markups–labor-hours, typical crews and equipment are included as well. When you need a detailed estimate, this publication has all the costs to help you come up with a complete, on the money, price you can rely on to win profitable work.

**$36.95 per copy**
**Over 300 pages, with charts and tables, 8-1/2 x 11**
**Catalog No. 60332   ISBN 0-87629-648-7**

## *Contractor's Pricing Guide:* Residential Repair & Remodeling Costs 2002

This book provides total unit price costs for every aspect of the most common repair & remodeling projects. Organized in the order of construction by component and activity, it includes demolition and installation, cleaning, painting, and more.

With simplified estimating methods; clear, concise descriptions; and technical specifications for each component, the book is a valuable tool for contractors who want to speed up their estimating time, while making sure their costs are on target.

**$36.95 per copy**
**Over 250 pages, illustrated, 8-1/2 x 11**
**Catalog No. 60342   ISBN 0-87629-655-X**

## Means Repair & Remodeling Estimating Methods

### 3rd Edition

By Edward B. Wetherill and R.S. Means Engineering Staff

This updated edition focuses on the unique problems of estimating renovations in existing structures—using the latest cost resources and construction methods. The book helps you determine the true costs of remodeling, and includes:

**Part I–The Estimating Process**
**Part II–Estimating by CSI Division**
**Part III–Two Complete Sample**
   **Estimates–Unit Price & Assemblies**
   **New Section on Disaster Reconstruction**

**$69.95 per copy**
**Over 450 pages, illustrated, hardcover**
**Catalog No. 67265A   ISBN 0-87629-454-9**

## Means Landscape Estimating Methods New 3rd Edition

By Sylvia H. Fee

**Professional Methods for Estimating and Bidding Landscaping Projects and Grounds Maintenance Contracts**

- Easy-to-understand text. Clearly explains the estimating process and how to use *Means Site Work & Landscape Cost Data.*
- Sample forms and worksheets to save you time and avoid errors.
- Tips on best techniques for saving money and winning jobs.
- **Two new chapters** help you control your equipment costs and bid landscape maintenance projects.

**$62.95 per copy**
**Over 300 pages, illustrated, hardcover**
**Catalog No. 67295A   ISBN 0-87629-534-0**

For more information
visit Means Web Site
at www.rsmeans.com

# Annual Cost Guides

## Means Building Construction Cost Data 2002

Offers you unchallenged unit price reliability in an easy-to-use arrangement. Whether used for complete, finished estimates or for periodic checks, it supplies more cost facts better and faster than any comparable source. Over 23,000 unit prices for 2002. The City Cost Indexes now cover over 930 areas, for indexing to any project location in North America.

$99.95 per copy
Over 700 pages, softcover
Catalog No. 60012   ISBN 0-87629-620-7

## Means Open Shop Building Construction Cost Data 2002

The open-shop version of the *Means Building Construction Cost Data*. More than 22,000 reliable unit cost entries based on open shop trade labor rates. Eliminates time-consuming searches for these prices. The first book with open shop labor rates and crews. Labor information is itemized by labor-hours, crew, hourly/daily output, equipment, overhead and profit. For contractors, owners and facility managers.

$99.95 per copy
Over 680 pages, softcover
Catalog No. 60152   ISBN 0-87629-628-2

## Means Plumbing Cost Data 2002

Comprehensive unit prices and assemblies for plumbing, irrigation systems, commercial and residential fire protection, point-of-use water heaters, and the latest approved materials. This publication and its companion, *Means Mechanical Cost Data*, provide full-range cost estimating coverage for all the mechanical trades.

$99.95 per copy
Over 570 pages, softcover
Catalog No. 60212   ISBN 0-87629-629-0

## Means Residential Cost Data 2002

Speeds you through residential construction pricing with more than 100 illustrated complete house square-foot costs. Alternate assemblies cost selections are located on adjoining pages, so that you can develop tailor-made estimates in minutes. Complete data for detailed unit cost estimates is also provided.

$87.95 per copy
Over 600 pages, softcover
Catalog No. 60172   ISBN 0-87629-623-1

## Means Electrical Cost Data 2002

Pricing information for every part of electrical cost planning: unit and systems costs with design tables; engineering guides and illustrated estimating procedures; complete labor-hour, materials, and equipment costs for better scheduling and procurement. With the latest products and construction methods used in electrical work. More than 15,000 unit and systems costs, clear specifications and drawings.

$99.95 per copy
Over 480 pages, softcover
Catalog No. 60032   ISBN 0-87629-635-5

## Means Repair & Remodeling Cost Data 2002

### Commercial/Residential

You can use this valuable tool to estimate commercial and residential renovation and remodeling. By using the specialized costs in this manual, you'll find it's not necessary to force fit prices for new construction into remodeling cost planning. Provides comprehensive unit costs, building systems costs, extensive labor data and estimating assistance for every kind of building improvement.

$87.95 per copy
Over 660 pages, softcover
Catalog No. 60042   ISBN 0-87629-622-3

## Means Site Work & Landscape Cost Data 2002

Hard-to-find costs are presented in an easy-to-use format for every type of site work and landscape construction. Costs are organized, described, and laid out for earthwork, utilities, roads and bridges, as well as grading, planting, lawns, trees, irrigation systems, and site improvements.

$99.95 per copy
Over 630 pages, softcover
Catalog No. 60282   ISBN 0-87629-624-X

## Means Light Commercial Cost Data 2002

Specifically addresses the light commercial market, which is an increasingly specialized niche in the industry. Aids you, the owner/designer/contractor, in preparing all types of estimates, from budgets to detailed bids. Includes new advances in methods and materials. Assemblies section allows you to evaluate alternatives in early stages of design/planning.

$87.95 per copy
Over 672 pages, softcover
Catalog No. 60182   ISBN 0-87629-626-6

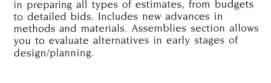

# Books for Builders

For more informatio
visit Means Web Sit
at www.rsmeans.com

## Builder's Essentials:
### Plan Reading & Material Takeoff

A complete course in reading and interpreting building plans—and performing quantity takeoffs to professional standards.

This book shows and explains, in clear language and with over 160 illustrations, typical working drawings encountered by contractors in residential and light commercial construction. The author describes not only how all common features are represented, but how to translate that information into a material list. Organized by CSI division, each chapter uses plans, details and tables, and a summary checklist.

**$35.95 per copy**
**Over 420 pages, illustrated, softcover**
**Catalog No. 67307   ISBN 0-87629-348-8**

## Builder's Essentials:
### Best Business Practices for Builders & Remodelers:
#### An Easy-to-Use Checklist System
By Thomas N. Frisby

A comprehensive guide covering all aspects of running a construction business, with more than 40 user–friendly checklists. This book provides expert guidance on: increasing your revenue and keeping more of your profit; planning for long-term growth; keeping good employees and managing subcontractors.

**$29.95 per copy**
**Over 220 pages, softcover**
**Catalog No. 67329    ISBN 0-87629-619-3**

## Means Estimating Handbook

This comprehensive reference is for use in the field and the office. It covers a full spectrum of technical data required for estimating, with information on sizing, productivity, equipment requirements, codes, design standards and engineering factors. It will help you evaluate architectural plans and specifications; prepare accurate quantity takeoffs; perform value engineering; compare design alternatives; prepare estimates from conceptual to detail; evaluate change orders.

**$99.95 per copy**
**Over 900 pages, hardcover**
**Catalog No. 67276   ISBN 0-87629-177-9**

## Builder's Essentials:
### Framing & Rough Carpentry, 2nd Edition

A complete, illustrated do-it-yourself course on framing and rough carpentry. The book covers wall, floors, stairs, windows, doors, and roofs, as well as nailing patterns and procedures. Additional sections are devoted to equipment and material handling, standards, codes, and safety requirements.

The "framer-friendly" approach includes easy-to-follow, step-by-step instructions. This practical guide will benefit both the carpenter's apprentice and the experienced carpenter, and sets a uniform standard for framing crews.

**$24.95 per copy**
**Over 125 pages, illustrated, softcover**
**Catalog No. 67298A   ISBN 0-87629-617-7**

## Interior Home Improvement Costs, New 7th Edition

Estimates for 66 interior projects, including:

• Attic/Basement Conversions
• Kitchen/Bath Remodeling
• Stairs, Doors, Walls/Ceilings
• Fireplaces
• Home Offices/In-law Apartments

**$19.95 per copy**
**Over 230 pages, illustrated, softcover**
**Catalog No. 67308C   ISBN 0-87629-576-6**

## Exterior Home Improvement Costs, New 7th Edition

Quick estimates for 64 projects, including:

• Room Additions/Garages
• Roofing/Siding/Painting
• Windows/Doors
• Landscaping/Patios
• Porches/Decks

**$19.95 per copy**
**Over 250 pages, illustrated, softcover**
**Catalog No. 67309C   ISBN 0-87629-575-8**

## Practical Pricing Guides for Homeowners and Contractors

These updated resources on the cost and complexity of the nation's most popular home improvement projects include estimates of materials quantities, total project costs, and labor hours. With costs localized to over 900 zip code locations.

# Books for Builders

For more information
visit Means Web Site
at www.rsmeans.com

## Means Illustrated
## Construction Dictionary
## (Condensed Edition)

Based on *Means Illustrated Construction Dictionary, New Unabridged Edition*, the condensed version features ,000 construction terms. If your work overlaps the construction business—from insurance, banking and real estate to building inspectors, attorneys, owners, and students—you will surely appreciate this valuable reference source.

$59.95 per copy
Over 500 pages, softcover
Catalog No. 67282   ISBN 0-87629-219-8

## Superintending for Contractors:
### *How to Bring Jobs in On-time, On-budget*
by Paul J. Cook

Today's superintendent has become a field project manager, directing and coordinating a large number of subcontractors, and overseeing the administration of contracts, change orders, and purchase orders. This book examines the complex role of the superintendent/field project manager, and provides guidelines for the efficient organization of this job.

$35.95 per copy
Over 220 pages, illustrated, softcover
Catalog No. 67233   ISBN 0-87629-272-4

## Means Forms for Contractors
### *The most-needed forms for contractors of various-size firms and specialties.*

Includes a variety of forms for each project phase — from bidding to punch list. With sample project correspondence. Includes forms for project administration, safety and inspection, scheduling, estimating, change orders, and personnel valuation. Blank forms are printed on heavy stock or easy photocopying. Each has a filled-in sample, with instructions and circumstances for use. 0 years of experience in construction project management, providing contractors with the tools they need to develop competitive bids.

$49.98 per copy
Over 400 pages, three-ring binder
Catalog No. 67288

## Estimating for Contractors:
### *How to Make Estimates that Win Jobs*
by Paul J. Cook

This widely used reference offers clear, step-by-step estimating instructions that lead to achieving the following goals: objectivity, thoroughness, and accuracy.

*Estimating for Contractors* is a reference that will be used over and over, whether to check a specific estimating procedure, or to take a complete course in estimating.

$35.95 per copy
Over 225 pages, illustrated, softcover
Catalog No. 67160   ISBN 0-87629-271-6

## Business Management for
## Contractors:
### *How to Make Profits in Today's Market*
By Paul J. Cook

Focuses on the manager's role in ensuring that the company fulfills contracts, realizes a profit, and shows steady growth. Offers guidance on planning company growth, financial controls, and industry relations.

New reduced price
Now $17.98 per copy; limited quantity
Over 230 pages, softcover
Catalog No. 67250   ISBN 0-87629-269-4

## Building Spec Homes Profitably
by Kenneth V. Johnson

The author offers a system to reduce risk and ensure profits in spec home building no matter what the economic climate. Includes:

* The 3 Keys to Success: location, floor plan and value
* Market Research: How to perform an effective analysis
* Site Selection: How to find and purchase the best properties
* Financing: How to select and arrange the best method
* Design Development: Combining value with market appeal
* Scheduling & Supervision: Expert guidance for improving your operation

$29.95 per copy
Over 200 pages, softcover
Catalog No. 67312   ISBN 0-87629-357-7

# MeansData™

## CONSTRUCTION COSTS FOR SOFTWARE APPLICATIONS
### Your construction estimating software is only as good as your cost data.

## Software Integration

A proven construction cost database is a mandatory part of any estimating package. We have linked MeansData™ directly into the industry's leading software applications. The following list of software providers can offer you MeansData™ as an added feature for their estimating systems. Visit them on-line at **www.rsmeans.com/demo/** for more information and free demos. Or call their numbers listed below.

**3D International**
713-871-7000 venegas@3di.com

**4Clicks-Solutions, LLC**
719-574-7721
mbrown@4clicks-solutions.com

**ACT**
**Applied Computer Technologies**
Facility Management Software
919-859-1335 info@srs.net

**AEPCO, Inc.**
301-670-4642 blueworks@aepco.com

**American Contractor/ Maxwell Systems**
800-333-8435 info@amercon.com

**ArenaSoft Estimating**
888-370-8806 info@arenasoft.com

**Ares Corporation**
650-401-7100
sales@arescorporation.com

**AssetWork CSI-Maximus**
Facility Management Software
800-659-9001 info@assetworks.com

**Benchmark, Inc.**
800-393-9193 sales@benchmark-inc.com

**BSD**
**Building Systems Design, Inc.**
888-273-7638 bsd@bsdsoftlink.com

**CProjects, Inc.**
203-262-6248 sales@cprojects.com

**cManagement**
800-945-7093 sales@cmanagement.com
**CDCI**
**Construction Data Controls, Inc.**
800-285-3929 sales@cdci.com

**CMS**
**Computerized Micro Solutions**
800-255-7407 cms@proest.com

**Conac Group**
800-663-2338 sales@conac.com

**Eagle Point Software**
800-678-6565 sales@eaglepoint.com

**Estimating Systems, Inc.**
800-967-8572 pulsar@capecod.net

**G2 Estimator**
**A Div. of Valli Info. Syst., Inc.**
800-657-6312 info@g2estimator.com

**G/C EMUNI, Inc.**
514-953-5148 rpa@gcei.ca

**Geac Commercial Systems, Inc.**
800-554-9865 info@geac.com

**Hard Dollar**
800-637-7496 sales@harddollar.com

**IQ Beneco**
801-565-1122 mdover@beneco.com

**Luqs International**
888-682-5573 info@luqs.com

**MC²**
**Management Computer Controls**
800-225-5622 vkeys@mc2-ice.com

**Prism Computer Corporation**
Facility Management Software
800-774-7622 famis@prismcc.com

**Quest Solutions, Inc.**
800-452-2342 info@questsolutions.com

**Sanders Software, Inc.**
800-280-9760 hsander@vallnet.com

**Sinisoft, Inc.**
877-669-4949 info.usa@sinistre.com

**Timberline Software Corp.**
800-628-6583
product.info@timberline.com

**TMA Systems, Inc.**
Facility Management Software
800-862-1130 sales@tmasys.com

**US Cost, Inc.**
800-372-4003
sales@uscost.com

**Vertigraph, Inc.**
800-989-4243
info-request@vertigraph.com

**Wendlware**
714-895-7222 sales@corecon.com

**Winestimator, Inc.**
800-950-2374 sales@winest.com

**DemoSource™**

**One-stop shopping for the latest cost estimating software for just $19.95.** This evaluation tool includes product literature and demo diskettes for ten or more estimating systems, all of which link to MeansData™. **Call 1-800-334-3509 to order.**

## FOR MORE INFORMATION ON ELECTRONIC PRODUCTS CALL 1-800-448-8182 OR FAX 1-800-632-6732.

MeansData™ is a registered trademark of R.S. Means Co., Inc., *CMD*.

For more information
visit Means Web Site
at www.rsmeans.com

# New Titles

# From Model Codes to the IBC: A Transitional Guide

## By Rolf Jensen & Associates, Inc.

**NEW!**

A time–saving resource for Architects, Engineers, Building Officials and Authorities Having Jurisdiction (AHJs), Contractors, Manufacturers, Building Owners, and Facility Managers.

Provides comprehensive, user-friendly guidance on making the transition to the International Building Code® from the model codes you're familiar with. Includes side-by-side code comparison of the IBC to the UBC, NBC, SBC, and NFPA 101®. Also features professional code commentary, quick-find indexes, and a Web site with regular code updates.

Also contains illustrations, abbreviations key, and an extensive resource section.

$114.95 per copy
580 pages, Softcover
Catalog No. 67328

# Historic Preservation: Project Planning & Estimating

## By Swanke Hayden Connell Architects

*Managing Historic Restoration, Rehabilitation, and Preservation Building Projects and Determining and Controlling Their Costs*

The authors explain:
- How to determine whether a structure qualifies as historic
- Where to obtain funding and other assistance
- How to evaluate and repair more than 75 historic building materials
- How to properly research, document, and manage the project to meet code, agency, and other special requirements
- How to approach the upgrade of major building systems

$99.95 per copy
Over 675 pages, Hardcover
Catalog No. 67323

# Means Illustrated Construction Dictionary, 3rd Edition

## New Updated Edition with interactive CD-ROM

Long regarded as the Industry's finest, the Means Illustrated Construction Dictionary is now even better. With the addition of over 1,000 new terms and hundreds of new illustrations, it is the clear choice for the most comprehensive and current information.

**The companion CD-ROM that comes with this new edition adds many extra features: larger graphics, expanded definitions, and links to both CSI MasterFormat numbers and product information.**

- 19,000 construction words, terms, phrases, symbols, weights, measures, and equivalents
- 1,000 new entries
- 1,200 helpful illustrations
- Easy-to-use format, with thumbtabs

$99.95 per copy
Over 790 pages, illustrated Hardcover
Catalog No. 67292A

| Qty. | Book No. | COST ESTIMATING BOOKS | Unit Price | Total |
|---|---|---|---|---|
| | 60062 | Assemblies Cost Data 2002 | $164.95 | |
| | 60012 | Building Construction Cost Data 2002 | 99.95 | |
| | 61012 | Building Const. Cost Data–Looseleaf Ed. 2002 | 124.95 | |
| | 63012 | Building Const. Cost Data–Metric Version 2002 | 99.95 | |
| | 60222 | Building Const. Cost Data–Western Ed. 2002 | 99.95 | |
| | 60112 | Concrete & Masonry Cost Data 2002 | 92.95 | |
| | 60142 | Construction Cost Indexes 2002 | 218.00 | |
| | 60142A | Construction Cost Index–January 2002 | 54.50 | |
| | 60142B | Construction Cost Index–April 2002 | 54.50 | |
| | 60142C | Construction Cost Index–July 2002 | 54.50 | |
| | 60142D | Construction Cost Index–October 2002 | 54.50 | |
| | 60342 | Contr. Pricing Guide: Resid. R & R Costs 2002 | 36.95 | |
| | 60332 | Contr. Pricing Guide: Resid. Detailed 2002 | 36.95 | |
| | 60322 | Contr. Pricing Guide: Resid. Sq. Ft. 2002 | 39.95 | |
| | 64022 | ECHOS Assemblies Cost Book 2002 | 164.95 | |
| | 64012 | ECHOS Unit Cost Book 2002 | 109.95 | |
| | 54002 | ECHOS (Combo set of both books) | 229.95 | |
| | 60232 | Electrical Change Order Cost Data 2002 | 99.95 | |
| | 60032 | Electrical Cost Data 2002 | 99.95 | |
| | 60202 | Facilities Construction Cost Data 2002 | 241.95 | |
| | 60302 | Facilities Maintenance & Repair Cost Data 2002 | 219.95 | |
| | 60162 | Heavy Construction Cost Data 2002 | 99.95 | |
| | 63162 | Heavy Const. Cost Data–Metric Version 2002 | 99.95 | |
| | 60092 | Interior Cost Data 2002 | 99.95 | |
| | 60122 | Labor Rates for the Const. Industry 2002 | 219.95 | |
| | 60182 | Light Commercial Cost Data 2002 | 87.95 | |
| | 60022 | Mechanical Cost Data 2002 | 99.95 | |
| | 60152 | Open Shop Building Const. Cost Data 2002 | 99.95 | |
| | 60212 | Plumbing Cost Data 2002 | 99.95 | |
| | 60042 | Repair and Remodeling Cost Data 2002 | 87.95 | |
| | 60172 | Residential Cost Data 2002 | 87.95 | |
| | 60282 | Site Work & Landscape Cost Data 2002 | 99.95 | |
| | 60052 | Square Foot Costs 2002 | 109.95 | |
| | | **REFERENCE BOOKS** | | |
| | 67147A | ADA in Practice | 59.98 | |
| | 67310 | ADA Pricing Guide | 59.98 | |
| | 67273 | Basics for Builders: How to Survive and Prosper | 34.95 | |
| | 67330 | Bldrs Essentials: Adv. Framing Techniques | 24.95 | |
| | 67329 | Bldrs Essentials: Best Bus. Practices for Bldrs | 29.95 | |
| | 67298A | Bldrs Essentials: Framing/Carpentry 2nd Ed. | 24.95 | |
| | 67298AS | Bldrs Essentials: Framing/Carpentry Spanish | 24.95 | |
| | 67307 | Bldrs Essentials: Plan Reading & Takeoff | 35.95 | |
| | 67261A | Bldg. Prof. Guide to Contract Documents-3rd Ed. | 64.95 | |
| | 67312 | Building Spec Homes Profitably | 29.95 | |
| | 67250 | Business Management for Contractors | 17.98 | |
| | 67146 | Concrete Repair & Maintenance Illustrated | 69.95 | |
| | 67278 | Construction Delays | 29.48 | |
| | 67255 | Contractor's Business Handbook | 21.48 | |
| | 67314 | Cost Planning & Est. for Facil. Maint. | 82.95 | |
| | 67317A | Cyberplaces: The Internet Guide-2nd Ed. | 59.95 | |
| | 67230A | Electrical Estimating Methods-2nd Ed. | 64.95 | |
| | 64777 | Environmental Remediation Est. Methods | 99.95 | |
| | 67160 | Estimating for Contractors | 35.95 | |
| | 67276 | Estimating Handbook | 99.95 | |

| Qty. | Book No. | REFERENCE BOOKS (Cont.) | Unit Price | Total |
|---|---|---|---|---|
| | 67249 | Facilities Maintenance Management | $ 86.95 | |
| | 67246 | Facilities Maintenance Standards | 69.95 | |
| | 67318 | Facilities Operations & Engineering Reference | 99.95 | |
| | 67301 | Facilities Planning & Relocation | 89.95 | |
| | 67231 | Forms for Building Const. Professional | 47.48 | |
| | 67288 | Forms for Contractors | 49.98 | |
| | 67328 | From Model Codes to IBC: Transitional Guide | 114.95 | |
| | 67260 | Fundamentals of the Construction Process | 34.98 | |
| | 67148 | Heavy Construction Handbook | 74.95 | |
| | 67323 | Historic Preservation: Proj. Planning & Est. | 99.95 | |
| | 67308C | Home Improvement Costs–Int. Projects 7th Ed. | 19.95 | |
| | 67309C | Home Improvement Costs–Ext. Projects 7th Ed. | 19.95 | |
| | 67324 | How to Estimate w/Means Data & CostWorks | 59.95 | |
| | 67304 | How to Estimate with Metric Units | 9.98 | |
| | 67306 | HVAC: Design Criteria, Options, Select.–2nd Ed. | 84.95 | |
| | 67281 | HVAC Systems Evaluation | 84.95 | |
| | 67282 | Illustrated Construction Dictionary, Condensed | 59.95 | |
| | 67292A | Illustrated Construction Dictionary, w/CD-ROM | 99.95 | |
| | 67295A | Landscape Estimating–3rd Ed. | 62.95 | |
| | 67299 | Maintenance Management Audit | 32.48 | |
| | 67302 | Managing Construction Purchasing | 19.98 | |
| | 67294 | Mechanical Estimating–2nd Ed. | 64.95 | |
| | 67245A | Planning and Managing Interior Projects-2nd Ed. | 69.95 | |
| | 67283A | Plumbing Estimating Methods-2nd Ed. | 59.95 | |
| | 67326 | Preventive Maint. Guidelines for School Facil. | 149.95 | |
| | 67236A | Productivity Standards for Constr.–3rd Ed. | 69.98 | |
| | 67247A | Project Scheduling & Management for Constr. | 64.95 | |
| | 67262 | Quantity Takeoff for Contractors | 17.98 | |
| | 67265A | Repair & Remodeling Estimating-3rd Ed. | 69.95 | |
| | 67322 | Residential & Light Commercial Const. Stds. | 59.95 | |
| | 67254 | Risk Management for Building Professionals | 15.98 | |
| | 67291 | Scheduling Manual–3rd Ed. | 32.48 | |
| | 67327 | Spanish/English Construction Dictionary | 22.95 | |
| | 67145B | Sq. Ft. & Assem. Estimating Methods–3rd Ed. | 69.95 | |
| | 67287 | Successful Estimating Methods | 64.95 | |
| | 67313 | Successful Interior Projects | 24.98 | |
| | 67233 | Superintending for Contractors | 35.95 | |
| | 67321 | Total Productive Facilities Management | 39.98 | |
| | 67284 | Understanding Building Automation Systems | 29.98 | |
| | 67303 | Unit Price Estimating Methods–2nd Ed. | 59.95 | |
| | 67319 | Value Engineering: Practical Applications | 79.95 | |

| | | |
|---|---|---|
| MA residents add 5% state sales tax | | |
| Shipping & Handling** | | |
| Total (U.S. Funds)* | | |

Prices are subject to change and are for U.S. delivery only. *Canadian customers may call for current prices. **Shipping & handling charges: Add 7% of total order for check and credit card payments. Add 9% of total order for invoiced orders.

**Send Order To:** ADDV-1001

Name (Please Print) _____

Company _____

☐ **Company**
☐ **Home**    Address _____

City/State/Zip _____

Phone # _____    P.O. # _____